Functional Cereals and Cereal Foods

Sneh Punia Bangar • Anil Kumar Siroha
Editors

Functional Cereals and Cereal Foods

Properties, Functionality and Applications

Springer

Editors
Sneh Punia Bangar
Department of Food, Nutrition
and Packaging Sciences
Clemson University
Clemson, SC, USA

Anil Kumar Siroha
Department of Food Science
and Technology
Chaudhary Devi Lal University
Sirsa, Haryana, India

ISBN 978-3-031-05613-0 ISBN 978-3-031-05611-6 (eBook)
https://doi.org/10.1007/978-3-031-05611-6

This Springer imprint is published by the registered company Springer Nature Switzerland AG
The registered company address is: Gewerbestrasse 11, 6330 Cham, Switzerland

Preface

Functional Cereals and Cereal Foods: Properties, Functionality and Applications is divided into two sections: Functional cereals and functional cereal foods.

Part I: Functional Cereals

This section will highlight the functional components of the various cereal grains. Chapter 1 contributed by Sneh Punia Bangar and Nita Kaushik discusses the functional components and health benefits of functional cereals. It highlights the major functional components present in whole grains, the role that whole grains may play in disease prevention, and applications of these grains for producing functional foods. A novel approach to improve the functional potential of cereals is presented in Chap. 2. This chapter elaborates on innovative technologies to improve the functional values and nutritional characteristics of cereals. Chapter 3 deals with improvement of genetic variation for nutrients and bioactive food components. This chapter discusses the components of nutritional genomics and the effect or role of nutrients and bioactive components in genome stability and various diseases. Chapter 4 by Amardeep Singh Virdi and Narpinder Singh covers functional cereals for gluten intolerance. Consumption of wheat-based (gluten) food products cause coeliac disease, so research is conducted on development of gluten-free food products. In this chapter, the characteristics of modified or improved products developed from alternative sources of wheat with improved functional, nutraceutical, texture, and sensory properties are discussed. Chapter 5 deals with functionality of resistant and slowly digesting starch in cereals. It elaborates on types of starch on the basis of digestion kinetics and their technological and beneficial physiological effects. The functionalities of β-Glucans and fibers in cereals are explored in Chap. 6, contributed by Mehnaza Manzoor and Sneh Punia Bangar. This chapter focuses on the physiological characteristics and their application in food formulations.

Part II: Functional Cereal Foods

Prebiotic and probiotic potential of cereals are explored in Chap. 7. In this chapter, cereal-based prebiotic, probiotic, and their health benefits will be discussed. Chapter 8 by Mishra and Panda deals with cereal-based fermented foods and non-alcoholic beverages. In this chapter, various indigenous fermented foods' and beverages' properties and their processing technologies are elaborated. Bakery products play an important role in our diet; these products are consumed as breakfast food, snacks, and bread. Functional cereal-based bakery products, breakfast cereals, and pasta products are discussed in Chap. 9, contributed by Maria Di Cairano, Roberta Tolve, NazarenaCela, Lucia Sportiello, Teresa Scarpa, and Fernanda Galgano. In this chapter, physical, chemical, nutritional, and sensory properties are covered. Plant-based milks or non-dairy milks are gaining much importance as a functional and specialty beverage all over the world. Cereal grain-based milks and their potential health properties are elaborated in Chap. 10. In Chap. 11, contributed by Patil, Usman, Mehmood, Ahmad, Haider, Zhang, Teng, and Li, cereal grain tea and its potential health properties are catalogued. Chapter 12 deals with low GI functional foods. High-fiber functional products are elaborated in Chap. 13, contributed by Aderonke Ibidunni Olagunju and Olufunmilayo Sade Omoba. This chapter throws light on the characteristics, types, and formulation, as well health benefits of high-fiber functional products. miRNA-based genetic engineering for crop improvement and production of functional foods are elaborated in Chap. 14.

Clemson, SC, USA — Sneh Punia Bangar
Sirsa, Haryana, India — Anil Kumar Siroha

Contents

About the Editors

Sneh Punia Bangar Ph.D., is a researcher at Clemson University, United States. Earlier, she worked as Assistant Professor (C) in the Department of Food Science and Technology, Chaudhary Devi Lal University, Sirsa. Her research interests include extraction and functional characterization of bioactive compounds, starches, functional foods, and nanocomposites films and coatings.She has presented her research in various national and international conferences and has published more than 100 research papers/book chapters in national and international journals/books. To date, she has authored or co-authored more than 80 (published/accepted) articles, 1 book series, 2 authored books (CRC, Taylor & Francis), 5 edited books (CRC, Taylor & Francis, Springer, Elsevier), 1 reference book (CRC, Taylor & Francis), 35 book chapters, and 20 conference proceedings/seminars. She has also served as Guest Editor of three special issues of the *International Journal of Food Science and Technology, Foods, Polymers, Coatings, and Materials.*

Anil Kumar Siroha Ph.D., is presently working as Assistant Professor (C) in the Department of Food Science and Technology at Chaudhary Devi Lal University, Sirsa. He received his doctorate in Food Science and Technology at Chaudhary Devi Lal University, Sirsa, India. His areas of interest include starch, starch modification, and the development of new products. He has published more than 25 research papers in national and international journals. Till date, he has written 1 authored book, 2 edited books, 1 reference book, and 20 book chapters. He also serves as a reviewer for various national and international journals. He is an active member of the Association of Food Scientists and Technologists (AFSTI), in Mysore, India. Dr. Siroha also co-supervised more than 20 research projects of M.Sc. students in the Department of Food Science and Technology at Chaudhary Devi Lal University, Sirsa, Haryana.

Contributors

Shabbir Ahmad Department of Food Science and Technology, MNS-University of Agriculture, Multan, Pakistan

Zeshan Ali Department of Life Sciences, COMSATS University Islamabad, Sahiwal Campus, Sahiwal, Pakistan

Nazarena Cela School of Agricultural, Forestry, Food and Environmental Sciences, University of Basilicata, Potenza, Italy

Maria Di Cairano School of Agricultural, Forestry, Food and Environmental Sciences, University of Basilicata, Potenza, Italy

Fernanda Galgano School of Agricultural, Forestry, Food and Environmental Sciences, University of Basilicata, Potenza, Italy

Neha Gautam Department of Basic Sciences, Dr. Y. S. Parmar University of Horticulture and Forestry, Nauni, Solan, HP, India

Junaid Haider Tianjin Institute of Industrial Biotechnology, Chinese Academy of Sciences, Tinajin, China

Ajinder Kaur School of Agricultural Biotechnology, College of Agriculture, Punjab Agricultural University, Ludhiana, Punjab, India

Ramandeep Kaur Department of Food Science and Technology, Punjab Agricultural University, Ludhiana, Punjab, India

Nita Kaushik Department of Food Technology, Guru Jambheshwar University of Science & Technology, Hisar, Harayana, India

Ajay Kumar Department of Law, Chaudhary Devi Lal University, Sirsa, Haryana, India

Anil Kumar Siroha Department of Food Science and Technology, Chaudhary Devi Lal University, Sirsa, Haryana, India

Satish Kumar College of Horticulture and Forestry, Thunag, Mandi. Dr YS Parmar University of Horticulture and Forestry, Nauni, Solan, HP, India

Vikas Kumar Department of Food Science and Technology, Punjab Agricultural University, Ludhiana, Punjab, India

Xiuting Li Beijing Advanced Innovation Center for Food Nutrition and Human Health, Beijing Technology and Business University, Beijing, China

School of Food and Health, Beijing Technology and Business University, Beijing, China

Beijing Engineering and Technology Research Center of Food Additives, School of Food and Chemical Technology, Beijing Technology and Business University, Beijing, China

Pooja Manchanda School of Agricultural Biotechnology, College of Agriculture, Punjab Agricultural University, Ludhiana, Punjab, India

Mehnaza Manzoor Division of Food Science and Technology, Sher-e-Kashmir University of Agriculture Science and Technology, Jammu, Jammu and Kashmir, India

Department of Food Technology and Nutrition, School of Agriculture, Lovely Professional University, Jalandhar, Punjab, India

Arshad Mehmood Beijing Advanced Innovation Center for Food Nutrition and Human Health, Beijing Technology and Business University, Beijing, China

School of Food and Health, Beijing Technology and Business University, Beijing, China

Beijing Engineering and Technology Research Center of Food Additives, School of Food and Chemical Technology, Beijing Technology and Business University, Beijing, China

Gayatri Mishra Centre for Applied Research, Innovation and Entrepreneurship, Lethbridge College, Lethbridge, AB, Canada

UniSA STEM, University of South Australia, Adelaide, SA, Australia

Sehrish Nasreen Doctor of Human Nutrition and Dietetics, Sargodha Medical College, Sargodha, Punjab, Pakistan

Aderonke Ibidunni Olagunju Department of Food Science and Technology, Federal University of Technology, Akure, Ondo State, Nigeria

Olufunmilayo Sade Omoba Department of Food Science and Technology, Federal University of Technology, Akure, Ondo State, Nigeria

Brajesh Kumar Panda Centre for Applied Research, Innovation and Entrepreneurship, Lethbridge College, Lethbridge, AB, Canada

Devashree N. Patil Department of Biotechnology, Shivaji University, Kolhapur, Maharashtra, India

Prasanna J. Patil Beijing Advanced Innovation Center for Food Nutrition and Human Health, Beijing Technology and Business University, Beijing, China

School of Food and Health, Beijing Technology and Business University, Beijing, China

Sneh Punia Bangar Department of Food, Nutrition and Packaging Sciences, Clemson University, Clemson, SC, USA

Pawan Kumar Rose Department of Energy and Environmental Sciences, Chaudhary Devi Lal University, Sirsa, Haryana, India

Teresa Scarpa School of Agricultural, Forestry, Food and Environmental Sciences, University of Basilicata, Potenza, Italy

Haroon Shah Beijing Advanced Innovation Center for Food Nutrition and Human Health, Beijing Technology and Business University, Beijing, China

School of Food and Health, Beijing Technology and Business University, Beijing, China

Deepansh Sharma College of Horticulture and Forestry, Punjab Agricultural University, Ludhiana, Punjab, India

Kartik Sharma Department of Biotechnology, Council of Scientific and Industrial Research-Institute of Himalayan Bioresource Technology (CSIR-IHBT), Palampur, HP, India

Arashdeep Singh Department of Food Science, College of Agriculture, Punjab Agricultural University, Ludhiana, Punjab, India

Narpinder Singh Department of Food Science and Technology, Guru Nanak Dev University, Amritsar, Punjab, India

Lucia Sportiello School of Agricultural, Forestry, Food and Environmental Sciences, University of Basilicata, Potenza, Italy

Khongsak Srikaeo Faculty of Food and Agricultural Technology, Pibulsongkram Rajabhat University, Muang Phitsanulok, Thailand

Chao Teng Beijing Advanced Innovation Center for Food Nutrition and Human Health, Beijing Technology and Business University, Beijing, China

School of Food and Health, Beijing Technology and Business University, Beijing, China

Beijing Engineering and Technology Research Center of Food Additives, School of Food and Chemical Technology, Beijing Technology and Business University, Beijing, China

Roberta Tolve Department of Biotechnology, University of Verona, Verona, Italy

Muhammad Usman Beijing Advanced Innovation Center for Food Nutrition and Human Health, Beijing Technology and Business University, Beijing, China

School of Food and Health, Beijing Technology and Business University, Beijing, China

Vanshika School of Agricultural Biotechnology, College of Agriculture, Punjab Agricultural University, Ludhiana, Punjab, India

Amardeep Singh Virdi Department of Food Science and Technology, Amity University Punjab, Mohali, Punjab, India

Syeda Mahvish Zahra Department of Environmental Design Health and Nutritional Sciences, Allama Iqbal Open University, Islamabad, Pakistan

Chengnan Zhang Beijing Advanced Innovation Center for Food Nutrition and Human Health, Beijing Technology and Business University, Beijing, China

School of Food and Health, Beijing Technology and Business University, Beijing, China

Beijing Engineering and Technology Research Center of Food Additives, School of Food and Chemical Technology, Beijing Technology and Business University, Beijing, China

Part I
Functional Cereals

Chapter 1
Functional Cereals: Functional Components and Benefits

Sneh Punia Bangar and Nita Kaushik

1.1 Introduction

Our regular food consumption plays a crucial part in our survival and the normal functioning of our body organs by providing fundamental nutrients. Cereals are the most common staple food around the world. Wheat, rice, barley, rye, oats, maize (corn), major and minor millets, and sorghum are the most common cereals. Cereals are the most widely farmed food on the planet; according to FAO (2020), global cereal production was 2,996,142,289 tonnes. The fact that cereal production is essential for global food security adds to the importance of cereals and cereal products (FAO, 2019). They are one of the most important sources of fundamental food nutrients, providing considerable amounts of energy, protein, and micronutrients in human nutrition (Đorđević et al., 2010). The nutritional value or utility of any grain as a human food is mostly determined by the quantity and quality of protein. Proteins are a class of bio-macromolecules that play a vital role in physiological processes. Because of their safety, great biocompatibility, nutritional value, and cost-effectiveness, natural plant proteins are useful resources (Sim et al., 2021).

The functionality of cereals depends mainly on the genetic composition and the impact of environmental factors on its main components such as carbohydrates (Damiri et al., 2020; Fouad et al., 2020; Bachra et al., 2020), proteins, vitamins, minerals and phenolic phytochemicals (Călinoiu & Vodnar, 2018). Whole grain cereals are rich sources of phenolic compounds such as benzoic and cinnamic acids, anthocyanidins, quinines, flavonols, chalcones, flavones, flavanones, and amino

S. P. Bangar (✉)
Department of Food, Nutrition and Packaging Sciences, Clemson University, Clemson, SC, USA

N. Kaushik
Department of Food Technology, Guru Jambheshwar University of Science & Technology Hisar, Hisar, Haryana, India

S. Punia Bangar, A. Kumar Siroha (eds.), *Functional Cereals and Cereal Foods*,
https://doi.org/10.1007/978-3-031-05611-6_1

phenolic compounds (Bellido & Beta, 2009; Hosseinian et al., 2008; Jang & Xu, 2009; Lloyd et al., 2000). The bioactive potential of phenolics have been linked to the reduced incidence of chronic diseases, including type-2 diabetes, cardiovascular disease, obesity and cancers (Shahidi & Yeo, 2018).

Many cereal crops might be weak in one component while being abundant in another. To combat this problem, researchers have focused on the extraction and usage of odd food plants, such as pseudocereals. Because of their nutritional value, phytochemical content, and application in gluten-free goods, pseudocereals are gaining a lot of attention (Boukid et al., 2018; Mir et al., 2018). This chapter summarizes the various functional components of cereal grains that have been shown to have health advantages in addition to basic nutrition.

1.2 Functional Components in Cereals

Cereals provide all of the macronutrients (proteins, lipids, and carbohydrates) that our bodies require for growth and maintenance. They are high in minerals, vitamins, and other micronutrients that are necessary for good health (Borneo & León, 2012). Cereals are made up of 65–75% carbohydrates, 7–12% protein, 2–6% fats, and 12–14% water in general (Baniwal et al., 2021). Cereals have a variety of useful polyphenolic components that are beneficial to our health in addition to providing nutrition, and they are necessary for our survival (Fig. 1.1).

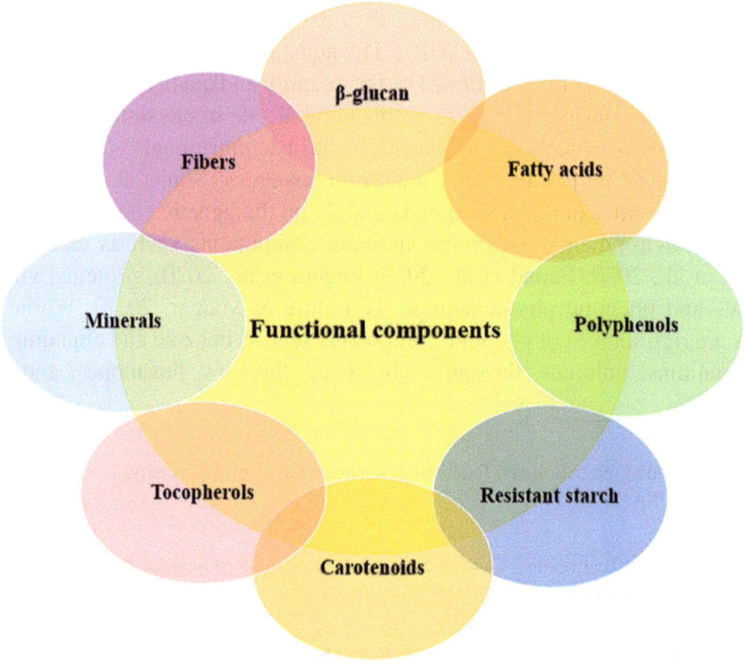

Fig. 1.1 Functional components in cereals

1.2.1 *Carbohydrate Functional Components*

1.2.1.1 Starch

Starch $(C_6H_{10}O_5)_n$ is a polymer made up of long chains of glucose molecules bound together by glycosidic bonds. It is made up of two main polymeric elements (amylose and amylopectin) that have different characteristics and ratios depending on the source. This macromolecule accounts for 60–70% of the weight of raw grains, and the processing/eating quality of cereal foods is mostly determined by starch properties (Li et al., 2020). Starch is deposited as partly crystalline and amorphous in granules that differ in shape and molecular structure from plant to plant (Blazek & Copeland, 2008). Starch is a major carbohydrate in the human diet in terms of quantity, as it contributes significantly to the external supply of glucose and total dietary energy consumption (Roder et al., 2005).

1.2.1.2 Inulin

Inulin is a fructosylfructose molecule with a β-(2→1) linkage that can contain either a β-D-fructose or a α-D-glucose molecule at the end (Gupta et al., 2019; Giri et al., 2021). Rose, a German chemist, originally isolated it from the roots of Inula helenium in 1804 and named it inulin in 1918. Despite the fact that inulin was previously thought to be a completely linear molecule with β-(2→1) linkages, permethylation study of inulin indicated a slight degree of branching. It also comes in a cyclic form with six, seven, or eight fructofuranose rings (Gupta et al., 2019). Inulin polysaccharides resist hydrolysis by gastric enzymes in the human digestive tract due to β-configurations of anomeric C2 in fructose monomers (Shoaib et al., 2016). As a result, non-digestible carbohydrates are inulin-type fructans. Long-chain inulin has a higher degree of polymerization (22–25) than native inulin, making it more thermostable, less soluble, and viscous. Inulin's general molecular structure is depicted in (Fig. 1.2). It provides less energy than other carbohydrate sources. And improves mineral absorption and intestinal health (Shoaib et al., 2016). Wheat, oat, dalia (Bulgur), rye (Secale cereale), and other common foods contain inulin, but in smaller amounts (Koruri et al., 2014).

1.2.1.3 Dietary Fiber (DF)

Dietary fiber is a nondigestible carbohydrate that is resistant to enzymatic digestion and absorption in the intestinal tract (Fig. 1.3). Dietary fibre is a group of compounds including various plant carbohydrate polymers (oligosaccharides and polysaccharides), such as cellulose, hemicelluloses, pectic substances, gums, resistant starch, and inulin, as well as lignin and other non-carbohydrate components (e.g., polyphenols, waxes, saponins, cutin, phytates, resistant protein) (Borneo & León, 2012; Kaur et al., 2021). In the small intestine, resistant starch (RS) and resistant protein (RP) withstand digestion. There are four types of resistant starch (RS1: physical

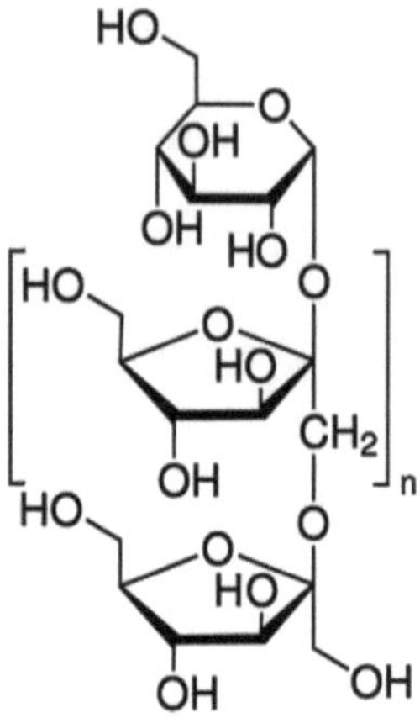

Fig. 1.2 Chemical structure of inulin

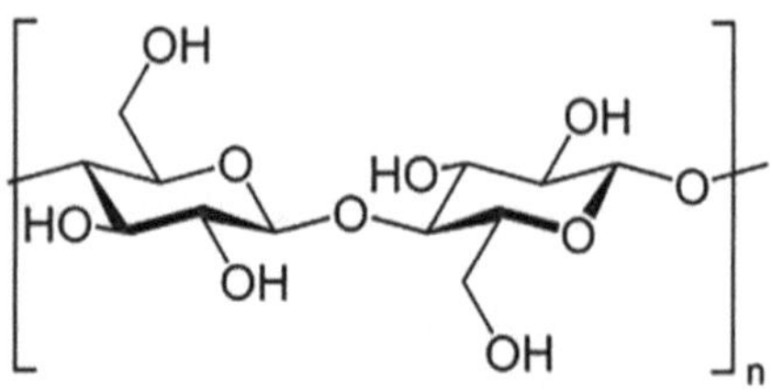

Fig. 1.3 Chemical structure of dietary fiber

inaccessible starch, RS2: ungelatinised starch granules, RS3: retrograded starch and RS4: chemically modified starch). RS increases faecal volume, promotes colonic fermentation, lowers postprandial blood glucose (insulin responses), and lowers preprandial cholesterol levels (Li et al., 2021). According to their water solubility, dietary fibers are divided into two categories: insoluble dietary fiber (IDF) and soluble dietary fiber (SDF) (Lattimer & Haub, 2010; Dhingra et al., 2012). Cellulose, hemicellulose, and lignins are examples of insoluble fibers. Pectins, gums, and mucilages are examples of soluble fiber. Insoluble fiber lengthens the transit time of the intestines, increases the volume of the feces, and so prevents constipation (Thebaudin et al., 1997). Soluble fiber has been shown to influence fat metabolism by binding to cholesterol and lowering its absorption. By producing a viscous film on the gut walls, soluble fiber also inhibits the release of glucose into the bloodstream (Anderson et al., 2009). Pearl millet (2.9–3.8%), wheat (0.79–0.93%), oat (3.5–5.8%), and barley (16.20%) had the good amount of crude fibre content (Siroha et al., 2016; Punia et al., 2019; Youssef et al., 2016; Aprodu & Banu, 2017).

1.2.1.4 β-Glucan

β-glucan is a linear polymer linked by a glycosidic bond between D-glucose monomer and a β-1,3 or β-1,4 glycosidic bond (Fig. 1.4). Some have β-1,6 branches and are found in fungi (such as mushrooms, yeast, and other yeasts), bacteria, seaweeds,

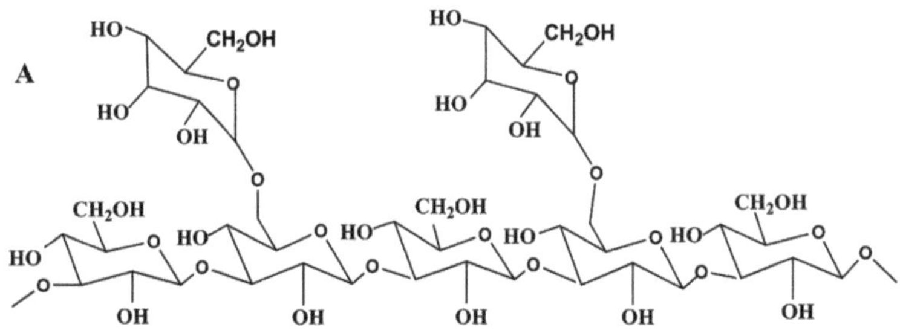

Fig. 1.4 Chemical structure of β-glucan

and cereals (Brennan & Cleary, 2005). The structure and description of β-glucan are given (Fig. 1.4). Due to a number of biological actions, such as hypocholesterolemia, hypoglycemia, antioxidant, anticancer, immunomodulatory, and anti-inflammatory effects, β–glucan is widely considered a useful food element (Adebowale et al., 2019). There are two types of β-glucans: water-soluble and water-insoluble. Soluble β-glucans offer several advantages over insoluble β-glucans. The solubility of β-glucans influences their functional activity, such as stability, emulsification, drug transport, and film forming ability.

Improving the solubility of β-glucan could expand its use in the pharmaceutical and functional food industries. (Liu et al., 2021). Although insoluble β-glucans have a strong immunostimulatory capacity, they also have a high risk of side effects due to hyperimmunity and can only be administered orally. β-glucan can be found in yeast, mushrooms, bacteria, algae, barley, and oats, among other natural sources (Zhu et al., 2015). Cereal β-glucans are naturally present in oat and barley, each containing around 4.5% β-glucans. These polymers also occur in rye (up to 2.5% in the whole flour) and in wheat, at lower concentrations (up to 2.5% in the bran fraction) (Cho and White, 1993; Li et al., 2006; Wood, 2010).

1.2.2 *Polyphenols*

The presence of one or more aromatic rings with one or more hydroxyl groups characterizes phenolic chemicals, which are secondary metabolites of plants (Borneo & León, 2012). Polyphenols are a diverse category of plant compounds with one or more benzene rings and various numbers of hydroxyl (OH), carbonyl (CO), and carboxylic acid (COOH) groups as functional components (Fig. 1.5). They are usually found in conjugated forms with one or more sugar residues attached. Polyphenolic compounds, in general, are active molecules that exhibit various physiological effects, such as antioxidant, anti-tumor, immunomodulation, anti-hyperglycemia, hypertension, antibacterial, anti-radiation, and anti-aging (Shi et al., 2015). Polyphenolic compounds are abundant in pigmented cereal grains. Flavonoids are the most frequent type of polyphenol. Catechins, thearubingens, theaflavins,

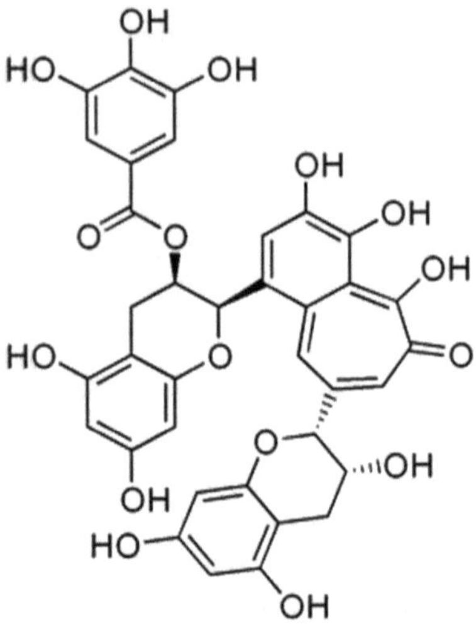

Fig. 1.5 Chemical structure of polyphenols

isoflavones, and over 8000 additional polyphenols are examples (Lobo et al., 2010). Chemically, they are benzoic acid and cinnamic acid derivatives that are mostly concentrated in the cortex and aleurone layer of grains and exist in two states: free and mixed (Mattila et al., 2005; Rao & Muralikrishna, 2002; Robbins, 2003). The phenolic content of different cereal grains is different, of which corn has the highest polyphenol content (15.55 μg/g), followed by wheat (7.99 μg/g), oats (6.53 μg/g), and rice (5.56 μg/g) (Adom and Lui, 2002).

Phenolic compounds have antioxidant characteristics and are thus beneficial to one's health since they may scavenge free radicals in the body, which are linked to cancer and cardiovascular disease (Dykes & Rooney, 2007). Because of their anti-apoptosis, antiaging, anticarcinogenic effects, and bioactive capacities, such as antioxidant activity, phenolics are becoming more popular (Hodzic et al., 2009).

1.2.2.1 Alkaloids

Alkaloids are basic organic chemicals with complex ring structures that are found primarily in dicotyledonous plants, such as the roots and fruits of the Ranunculaceae, Rutaceae, and Leguminosae families. Free bases, salts, amides (colchicine), N-oxide (matrine), and nitrogen-containing complex aldehydes are some of the chemical forms they take (Liu et al., 2019). Alkaloids exhibit substantial biological activity due to their complex ring structure and nitrogen content, and are one of the most important and effective components in Chinese herbal medicine as well as Western medicine. The most frequent alkaloids includes indole, avenanthramides (AVNs) are plentiful in the human food chain and constitute an integral part of the ordinary Western diet (Singh et al., 2013). Researchers have revealed that oats, a versatile grain with a distinct mix of natural phenols, possess around 40 different oat alkaloids when compared to other cereals. The amount of AVNs in oats ranges

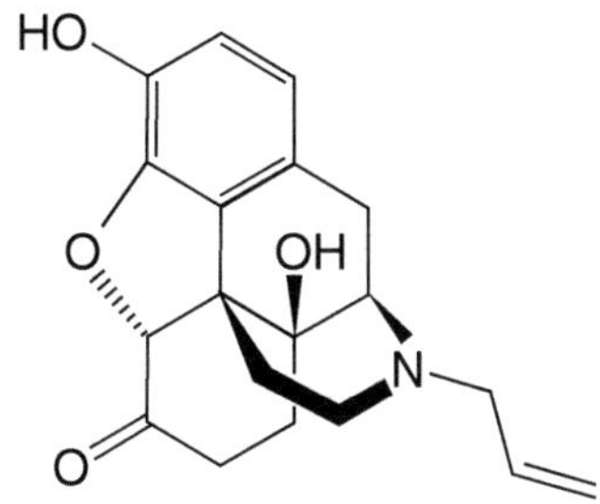

Fig. 1.6 Chemical structure of alkaloids

from 2mg/kg-53mg/kg. AVNs are created by forming amide bonds between a series of anthranilic acid and its derivatives and a sequence of cinnamic acid and its derivatives (Turrini et al., 2019) (Fig. 1.6).

1.2.2.2 Tocopherols and Tocotrienols

Tocopherols and tocotrienols are fat-soluble functional compounds with a phenolic-chromanol ring coupled to a saturated (tocopherols) or unsaturated (tocotrienols) isoprenoid side chain (Fig. 1.7). Alpha, beta, gamma, and delta are the four primary types of tocopherols and tocotrienols, which differ in the number and position of methyl groups on the phenolic-chromanol rings (Lobo et al., 2010; Srividya et al., 2010). In terms of health advantages, a plethora of evidence has been documented on tocopherols and tocotrienols' possible protective effects against cardiovascular illnesses, certain types of cancer, metabolic disorders, neurodegeneration, and oxidative stress (de Camargo et al., 2019). Immune modulation and anti-inflammatory properties have been highlighted as well (Meganathan & Fu, 2016). Most monocots, including agronomically important cereal grains like wheat and rice, include tocotrienols as the predominant form of vitamin E in their seed endosperm (Sen et al., 2007).

1.2.2.3 Carotenoids (E.G. Lycopene, Lutein)

Carotenolds represent the primary pigments in cereals and have potent antioxidant properties (Fig. 1.8). They are lipid-soluble plant pigments made up of oxygenated or non-oxygenated hydrocarbons with at least 40 carbon atoms and a complex conjugated double bond structure. The non-polar functional carotenoids alpha-carotene, beta-carotene, and lycopene are the most common, while lutein is the most common polar carotenoid. They give fruits and vegetables their red, yellow, and orange colors, and their antioxidant qualities have recently gotten a lot of attention. In plant tissues, carotenoids can be found esterified to fatty acids or unesterified. Lycopene is the most powerful oxygen neutralizer, with chemopreventive properties (Parker, 2000).

The carotenoids α- and β-carotene, β-cryptoxanthin, zeaxanthin, and lutein are all common carotenoids found in whole grains. While vegetables and fruits are the

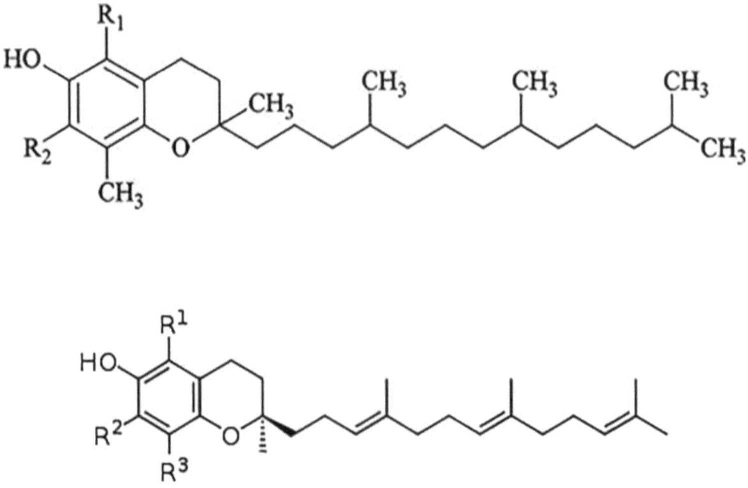

Fig. 1.7 Chemical structure of tocopherols and tocotrienols

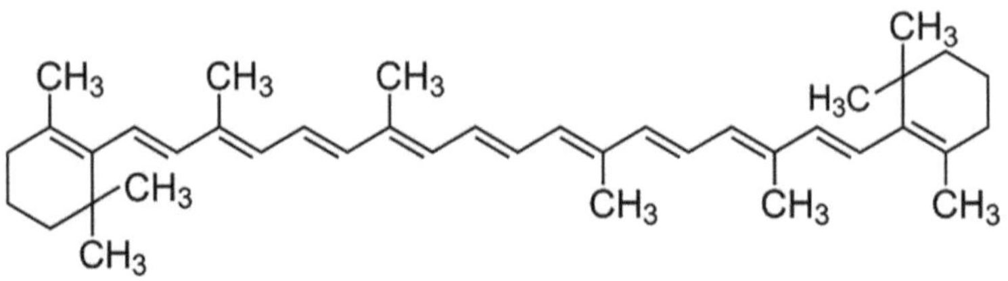

Fig. 1.8 Chemical structure of carotenoids

most common sources of carotenoids in human diets, whole grains are becoming more widely acknowledged as a significant source of these phytonutrients. Carotenoids have been linked to a variety of biological actions in the protection of chronic diseases, including cardiovascular disease and many types of cancer (Borneo & León, 2012). According to Ma et al. (2019), black wheat has a greater average total carotenoid content than white and red wheat. Suriano et al. (2020) discovered that barley contains lutein, zeaxanthin, and β-carotene.

1.2.2.4 Flavonoids

Flavonoids are a class of polyphenolic chemicals that are formed as a secondary metabolite in plants (Fig. 1.9). Flavonoids are compounds with C6-C3-C6 skeleton that consists of two aromatic rings joined by three carbon link including anthocyanins, flavanols, flavones, flavanones and flavonols (Van Hung, 2016). They are an important component of many nutraceutical products. In wheat, barley, and millet, total flavonoids content is 75–106 g CE/g, 1968–2198 g FAE/g, and

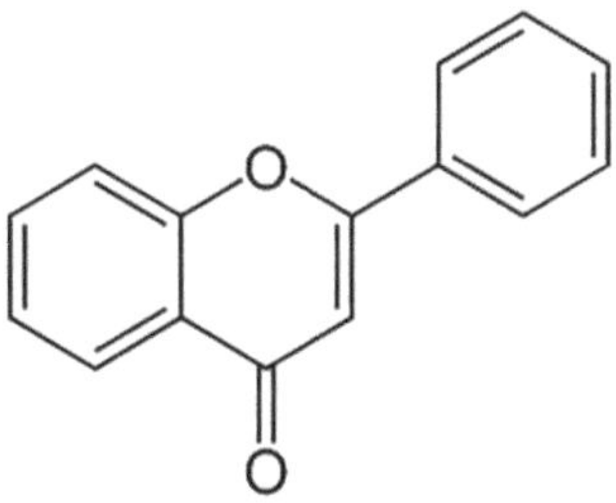

Fig. 1.9 Chemical structure of flavonoids

1721–2484 g CE/g, respectively (Punia et al., 2019; Sandhu & Punia, 2017; Siroha et al., 2016). Flavonols, anthocyanins and proanthocyanidins are the major types of flavonoids found in cereal grains; the first two compounds are the main flavonoids located in the pericarp and contribute to grain pigmentation and protection against UV-B radiation. Flavonols and anthocyanins frequently exist in cereal pigmented grains as glycoside derivatives, including cyanidin-3 glucoside, peonidin-3 glucoside and delphinidin-3 glucoside. Anthocyanins are responsible for pigmentation in cereals and have been identified as one of the most active compounds in terms of their antioxidant activity (Rao et al., 2018).

1.2.2.5 γ- Oryzanol

γ-oryzanol, a mixture of 10 esters of triterpene alcohols is mainly composed of esters of transferulic acid (trans-hydroxycinnamic acid) with phytosterols (Fig. 1.10). Among these phytosterols, cycloartenol, β-sitosterol, 24-methylenecycloartenol and campesterol are the major components in γ-oryzanol. γ-oryzanol is anantioxidant compound and is associated with decreasing plasma cholesterol, lowering serum cholesterol, decreasing cholesterol absorption and decreasing platelet aggregation. The beneficial effects of γ-oryzanol on human health have generated global interest in developing simple methods for its separation from natural sources, such as crude rice bran oil, rice bran oil soap stock, rice bran acid oil, or biodiesel residue from rice bran (Zullaikah et al., 2009). Rice bran is a rich source of steryl ferulate esters, commonly referred to as oryzanols (Xu & Godber, 1999). Kim et al. (2015) also concluded that High Hydrostatic Pressure Treatment (HPT) has positive effect on γ-oryzanol content in germinated rough rice.

1.2.2.6 Phytosterols

Phytosterols are the plant equivalent of cholesterol, which is found in animals. They have similar structures. Plant sterols, on the other hand, have more double bonds and methyl and/or ethyl groups on their side chains. Beta-sitosterol, campesterol, and stigmasterol are the most frequent bioactive phytosterols. Plant stanols, such as sitostanol, are saturated derivatives of plant sterols (Swanson, 2003 and Anon, 2013).

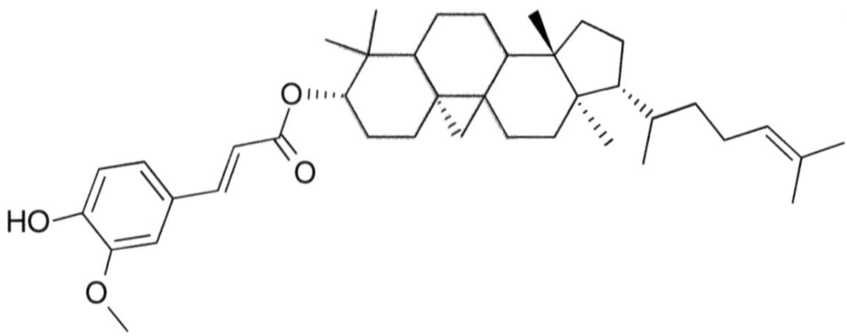

Fig. 1.10 Chemical structure of γ-oryzanol

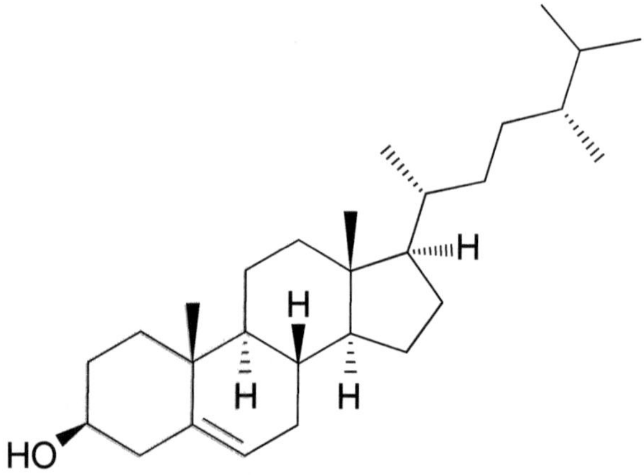

Fig. 1.11 Chemical structure of phytosterol

Phytosterols are found in legumes, cereals, and seeds, as well as oils, nuts, fruits, and greens. Many variations in plant sterol type and content have been observed for a variety of items in each food group (Jiménez-Escrig et al., 2006). The ability of phytosterols to decrease cholesterol is one of their most well-known medicinal properties (Gylling et al., 2014) (Fig. 1.11).

1.2.2.7 Anthocyanins

Due to their appealing nutritional values, colored cereal grains (black, purple, blue, pink, red, and brown) have recently attracted a lot of attention. Anthocyanins are a type of pigment that is responsible for the color of cereals as well as their health benefits. They are a group of highly pigmented pigments that give many plants their orange, brown, red, blue, and purple hues (Escribano-Bailón et al., 2004). Anti-oxidation, anti-cancer, retinal protection, hypolipidaemia, anti-ageing, and promoting gut health are all qualities of anthocyanins, which are glycosides of polymethoxy

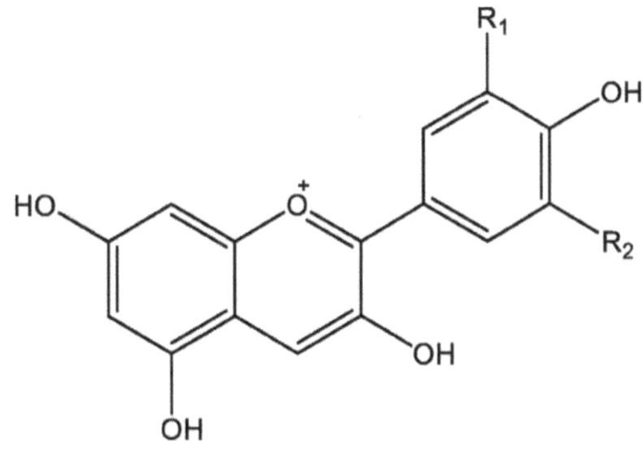

Fig. 1.12 Chemical structure of anthocyanins

and/or polyhydroxy derivatives of flavylium or 2-phenylbenzopyrylium salts (Tsuda, 2012; Kamiloglu et al., 2015; Olivas-Aguirre et al., 2016). As a result, eating anthocyanin-rich foods on a daily basis may help to avoid many chronic diseases and enhance overall health (Tsuda, 2012). Natural coloured grains include black, purple, blue, pink, red, and brown varieties (Escribano-Bailón et al., 2004). Anthocyanins can be found in abundance in these colorful grains. They can be transformed into functional components that can be used to create unique food products with health benefits (Ficco et al., 2016) (Fig. 1.12).

1.2.3 Minerals

The proper functioning of important metabolic processes in the human body necessitates the use of vitamins and minerals. They are cofactors for numerous enzymes, and they also aid in osmotic pressure regulation. Cereals are high in vitamins B and E, as well as minerals including Ca, Mg, Mg, Fe, and Zn (Hübner & Arendt, 2013). Millets are a good source of nutrients including magnesium, manganese, and phosphorus, according to nutritionists (Shashi et al., 2007). White maize has high mineral content, such as Fe and Zn (0.32 mg/100 g and 0.10 mg/100 g, respectively) (Oboh et al., 2010). Chia seeds have been found to have higher levels of calcium, phosphorus, magnesium, zinc, iron, and copper than many other seeds (Capitani et al., 2012).

1.2.4 Protein

Cereal crops are the most important source of protein in the human diet nutritionally. Cereals can provide more than half of the body's daily protein requirements, according to "The Pagoda of Balanced Diet for Chinese Residents." The majority of proteins are present in the endosperm of seeds and are referred to as storage proteins. Cereal storage proteins are divided into three categories: water soluble (albumin; 60.27%), dilute saline soluble (globulin; 3.68%), and dilute acids/bases soluble (globulin; 3.68%) (glutenin; 3.05%, gliadin; 2.91%). Other proteins make up 30.09% of the total. Wheat, for example, is abundant in gliadin (40–50%) and glutenin (30–45%), and buckwheat protein is comparatively high in albumin (20.99–30.30%). Further, rice, barley, and finger millet are reported to be low, and wheat, oats, pearl millet, and proso millet are high in total protein content. Wheat is the most significant grain, and the availability of protein is one of the key reasons for its popularity as a dietary ingredient. The viscoelastic nature of glutenin and gliadin proteins of wheat plays a tremendous role in baking process. Oat has the highest protein level (12–20%) among cereals with a superior amino acid profile due to higher amounts of limiting amino acids lysine and threonine. Oat groats contain protein content of (15–20 %) followed by wheat (10.69–13.68 %), brown rice (7.50 %), barley (9.91 %), sorghum (10.62 %), millet (11.02 %) and rye (10.34 %) (Kumar et al. 2021).

1.3 Health Benefits of Functional Cereals

Numerous bioactive components generated from plants have been demonstrated to have beneficial impacts on human health and disease prevention. Cereals are the most widely consumed supplementary meals worldwide, and there is now a fresh opportunity to add functional components to address health issues that are not caused by a basic nutritional deficiency. Extensive study has shown that grains are high in fermentable carbohydrates that are normally not destroyed by human digestive enzymes, but are selectively degraded and fermented into short-chain fatty acids by the intestinal microbial community. These short-chain fatty acids are absorbed and used by the host's intestinal wall, as well as by helpful bacteria in the intestine, such as bifidobacteria and lactobacilli, increasing the intestinal microbial ecology, reducing the intestinal pH, and boosting the body's immunity. For example, dietary fiber-rich barley and oats boost probiotic growth and improve intestinal microecology (Dvoncova et al., 2010).

1.3.1 Weight Management

It's difficult to pinpoint a precise method by which whole grains might help with weight loss or control. Body weight management is a complicated topic. Consumption of dietary fibers and whole grains has been linked to a lower risk of obesity, overweight, and a high waist-to-hip ratio. Dietary fiber increases volume and easily causes satiation, resulting in a reduction in food consumption. Soluble dietary fiber has a great ability to absorb water, increasing volume and weight by 10 to 15 times, resulting in improved satiety and lower fat absorption in food, allowing for successful weight management. Certain fibers can also limit the rate of glucose absorption by slowing stomach emptying. Some fibers can also help to increase intestinal satiety by slowing stomach emptying and lowering glucose absorption in the small intestine (Lattimer & Haub, 2010).

1.3.2 Cardiovascular Disease

According to estimates from the World Health Organization (WHO), 17.5 million people died of cardiovascular illnesses in 2005, accounting for 30% of all fatalities worldwide (AHA, 2007). CVD is a widespread disease with a high rate of morbidity and mortality that poses a severe danger to human health worldwide. Cereals are high in antioxidants and have potential to decrease cholesterol and lessening the risk of coronary heart disease (CHD). Anti-cholesterolemic qualities are found in fibers, phytosterols, β–glucans, and policosanols, while antioxidant properties are found in flavonoids and anthocyanins. Furthermore, high molecular weight β-glucan has potential to improve the composition and function of gut flora, lowering cardiovascular and cerebrovascular risk markers in patients with mild hypercholesterolemia (Wang et al., 2017).

Numerous epidemiological studies have linked dietary fiber intake to a lower risk of cardiovascular disease, mostly due to a sustained drop in LDL levels (Soliman, 2019). Whole grains have more dietary fibre, protein, vitamins, and inorganic salts than refined grains. Increasing whole grain intake can minimize the risk of coronary disease and stroke and lowering the risk of CVD by regulating blood lipids and controlling blood pressure, increasing Oat β-glucan has been demonstrated in studies to considerably lower serum total triglycerides and serum cholesterol in mice that had consumed a high-fat diet. The cholesterol levels of Syrian hamsters were also decreased by feeding them barley and insoluble fiber (Wilson et al., 2004).

Studies have also shown that eating more whole-grain foods, such as whole-grain morning cereal, oatmeal, brown rice, and wheat germ, is associated with a lower risk of diabetes. Furthermore, increasing dietary fiber intake can lower the food glycemic index, lowering the effect of dietary sugar (Yang, Ding, et al., 2020).

1.3.3 Modulating Intestinal Flora

Long-term fixed eating habits may have a negative impact on gut microbes. Obesity, diabetes, hypertension, coronary heart disease, various cardiovascular and cerebrovascular disorders, and colon cancer have all been linked to a structural disturbance of the gut flora, according to microecological and nutritional research. Healthy gut microbiota produces a number of beneficial metabolites, including short-chain fatty acids like acetic acid, propionic acid, and butyric acid (Tang & Hazen, 2017), which are favorable to human health. Probiotics such as bifidobacterium, lactobacillus, and some enterococci have been shown to offer nutrition for intestinal cells and increase nutrient absorption, resulting in a healthy intestinal environment (Ma et al., 2019).

Extensive study has shown that grains are high in fermentable carbohydrates that are normally not destroyed by human digestive enzymes, but are selectively degraded and fermented into short-chain fatty acids by the intestinal microbial community. These short chain fatty acids are absorbed and used by the host's intestinal wall, as well as by helpful bacteria in the intestine, such as bifidobacteria and lactobacilli, increasing the intestinal microbial ecology, reducing the intestinal pH, and boosting the body's immunity. Undigested carbohydrates also raise the water content of the feces and speed up the intestine's peristalsis. Given this, researchers hypothesized that increasing cereal intake appropriately could result in an abundance of probiotics and improved intestinal function, with multiple good impacts on a variety of disorders (Samantha et al., 2020). Dietary fiber-rich foods like barley and oats, for example, boost probiotic development and improve gut microecology. In vivo and in vitro studies have shown that oat foods, barley husks and rye bran, wheat germ, whole wheat flakes, and wheat bran selectively increase the growth and number of the probiotic bifidobacterium or lactobacillus, thereby exerting antitumor potential and enhancing the formation of short-chain fatty acids such as acetate and butyrate (Kiarie et al., 2014).

1.3.4 Cancer

Fermentable carbohydrates are abundant in whole grains (dietary fiber, resistant starch, and oligosaccharides). Dietary fiber increases fecal volume and reduces transit time, lowering the risk of mutagens interacting with gut epithelial cells. A number of phytochemicals found in grains have been shown to have anticancer properties. The extracts of buckwheat, barley, rice, and wheat, for example, demonstrated a clear inhibitory effect on Caco-2 cell proliferation, with buckwheat showing the highest effect, with a 55.62% inhibition rate on Caco-2 cell growth. Furthermore, following digestion in the gastrointestinal tract, insoluble phenolic chemicals contained in grains reach the colon in intact form and protect the colon after fermentation and microbe release, preventing colon cancer (Kim et al., 2009). Because whole grains are high in antioxidant chemicals, it's been suggested that they may have direct protective effects on colonic cells against reactive oxygen species and other harmful radicals. Carcinogen production from precursor compounds may be inhibited by phytic acid, vitamin E, and phenolics.

1.3.5 Hypertension

Whole grains are high in fiber, which aids in the improvement of lipid profiles and the reduction of systolic and diastolic blood pressure in obese men and women. Many other ingredients in cereals, such as magnesium, potassium, and various proteins, have also been shown to help lower blood pressure. Consumption of whole grains may improve blood pressure regulation by increasing vascular reactivity response (Kochar et al., 2011).

1.3.6 Diabetes

Numerous studies have connected a high intake of cereal fiber to a lower risk of diabetes. Satiety is induced by low-GI meals, which assists weight loss. Despite the fact that one study demonstrated a similar level of weight loss with a high-GI diet as with a low-GI diet, several intervention studies have found that energy-restricted diets based on low-GI meals cause greater weight loss than those based on high-GI foods. Epidemiological studies strongly support the suggestion that high intakes of whole grain foods protect against the development of type II diabetes mellitus (T2DM). People who consume ~3 servings per day of whole grain foods are less likely to develop T2DM than low consumers (Venn, & Mann, 2004).

1.3.7 Fighting Obesity/Anti-Obesity

Obesity has become a worldwide problem. Obesity is a risk factor for various chronic diseases, including type 2 diabetes, cardiovascular disease, and cancer, according to a body of epidemiological data. Each year, almost 2.8 million individuals die as a result of being overweight or obese. Because obesity has reached epidemic proportions, it is expected to overtake heart disease as the leading cause of death in the near future. Obesity is strongly linked to our everyday behaviors, particularly our nutrition, which has changed dramatically. Consumption of dietary fibers and whole grains has been linked to a lower risk of obesity, overweight, and a high waist-to-hip ratio. Dietary fiber increases volume and easily causes satiation, resulting in a reduction in food consumption. Soluble dietary fiber has a great ability to absorb water, increasing volume and weight by 10 to 15 times, resulting in increased satiety and lower fat absorption in food, allowing for efficient weight control. Certain fibre can also help to increase intestinal satiety by slowing stomach emptying and lowering glucose absorption in the small intestine (Gadde et al., 2018).

In short, it has been proposed that a high fiber diet can help to regulate obesity and its related abnormalities by avoiding overeating and fat buildup in the body. Functional cereals are employed in a weight-loss strategy that involves lowering nutritional activity energy consumption through energy reduction, dietary density, and hunger suppression. (Lattimer & Haub, 2010).

1.3.8 Anti-Inflammatory and Antibacterial

Inflammation is the body's powerful response to infection, injury, and disease, and it's necessary for the host to eliminate pestilent stimuli and repair the tissue damage that results (Song et al., 2019). Macrophages, which are tissue-resident immune cells, start the inflammatory response by generating inducible nitric oxide synthase, cyclooxygenase-2, and a slew of proinflammatory cytokines like tumor necrosis factor (TNF)-a, interleukin (IL)-6, and interleukin (IL)-12. Uncontrolled and improper macrophage responses, on the other hand, might stymie the host's healing process, leading to chronic inflammation that can eventually damage healthy tissues. Cereal active compounds and cereal-derived active components have been shown to have anti-inflammatory properties. Yao et al. (2014) extracted and identified 11 monomeric saponins from quinoa grains, then tested their impact on the production of nitric oxide, TNF-a, and IL-6. To minimize chronic inflammation, quinoa saponin blocked the release of inflammatory mediators and downstream pathways. In Raw264.7 cells, millet polyphenols strongly reduced the release of pro-inflammatory mediators TNF-a and IL-6, confirming previous findings (Hosoda et al., 2012).

On the other hand, p-coumaric acid and ferulic acid in polyphenol extracts of cereals have antibacterial properties. They work by lowering the oxidation of microbial enzymes and microbial membranes, which inhibits bacterial cell development. The antibacterial and antifungal activity of the millet seed coat epidermal extract was strong, especially caffeic acid, p-coumaric acid, ferulic acid, and protocatechuic acid (Shahidi & Chandrasekara, 2013).

1.3.9 Antioxidants

Reactive oxygen species (ROS) including superoxide anion, hydrogen peroxide, peroxyl radical, hydroxyl radical, singlet oxygen and peroxynitrite are produced during cellular metabolic processes. Several studies suggested a strong association between elevated levels of these molecules and the pathogenesis of several chronic diseases through the process of oxidative stress. The process of oxidation in cells is regulated by antioxidants, which delay or prevent cellular damage. Antioxidant protection is normally achieved through a balance between pro-oxidants and endogenous and/or dietary antioxidants. Polyphenols' physiological action is linked to their powerful free radical scavenging and antioxidant capabilities (Shi et al., 2015). Phenols interact with enzymes that create free radicals to complex excess metal ions that drive oxidation, so indirectly eliminating free radicals (Chandrasekara & Shahidi, 2011a; 2011b). Cereal is a good example of antioxidants because it is high in phenols. According to new research, dietary and age-related disorders such as metabolic syndrome, cardiovascular disease, type 2 diabetes, and cancer, all of which are linked to increased oxidative stress, can be effectively averted by eating grains on a regular basis (Shah et al., 2015).

1.4 Conclusion

Consumption of cereal grains has been associated with the prevention of widespread health issues. The majority of the health-promoting compounds present in whole grains are con-centrated in the germ and bran, which are discarded during white flour manufacturing. Epidemiological studies demonstrate a protective impact of a whole grain diet against cardiovascular disease, diabetes, and cancer, as well as in weight control, and there is accumulating evidence to support the benefits of whole grain consump-tion beyond a basic diet. European and global health authorities urge the use of whole grain foods as part of a balanced diet. Knowledge of the biochemistry, biological activity, bioavailability and genetics of whole grain phenolic acids is very vast. Bioprocessing and encapsulation techniques support the concept of functional health-related whole grain products by increasing bioaccessibility of bioactive chemicals and nutrients while maintaining sensory qualities. Further, more research is needed regarding the impact of grain phenolic acids on the gut microbiota and their mechanisms of action in humans.

References

Adebowale, T. O., Yao, K., & Oso, A. O. (2019). Major cereal carbohydrates in relation to intestinal health of monogastric animals: A review. *Animal Nutrition, 5*(4), 331–339.

Adom, K. K., & Liu, R. H. (2002). Antioxidant activity of grains. *Journal of agricultural and food chemistry, 50*(21), 6182–6187.

American Heart Association (AHA). (2007). Heart disease and stroke statistics-2007 update. *Circulation, 115*, 69–171.

Anderson, J. W., Baird, P., Davis, R. H., Ferreri, S., Knudtson, M., Koraym, A., … Williams, C. L. (2009). Health benefits of dietary fiber. *Nutrition Reviews, 67*(4), 188–205.

Anon. (2013). *Nutrition*. Wikipedia, the Free Encyclopaedia. http://en.wikipedia.org/wiki/Nutrition. Accessed 26 March 2013.

Aprodu, I., & Banu, I. (2017). Milling, functional and thermo-mechanical properties of wheat, rye, triticale, barley and oat. *Journal of Cereal Science, 77*, 42–48.

Aspinall, G. O. (1970). *Arabinans and xylans. Polysaccharides*. Pergamon Elmsford, New York, EU, 103–115.

Bachra, Y., Grouli, A., Damiri, F., Bennamara, A., & Berrada, M. (2020). A new approach for assessing the absorption of disposable baby diapers and superabsorbent polymers: A comparative study. *Results in Materials, 8*, 100156.

Backer, E. D., Gebruers, K., Van den Ende, W., Courtin, C. M., & Delcour, J. A. (2010). Post-translational processing of β-D-xylanases and changes in extractability of arabinoxylans during wheat germination. *Plant Physiology and Biochemistry, 48*(2–3), 90–97.

Baniwal, P., Mehra, R., Kumar, N., Sharma, S., & Kumar, S. (2021). Cereals: Functional constituents and its health benefits. *Pharma Innov, 10*(3), 01–07.

Bellido, G. G., & Beta, T. (2009). Anthocyanin composition and oxygen radical scavenging capacity (ORAC) of milled and pearled purple, black, and common barley. *Journal of Agricultural and Food Chemistry, 57*(3), 1022–1028.

Blazek, J., & Copeland, L. (2008). Pasting and swelling properties of wheat flour and starch in relation to amylose content. *Carbohydrate Polymers, 71*, 380–387.

Borneo, R., & León, A. E. (2012). Whole grain cereals: Functional components and health benefits. *Food & Function, 3*(2), 110–119.

Boukid, F., Folloni, S., Sforza, S., Vittadini, E., & Prandi, B. (2018). Current trends in ancient grains-based foodstuffs: Insights into nutritional aspects and technological applications. *Comprehensive Reviews in Food Science and Food Safety, 17*(1), 123–136.

Brennan, C. S., & Cleary, L. J. (2005). The potential use of cereal (1→ 3, 1→ 4)-β-D-glucans as functional food ingredients. *Journal of Cereal Science, 42*(1), 1–13.

Călinoiu, L. F., & Vodnar, D. C. (2018). Whole grains and phenolic acids: A review on bioactivity, functionality, health benefits and bioavailability. *Nutrients, 10*(11), 1615.

de Camargo, A. C., Franchin, M., & Shahidi, F. (2019). Tocopherols and tocotrienols: Sources, analytical methods, and effects in food and biological systems. *Encyclopedia of Food Chemistry*, 561–570.

Capitani, M. E. A., Spotorno, V., Nolasco, S. M., & Tomás, M. C. (2012). Physicochemical and functional characterization of by-products from chia (*Salvia hispanica* L.) seeds of Argentina. *LWT-Food Science and Technology, 45*(1), 94–102.

Chandrasekara, A., & Shahidi, F. (2011a). Bioactivities and antiradical properties of millet grains and hulls. *Journal of Agricultural and Food Chemistry, 59*(17), 9563–9571.

Chandrasekara, A., & Shahidi, F. (2011b). Determination of antioxidant activity in free and hydrolyzed fractions of millet grains and characterization of their phenolic profiles by HPLC-DAD-ESI-MSn. *Journal of Functional Foods, 3*(3), 144–158.

Cho, K. C., & White, P. J. (1993). Enzymatic analysis of β-glucann content in different oat genotypes. *Cereal Chemistry*, 70(5), 539–542

Damiri, F., Bachra, Y., Bounacir, C., Laaraibi, A., & Berrada, M. (2020). Synthesis and characterization of lyophilized chitosan-based hydrogels cross-linked with benzaldehyde for controlled drug release. *Journal of Chemistry*. https://doi.org/10.1155/2020/8747639

Dhingra, D., Michael, M., Rajput, H., & Patil, R. T. (2012). Dietary fiber in foods: A review. *Journal of Food Science & Technology, 49*, 255–266.

Đorđević, T. M., Šiler-Marinković, S. S., & Dimitrijević-Branković, S. I. (2010). Effect of fermentation on antioxidant properties of some cereals and pseudo cereals. *Food Chemistry, 119*(3), 957–963.

Dvoncova, D., Havrlentova, M., Hlinkova, A., & Hozlar, P. (2010). Effect of fertilization and variety on the beta-glucan content in the grain of oats. *Żywność Nauka Technologia Jakość, 17*(3), 108.

Dykes, L., & Rooney, L. W. (2007). Phenolic compounds in cereal grains and their health benefits. *Cereal Foods World, 52*(3), 105–111.

Escribano-Bailón, M. T., Santos-Buelga, C., & Rivas-Gonzalo, J. C. (2004). Anthocyanins in cereals. *Journal of Chromatography, 1054*, 129–141.

FAO. (2020). https://www.fao.org/faostat/en/#data/QCL. Accessed 8 March 2022.

FAO Cereal Supply and Demand Brief. World Food Situation. Cereal production and inventories to decline but overall supplies remain adequate. Food and Agriculture Organization of the United Nations. Available online: http://www.fao.org/worldfoodsituation/csdb/en/. Accessed 29 January 2019.

Ficco, D. B. M., De Simone, V., De Leonardis, A. M., Giovanniello, V., Del Nobile, M. A., et al. (2016). Use of purple durum wheat to produce naturally functional fresh and dry pasta. *Food Chemistry, 205*, 187–195.

Fouad, D., Bachra, Y., Ayoub, G., Ouaket, A., Bennamara, A., Knouzi, N., & Berrada, M. (2020). A novel drug delivery system based on nanoparticles of magnetite Fe 3 O 4 embedded in an auto cross-linked chitosan. In *Chitin and chitosan-physicochemical properties and industrial applications* (pp. 1–24). IntechOpen.

Gadde, K. M., Martin, C. K., Berthoud, H. R., & Heymsfield, S. B. (2018). Obesity: Pathophysiology and management. *Journal of the American College of Cardiology, 71*(1), 69–84.

Ghoneum, M., & Abedi, S. (2004). Enhancement of natural killer cell activity of aged mice by modified arabinoxylan rice bran (MGN-3/biobran). *Journal of Pharmacy and Pharmacology, 56*(12), 1581–1588.

Giri, S., Dutta, P., & Giri, T. K. (2021). Inulin-based carriers for colon drug targeting. *Journal of Drug Delivery Science and Technology*, 102595.

Gupta, N., Jangid, A. K., Pooja, D., & Kulhari, H. (2019). Inulin: A novel and stretchy polysaccharide tool for biomedical and nutritional applications. *International Journal of Biological Macromolecules, 132*, 852–863.

Gylling, H., Plat, J., Turley, S., Ginsberg, H. N., Ellegård, L., Jessup, W., … Chapman, M. J. (2014). Plant sterols and plant stanols in the management of dyslipidaemia and prevention of cardiovascular disease. *Atherosclerosis, 232*(2), 346–360.

Hodzic, Z., Pasalic, H., Memisevic, A., Srabovic, M., Saletovic, M., & Poljakovic, M. (2009). The influence of total phenols content on antioxidant capacity in the whole grain extracts. *European Journal of Scientific Research, 28*(3), 471–477.

Hosoda, A., Okai, Y., Kasahara, E., Inoue, M., Shimizu, M., Usui, Y., … Higashi-Okai, K. (2012). Potent immunomodulating effects of bran extracts of traditional Japanese millets on nitric oxide and cytokine production of macrophages (RAW264. 7) induced by lipopolysaccharide. *Journal of UOEH, 34*(4), 285–296.

Hosseinian, F. S., Li, W. D., & Beta, T. (2008). Measurement of anthocyanins and other phytochemical in purple wheat. *Food Chemistry, 109*(4), 916–924.

Hübner, F., & Arendt, E. K. (2013). Germination of cereal grains as a way to improve the nutritional value: A review. *Critical Reviews in Food Science and Nutrition, 53*(8), 853–861.

Izydorczyk, M. S., & Biliaderis, C. G. (2000). *Structural and functional aspects of cereal arabinoxylans and β-glucans*. In Developments in Food Science (Vol. 41, pp. 361–384). Elsevier.

Jang, S., & Xu, Z. (2009). Lipophilic and hydrophilic antioxidants and their antioxidant activity in purple rice bran. *Journal of Agricultural and Food Chemistry, 57*(3), 858–862.

Jiménez-Escrig, A., Santos-Hidalgo, A. B., & Saura-Calixto, F. (2006). Common sources and estimated intake of plant sterols in the Spanish diet. *Journal of Agricultural and Food Chemistry, 54*(9), 3462–3471.

Kamiloglu, S., Capanoglu, E., Grootaert, C., & Van Camp, J. (2015). Anthocyanin absorption and metabolism by human intestinal Caco-2 cells—A review. *International Journal of Molecular Sciences, 16*, 21555–21574.

Kaur, H., Singh, B., & Singh, A. (2021). Comparison of dietary fibers obtained from seven Indian cereal grains. *Journal of Cereal Science, 102*, 103331.
Kiarie, E., Romero, L. F., & Ravindran, V. (2014). Growth performance, nutrient utilization, and digesta characteristics in broiler chickens fed corn or wheat diets without or with supplemental xylanase. *Poultry Science, 93*(5), 1186–1196.
Kim, M. J., Hong, S. Y., Kim, S. K., Cheong, C., Park, H. J., Chun, H. K., … Kang, S. A. (2009). β-Glucan enhanced apoptosis in human colon cancer cells SNU-C4. *Nutrition Research and Practice, 3*(3), 180–184.
Kim, M. Y., Lee, S. H., Jang, G. Y., Park, H. J., Li, M., Kim, S., … Jeong, H. S. (2015). Effects of high hydrostatic pressure treatment on the enhancement of functional components of germinated rough rice (Oryza sativa L.). *Food Chemistry, 166*, 86–92.
Kochar, J., Gaziano, J. M., & Djoussé, L. (2011). Breakfast cereals and risk of hypertension in the physicians' health study I. *Clinical Nutrition, 31*(1), 89–92.
Koruri, S. S., Banerjee, D., Chowdhury, R., & Bhattacharya, P. (2014). Studies on prebiotic food additive (inulin) in indian dietary fibre sources-garlic (Allium sativum), wheat (Triticum spp.), oat (Avena sativa) and dalia (bulgur). *International Journal of Pharmacy and Pharmaceutical Sciences, 6*(9), 278–282.
Lattimer, J. M., & Haub, M. D. (2010). Effects of dietary fiber and its components on metabolic health. *Nutrients, 2*(12), 1266–1289.
Li, W., Cui, S. W., & Kakuda, Y. (2006). Extraction, fractionation, structural and physical characterization of wheat β-D-glucans. *Carbohydrate Polymers, 63*(3), 408–416.
Li, H., Wang, R., Liu, J., Zhang, Q., Li, G., Shan, Y., & Ding, S. (2020). Effects of heatmoisture and acid treatments on the structural, physicochemical, and in vitro digestibility properties of lily starch. *International Journal of Biological Macromolecules, 148*, 956–968.
Li, H., Yin, J., Tan, B., Chen, J., Zhang, H., Li, Z., & Ma, X. (2021). Physiological function and application of dietary fiber in pig nutrition: A review. *Animal Nutrition, 7*, 259–267.
Liu, Y., Wu, Q., Wu, X., Algharib, S. A., Gong, F., Hu, J., … Wang, Y. (2021). Structure, preparation, modification, and bioactivities of β-glucan and mannan from yeast cell wall: A review. *International Journal of Biological Macromolecules, 173*, 445–456.
Liu, C., Yang, S., Wang, K., Bao, X. M., Liu, Y. M., Zhou, S. Y., Liu, H. W., Qiu, Y. L., Wang, T., & Yu, H. Y. (2019). Alkaloids from traditional Chinese medicine against hepatocellular carcinoma. *Biomedicine & Pharmacotherapy, 120*, 109543.
Lloyd, B. J., Siebenmorgen, T. J., & Beers, K. W. (2000). Effect of commercial processing on antioxidants in rice bran. *Cereal Chemistry, 77*(5), 551–555.
Lobo, V., Patil, A., Phatak, A., & Chandra, N. (2010). Free radicals, antioxidants and functional foods: Impact on human health. *Pharmacognosy Reviews, 4*(8), 118.
Lu, J., & Li, Y. (2006). Effects of arabinoxylan solubilization on wort viscosity and filtration when mashing with grist containing wheat and wheat malt. *Food Chemistry, 98*(1), 164–170.
Ma, Q. T., Li, Y. Q., Li, P. F., Wang, M., Wang, J. K., Tang, Z. Y., Wang, T., Luo, L. L., Wang, C. G., & Wang, T. (2019). Research progress in the relationship between type 2 diabetes mellitus and intestinal flora. *Biomedicine & Pharmacotherapy, 117*, 109138.
Mattila, P., Pihlava, J. M., & Hellström, J. (2005). Contents of phenolic acids, alkyl-and alkenylresorcinols, and avenanthramides in commercial grain products. *Journal of Agricultural and Food Chemistry, 53*(21), 8290–8295.
McKevith, B. (2004). Nutritional aspects of cereals. *Nutrition Bulletin, 29*(2), 111–142.
Meganathan, P., & Fu, J. Y. (2016). Biological properties of tocotrienols: Evidence in human studies. *International Journal of Molecular Sciences, 17*(11), 1682.
Mir, N. A., Riar, C. S., & Singh, S. (2018). Nutritional constituents of pseudo cereals and their potential use in food systems: A review. *Trends in Food Science & Technology, 75*, 170–180.

Nelson, K., Stojanovska, L., Vasiljevic, T., & Mathai, M. (2013). Germinated grains: A superior whole grain functional food? *Canadian Journal of Physiology and Pharmacology, 91*(6), 429–441.

Oboh, G., Ademiluyi, A. O., & Akindahunsi, A. A. (2010). The effect of roasting on the nutritional and antioxidant properties of yellow and white maize varieties. *International Journal of Food Science & Technology, 45*(6), 1236–1242.

Olivas-Aguirre, F. J., Rodrigo-García, J., Martínez-Ruiz, N. D. R., Cárdenas-Robles, A. I., Mendoza-Díaz, S. O., et al. (2016). Cyanidin-3-O-glucoside: Physical-chemistry, foodomics and health effects. *Molecules, 21*, 1264.

Parker, R. S. (2000). Phytochemicals: Carotenoids. *Wiley Encyclopedia of Food Science and Technology, 3*, 909–915.

Punia, S., Sandhu, K. S., & Siroha, A. K. (2019). Difference in protein content of wheat (*Triticum aestivum* L.): Effect on functional, pasting, color and antioxidant properties. *Journal of the Saudi Society of Agricultural Sciences, 18*(4), 378–384.

Rao, M. V. S. S. T., & Muralikrishna, G. (2002). Evaluation of the antioxidant properties of free and bound phenolic acids from native and malted finger millet (Ragi, Eleusine coracana Indaf-15). *Journal of Agricultural and Food Chemistry, 50*(4), 889–892.

Rao, S., Schwarz, L. J., Santhakumar, A. B., Chinkwo, K. A., & Blanchard, C. L. (2018). Cereal phenolic contents as affected by variety and environment. *Cereal Chemistry, 95*(5), 589–602.

Roder, N., Ellis, P. R., & Butterworth, P. J. (2005). Starch molecular and nutritional properties: A review. *Advances in Molecular Medicine, 1*(12), 5–14.

Robbins, R. J. (2003). Phenolic Acids in Foods: An Overview of Analytical Methodology. *Journal of Agricultural and Food Chemistry, 51*, 2866–2887.

Samantha, K. G., Megan, R., Balazs, B., & Kevin, W. (2020). Dietary fibre in gastrointestinal health and disease. *Nature Reviews Gastroenterology & Hepatology, 18*, 101–116.

Sandhu, K. S., & Punia, S. (2017). Enhancement of bioactive compounds in barley cultivars by solid substrate fermentation. *Journal of Food Measurement and Characterization, 11*(3), 1355–1361.

Sen, C. K., Khanna, S., & Roy, S. (2007). Tocotrienols in health and disease: The other half of the natural vitamin E family. *Molecular Aspects of Medicine, 28*(5–6), 692–728.

Shahidi, F., & Yeo, J. D. (2018). Bioactivities of phenolics by focusing on suppression of chronic diseases: A review. *International Journal of Molecular Sciences, 19*(6), 1–16.

Shashi, B. K., Sharan, S., Shittalamani, S., Shankar, A. G., & Nagarathna, T. K. (2007). Micronutrient composition, antinutritional factors and bioaccessibility of iron in different finger millet (*Eleusine coracana*) genotypes. *Karnataka Journal of Agricultural Sciences, 20*(3), 583–585.

Shi, J., Shan, S., Li, Z., Li, H., Li, X., & Li, Z. (2015). Bound polyphenol from foxtail millet bran induces apoptosis in HCT-116 cell through ROS generation. *Journal of Functional Foods, 17*, 958–968.

Shoaib, M., Shehzad, A., Omar, M., Rakha, A., Raza, H., Sharif, H. R., … Niazi, S. (2016). Inulin: Properties, health benefits and food applications. *Carbohydrate Polymers, 147*, 444–454.

Shahidi, F., & Chandrasekara, A. (2013). Millet grain phenolics and their role in disease risk reduction and health promotion: A review. *Journal of Functional Foods, 5*(2), 570–581.

Shah, A. D., Langenberg, C., Rapsomaniki, E., Denaxas, S., Pujades-Rodriguez, M., Gale, C. P., ... & Hemingway, H. (2015). Type 2 diabetes and incidence of cardiovascular diseases: a cohort study in 1· 9 million people. *The lancet Diabetes & endocrinology, 3*(2), 105–113.

Sim, S. Y. J., Srv, A., Chiang, J. H., & Henry, C. J. (2021). Plant proteins for future foods: A roadmap. *Food, 10*(8), 1967.

Singh, R., De, S., & Belkheir, A. (2013). Avena sativa (oat), a potential neutraceutical and therapeutic agent: An overview. *Critical Reviews in Food Science and Nutrition, 53*(2), 126–144.

Siroha, A. K., Sandhu, K. S., & Kaur, M. (2016). Physicochemical, functional and antioxidant properties of flour from pearl millet varieties grown in India. *Journal of Food Measurement and Characterization, 10*(2), 311–318.

Soliman, G. A. (2019). Dietary fiber, atherosclerosis, and cardiovascular disease. *Nutrients, 11*(5), 1155.

Song, Y., Jeong, H. Y., Lee, J. K., Choi, Y. S., Kim, D. O., Jang, D., … Kang, H. (2019). Enzyme treatment alters the anti-inflammatory activity of the water extract of wheat germ in vitro and in vivo. *Nutrients, 11*(10), 2490.

Srividya, A. R., Venkatesh, N., & Vishnuvarthan, V. J. (2010). Nutraceutical as medicine. *International Journal of Pharmacy and Pharmaceutical Sciences, 1*, 132–145.

Suriano, S., Iannucci, A., Codianni, P., Fares, C., Menga, V., Russo, M., … Troccoli, A. (2020). Carotenoids and tocols content in genotypes of colored barley. *Journal of Cereal Science, 96*, 103110.

Swanson, J. E. (2003). *Bioactive food components.* Encycl food cult, gale cengage. http://www.enotes.com/bioactive-food-components-reference/bioactive-foodcomponents.

Tang, W. H. W., & Hazen, S. L. (2017). Gut microbiome and its role in cardiovascular diseases. *Current Opinion in Cardiology, 32*, 761–766.

Thebaudin, J. Y., Lefebvre, A. C., Harrington, M., & Bourgeois, C. M. (1997). Dietary fibres: Nutritional and technological interest. *Trends in Food Science & Technology, 8*(2), 41–48.

Ti, H., Zhang, R., Zhang, M., Li, Q., Wei, Z., Zhang, Y., … Ma, Y. (2014). Dynamic changes in the free and bound phenolic compounds and antioxidant activity of brown rice at different germination stages. *Food Chemistry, 161*, 337–344.

Tsuda, T. (2012). Dietary anthocyanin-rich plants: Biochemical basis and recent progress in health benefits studies. *Molecular Nutrition & Food Research, 56*, 159–170.

Turrini, E., Maffei, F., Milelli, A., Calcabrini, C., & Fimognari, C. (2019). Overview of the anticancer profile of avenanthramides from oat. *International Journal of Molecular Sciences, 20*(18), 4536.

Van Hung, P. (2016). Phenolic compounds of cereals and their antioxidant capacity. *Critical reviews in food science and nutrition, 56*(1), 25–35.

Venn, B. J., & Mann, J. I. (2004). Cereal grains, legumes and diabetes. *European journal of clinical nutrition, 58*(11), 1443–1461.

Wang, Y., Harding, S. V., Thandapilly, S. J., Tosh, S. M., Jones, P. J., & Ames, N. P. (2017). Barley β-glucan reduces blood cholesterol levels via interrupting bile acid metabolism. *British Journal of Nutrition, 118*(10), 822–829.

Wilson, T. A., Nicolosi, R. J., Delaney, B., Chadwel, K., Moolchandani, V., Kotyla, T., Ponduru, S., Zheng, G. H., Hess, R., Knutson, N., et al. (2004). Reduced and high molecular weight barley beta-glucans decrease plasma total and non-HDL-cholesterol in hypercholesterolemic Syrian golden hamsters. *Journal of Nutrition, 134*, 2617–2622.

Wood, P. J. (2010). Oat and rye β-glucan: properties and function. Cereal Chemistry, 87(4), 315–330..

Xu, Z., & Godber, J. S. (1999). Purification and identification of components of γ-oryzanol in rice bran oil. *Journal of Agricultural and Food Chemistry, 47*(7), 2724–2728.

Yang, H., Ding, M., Sampson, L., Willett, W. C., Manson, J. E., Wang, M., … Sun, Q. (2020). Intake of whole grain foods and risk of type 2 diabetes: Results from three prospective cohort studies. *BMJ, 370*, 2206–2218.

Yao, Y., Shi, Z., & Ren, G. (2014). Antioxidant and immunoregulatory activity of polysaccharides from quinoa (*Chenopodium quinoa Willd.*). *International Journal of Molecular Sciences, 15*(10), 19307–19318.

Youssef, M. K. E., Nassar, A. G., El Fishawy, F. A., & Mostafa, M. A. (2016). Assessment of proximate chemical composition and nutritional status of wheat biscuits fortified with oat powder. Assiut Journal of Agricultural Sciences 47(5), 83–94.

Zhu, F., Du, B., Bian, Z., & Xu, B. (2015). Beta-glucans from edible and medicinal mushrooms: Characteristics, physicochemical and biological activities. *Journal of Food Composition and Analysis, 41*, 165–173.

Zullaikah, S., Melwita, E., & Ju, Y. H. (2009). Isolation of oryzanol from crude rice bran oil. *Bioresource Technology, 100*(1), 299–302.

Chapter 2
Novel Approaches to Improve Functional Potential of Cereals

Anil Kumar Siroha, Sneh Punia Bangar, Pawan Kumar Rose, and Ajay Kumar

2.1 Introduction

As the population grows, food insecurity such as chronic malnutrition and potential food shortages are becoming a major threat to humanity (Mensi & Udenigwe, 2021; Mir et al., 2018). In response, the United Nations has proposed "Zero Hunger" under Sustainable Development Goal 2 to end hunger, achieve food security and improved nutrition and promote sustainable agriculture (Fischer et al., 2021). Cereal grains are universally recognized as stable food that is the primary source of energy, carbs, protein, and some minerals, vitamins, bioactive nutrients, as well as antinutrients (Swaminathan, 2021). There are seven major cereals: wheat, rice, corn, rye, oats, millet, barley and sorghum, among that wheat, rice, and maize, dominate global agricultural output. The human diet was already dominated by cereal-based bread and porridge-like foods in prehistoric times (Valamoti et al., 2019). At the household level, several traditional methods of processing and preparing foods can enhance the bioavailability of the micronutrients in plant-based diets (Lakshmi et al., 2015). The nutritional and organoleptic characteristics of several cereal-based products have been improved through conventional processes like fermentation and

A. K. Siroha (✉)
Department of Food Science and Technology, Chaudhary Devi Lal University, Sirsa, Haryana, India

S. P. Bangar
Department of Food, Nutrition and Packaging Sciences, Clemson University, Clemson, SC, USA

P. K. Rose
Department of Energy and Environmental Sciences, Chaudhary Devi Lal University, Sirsa, Haryana, India

A. Kumar
Department of Law, Chaudhary Devi Lal University, Sirsa, Haryana, India

S. Punia Bangar, A. Kumar Siroha (eds.), *Functional Cereals and Cereal Foods*,
https://doi.org/10.1007/978-3-031-05611-6_2

malting in monitored conditions (Marsh et al., 2014; Ochanda et al., 2010). In recent studies, germination has been found to enhance further the nutritional and medicinal benefits of edible grains (Gan et al., 2017). From the findings, it is evident that fermentation is a viable method that increases the nutritional and functional aspects of the cereal-based product (Punia et al., 2021). Additionally, these processes have been reported to reduce antinutrients and boost the bioavailability of the nutrients (Ochanda et al., 2010; Onyango et al., 2013). An overview of different methods for improving the functional potential of cereals is presented in this chapter.

2.2 Germination

The process of germination is relatively inexpensive, easy, biologically safe that triggered by the actions of enzymes, external stimuli, and phytohormones (Fig. 2.1; Table 2.1). Furthermore, germination increases cereal nutrition value and reduces amount of phytic acid, a substance that interferes with mineral absorption and protein digestion (Albarracín et al., 2019). In germination, seeds emerge from their latency stage, causing an increase in weight and minerals bioavailability and these seeds contain an appreciable amount of ascorbic acid, riboflavin, choline, thiamine, tocopherols, and pantothenic acid (Sangronis & Machado, 2007). The amount of metabolic changes that grains undergo during germination is affected by genetic diversity, the presence or absence of hulls, and the conditions of germination (Caceres et al., 2014; Swanston & Middlefell-Willliams, 2012). Selection of the right cultivar and optimization of sprouting conditions are therefore vital for improving desirable and reducing detrimental changes in sprouted cereal flour quality (Aparicio-García et al., 2020). Karwasra et al. (2018) reported an increase in the antioxidant potential in various wheat flour with germination. Wang et al. (2022) observed increased γ-aminobutyric acid (about 194 mg/kg), ferulic acid (8.07 mg/kg), and total phenolic compounds (5370 mg/kg) in germinated rice flour. During

Fig. 2.1 Germination of pearl millet grain

Table 2.1 Effect of different processing's on nutritional profile of cereals

Sample	Major results	Reference
Germination		
Oat	Enhanced free phenolics, antioxidant capacity, protein hydrolysis and α-amylase activity in oat flours, β-glucan content reduced after 96 h	Aparicio-García et al. (2020)
Red rice	Phenolic acids and flavonoids, and GABA content increased with germination	Müller et al. (2021)
Rice	As rice germination progressed, the reducing sugar content and the amylase activity increased	Veluppillai et al. (2009)
Barley, wheat, rice, oat, maize	Protein content improved with germination	Youssef et al. (2013); Hung et al. (2012); Moongngarm and Saetung (2010); Tian et al. (2010); Hiran et al. (2016)
Wheat	Significant increase in antioxidant activity, solubility, water absorption capacity and oil absorption capacity	Karwasra et al. (2018)
Foxtail and kodo millet	Functional properties of β-glucan extracted from germinated millet flour improved	Sharma et al. (2018)
Fermentation		
Wheat	Antioxidant activity increased with fermentation process	Sandhu and Punia (2017)
Wheat, rice, oat, sorghum, maize	Polyphenols and antioxidants increased after fermentation	Saharan et al. (2017)
Rice, seim seeds	Fermented flour showed higher total phenolic content and DPPH radical-scavenging activity	Sadh et al. (2017)
Barley	Total phenolic content, antioxidant activity, total flavonoids content, and metal chelating activity increased after fermentation process	Sandhu and Punia (2017)
Pearl millet	Bioactive compounds (ascorbic acid, gallic acid and p-Coumaric acid) were significantly increased	Salar et al. (2017)
Sorghum, maize, millet	BD decreased and water and oil absorption capacity increased after fermentation	Atuna et al. (2022)
Barley	Barley bran showed higher content of ascorbic acid, gallic acid, catechin, vanillin and resorcinol compared to unfermented sample	Bangar et al. (2021a)
Rice	Rice bran showed higher content of ascorbic acid, and gallic acid while lesser value for catechin, and vanillin acid	Punia et al. (2021)

(continued)

Table 2.1 (continued)

Sample	Major results	Reference
Extrusion		
Barley	Extrusion improves the antioxidant activity and resistant starch content. Glycemic index values reduced with the extrusion process	Bangar et al. (2022)
Barley	Antioxidant properties increased whereas total phenolic content decreased	Sharma et al. (2012)
Rice	RS content improve with extrusion process. Functional properties gelation capacity, swelling power, emulsion and foaming capacity decreased	Gulzar et al. (2021)
Rice	Antioxidant properties reduced after extrusion process	Gujral et al. (2012)
Maize	RDS content increase whereas SDS and RS content decrease after treatment. Pasting and gelatinization behaviour also affected.	Zhang et al. (2016)
Wheat	Antioxidant properties reduced after extrusion process and extruded product showed acceptable nutritional value	Bhat et al. (2019)
Brown rice, wheat, and oat	Free phenolics reduced while bound phenolics increase	Zeng et al. (2016)

germination, binding phenolic acids in the cell wall are transformed and decomposed, which increases the amount of free ferulic acid. During germination of rice, amylose and amylopectin are broken down into maltose, dextrin, and glucose as well as proteins into amino acids and peptides through the action of activated hydrolytic enzymes (Singh et al., 2015). Recently, scientists have studied the germination of several types of grains and reported improvement in bioactive compounds and protein digestibility (Müller et al., 2021). Aparicio-García et al. (2020) demonstrated that germination mainly at elevated temperatures and extended periods (12–20 °C and 24–216 h) enhanced the free phenolics, antioxidant capacity, protein hydrolysis, and α-amylase activity in oat flours. The germination for short periods (24–96 h) retained β-glucan content but decreased beyond this time period. Compared with raw grain, sprouted barley had a 38.6% higher protein content (Ortiz et al., 2021). Hiran et al. (2016) observed significant improvement in crude protein (15.7% in maize) after germination. After germination, the increase in protein content attributed to the production of enzymes, modification in composition caused by the degradation of other constituents, and newly synthesized protein (Xu et al., 2019b). Germination also increased phenols level and antioxidant activity in cereal grain which attributed to their improved nutraceutical properties (Hiran et al., 2016).

The hydrolytic and amylolytic enzymes are activates during malting and germination process leads to a starch reduction and simple sugars improvement with time as a determining factor (Nkhata et al., 2018). The concentration of maltose has been high in sprouted rice, sorghum, and millet in comparison to glucose, however, the sugar profile of sprouted grain is mainly dependent on the species (Agu et al., 2012; You et al., 2016). El-Refai et al. (2012) demonstrated that the germination increased the total sugar content and decreased the starch content in barley and oat grain. Qin et al. (2021) found that total sugar content of barley decreased with germination, except GAN-7 variety which increases at 5th day of germination. Maltotetraose ratio was found to decrease with increasing germination time, while the ratios of maltose and sucrose, fructose, glucose were increased during germination. Yang et al. (2021) observed an increased level of crude fibre, soluble sugars, free amino acids, bioactive components, and α-amylase activity in germinated proso millet flours. Germination improves protein and starch digestibility. As brown rice germinates, its outer bran layer softens, allowing the rice to absorb more water and become more easily cooked. The enzymes liberated during the sprouting of rice cause break down of grain carbohydrate and protein, which attribute sweet flavour to cooked sprouted rice (Kayahara & Tsukahara, 2000).

Elkhalifa and Bernhardt (2010) evaluated the functional aspects of germinated sorghum flour and observed enhancement in protein solubility, water absorption capacity, oil absorption capacity after the germination process as compared to the control sample while bulk density and least gelation capacity of flour reduced. Ocheme et al. (2015) observed an increase in foaming and emulsion capacity while a decrease in bulk density in germinated sorghum flour. The decrease in bulk density (loose and packed) value might be associated with the breakdown of starch and protein during germination. Karwasra et al. (2018) stated that the swelling power was reduced with germination while other properties (solubility index, water holding capacity, oil absorption capacity) improved with germination in wheat. Hussain and Uddin (2012) observed that the germination process reduces the bulk density and oil absorption capacity while increasing the water absorption and foaming capacity. Carbohydrates and proteins are the main constituents that enhance water uptake and solubility of the flours due to the presence of hydrophilic parts (Wani et al., 2013). Sharma et al. (2018) reported improvement in foaming capacity, foaming stability, water binding capacity, and swelling power in foxtail and kodo millet flour after germination. However, rheological properties (storage modulus, loss modulus, apparent viscosity) of β-glucan decreased with germination.

2.3 Fermentation

Fermentation is accomplished through the action of microorganisms or enzymes that causes desirable modifications in flavour and texture (Heiniö et al., 2003). Fermentation is a important steps in the food processing (Fig. 2.2; Table 2.1). It offers several advantages, including the ability to preserve food, improve food safety, enhance flavor and acceptability, increase diversity in the diet, improve nutritional content, reduce antinutrients, and enhanced functional properties (Westby

Fig. 2.2 Fermented and non-fermented pearl millet grain

et al., 1997). The fermentation technology accounts for about one-third of the total human food consumption (Nout & Kiers, 2005). Solid-state (SSF) and submerged method (SMF) are commonly used fermentation processes. In SMF, microorganisms grow in an enriched liquid culture of nutrients, water, and oxygen which allow rapid consumption of substrates (Liu & Kokare, 2017; Rose, 2022; Subramaniyam & Vimala, 2012). While, SSF technique permits microorganisms to grow on solid substrates surrounded by a continuous gaseous phase (Pandey, 2003). Microorganisms such as *Saccharomyces cerevisiae* (Yeast) (Rani et al., 2018), *Helvella lacunose* X1, *Agaricus bisporus* AS2796, *Fomitiporia yanbeiensis* G1, *Aspergillus oryzae, Cordyceps militaris, Rhizopus oligosporus, Lentinula edode* (Fungus) (Saharan et al., 2017; Sánchez-Magaña et al., 2019; Stoffel et al., 2019; Xu et al., 2018, 2019a), and *Bifido bacterium spp., B. animalis, B. breve B. longum* etc. (Bacteria) are generally used for fermenting cereal grains (Ayyash et al., 2018).

Fermentation is an ancient technique that has been used to improve nutritional quality (Frias et al., 2005) as well as to raise phenolic content in foods (Martins et al., 2011). Sandhu and Punia (2017) studied role of SSF with time on the bioactive compound of wheat and reported until the 5th day, fermentation increased bioactive properties, and afterward, they decreased. Saharan et al. (2017) assessed the potential of *Aspergillus oryzae* fermentation on the bioactive profile of different cereals grains. It was observed that fermentation enhances the bioactive profile of cereal grains. After fermentation, *Oryza sativa* and *Triticum aestivum* showed the highest levels of polyphenols and antioxidants followed by > *Sorghum bicolour* > *A. sativa* > *Zea mays* because, during fermentation, they exhibited high enzyme activities. Sadh et al. (2017) determined the effect of two fungal strains (*Aspergillus oryzae* and *Aspergillus awamori*) on the antioxidant activity of rice and seim seeds. Total phenolic content and DPPH radical scavenging of fermented flours were higher than non-fermented flour. In another study, Sandhu and Punia (2017) evaluated the bioactive profile of different barley cultivars fermented with *Aspergillus awamori nakazawa* (MTCC-548) and observed that bioactive properties were significantly increased during fermentation up to day 5, but decreased thereafter. The SSF process involves the production of α-amylase and β-glucosidase (hydrolytic enzymes), which soften the kernel structure, dissolve the cell walls and liberate bound nutrients. As a result of enzyme hydrolysis of phenolic glucosides and the free polyphenol content improved which enhanced the nutraceutical properties of food (Cai et al., 2012; McCue & Shetty, 2003). Salar et al. (2017) fermented the pearl millet using *Aspergillus sojae* to evaluate bioactivity and DNA damage protection activity. During SSF, the productions of enzymes (mainly α-amylase, β-glucosidase, xylanase) and specific bioactive compounds (mainly ascorbic acid, gallic acid, p-Coumaric acid) were significantly increased. The authors concluded that fermented flours could be used to make functional and nutraceutical foods. Bangar et al. (2021a) reported a high concentration of ascorbic acid (107.15 μg/g), gallic acid (405.5 μg/g), catechin (88.3 μg/g), vanillin (40.89 μg/g), and resorcinol (20.7 μg/g) in barley bran fermented with *Aspergillus oryzae* in comparison to non-fermented barley bran i.e. ascorbic acid (20.44 μg/g), gallic acid (12.75 μg/g), and catechin (9.9 μg/g). In another study, Punia et al. (2021) fermented rice bran with *Aspergillus oryzae* and observed fermentation improved the level of ascorbic acid and gallic acid from 11.1 μg/g to 12.7 μg/g, and 14.8 μg/g to 23.3 μg/g, respectively. However, catechin and vanillin decreased after the fermentation process. After fermentation, cereal-based food becomes rich in protein by eliminating carbohydrates portion and lowering energy consumption during cooking (Asensio-Grau et al., 2020; Petrova & Petrov, 2020).

Additionally, fermentation affects the properties of cereal grains on a nutritional and functional level. Atuna et al. (2022) studied the fermentation effect on the functional properties of different cereals and observed increased water absorption capacity and oil absorption capacity of cereal flours while reduced bulk density. Sorghum and maize flours exhibited slightly higher peak viscosity (PV) whereas millet flours had a lower value compared with untreated flour. Abd Elmoneim et al. (2005) observed higher protein solubility, OAC, FC, emulsion capacity (EC) and emulsion stability (ES) content while lower WAC in fermented sorghum flour than in the non-fermented sample. Protein solubility may enhance the EC of fermented sorghum flour.

2.4 Extrusion

Extrusion cooking is the preferred method of making breakfast cereals, ready-to-eat snacks, and other textured foods, since mid-1930s (Brennan et al., 2011). The word 'extrusion' comes from the Latin word 'extrude' which means 'thrust out' or 'force out'. Extrusion is a complex process that involves mixing, mechanical shearing, heating, pressurization and of two types (basis of degree of heat treatment), conventional cold extrusion (usually between 25 °C and 45 °C) and thermoplastic extrusion (extrusion cooking/puffing) (Menis-Henrique et al., 2020). Cold extrusion forms desired shapes by forcing hydrated flour through a die. The material becomes compacted by applied pressure during the passage through along with slight heating at 35–55 °C which strengthen the structure and prevents further expansion (Robin & Palzer, 2015). Thermoplastic extrusion is a thermo-mechanical process that alters the microstructure and chemical properties of the materials, as well as the macroscopic characteristics, such as shape and appearance (Spotti & Campanella, 2017). During extrusion, covalent and non-covalent bonds between components in complex food matrixes are disrupted, which alters functional properties. Extrusion is applied to develop ready-to-eat cereals, confectionery goods, and sweet and salty snacks, which has sparked an interest in studying its function and nutritional properties (Gulati et al., 2020). Extrusion's advantages include being economical, high-productivity, energy-efficient, and ability to synthesize various products (Faraj et al., 2004; Kim et al., 2006; Ye et al., 2018). The physicochemical properties of extruded products are influenced by several variables including raw material attributes, mixing and conditioning the raw material, barrel temperature, pressure, screw speed, moisture content, flow rate, energy input, residence type, screw configuration, etc. (Brennan et al., 2011). Considering the final product's nutritional value, extrusion cooking offers several advantages, including deactivation of antinutrients, the annihilation of aflatoxins, and enhanced digestion of fiber (Saalia & Phillips, 2011; Singh et al., 2007). Higher plants store starch in partially crystalline granules form which can vary in shape and molecular structure between species and within plant species (Blazek & Copeland, 2008). Due to its numerous functional properties, starch is vital in food processing, especially as it allows products to be modified in texture. Extrusion can change starch properties as it becomes retrograde when frozen and thawed. Through IECT (Improved Extrusion Cooking Technology), freeze-thaw stability can be improved by reducing syneresis and retrograding the interaction of starch chains (Ye et al., 2016). The extrusion process increased resistant starch (RS) and reduces Glycaemic index (GI) values for barley flour (Bangar et al., 2022). Gulzar et al. (2021) reported an improvement of 4.91–6.83% in the resistant starch content of extruded rice flour. The combination of heat and moisture in extrusion modifies the starch characteristics simultaneously, their interaction with non-starches, resulting in retrograded starch production. Extrusion is a broadly used process to modify the functional aspects of materials rich in starch content through pregelatinization, followed by retrogradation, which eventually produces retrograded resistant starch (RS3) and such RS synthesis is significantly influenced by variation in the extrusion process parameters (Ai et al., 2016). Additionally, extrusion may alter the molecular weight distribution and solubility of additional polysaccharides in cereal grains, which creates variability in digestion characteristics of starch (Rathod & Annapure, 2016; Robin et al., 2016; von Borries-Medrano et al., 2016; Zhang et al., 2016). RDS content increase whereas SDS and RS content decrease after extrusion treatment of maize flour (Zhang et al., 2016). Vasanthan et al. (2002) determined the impact of extrusion conditions on total (TDF), soluble (SDF), and insoluble dietary fiber (IDF) in barley flour and observed that the extrusion cooking increased SDF and TDF content in both types of barley flours. Protein, lipid, and ash contents of CDC-Candle and Phoenix flours were quite similar, but starch, IDF, SDF, and β-glucan contents were significantly distinct. Further, extrudates exhibit high protein digestibility value than non-extruded products attributed to the inactivation of anti-nutritional factors and denaturation of proteins (Singh et al., 2007). Zhang et al. (2016) observed that the protein content of maize flour after the extrusion process showed no significant changes.

After extrusion cooking, DPPH free radical scavenging activity significantly increased, but total phenolic content decreased (Sharma et al., 2012). Thermal processing of foods results in the production of dark colour pigments (brown) due to the Maillard reaction (Sharma & Gujral, 2011; Xu & Chang, 2008). These pigments (especially melanoidins) are known to act as antioxidants (Manzocco et al., 2001). Gujral et al. (2012) stated improvement in the antioxidant activity in the rice flour after germination while reduction after the extrusion process. Bhat et al. (2019) evaluate the extrusion impact on the antioxidant activity of extruded snacks from wheat flour. The results demonstrated a reduction in total phenolic content (TPC), DPPH scavenging activity and pasting properties, and improvement in reducing power and suppression of lipid peroxidation. In general, extruded snacks were acceptable in terms of their sensory properties. Liu et al. (2020a) observed the effect of pre-treatment (extrusion) on the antioxidant properties of oil extracted from oat bran. According to the study, extrusion pre-treatment had no adverse impact on the antioxidant activity of polar antioxidants in oat bran oil, such as phenols.

The higher barrel temperature along with high moisture content may decarboxylate phenolic compounds during extrusion, which encourage polymerisation of phenols and tannins ultimately reducing extractability and antioxidant activity (Dlamini et al., 2007; Repo-Carrasco-Valencia et al. 2009a, b). Zeng et al. (2016) applied IECT technique (moisture 30%, extrusion temperature 120 °C, drying 40 °C) to evaluate the phenolic profile of different cereals. There were significant decreases in total free phenolic acids in brown rice (5.88%) and wheat (45.66%), while increases in total bound phenolic acids were noted in brown rice (6.45%), wheat (8.78%) and oat (9.10%).

After extrusion treatment, water absorption capacity was improved while gelation capacity, SP, EC and FC decreased in rice flour (Gulzar et al., 2021). Martínez et al. (2014) demonstrated that wheat and rice flour hydrated more rapidly after extrusion processing. Zhang et al. (2016) noted a change in the gelatinization behaviour and pasting properties of maize flours after extrusion. PV and gelatinization temperature increases after the treatment. According to Gujral et al. (2012), the water solubility, WAC, and percent expansion of rice extrudates increased significantly after extrusion. Liu et al. (2020a, b) evaluated the impact of different extrusion parameters (temperature, moisture, speed). High moisture content and low screw speed were observed increased in BD and total starch content (TSC) and decreased in water solubility index (WSI) in oat. Furthermore, as the extrusion temperature and moisture content increase, WSI and starch digestibility initially increased and later decreased, and reached their maximum values at 130 °C and 26%, respectively

2.5 Thermal Processing

In cereals, thermal processing is a widely used method to enhance texture, palatability, and nutritional value through gelatinization of starch, denature proteins, increase nutrient availability, inactivate heat-labile toxic compounds and other enzyme inhibitors (Bakr & Gawish, 1991). Toasting is a rapid food processing technique involving dry heat for a short period of time. Sandhu et al. (2017) reported an increase in the total phenolic content by 11.5–27.1%, antioxidant activity by 29.1–53.6%, and metal chelating activity by 33.9–74.4%, respectively, and decreased in total flavonoid content by 23–40.1% in oat. Siroha and Sandhu (2017) compared impact of cooking and toasting on the bioactive properties of different pearl millet cultivars. The results demonstrated the superiority of toasting over cooking in terms of TPC, total flavonoids content (TFC), antioxidant activity, and $ABTS^{+}$ scavenging activity. Further, all the antioxidant characteristics were decreased in the cooking process. Sharma and Gujral (2011) observed an increase in antioxidant activity after sand roasting and microwave cooking. When compared with sand roasting, microwave cooking can effectively inactivate polyphenol oxidase activity. Heat treatment increases antioxidant activity by forming non-enzymatic browning products, generally melanoids at high temperature, which have antioxidant properties (Sharma & Gujral, 2011). Schlörmann et al. (2020) reported a slight variation in dietary fibre fractions after roasting however, fat, protein, starch, and β-glucan contents did not change. According to the results, roasting up to 160°C can yield oat products with enhanced sensory qualities and promising nutritional compositions. Further, as the roasting temperature increased, the viscosities decreased significantly. Sandhu et al. (2017) also assessed the functional properties of toasted oat flour and observed elevated value of WAC and OAC after the toasting process. Color parameters L* (Light) decreased whereas a* (Redness) value increased. In comparison to microwave cooking, sand roasting resulted in greater bulk density reduction and more puffing (Sharma & Gujral, 2011). The decrease in BD, EC and FC while the increase in WAC and OAC was reported in pigmented wheat after roasting treatment (Dhua et al. (2021). Zhao et al. (2020) reported that the roasting treatment of barley starch slightly decreased the birefringence intensity, SP, and solubility along with the presence of polarized cross, but with a slight increase in relative crystallinity and unchanged crystal type.

2.6 Ultrasound Processing

In the food processing industry, ultrasound technology enhances the rate of numerous processes and their efficiency. Additionally, can applied in conjunction with pressure (manosonication) and temperature (thermosonication) to further increases its efficacy (Condón-Abanto et al., 2016). Zhu and Li (2019) reported a decrease in antioxidant activity in quinoa flour after ultra-sonication treatment. Xia et al. (2020)

studied the impact of ultrasound treatment on DPPH radical scavenging activity of wholegrain rice. The results indicated that the ultrasound treatment liberated various antioxidants compounds from wholegrain or some neutralized by free radicals generated due to cavitation phenomena as indicated through a significantly low value of EC50. İşçimen and Hayta (2018) reported positive correlation between DPPH scavenging activity and ultrasound power for rice protein isolate. Additionally, a study that evaluated the extraction of antioxidants from rice using ultrasound reported an increase in percent scavenging activity with the power of ultrasound increased (Zhang et al., 2003).

Ultrasound can be used to physically modify food components to give them a new structure and function. The ultrasound treatment improved the antioxidant capacity and protein digestibility of sorghum flour by producing low molecular weight peptides (Sullivan et al., 2018). The ultrasound treatment decreased the hydrophobicity of wheat flour dough and the secondary structure of gliadin, gluten, and glutenin was also brought into order (Zhang et al., 2020). Sonication treatment may be utilized for the fortification process of cereals. In the food matrix, ultrasound creates micro-channels and cavities that improve hydration and micronutrient uptake through capillarity and sponge effect. In the food matrix, ultrasound created micro-channels and cavities that facilitated hydration and micronutrient absorption through capillarity and sponge effect (Miano & Augusto, 2018). Bangar et al. (2021b) evaluated the ultra-sonication treatment effect on pearl millet starch properties. The results revealed that the ultra-sonication treatment improves peak viscosity, gelatinization temperature, storage and loss modulus of pearl millet starch compare to native starch. Further, no significant effects of ultra-sonication treatment were observed on morphological properties of pearl millet starch.

2.7 Pulsed Electric Fields (PEF)

The first application of electric field technology in food processing dates to the early 1920s when milk was pasteurized with a low voltage electric field (Prescott, 1927). In PEF treatment, a high voltage pulse is applied across food placed between conducting electrodes for a short period (ranging from milliseconds to microseconds) (Duque et al., 2020). Wanyo et al. (2014) found that infrared treatment for 2 h at an intensity of 2 kW/m^2 and a temperature of 40 °C enhanced rice bran phenolic content and antioxidant activity. However, rice bran and husk antioxidant activity related to hot air and enzymatic treatment using cellulose. Duque et al. (2020) observed that the PEF-treated thermally processed oat flour (TPOF) exhibited a enhanced β-glucan content compared to untreated TPOF. The relative amounts of starch, protein, and fat in each fraction didn't differ significantly from one another. A significantly lower gelatinisation enthalpy and percentage relative crystallinity were found in the treated samples. PEF-treated wheat seeds plantlets juice exhibited significant rise in total phenolic contents, DPPH, chlorophylls, carotenoids, soluble proteins, minerals, and amino acids content than untreated sample. Results showed that PEF treatment may improve metabolism, optimize the nutrients, and enhance the strength of the wheat kernels plantlets (Ahmed et al., 2020). Wu et al. (2019a) reported decreased in gelatinization enthalpy (ΔH), retrogradation value, slowly digestible starch value while increased in rapid digestible starch value in rice after PEF treatment. Han et al. (2009) reported that at higher electric fields strength, gelatinization temperatures, enthalpies, pasting parameters peak, breakdown and final viscosity decreased.

2.8 Biotechnology Approaches

The use of biotechnology approaches to increase micronutrient (vitamins and minerals) availability is technically feasible without affecting agronomic performance. Genetic bio-fortification can be accomplished using a single or a combination of approaches. Polishing is essential to avoid seed rancid upon storage but it removes the essential micronutrients that are present in the husk, aleurone, and embryo of rice (Bhullar & Gruissem, 2013; Dubock, 2019). The immature endosperm can produce geranylgeranyl diphosphate which transforms into carotene phytoene via the expression of the enzyme phytoene synthase. Further, the complementation with Daffodil (*Narcissus pseudonarcissus*) phytoene synthase (PSY gene) and bacterial carotene desaturase (CrtI) (from *Erwinia uredovora*) increased β-carotene (vitamin A source for humans) by four-fold in golden rice (Burkhardt et al., 1997; Ye et al., 2000). The application of maize phytoene synthase instead of Daffodil phytoene synthase and *E. uredovora* carotene desaturase improved the β-carotene level 23-fold in golden rice variety 2 (Paine et al., 2005). Expression of *Aspergillus thaliana* GTP-cyclohydrolase I and aminodeoxychorismate synthase increased the folate (Vitamin B9) content by 15–100 fold in comparison to polished rice grain (Storozhenko et al., 2007). The over-expression of 4-amino-2-methyl-5-hydroxy methylpyrimidine monophosphate (HMP-P) synthase (THIC) and HMP-P kinase (THID) increased thiamin (B1) level about 5-fold in unpolished rice seeds (Goyer, 2017). Blancquaert et al. (2015) applied metabolic engineering to increase vitamin B9 by 150-fold by adding folate-binding protein along with enhancing the activity of GTPCH1 (GTP cyclohydrolase I) and ADCS (aminodeoxychorismate synthase). The processing reduced the initial amount of iron i.e. 38 parts per million (ppm) in paddy (rough rice) to 8.8 ppm in brown rice, then to 4.1 ppm in milled rice (Dexter, 1998). The first attempt to enrich the rice with iron was reported by Goto et al. (1999), by ectopic overexpression of SoyFerH1 (soybean iron storage protein ferritin gene) and *GluB-1* (promoter for the rice seed storage protein glutelin) in the endosperm that increased iron content up to 38 μg/g in brown rice than wildtype *Japonica cv. Kitaake* (about 14.3 μg/g). Ferritin has the ability to accommodate up to 4000 iron atoms per protein molecule and due to the solubility and bioavailability of ferritin iron various strategies can be applied for overexpressing ferritin in crops (Bhullar & Gruissem, 2013). The endosperm-specific expression of soybean ferritin

(SoyferH1) increased iron content by about 37 μg/g in rice grain (Indica cv. IR68144) (Vasconcelos et al., 2003). However, iron content in rice endosperm cannot be meet the recommended dietary level through the expression of exogenous ferritin or single genes. But, the combinatorial approach (fusion of genes) improved both uptake and storage of iron in rice. Wirth et al. (2009) evaluated the expression of *Aspergillus thaliana* Nicotianamine synthase gene (AtNAS) along with endosperm-specific expression of *Phaseolus vulagaris* Ferritin and *Aspergillus fumigatus* Phytase and observed significantly increased (above six-fold) in the iron concentration in the polished rice grains. Masuda et al. (2012) applied three transgenic approaches: expression of the iron storage protein ferritin (Soybean ferritin gene) using endosperm-specific promoters (the OsGlb1 and 2.3-kb OsGluB1) to improve iron storage in grains, overproduction of the natural metal chelator nicotianamine (barley nicotianamine synthase 1 gene, HvNAS1) to improve iron translocation, and expression of rice nicotianamine–metal transporter gene (OsYSL2) in the presence of endosperm-specific promoter (OsGlb1) and sucrose transporter promoter (OsSUT1) to improve iron flux into the endosperm. The results reported 6-fold and 4.4-fold increased iron concentration in rice grain that grown in greenhouse and paddy field conditions, respectively in comparison to non-transgenic seeds. Further, transgenic rice grain also stored zinc up to 1.6-times. Rice grain iron content was increased significantly by introducing three barley genes that involved mugineic acid biosynthesis with soybean ferritin genes (Masuda et al., 2013). Wu et al. (2019b) applied multigene overexpression approach using endosperm storage gene PvFER, the chelator AtNAS1gene, and an intracellular iron stores AtNRAMP3 and observed significant iron concentration (13.65 μg/g) in rice grain in the greenhouse condition. In wheat, β-carotene content can increase by metabolic engineering through Endosperm-specific silencing of the carotenoid hydroxylase gene (TaHYD) that improved β-carotene concentration 10.5-fold. Overexpression of phytoene synthase (CrtB) further improved β-carotene concentration about 14.6-fold. Further, the combination of both silencing of TaHYD (block strategy) and overexpressing of CrtB (push strategy) significantly improved β-carotene concentration up to 31-fold (Zeng et al., 2015). Liang et al. (2019) applied metabolic engineering to improved folate content in wheat through two Gm8gGCHI and Gm3gGCHI gene (cloned gene) and one GmADCS (aminodeoxychorismate synthase) gene from soybean as they are involved in the synthesis of pterin and p-aminobenzoate (folate precursors), respectively. The results reported superiority of Gm8gGCHI over Gm3gGCHI and co-expression of Gm8gGCHI and GmADCS along with endosperm-specific promoters which boost folate level 2.3-fold in transgenic wheat grains. Moreover, codon-optimized Gm8gGCHI, tomato LeADCS genes and wheat endosperm-specific glutenin promoter (1Dx5) increased folate yield by 5.6-fold. The whole grain of wheat is estimated to contain 30 ppm iron, of which around 5% is bioavailable, meaning that an additional 22 ppm iron is required, for a total of 52 ppm (Yashveer et al., 2014). Borg et al. (2012) first time applied the transgenic approach to the increased iron concentration in wheat (*Triticum aestivum*) under greenhouse conditions through overexpression of wheat ferritin gene

(TaFer1-A) that increased iron concentration up to 1.5 to 1.9-fold in wheat grains. The endosperm-specific expression of soybean ferritin increased iron content 1.1 to 1.6-fold (Xiaoyan et al., 2012). The overexpression of OsNAS2 (single gene expression) produced iron content up to 80 μg/g in the field condition (Beasley et al., 2019). Moreover, overexpression of OsNAS2 and PvFERRITIN generated iron content up to 93.1 μg/g in greenhouse conditions (Singh et al., 2017). A study in the University of Melbourne, Australia, based on overexpression of rice nicotianamine synthase gene (OsNAS2) along with maize ubiquitin promoter (*Ubi1)* yielded transgenics wheat grain with 52 ppm of iron content (Yashveer et al., 2014). Iron level in grain are controlled by several genes associated with uptake, transportation, and storage of iron (Shi et al., 2020) (Fig. 2.3). Iron concentration in wheat grain increased remarkably through transgenic manipulation, although transgenic wheat development is in its infancy (Li et al., 2018; Shi et al., 2020).

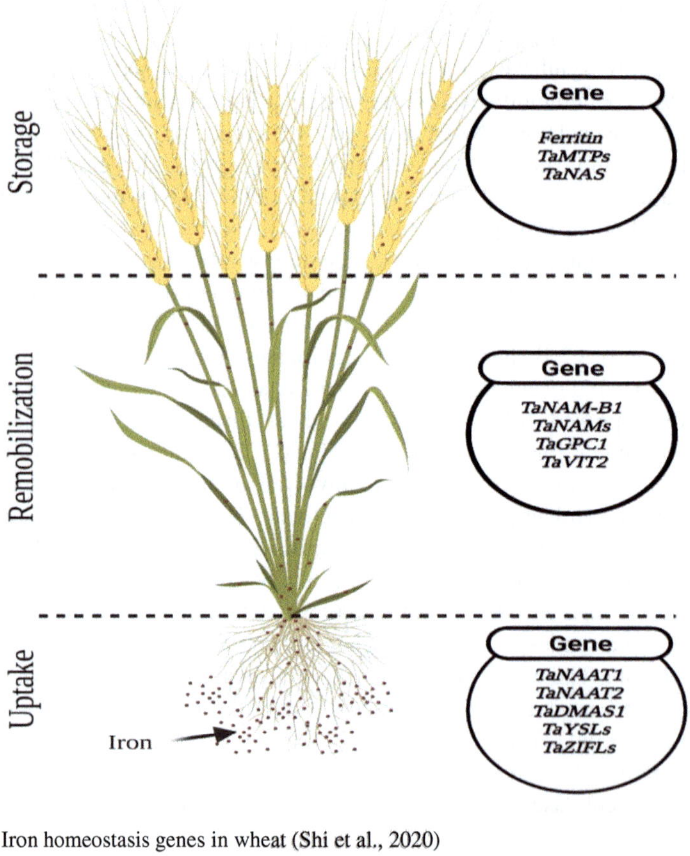

Fig. 2.3 Iron homeostasis genes in wheat (Shi et al., 2020)

2.9 Conclusion

Processing's are novel techniques which are used to improve the functionalities of cereals and its products. To produce more natural food components, there is an interest to improve the properties with physical treatments like germination, extrusion, heat processing's and ultra-sonication treatment. These treatments improve the functional and nutritional properties of cereals. Processing treatments improves the TPC, DPPH scavenging activity and antioxidant properties of cereal foods. Thermal processing's reduced the anti-nutritional components of cereals, and improves the digestibility of cereal components. Extrusion, ultra-sonication and PEF are novel techniques which are used to increase specific functional properties of cereals. Ultrasound produced micro-channels and cavities, this treatment may be utilised for fortification process of cereals because this treatment increase the absorption power of food. Biotechnology approach showed specific role in food production as well as in nutritional quality. Micronutrients i.e. β-carotene, folate, thiamine, iron etc. are significantly improved by using the biotechnological approach. It is concluded that processing treatments are basic requirement to increase the functionality and nutritional value of cereals.

References

Abd Elmoneim, O. E., Schiffler, B., & Bernhardt, R. (2005). Effect of fermentation on the functional properties of sorghum flour. *Food Chemistry, 92*(1), 1–5.

Agu, R. C., Chiba, Y., Goodfellow, V., Mackinlay, J., Brosnan, J. M., Bringhurst, T. A., et al. (2012). Effect of germination temperatures on proteolysis of the gluten-free grains rice and buckwheat during malting and mashing. *Journal of Agricultural and Food Chemistry, 60*(40), 10147–10154.

Ahmed, Z., Manzoor, M. F., Ahmad, N., Zeng, X. A., Din, Z. U., Roobab, U., et al. (2020). Impact of pulsed electric field treatments on the growth parameters of wheat seeds and nutritional properties of their wheat plantlets juice. *Food Science and Nutrition, 8*(5), 2490–2500.

Ai, Y., Cichy, K. A., Harte, J. B., Kelly, J. D., & Ng, P. K. (2016). Effects of extrusion processing on the chemical composition and functional properties of dry common bean powders. *Food Chemistry, 211*, 538–545.

Albarracín, M., Dyner, L., Silvia, M., Adriana, G., Zuleta, A., & Drago, S. R. (2019, March 1–8). Modification of nutritional properties of whole rice flours (*Oryza sativa* L.) by soaking, germination, and extrusion. *Journal of Food Biochemistry, 43*(7), e12854. https://doi.org/10.1111/jfbc.12854.

Aparicio-García, N., Martínez-Villaluenga, C., Frias, J., & Peñas, E. (2020). Changes in protein profile, bioactive potential and enzymatic activities of gluten-free flours obtained from hulled and dehulled oat varieties as affected by germination conditions. *LWT-Food Science and Technology, 134*, 109955.

Asensio-Grau, A., Calvo-Lerma, J., Heredia, A., & Andrés, A. (2020). Enhancing the nutritional profile and digestibility of lentil flour by solid state fermentation with Pleurotus ostreatus. *Food and Function, 11*(9), 7905–7912.

Atuna, R. A., Ametei, P. N., Bawa, A. A., & Amagloh, F. K. (2022). Traditional processing methods reduced phytate in cereal flour, improved nutritional, functional and rheological properties. *Scientific African, 15*, e01063.

Ayyash, M., Johnson, S. K., Liu, S., Al-Mheiri, A., & Abushelaibi, A. (2018). Cytotoxicity, antihypertensive, antidiabetic and antioxidant activities of solid-state fermented lupin, quinoa and wheat by Bifidobacterium species: In-vitro investigations. *Food Science and Technology, 95*, 295–302.

Bakr, A. A., & Gawish, R. A. (1991). Nutritional evaluation and cooking quality of dry cowpea (Vigna sinensis L.) grown under various agricultural conditions. I: Effect of soaking and cooking on the chemical composition and nutritional quality of cooked seeds. *Journal of Food Science and Technology, 28*(5), 312–316.

Bangar, S. P., Sandhu, K. S., Purewal, S. S., Kaur, M., Kaur, P., Siroha, A. K., et al. (2021a). Fermented barley bran: An improvement in phenolic compounds and antioxidant properties. *Journal of Food Processing and Preservation, e15543,* 1–8.

Bangar, S. P., Nehra, M., Siroha, A. K., Petrů, M., Ilyas, R. A., Devi, U., & Devi, P. (2021b). Development and characterization of physical modified pearl millet starch-based films. *Food, 10*(7), 1609.

Bangar, S. P., Sandhu, K. S., Trif, M., Rusu, A., Pop, I. D., & Kumar, M. (2022). Enrichment in different health components of barley flour using twin-screw extrusion technology to support nutritionally balanced diets. *Frontiers in Nutrition, 8*, 823148.

Beasley, J. T., Bonneau, J. P., Sánchez-Palacios, J. T., Moreno-Moyano, L. T., Callahan, D. L., Tako, E., et al. (2019). Metabolic engineering of bread wheat improves grain iron concentration and bioavailability. *Plant Biotechnology Journal, 17*(8), 1514–1526.

Bhat, N. A., Wani, I. A., Hamdani, A. M., & Gani, A. (2019). Effect of extrusion on the physicochemical and antioxidant properties of value added snacks from whole wheat (*Triticum aestivum* L.) flour. *Food Chemistry, 276*, 22–32.

Bhullar, N. K., & Gruissem, W. (2013). Nutritional enhancement of rice for human health: The contribution of biotechnology. *Biotechnology Advances, 31*(1), 50–57.

Blancquaert, D., Van Daele, J., Strobbe, S., Kiekens, F., Storozhenko, S., De Steur, H., et al. (2015). Improving folate (vitamin B9) stability in biofortified rice through metabolic engineering. *Nature Biotechnology, 33*(10), 1076–1078.

Blazek, J., & Copeland, L. (2008). Pasting and swelling properties of wheat flour and starch in relation to amylose content. *Carbohydrate Polymers, 71*, 380–387.

Borg, S., Brinch-Pedersen, H., Tauris, B., Madsen, L. H., Darbani, B., Noeparvar, S., & Holm, P. B. (2012). Wheat ferritins: Improving the iron content of the wheat grain. *Journal of Cereal Science, 56*(2), 204–213.

Brennan, C., Brennan, M., Derbyshire, E., & Tiwari, B. K. (2011). Effects of extrusion on the polyphenols, vitamins and antioxidant activity of foods. *Trends in Food Science and Technology, 22*(10), 570–575.

Burkhardt, P. K., Beyer, P., Wünn, J., Klöti, A., Armstrong, G. A., Schledz, M., et al. (1997). Transgenic rice (*Oryza sativa*) endosperm expressing daffodil (Narcissus pseudonarcissus) phytoene synthase accumulates phytoene, a key intermediate of provitamin A biosynthesis. *The Plant Journal, 11*(5), 1071–1078.

Caceres, P. J., Martínez-Villaluenga, C., Amigo, L., & Frias, J. (2014). Maximising the phytochemical content and antioxidant activity of Ecuadorian brown rice sprouts through optimal germination conditions. *Food Chemistry, 152*, 407–414.

Cai, S., Wang, O., Wu, W., Zhu, S., Zhou, F., Ji, B., et al. (2012). Comparative study of the effects of solid-state fermentation with three filamentous fungi on the total phenolics content (TPC), flavonoids, and antioxidant activities of subfractions from oats (*Avena sativa* L.). *Journal of Agricultural and Food Chemistry, 60*(1), 507–513.

Condón-Abanto, S., Arroyo, C., Álvarez, I., Condón, S., & Lyng, J. G. (2016). Application of ultrasound in combination with heat and pressure for the inactivation of spore forming bacteria isolated from edible crab (*Cancer pagurus*). *International Journal of Food Microbiology, 223*, 9–16.

Dexter, P. B. (1998). Rice fortification for developing countries; OMNI/USAID:Washington, DC, USA, https://pdf.usaid.gov/pdf_docs/PNACD279.pdf

Dhua, S., Kheto, A., Sharanagat, V. S., Singh, L., Kumar, K., & Nema, P. K. (2021). Quality characteristics of sand, pan and microwave roasted pigmented wheat (*Triticum aestivum*). *Food Chemistry, 365*, 130372.

Dlamini, N. R., Taylor, J. R. N., & Rooney, L. W. (2007). The effect of sorghum type and processing on the antioxidant properties of African sorghum-based foods. *Food Chemistry, 105*(4), 1412–1419.

Dubock, A. (2019). Golden rice: To combat vitamin A deficiency for public health. *Vitamin A*, 1–21. https://doi.org/10.5772/intechopen.84445

Duque, S. M. M., Leong, S. Y., Agyei, D., Singh, J., Larsen, N., & Oey, I. (2020). Modifications in the physicochemical properties of flour "fractions" after pulsed electric fields treatment of thermally processed oat. *Innovative Food Science and Emerging Technologies, 64*, 102406.

Elkhalifa, A. E. O., & Bernhardt, R. (2010). Influence of grain germination on functional properties of sorghum flour. *Food Chemistry, 121*(2), 387–392.

El-Refai, A. A., El-Bastawesy, A. M., & El-Ashaal, E. S. (2012). Effect of germination process on the chemical and biological active compounds of barley and oat grains. *Journal of Food and Dairy Sciences, 3*(10), 553–564.

Faraj, A., Vasanthan, T., & Hoover, R. (2004). The effect of extrusion cooking on resistant starch formation in waxy and regular barley flours. *Food Research International, 37*(5), 517–525.

Fischer, S., Hilger, T., Piepho, H.-P., Jordan, I., & Cadisch, G. (2021). Missing association between nutrient concentrations in leaves and edible parts of food crops – A neglected food security issue. *Food Chemistry, 345*, 128723.

Frias, J., Miranda, M. L., Doblado, R., & Vidal-Valverde, C. (2005). Effect of germination and fermentation on the antioxidant vitamin content and antioxidant capacity of Lupinus albus L. var. Multolupa. *Food Chemistry, 92*(2), 211–220.

Gan, R. Y., Lui, W. Y., Wu, K., Chan, C. L., Dai, S. H., Sui, Z. Q., & Corke, H. (2017). Bioactive compounds and bioactivities of germinated edible seeds and sprouts: An updated review. *Trends in Food Science and Technology, 59*, 1 14.

Goto, F., Yoshihara, T., Shigemoto, N., Toki, S., & Takaiwa, F. (1999). Iron fortification of rice seed by the soybean ferritin gene. *Nature Biotechnology, 17*(3), 282–286.

Goyer, A. (2017). Thiamin biofortification of crops. *Current Opinion in Biotechnology, 44*, 1–7.

Gujral, H. S., Sharma, P., Kumar, A., & Singh, B. (2012). Total phenolic content and antioxidant activity of extruded brown rice. *International Journal of Food Properties, 15*(2), 301–311.

Gulati, P., Brahma, S., & Rose, D. J. (2020). Impacts of extrusion processing on nutritional components in cereals and legumes: Carbohydrates, proteins, lipids, vitamins, and minerals. In *Extrusion cooking* (pp. 415–443). Woodhead Publishing.

Gulzar, B., Hussain, S. Z., Naseer, B., & Naik, H. R. (2021). Enhancement of resistant starch content in modified rice flour using extrusion technology. *Cereal Chemistry, 98*(3), 634–641.

Han, Z., Zeng, X. A., Zhang, B. S., & Yu, S. J. (2009). Effects of pulsed electric fields (PEF) treatment on the properties of corn starch. *Journal of Food Engineering, 93*(3), 318–323.

Heiniö, R. L., Katina, K., Wilhelmson, A., Myllymäki, O., Rajamäki, T., Latva-Kala, K., et al. (2003). Relationship between sensory perception and flavour-active volatile compounds of germinated, sourdough fermented and native rye following the extrusion process. *LWT-Food Science and Technology, 36*(5), 533–545.

Hiran, P., Kerdchoechuen, O., & Laohakunjit, N. (2016). Combined effects of fermentation and germination on nutritional compositions, functional properties and volatiles of maize seeds. *Journal of Cereal Science, 71*, 207–216.

Hung, P. V., Maeda, T., Yamamoto, S., & Morita, N. (2012). Effects of germination on nutritional composition of waxy wheat. *Journal of the Science of Food and Agriculture, 92*(3), 667–672.

Hussain, I., & Uddin, M. B. (2012). Optimization effect of germination on functional properties of wheat flour by response surface methodology. *International Research Journal of Plant Science, 3*(3), 31–37.

İşçimen, E. M., & Hayta, M. (2018). Optimisation of ultrasound assisted extraction of rice bran proteins: Effects on antioxidant and antiproliferative properties. *Quality Assurance and Safety of Crops and Foods, 10*(2), 165–174.

Karwasra, B. L., Gill, B. S., Kaur, M., & Kaur, H. (2018). Influence of germination period on physicochemical, pasting and antioxidant properties of Indian wheat cultivars. *Journal of Food Measurement and Characterization, 12*(1), 68–77.

Kayahara, H., & Tsukahara, K. (2000). Flavor, health and nutritional quality of pre-Spruted Brown Rice. In *Proceedings of International Chemical Congress of Pacific Basin Societies*.

Kim, J. H., Tanhehco, E. J., & Ng, P. K. W. (2006). Effect of extrusion conditions on resistant starch formation from pastry wheat flour. *Food Chemistry, 99*(4), 718–723.

Lakshmi, K. B., Kuna, V., & Prasuna, M. (2015). Traditional processes and methods to enhance nutritional value in foods. *Health Action, 28*, 32–35.

Li, A., Liu, D., Yang, W., Kishii, M., & Mao, L. (2018). Synthetic hexaploid wheat: Yesterday, today, and tomorrow. *Engineering, 4*(4), 552–558.

Liang, Q., Wang, K., Liu, X., Riaz, B., Jiang, L., Wan, X., et al. (2019). Improved folate accumulation in genetically modified maize and wheat. *Journal of Experimental Botany, 70*(5), 1539–1551.

Liu, X., & Kokare, C. (2017). Microbial enzymes of use in industry. In G. Brahmachari, A. L. Demain, & J. L. Adro (Eds.), *Biotechnology of microbial enzymes* (pp. 267–298). Elsevier. https://doi.org/10.1016/B978-0-12-803725-6.00011-X

Liu, J., Jin, S., Song, H., Huang, K., Li, S., Guan, X., & Wang, Y. (2020a). Effect of extrusion pretreatment on extraction, quality and antioxidant capacity of oat (Avena sativa L.) bran oil. *Journal of Cereal Science, 95*, 102972.

Liu, P., Cheng, J., Li, M., Li, J., Zhu, H., Wang, S., & Yu, G. (2020b). Effects of extrusion conditions on physicochemical properties of finger millet (*Gramineae Setaria*). *Journal of Agricultural Science, 12*(9), 149–158.

Manzocco, L., Calligaris, S., Masrrocola, D., Nicoli, M., & Lerici, C. R. (2001). Review of non-enzymatic browning and antioxidant capacity in processed foods. *Trends in Food Science and Technology, 11*, 340–346.

Marsh, A. J., Hill, C., Ross, R. P., & Cotter, P. D. (2014). Fermented beverages with health-promoting potential: Past and future perspectives. *Trends in Food Science and Technology, 38*(2), 113–124.

Martínez, M. M., Calviño, A., Rosell, C. M., & Gómez, M. (2014). Effect of different extrusion treatments and particle size distribution on the physicochemical properties of rice flour. *Food and Bioprocess Technology, 7*(9), 2657–2665.

Martins, S., Mussatto, S. I., Martínez-Avila, G., Montañez-Saenz, J., Aguilar, C. N., & Teixeira, J. A. (2011). Bioactive phenolic compounds: Production and extraction by solid-state fermentation. A review. *Biotechnology Advances, 29*(3), 365–373.

Masuda, H., Ishimaru, Y., Aung, M. S., Kobayashi, T., Kakei, Y., Takahashi, M., et al. (2012). Iron biofortification in rice by the introduction of multiple genes involved in iron nutrition. *Scientific Reports, 2*(1), 1–7.

Masuda, H., Kobayashi, T., Ishimaru, Y., Takahashi, M., Aung, M. S., Nakanishi, H., et al. (2013). Iron-biofortification in rice by the introduction of three barley genes participated in mugineic acid biosynthesis with soybean ferritin gene. *Frontiers in Plant Science, 4*, 132.

McCue, P., & Shetty, K. (2003). Role of carbohydrate-cleaving enzymes in phenolic antioxidant mobilization from whole soybean fermented with Rhizopus oligosporus. *Food Biotechnology, 17*(1), 27–37.

Menis-Henrique, M. E. C., Scarton, M., Piran, M. V. F., & Clerici, M. T. P. S. (2020). Cereal fiber: Extrusion modifications for food industry. *Current Opinion in Food Science, 33*, 141–148.

Mensi, A., & Udenigwe, C. C. (2021). Emerging and practical food innovations for achieving the Sustainable Development Goals (SDG) target 2.2. *Trends in Food Science and Technology, 111*, 783–789.

Miano, A. C., & Augusto, P. E. D. (2018). The ultrasound assisted hydration as an opportunity to incorporate nutrients into grains. *Food Research International, 106*, 928–935.

Mir, N. A., Riar, C. S., & Singh, S. (2018). Nutritional constituents of pseudo cereals and their potential use in food systems: A review. *Trends in Food Science and Technology, 75*, 170–180.

Moongngarm, A., & Saetung, N. (2010). Comparison of chemical compositions and bioactive compounds of germinated rough rice and brown rice. *Food Chemistry, 122*(3), 782–788.

Müller, C. P., Hoffmann, J. F., Ferreira, C. D., Diehl, G. W., Rossi, R. C., & Ziegler, V. (2021). Effect of germination on nutritional and bioactive properties of red rice grains and its application in cupcake production. *International Journal of Gastronomy and Food Science, 25*, 100379.

Nkhata, S. G., Ayua, E., Kamau, E. H., & Shingiro, J. B. (2018). Fermentation and germination improve nutritional value of cereals and legumes through activation of endogenous enzymes. *Food Science and Nutrition, 6*(8), 2446–2458.

Nout, M. R., & Kiers, J. L. (2005). Tempe fermentation, innovation and functionality: Update into the third millenium. *Journal of Applied Microbiology, 98*(4), 789–805.

Ochanda, S. O., Akoth, O. C., Mwasaru, A. M., Kagwiria, O. J., & Mathooko, F. M. (2010). Effects of malting and fermentation treatments on group B-vitamins of red sorghum, white sorghum and pearl millets in Kenya. *Journal of Applied Biosciences, 34*, 2128–2134.

Ocheme, O. B., Adedeji, O. E., Lawal, G., & Zakari, U. M. (2015). Effect of germination on functional properties and degree of starch gelatinization of sorghum flour. *Journal of Food Research, 4*(2), 159.

Onyango, C. A., Ochanda, S. O., Mwasaru, M. A., Ochieng, J. K., Mathooko, F. M., & Kinyuru, J. N. (2013). Effects of malting and fermentation on antinutrient reduction and protein digestibility of red sorghum, white sorghum and pearl millet. *Journal of Food Research, 2*(1), 41.

Ortiz, L. T., Velasco, S., Treviño, J., Jiménez, B., & Rebolé, A. (2021). Changes in the nutrient composition of barley grain (Hordeum vulgare L.) and of morphological fractions of sprouts. *Scientifica, 2021*, 9968864.

Paine, J. A., Shipton, C. A., Chaggar, S., Howells, R. M., Kennedy, M. J., Vernon, G., et al. (2005). Improving the nutritional value of Golden Rice through increased pro-vitamin A content. *Nature Biotechnology, 23*(4), 482–487.

Pandey, A. (2003). Solid-state fermentation. *Biochemical Engineering Journal, 13*(2–3), 81–84.

Petrova, P., & Petrov, K. (2020). Lactic acid fermentation of cereals and pseudocereals: Ancient nutritional biotechnologies with modern applications. *Nutrients, 12*(4), 1118.

Prescott, S. C. (1927). The treatment of milk by an electrical method. *American Journal of Public Health (New York, N.Y.: 1912), 17*(3), 221–223. https://doi.org/10.2105/AJPH.17.3.221

Punia, S., Sandhu, K. S., Grasso, S., Purewal, S. S., Kaur, M., Siroha, A. K., et al. (2021). Aspergillus oryzae fermented rice bran: A byproduct with enhanced bioactive compounds and antioxidant potential. *Food, 10*(1), 70.

Qin, Q., Liu, J., Hu, S., Dong, J., Yu, J., Fang, L., et al. (2021). Comparative proteomic analysis of different barley cultivars during seed germination. *Journal of Cereal Science, 102*, 103357.

Rani, P., Kumar, A., Purohit, S. R., & Rao, P. S. (2018). Impact of fermentation and extrusion processing on physicochemical, sensory and bioactive properties of rice-black gram mixed flour. *LWT – Food Science and Technology, 89*, 155–163.

Rathod, R. P., & Annapure, U. S. (2016). Effect of extrusion process on antinutritional factors and protein and starch digestibility of lentil splits. *LWT-Food Science and Technology, 66*, 114–123.

Repo-Carrasco-Valencia, R., de La Cruz, A. A., Alvarez, J. C. I., & Kallio, H. (2009a). Chemical and functional characterization of kaiwa (*Chenopodium pallidicaule*) grain, extrudate and bran. *Plant Foods for Human Nutrition, 64*(2), 94–101.

Repo-Carrasco-Valencia, R., Pena, J., Kallio, H., & Salminen, S. (2009b). Dietary fiber and other functional components in two varieties of crude and extruded kiwicha *(Amaranthus caudatus)*. *Journal of Cereal Science, 49*(2), 219–224.

Robin, F., & Palzer, S. (2015). Texture of breakfast cereals and extruded products. In J. Chen & A. Rosenthal (Eds.), *Modifying food texture: Novel ingredients and processing techniques* (pp. 201–230). Elsevier.

Robin, F., Heindel, C., Pineau, N., Srichuwong, S., & Lehmann, U. (2016). Effect of maize type and extrusion-cooking conditions on starch digestibility profiles. *International Journal of Food Science & Technology, 51*(6), 1319–1326.

Rose, P. K. (2022). Bioconversion of agricultural residue into biofuel and high-value biochemicals: Recent advancement. In Y. K. Nandabalan, V. K. Garg, N. K. Labhsetwar, & A. Singh (Eds.), *Zero waste biorefinery. Energy, environment, and Sustainability*. Springer. https://doi.org/10.1007/978-981-16-8682-5_9

Saalia, F. K., & Phillips, R. D. (2011). Degradation of aflatoxins by extrusion cooking: Effects on nutritional quality of extrudates. *LWT-Food Science and Technology, 44*(6), 1496–1501.

Sadh, P. K., Saharan, P., Duhan, S., & Duhan, J. S. (2017). Bio-enrichment of phenolics and antioxidant activity of combination of Oryza sativa and Lablab purpureus fermented with GRAS filamentous fungi. *Resource-Efficient Technologies, 3*(3), 347–352.

Saharan, P., Sadh, P. K., & Duhan, J. S. (2017). Comparative assessment of effect of fermentation on phenolics, flavanoids and free radical scavenging activity of commonly used cereals. *Biocatalysis and Agricultural Biotechnology, 12*, 236–240.

Salar, R. K., Purewal, S. S., & Sandhu, K. S. (2017). Fermented pearl millet (Pennisetum glaucum) with in vitro DNA damage protection activity, bioactive compounds and antioxidant potential. *Food Research International, 100*, 204–210.

Sánchez-Magaña, L. M., Reyes-Moreno, C., Milán-Carrillo, J., Mora-Rochín, S., León-López, L., Gutiérrez-Dorado, R., & Cuevas-Rodríguez, E. O. (2019). Influence of solid state bioconversion by rhizopusoligosporus on antioxidant activity and phenolic compounds of Maize (*Zea Mays* L.). *Agrociencia, 53*, 45–57.

Sandhu, K. S., & Punia, S. (2017). Enhancement of bioactive compounds in barley cultivars by solid substrate fermentation. *Journal of Food Measurement and Characterization, 11*(3), 1355–1361.

Sandhu, K. S., Godara, P., Kaur, M., & Punia, S. (2017). Effect of toasting on physical, functional and antioxidant properties of flour from oat (Avena sativa L.) cultivars. *Journal of the Saudi Society of Agricultural Sciences, 16*(2), 197–203.

Sangronis, E., & Machado, C. J. (2007). Influence of germination on the nutritional quality of Phaseolus vulgaris and Cajanus cajan. *LWT-Food Science and Technology, 40*(1), 116–120.

Schlörmann, W., Zetzmann, S., Wiege, B., Haase, N. U., Greiling, A., Lorkowski, S., et al. (2020). Impact of different roasting conditions on sensory properties and health-related compounds of oat products. *Food Chemistry, 307*, 125548.

Sharma, P., & Gujral, H. S. (2011). Effect of sand roasting and microwave cooking on antioxidant activity of barley. *Food Research International, 44*, 235–240.

Sharma, P., Gujral, H. S., & Singh, B. (2012). Antioxidant activity of barley as affected by extrusion cooking. *Food Chemistry, 131*(4), 1406–1413.

Sharma, S., Saxena, D. C., & Riar, C. S. (2018). Characteristics of β-glucan extracted from raw and germinated foxtail (*Setaria italica*) and kodo (Paspalum scrobiculatum) millets. *International Journal of Biological Macromolecules, 118*, 141–148.

Shi, Y., Li, J., & Sun, Z. (2020). Success to iron biofortification of wheat grain by combining both plant and microbial genetics. *Rhizosphere, 15*, 100218.

Singh, S., Gamlath, S., & Wakeling, L. (2007). Nutritional aspects of food extrusion: A review. *International Journal of Food Science & Technology, 42*(8), 916–929.

Singh, A. K., Rehal, J., Kaur, A., & Jyot, G. (2015). Enhancement of attributes of cereals by germination and fermentation: A review. *Critical Reviews in Food Science and Nutrition, 55*(11), 1575–1589.

Singh, S. P., Keller, B., Gruissem, W., & Bhullar, N. K. (2017). Rice NICOTIANAMINE SYNTHASE 2 expression improves dietary iron and zinc levels in wheat. *Theoretical and Applied Genetics, 130*(2), 283–292.

Siroha, A. K., & Sandhu, K. S. (2017). Effect of heat processing on the antioxidant properties of pearl millet (*Pennisetum glaucum* L.) cultivars. *Journal of Food Measurement and Characterization, 11*(2), 872–878.
Spotti, M. J., & Campanella, O. H. (2017). Functional modifications by physical treatments of dietary fibers used in food formulations. *Current Opinion in Food Science, 15*, 70–78.
Stoffel, F., Santana, W. D. O., Fontana, R. C., Gregolon, J. G. N., Kist, T. B. L., De Siqueira, F. G., Mendonça, S., & Camassola, M. (2019). Chemical features and bioactivity of grain flours colonized by macro fungi as a strategy for nutritional enrichment. *Food Chemistry, 297*, 124988.
Storozhenko, S., De Brouwer, V., Volckaert, M., Navarrete, O., Blancquaert, D., Zhang, G. F., et al. (2007). Folate fortification of rice by metabolic engineering. *Nature Biotechnology, 25*(11), 1277–1279.
Subramaniyam, R., & Vimala, R. (2012). Solid state and submerged fermentation for the production of bioactive substances: A comparative study. *International Journal of Science and Nature, 3*(3), 480–486.
Sullivan, A. C., Pangloli, P., & Dia, V. P. (2018). Impact of ultrasonication on the physicochemical properties of sorghum kafirin and in vitro pepsin-pancreatin digestibility of sorghum gluten-like flour. *Food Chemistry, 240*, 1121–1130.
Swaminathan, S. (2021). Cereal grains. *Reference Modules in Food Science*. https://doi.org/10.1016/B978-0-12-821848-8.00017-2
Swanston, J. S., & Middlefell-Willliams, J. E. (2012). The influence of steep regime and germination period on the mating properties of some hull-less barley lines. *Journal of the Institute of Brewing, 118*, 186–191.
Tian, B., Xie, B., Shi, J., Wu, J., Cai, Y., Xu, T., et al. (2010). Physicochemical changes of oat seeds during germination. *Food Chemistry, 119*(3), 1195–1200.
Vasanthan, T., Gaosong, J., Yeung, J., & Li, J. (2002). Dietary fiber profile of barley flour as affected by extrusion cooking. *Food Chemistry, 77*(1), 35–40.
Vasconcelos, M., Datta, K., Oliva, N., Khalekuzzaman, M., Torrizo, L., Krishnan, S., et al. (2003). Enhanced iron and zinc accumulation in transgenic rice with the ferritin gene. *Plant Science, 164*(3), 371–378.
Veluppillai, S., Nithyanantharajah, K., Vasantharuba, S., Balakumar, S., & Arasaratnam, V. (2009). Biochemical changes associated with germinating rice grains and germination improvement. *Rice Science, 16*(3), 240–242.
von Borries-Medrano, E., Jaime-Fonseca, M. R., & Aguilar-Mendez, M. A. (2016). Starch–guar gum extrudates: Microstructure, physicochemical properties and in-vitro digestion. *Food Chemistry, 194*, 891–899.
Valamoti, S. M., Marinova, E., Heiss, A. G., Hristova, I., Petridou, C., Popova, T., ... & Ruseva, T. K. (2019). Prehistoric cereal foods of southeastern Europe:an archaeobotanical exploration. *Journal of Archaeological Science, 104*, 97–113.
Wang, Y., Liu, S., Yang, X., Zhang, J., Zhang, Y., Liu, X., et al. (2022). Effect of germination on nutritional properties and quality attributes of glutinous rice flour and dumplings. *Journal of Food Composition and Analysis, 44*, 104440.
Wani, I. A., Sogi, D. S., Wani, A. A., & Gill, B. S. (2013). Physico-chemical and functional properties of flours from Indian kidney bean (*Phaseolus vulgaris* L.) cultivars. *LWT-Food Science and Technology, 53*(1), 278–284.
Wanyo, P., Meeso, N., & Siriamornpun, S. (2014). Effects of different treatments on the antioxidant properties and phenolic compounds of rice bran and rice husk. *Food Chemistry, 157*, 457–463.
Westby, A., Reilly, A., & Bainbridge, Z. (1997). Review of the effect of fermentation on naturally occurring toxins. *Food Control, 8*, 329–339.
Wirth, J., Poletti, S., Aeschlimann, B., Yakandawala, N., Drosse, B., Osorio, S., et al. (2009). Rice endosperm iron biofortification by targeted and synergistic action of nicotianamine synthase and ferritin. *Plant Biotechnology Journal, 7*(7), 631–644.

Wu, C., Wu, Q. Y., Wu, M., Jiang, W., Qian, J. Y., Rao, S. Q., et al. (2019a). Effect of pulsed electric field on properties and multi-scale structure of japonica rice starch. *LWT-Food Science and Technology, 116*, 108515.

Wu, T. Y., Gruissem, W., & Bhullar, N. K. (2019b). Targeting intracellular transport combined with efficient uptake and storage significantly increases grain iron and zinc levels in rice. *Plant Biotechnology Journal, 17*(1), 9–20.

Xia, Q., Tao, H., Li, Y., Pan, D., Cao, J., Liu, L., et al. (2020). Characterizing physicochemical, nutritional and quality attributes of wholegrain Oryza sativa L. subjected to high intensity ultrasound-stimulated pre-germination. *Food Control, 108*, 106827.

Xiaoyan, S., Yan, Z., & Shubin, W. (2012). Improvement Fe content of wheat (*Triticum aestivum*) grain by soybean ferritin expression cassette without vector backbone sequence. *Journal of Agricultural Biotechnology, 20*, 766–773.

Xu, B., & Chang, S. K. C. (2008). Total phenolics, phenolic acids, isoflavones, and anthocyanins and antioxidant properties of yellow and black soybeans as affected by thermal processing. *Journal of Agricultural and Food Chemistry, 56*, 7165–7175.

Xu, L. N., Guo, S., & Zhang, S. W. (2018). Effects of solid-state fermentation with three higher fungi on the total phenol contents and antioxidant properties of diverse cereal grains. *FEMS Microbiology Letters, 365*(16), 1–8.

Xu, L. N., Guo, S., & Zhang, S. W. (2019a). Effects of solid-state fermentation on the nutritional components and antioxidant properties from quinoa. *Emirates Journal of Food and Agriculture, 31*(1), 39–45.

Xu, M., Jin, Z., Simsek, S., Hall, C., Rao, J., & Chen, B. (2019b). Effect of germination on the chemical composition, thermal, pasting, and moisture sorption properties of flours from chickpea, lentil, and yellow pea. *Food Chemistry, 295*, 579–587.

Yang, Q., Luo, Y., Wang, H., Li, J., Gao, X., Gao, J., & Feng, B. (2021). Effects of germination on the physicochemical, nutritional and in vitro digestion characteristics of flours from waxy and nonwaxy proso millet, common buckwheat and pea. *Innovative Food Science and Emerging Technologies, 67*, 102586.

Yashveer, S., Singh, V., Kaswan, V., Kaushik, A., & Tokas, J. (2014). Green biotechnology, nanotechnology and bio-fortification: Perspectives on novel environment-friendly crop improvement strategies. *Biotechnology and Genetic Engineering Reviews, 30*(2), 113–126.

Ye, X., Al-Babili, S., Kloti, A., Zhang, J., Lucca, P., Beyer, P., & Potrykus, I. (2000). Engineering the provitamin A (β-carotene) biosynthetic pathway into (carotenoid-free) rice endosperm. *Science, 287*(5451), 303–305.

Ye, J., Hu, X., Zhang, F., Fang, C., Liu, C., & Luo, S. (2016). Freeze-thaw stability of rice starch modified by improved extrusion cooking technology. *Carbohydrate Polymers, 151*, 113–118.

Ye, J., Hu, X., Luo, S., Liu, W., Chen, J., Zeng, Z., & Liu, C. (2018). Properties of starch after extrusion: A review. *Starch-Stärke, 70*(11–12), 1700110.

You, S. Y., Oh, S. G., Han, H. M., Jun, W., Hong, Y. S., & Chung, H. J. (2016). Impact of germination on the structures and in vitro digestibility of starch from waxy brown rice. *International Journal of Biological Macromolecules, 82*, 863–870.

Youssef, M. K. E., El-Fishawy, F. A. E. K., Ramadan, E. S. A. E. N., & Abd El-Rahman, A. M. (2013). Nutritional assessment of barley, talbina and their germinated products. *Frontiers in Science, 3*(2), 56–65.

Zeng, J., Wang, X., Miao, Y., Wang, C., Zang, M., Chen, X., et al. (2015). Metabolic engineering of wheat provitamin A by simultaneously overexpressing CrtB and silencing carotenoid hydroxylase (TaHYD). *Journal of Agricultural and Food Chemistry, 63*(41), 9083–9092.

Zeng, Z., Liu, C., Luo, S., Chen, J., & Gong, E. (2016). The profile and bioaccessibility of phenolic compounds in cereals influenced by improved extrusion cooking treatment. *PLoS One, 11*(8), e0161086.

Zhang, Z., Wei, T., Hou, J., Li, G., Yu, S., & Xin, W. (2003). Iron-induced oxidative damage and apoptosis in cerebellar granule cells: Attenuation by tetramethylpyrazine and ferulic acid. *European Journal of Pharmacology, 467*(1–3), 41–47.

Zhang, X., Chen, Y., Zhang, R., Zhong, Y., Luo, Y., Xu, S., et al. (2016). Effects of extrusion treatment on physicochemical properties and in vitro digestion of pregelatinized high amylose maize flour. *Journal of Cereal Science, 68*, 108–115.
Zhao, B., Shang, J., Liu, L., Tong, L., Zhou, X., Wang, S., et al. (2020). Effect of roasting process on enzymes inactivation and starch properties of highland barley. *International Journal of Biological Macromolecules, 165*, 675–682.
Zhu, F., & Li, H. (2019). Modification of quinoa flour functionality using ultrasound. *Ultrasonics Sonochemistry, 52*, 305–310.
Zhang, Y., Li, Y., Wang, H., Oladejo, A. O., Zhang, H., & Liu, X. (2020). Effects of ultrasound-assisted freezing on the water migration of dough and the structural characteristics of gluten components. *Journal of Cereal Science, 94*, 102893.

Chapter 3
Improvement of Genetic Variation for Nutrients and Bioactive Food Components in Cereal Crops

Pooja Manchanda, Deepansh Sharma, Ajinder Kaur, and Arashdeep Singh

3.1 Introduction

Genetic variation signifies differences that prevail in the genetic make-up of individuals or plants within the population. It occurs because of differences in DNA (Deoxyribonucleic acid) or RNA (Ribonucleic acid) sequences and variant alleles of a gene. It occurs in the form of different morphological traits in plants. The genetic variation existing among genotypes of one species or different species is termed as genetic diversity. We can measure genetic diversity by counting the number of different genes in a gene pool. Genetic variation thus, forms the building block of genetic diversity (Barton & Keightley, 2002).

Plant genetics is a field that focuses on the systematic and continual generation of new cultivars. It takes advantage of genetic variation among individuals within a plant species to combine desirable features into new and improved varieties. Besides existing variation, new variation is critical for introducing new features for genetic improvement of plants. Generally, genetic variation is a driving force in evolution to carry out natural selection that triggers alteration in the frequency of allelic expression within the population. The genetic variation generates diversity within and between populations, thereby affecting their phenotypic characteristics. The high extent of genetic variability imparts adaptation features to plants for their survival under unfavorable conditions (Holme et al., 2019).

P. Manchanda · A. Kaur
School of Agricultural Biotechnology, College of Agriculture, Punjab Agricultural University, Ludhiana, Punjab, India

D. Sharma
College of Horticulture and Forestry, Punjab Agricultural University, Ludhiana, Punjab, India

A. Singh (✉)
Department of Food Science, College of Agriculture, Punjab Agricultural University, Ludhiana, Punjab, India

S. Punia Bangar, A. Kumar Siroha (eds.), *Functional Cereals and Cereal Foods*,
https://doi.org/10.1007/978-3-031-05611-6_3

The genetic variation for important agricultural traits in different crop plants can be explored due to advances in genomic technologies. If a specific genetic feature is not immediately available for introgression into breeding materials, genetic variation in a crop species can be increased through alternative techniques.

Today, millions of individuals suffer from a variety of ailments because of lack of adequate nutrition and vital amino acids, and it becomes critical to improve the nutrient makeup of key staple crops (FAO, 2013). Over the last decade, significant progress has been made in the applications of biotechnology to produce nutritionally superior food crops. Cereal crops, one of the most widely used crops for food, have a diverse range of nutrients and physiologically active substances that benefit human health. The eight cereal grains of the family Gramineae (*Poaceae*) i.e. wheat, rice, barley, oats, maize, rye, sorghum and pearl millet are the most significant, accounting for 52% of total food calories (Vasil, 1995). Fortunately, crop biofortification has emerged as a potential technique for improving nutritional content of the cereals. Supplementation or biofortification procedures have traditionally been used to add vitamins and minerals to food crops. Crops can be biofortified in a variety of ways, such as addition of proper mineral or inorganic chemical using conventional plant breeding and biotechnological techniques (Bhutta et al., 2013).

In addition, targeted genome engineering with artificial nucleases offers the opportunity for speeding up the process for both basic research and plant improvement by allowing precise, quick and exact changes to genomes. It entails utilising an engineered nuclease to introduce targeted DNA double-strand breaks (DSBs), which stimulate cellular DNA repair system. Depending on the repair pathway as well as accessibility of a repair template, different genomic alterations can be performed (Cristea et al., 2013). Moreover, loss-of-function screening is now feasible and economical on a genomic scale, due to the simplicity and accessibility of the CRISPR/Cas9 technological platform that has many merits over alternative genome editing technologies (Shalem et al., 2014).

3.2 Advantages of Genetic Variation

Genetic variation offers a series of advantages as

1. Genetic variation serves as raw material for evolution
2. They provide each plant its own unique identity
3. Plant traits, like biotic or abiotic stress tolerance can be improved by crop breeders through genetic variation
4. Variations help plants to acclimatize themselves according to the changing climate, and for their survival against sudden fluctuations in environment
5. Generation of new traits or phenotypes in the plants is possible due to mutations or discontinuous variation
6. Specific phenotypes provide some plants an opportunity to better fit in the struggle for survival or existence

The various factors affecting genetic variation include mutation, gene flow, selection, genetic drift, migration, mating and crossing over; these are explained in the following section for successful grain improvement in cereals. Genomic technologies are rapidly advancing, allowing plant geneticists to access the genetic diversity in this key agricultural species. Genome-wide association studies are being used to build and mine diverse collections of cereal germplasm, and the variation so found can be linked to the sequence assembly in cereal crops (Muñoz-Amatriaín et al., 2014).

3.3 Methods of Genetic Variation

3.3.1 Mutation

Mutation results in changes in an organism's DNA leading to diversity in population. The accumulation of mutations with time lead to evolution of species. The novel mutations have led to the generation of novel genotypic and phenotypic variability. The unfavourable mutations are eliminated from population through natural selection. On the contrary, the beneficial mutations are passed on to the subsequent generation. The extent of mutations i.e. harmful or beneficial determines the survival of an organism with respect to sexual maturity and reproduction. The mutations may put a dramatic effect on a gene and the resulting phenotype (Slarkin, 1985).

Mutations result in chromosomal aberrations leading to genetic variation and diversity in the form of changed phenotype. Apart from this, plant parts/seeds are treated with different mutagens like gamma rays, chemicals (e.g. sodium azide) to generate mutations. Many varieties released in cereal crops (like rice, barley, wheat), legumes (soybean), vegetables (potatoes, onions) are an outcome of mutation breeding (ISAAA, 2006).

3.3.2 Genetic Drift

The change in frequency of an allele in a population due to random sampling and random event is referred to as genetic drift, which determines survival and reproduction of individuals. The two major types of genetic drift include population bottleneck (loss of individuals in a population due to natural disaster) and the founder effect (group of individuals separating from a population). The genetic variation among these populations will decrease with significant decrease in the number of mating individuals (Masel, 2011).

3.3.3 Gene Flow

Gene flow is an evolutionary force that represents the flow of alleles in and out of a population due to migration of individuals or gametes. The gene flow leads to stability and fluctuations in the population. The steady gene flow between two populations results in the reduction of genetic variation between two groups. Furthermore, introduction of new alleles through gene flow increases variability within a population and makes possible new combination of traits. The antagonistic feature of gene flow is the creation of new species through recombination of gene pools in the group of organisms (Cole, 2003).

3.3.4 Mating

Inbreeding and outbreeding are two types of mating strategies in crop plants, which increase and decrease genetic variation, respectively. In addition, interspecific hybridization results in creation of new phenotypes, thereby, enhancing genetic variation, whereas intraspecific hybridization lowers genetic variation within a population (Osawaru et al., 2015). The existing genes can also be arranged in new ways using chromosomal crossing over and recombination techniques, occurring during sexual reproduction.

3.3.5 Natural Selection

Natural selection leads to the survival and reproduction of individuals with certain genotypes that pass on their alleles to the next generation. According to Charles Darwin, natural selection is based on the following conditions:

1. There is variation among individuals within a population for some trait
2. The variation must be heritable
3. Variation of the trait is linked with variation in fitness

Selection favors a few among total alleles that lead to increase in number of selected alleles. Thus, selection reduces genetic variation.

3.4 Molecular Bases for Genetic Variation

Single nucleotide polymorphisms (SNPs), minor insertion/deletion polymorphisms (INDELs) and copy-number variations (CNVs) are all examples of genetic variations across individuals of the same species. SNPs have undoubtedly been the most studied intraspecific genomic variation. The development of genomic tools has provided the capabilities needed to find crucial trait controlling loci (Russell et al., 2011).

3.4.1 *Single Nucleotide Polymorphisms*

SNPs represent single nucleotide changes through transition (purine to purine or pyrimidine to pyrimidine) and transversion (purine to pyrimidine or vice versa). The observation of single nucleotide variation frequency greater than or equal to 1% is known as single nucleotide polymorphism. The presence of SNPs within coding and non-coding regions of genes and in intergenic regions cause genetic variations within the population. If a SNP occurs within a gene, then the gene is described as having more than one allele. In such cases, SNPs may lead to variations in the amino acid sequence. SNPs have been used in cereal crops to improve amylose content in rice (Bao et al., 2006).

3.4.2 *Insertions and Deletions*

Indels cause insertion and deletion of one or more base pairs leading to DNA mutations. These changes result in addition and deletion of nucleotide bases causing frameshift mutation, which leads to change in the translated product of a gene. For instance, the deletion of 3 base pair nucleotide sequence (DF508) in coding region of *CFTR* (cystic fibrosis transmembrane conductance regulator) gene is the cause of cystic fibrosis.

3.4.3 *Structural Variations*

Structural variations cause large-scale structural differences in DNA due to deletion, duplication, inversion and balanced translocation. These include copy number variations, variable number of tandem repeats and chromosomal re-arrangement.

3.4.4 *Copy Number Variations*

The genomic structural variations leading to change in the number of copies of a particular segment are called as copy number variations.

3.4.5 Variable Number Tandem Repeats (VNTRs)

The changes in the number of repeated DNA sequence arranged in tandem arrays are referred to as VNTRs. These are classified according to the size and number of their repeat units i.e. microsatellite, minisatellite, satellite and telomere repeat arrays.

3.5 Advantages of Genetic Variability in Cereal Crop Improvement

Genetic variability is the foundation for crop improvement, which allows crop breeders to develop cultivars with desired features, such as high yield, higher 100-seed weight, disease resistance, high nutrients, and enhanced bioactive components and so on. Initially, genetic variability within crop species was used to meet up subsistence food requirement. Later, the need arose for production of surplus food to meet food requirements for growing population. Currently, the focus is on the high yield as well as good quality aspects of staple cereal crops, such as rice, wheat, maize and barley, so as to provide balanced nutrition to human beings. Moreover, breeding for climate robust cultivars is becoming increasingly important as a result of climate change. The presence of genetic variation and diversity in similar species, wild species, germplasm lineages and mutant lines potentially enable breeders to develop climate tolerant cultivars as well as cultivars with enhanced agronomic and nutrient quality features. Thus, diverse genes in the form of variant crossable germplasm resources must be preserved to satisfy ever-changing breeding objectives.

Since, there is genetic heterogeneity within and between crop species, researchers can choose superior genotypes that can be released as a new variety or used as a parent in hybridization programs. Heterosis in F_1 plants and transgressive segregants in the F_2 generation need the presence of contrasting characteristics between two parents. The importance of genetic variation in cereal crop improvement has been documented in barley. Genomic technologies are rapidly advancing, allowing researchers to access the genetic diversity of this major agricultural species. Genome-wide association studies are being used to build and mine diverse collections of barley germplasm, and the variability so found can be linked to the barley sequence assembly. Interestingly, single nucleotide polymorphism platforms are now available. So, introducing advantageous alleles *via* marker-assisted selection has become quick and effective (Muñoz-Amatriaín et al., 2014).

3.6 Composition of Cereals: Nutrients and Bioactive Components

Since the dawn of civilization, cereals have supplied a multitude of health benefits to mankind. These are staple foods and essential components of human diet that have played an important role in shaping human civilisation. Cereals are necessary for the survival of billions of people on daily basis because they are abundant source of proteins, carbohydrates, minerals, lipids, B vitamins and various essential bioactive compounds (Goldberg, 2003).

3.6.1 Macronutrients

3.6.1.1 Carbohydrates

Cereals are commonly classified as carbohydrate-rich foods because they contain about 75% carbohydrate. The endosperm of cereal grains contains starch granules, which vary in size and form and are a primary component of cereals.

3.6.1.2 Starch

Cereal starch is a key storage carbohydrate and an important part of our diet (Goesaert et al., 2005). Starch in its native form is a versatile product that is often utilised as a raw material in the manufacture of sweeteners (Zeeman et al., 2010), bioethanol and biogas (Baghurst et al., 1996).

3.6.1.3 Amylose and Amylopectin

The endosperm contains granular starch that is made up of water insoluble homoglucans: amylopectin and amylose, that make upto 72–75% and 25–28% content, respectively in grain starches (Colonna & Buleon, 1992). In mutant genotypes, the ratio of amylopectin to amylose has been modified. A change in ratio has also been observed to impact the technical features of a variety (Van Hung et al., 2006). Wheat with high amylose content has been proposed as a raw material for enzyme-resistant starch synthesis (Topping et al., 2010).

Amylose is a linear molecule composed of alpha-(1,4)-linked D-glucopyranosyl units with a polymerization ratio ranging from 500 to 6000 glucose units, whereas amylopectin that gives starch its granular shape, has a polymerization ratio ranging between 30,000–30,00,000 glucose units (Hizukuri et al., 1981). The molecular weight of amylopectin is greater than that of amylose (Shibanuma et al., 1994).

3.6.1.4 Proteins

Cereals have a protein content of 6–15% (Goldberg, 2003). Gliadin and glutenin are the principal storage proteins in wheat, whereas other important cereal crops contain oryzenin, prolamin (zein), hordein, albumins and globulins (Kulp & Ponte, 2000). Cereals contain a wide variety of amino acids, but in limited amounts. Dietary essential amino acids must be consumed, so that, human body can synthesise other (non-essential) amino acids from them. The limiting amino acid is the essential amino acid that is in shortest supply in relation to demand. Rice, barley, rye and high lysine cultivars (e.g. sorghum, maize and barley) have more favourable essential amino acid compositions (Macrae et al., 1993).

The percentage of a few vital amino acids in cereal crops is listed below:

1. Leucine 7–14%
2. Alanine 4–11%
3. Methionine 1.3–2.9%
4. Tryptophan 0.2–1.0%
5. Histidine 1.8–2.2%
6. Lysine 1.4–3.3%

3.6.1.5 Lipids

Seed oils are the largest source of lipids. On dry matter basis, lipid content of cereal grains varies from 1% to 3% in rice, barley, wheat and rye, 5–10% in oats, and 5–9% in corn. Linoleic acid, an important fatty acid, is abundant in lipid fraction (Southgate, 1993).

3.6.2 Micronutrients

Cereals can help humans to fulfil their daily requirements of vitamins and minerals, but the number of micronutrients that we get depends on how much seed, fibre and kernel is consumed. Unfortunately, the refined grain products lose some of the pericarp, seed and parenchymatous layer that are high in vitamins and minerals.

3.6.2.1 Vitamins and Minerals

Vitamin concentration in cereals range from less than 1 to about 50 mg/kg in seeds (Koehler & Wieser, 2013). Cereals lack vitamin B_{12} and C, as well as vitamin A and β-carotene with exception of yellow corn (Cordain, 1999). Cereals, on the other hand, provide a valuable supply of most B vitamins, including riboflavin, thiamine and niacin (Kulp & Ponte, 2000). These also have significant levels of minerals and vitamin E. Cereals, like most plant meals, have a low salt content and high potassium content. Iron, zinc and magnesium are abundant minerals in whole-grain cereals. However, numerous trace elements, such as selenium are found in meager quantities. Rice has the greatest selenium content (10–13 mg/100 mg) as compared to any other cereal grain. The selenium concentration of a cereal varies, based on the amount of selenium in the soil; for example, the selenium content of wheat grain could range from 0.001 to 30 mg/100 g (Lyons et al., 2003).

3.6.2.2 Bioactive Compounds

Cereals have a variety of molecules known as phytochemicals or plant bioactive substances that may have nutrition properties (Goldberg, 2003). Apart from flavonoids that are only found in trace amounts in grains, other antioxidants, such as tocotrienols, tocopherols and carotenoids are also present in fair amounts. Whole grain breakfast cereals have been discovered to have an antioxidant content comparable to fruits and veggies (Miller et al., 2000). However, bound phytochemicals in cereals are found to be the largest contributors in providing various antioxidants (Adom & Liu, 2002). Lignans are a form of phytoestrogen compounds present in cereals and their quantity in grains is minimal. Nonetheless, grains may be a major source of lignans due to the huge amounts consumed.

Phenols

Phenolic compounds, often known as plant secondary metabolites, are widely distributed throughout the plant kingdom. The phenolic compounds with one functional carboxylic acid are referred to as phenols (Croteau et al., 2000). Based on the phenol subunits, there are two basic categories: simple phenols and polyphenols. Flavonoids are polyphenols with at least two phenol subunits, while tannins have three or more phenol subunits (Clifford, 2001).

The different studies on phenolic acids are focused on elucidating their functions in plant life. Appearance, sensory attributes, nutritional and antioxidant capabilities of foods have all been linked to phenolic acids (Maga & Katz, 1978). The function of phenolics in organoleptic qualities, such as flavour, astringency and hardness has sparked analytical analyses. Phenolics act as antioxidants because of redox potential of the phenol molecule. Although, there are various methods for antioxidant activity, radical scavenging *via* hydrogen atom donation is thought to be the most common. Electron donation and dioxygen quenching are two other well-known antioxidant and radical-suppressing processes (Shahidi et al., 1992).

Furthermore, both bound and free forms of phenolic acids have been found in cereal grains. The majority of phenolic acids may be found in sorghum and millet (Hahn et al., 1983). Ferulic acid and p-coumaric acid are the most common phenolic acids found in cereals. The most prevalent hydroxycinnamic acid found in wheat grains is ferulic acid. It is the most abundant polyphenol in cereals, where it is esterified to arabinoxylans in the grain cell wall. Wheat bran contains ferulic acid, which is esterified to hemicelluloses in cell wall (Dewanto et al., 2002). Grain types, cultivars and the section of the grain examined, all influence the content of phenolic chemicals in whole-grain cereals (Holtekjølen et al., 2006).

Lignans

Lignan, an essential product, can be found in both fungi and plants. Various lignans have been shown to have anti-tumour, anti-proliferative and anti-microbial activities, as well as the ability to block certain enzymes. Talking about the structure, lignans are dimers of phenylpropanoid units connected by core carbon molecules (Hano et al., 2021).

Lignan, a polyphenolic active molecule, is a class of phytoestrogen chemicals found in a variety of plant crops, particularly cereals including wheat, oats and rye (Hooper & Cassidy, 2006). The proportion of lignans has been determined to be between 10–15% in the bran layer of rye, wheat and oat grains. The amount of total lignans in whole grain wheat relative to bran was just 9.6%, but oat, rye, spelt wheat and barley lignan content ranged between 25–40%. However, lignans did not appear to be concentrated in the bran layer of buckwheat, triticale or millet. In these species, the percentage of lignan in whole grain and bran was between 62–78% (Mazur & Adlercreutz, 1998).

The cereal species (in descending order of their lignan content) can be arranged as follows:

Rye > wheat > triticale > oat > spelt wheat > Japanese rice > wild rice > buckwheat> barley >amaranth > corn > millet > quinoa > red rice > brown rice > dhurra. However, when the alkaline-extractable component of wheat is considered, its lignan concentration exceeds that of rye (Smeds et al., 2007).

Carotenoids

Carotenoids, having orange, yellow and red colour, are the most widely distributed pigments in nature and have attracted a lot of attention for their antioxidant and provitamin properties. The pro-vitamin A activity of β-cryptoxanthin, β-carotene and others with at least one intact non-oxygenated β-ionone ring contributes significantly to the nutritional value of carotenoids (Palozza & Krinsky, 1992).

Isoprene units make up the 40-carbon backbone of carotenoids. The long line of conjugated double bonds constituting the centre section of the molecule is the most distinguishing property of carotenoids. This is what gives them their structure, chemical reactivity and ability to absorb light.

Lutein is a prominent carotenoid found in high amounts in cereals, particularly wheat, while zeaxanthin and lutein are present in rice, both of which help to improve eyesight (Adom et al., 2003). On a dry weight basis, maize is an excellent source of carotenoids, with a weight of 11 μg/kg of seeds (Saikia & Deka, 2011). In contrast to other micronutrients like minerals, trace elements and polyphenols, carotenoids are more equally distributed throughout the grain, with large amounts in the endosperm (Panfili et al., 2004).

Phytosterols

Phytosterols, that belong to the triterpene family of plant metabolites, are necessary macromolecules for body health. They are present in cereals as glycosides, essential fats, free alcohols and as esters when linked with ferulic acid. Campesterol (C-28, carbon skeleton), stigmasterol (C-29), and β-sitosterol (C-29) are the few primary sterols found in plant world, accounting for 98% of all phytosterols (Miras-Moreno et al., 2016). Cereal grains, such as wheat, oats, rye, corn and triticale contain sterols. Non-saponifiable extracts of wheat flour and whole grains contain campesterol and sitosterol (Seitz, 1989).

Flavonoids

Flavonoids are polyphenolic secondary metabolic products with a low molecular weight. Flavonols have long been recognised to be formed in specific places in plants, and they are responsible for flower colour, aroma and fruit dispersion, which aid in seed and seedling germination, growth and development. Table 3.1 shows the flavonoid profile of a few cereal crops. Flavonoids may serve a functional role in freezing tolerance and heat acclimatisation in plants, as well as frost resilience and drought tolerance (Samanta et al., 2011).

Tocols: Tocopherols and Tocotrienols

Tocols are naturally occurring antioxidants found in plant-based foods, such as cereals. These lipid-soluble antioxidants are further divided into tocopherols and tocotrienols, which are highly acknowledged for their bioactivities and are abundant

Table 3.1 Flavonoid profile in some cereal crops

S.No.	Cereal crop	Important flavonoid compounds	Position	Colour	Reference
1	Barley (*Hordeum vulgare*)	Quercetin and kaempferol	Pericarp	Purple	Yang et al. (2013)
2	Maize (*Zea mays*)	Pelargonidin 3-glucoside	Aleurone and pericarp	Pink	Abdel-Aal et al. (2006)
3	Wheat (*Triticum aestivum*)	Cyanidin-3-rutinoside and cyanidin-3-glucoside	Pericarp	Purple	Knievel et al. (2009)
4	Rice (*Oryza sativa*)	Apigenin	Pericarp	Red	Kim et al. (2010)

in cereal grains (Nielsen & Hansen, 2008). Owing to their ability to decrease free radicals, tocols serve a variety of functions in plants, including non-photosynthetic and photosynthetic activities. They safeguard photosynthetic machinery from lipid peroxidation in the chloroplasts (Yamauchi & Matsushita, 1979).

Tocols can be found in large quantities in cereal grains. In oats, total tocol concentration vary from 20 to 30 mg per kg, with α-tocotrienol being the most prevalent (Peterson, 1995). The presence of tocotrienols and tocopherols in cereal grains was demonstrated in a study involving seven *Triticum aestivum* and eight *Triticum turgidum* cultivars. The results showed abundant accumulation of tocols, with α-tocotrienol, β-tocotrienol, β-tocopherol and α-tocopherol having high accumulation in grains, and the mean tocotrienol: tocopherol ratio as 3.68 (Hidalgo et al., 2006).

β-Glucans

All the genotypic and environmental variables influence the β-glucan content of cereal grains. Large amounts of β-glucans are found in barley and oats. They are contained in the endosperm of oats and the aleurone layer of barley (Bhatty, 1993). The total β-glucan content of barley grain varies from 2.5% to 11.3% by weight of the seed, but it is most commonly between 4–7% , whereas in oats, β-glucan content is highly variable, ranging from 2.2% to 7.8%. Furthermore, carbohydrate content analysis of wild barley (*Hordeum spontaneum*) genotypes indicated β-glucan levels of upto 13.2% (Lazaridou et al., 2007).

Avenanthramides

Avenanthramides are oat-derived polyphenols. There are at least 25 different substances that are substituted cinnamic acid amides of anthranilic acids. Avenanthramide 1, 3 and 4, commonly known as avenanthramides B, A and C are the three primary avenanthramides found in oats (Collins, 1989).

3.7 Role of Genetic Variation in Food Security

Genetic variability and genetic diversity contribute towards following three dimensions of food security: food availability, food access and food utilization.

Millions of people world-wide suffer from micronutrient deficiency, this kind of food insecurity along with low dietary quality cause various physical, mental disorders, diseases and a number of untimely deaths. Achieving Sustainable Development Goal 2, "zero hunger and improved nutrition," requires huge changes in prevailing global food systems. Among other approaches, agricultural technologies have a key role to perform. An increase in agricultural production due to intensive crop cultivation, high use of pesticides and insecticides has contributed to environmental extremes and climate change (increased temperature, drought stress etc.) having a negative effect on agricultural production. Poor people living in Africa, Asia suffer the most by climate change as they rely purely on agriculture for earning their livelihoods. Thus, there is a need to adopt new types of agricultural technologies, so that food security is achieved through sustainable agriculture (Poole et al., 2021).

Different development organizations, e.g. The Rockefeller Foundation were involved in launching various research programs that aimed at making available new agricultural technologies to farmers in the developing countries. The high-yielding varieties of wheat, rice and maize, developed by virtue of efforts of these research programs were widely grown by farmers, these varieties along with increased use of agrochemicals and fertilizers resulted in bumper yields in short time period. This increase in food production became popularly known as the Green Revolution. Upon simulation, it was demonstrated that mean calorie consumption in developing countries had been 10–15% less if high yielding wheat, rice and maize varieties were not introduced. Therefore, Green Revolution played a significant role to lessen hunger in developing countries. Bumper yields increased availability of food, agricultural profits, and incomes of poor people belonging to small farm sector. However, in Africa, growth in agricultural productivity was much slower than other countries as it could not keep pace with the growing population (Stevenson et al., 2013).

The growth in yields of cereal crops was due to cultivation of new varieties, high use of pesticides, fertilizers, water etc. Further hike in fertilizer use and other inputs are connected with marginal yield enhances, as there is climate change as well as continuous growth of population. So, there is a need to use new tools (to accelerate crop breeding) that are safe and speed up the work towards food security (ISAAA, 2021). In this regard, new plant breeding technologies (NPBTs) encompassing GMOs (genetically modified organisms/transgenic crops) and gene/genome edited crops could contribute towards production of plants with higher yields, decreased application of herbicides/insecticides, improved quality and shelf life of products. The GMOs are seen very critically, which undergo rigorous testing before their release for commercial cultivation. Although, extensive years of research have demonstrated that transgenic crops are not more risky than traditionally bred crops, yet there are still widespread concerns about their possible negative consequences on human health and environment. These issues have resulted in outlining of strict safety regulations for transgenic crops. Since majority of the transgenic crops commercialized up till now have been developed by multinational companies, their economic and social concerns are still prevalent.

3.8 Genetic Improvement in Cereal Crops: Techniques/Approaches

Various examples of cereal crops have been summarized in Table 3.2, using techniques/approaches for the improvement of nutrients and bioactive compounds.

Table 3.2 A representative list of cereal crops where nutrients and bioactive compounds have been improved using different approaches

S. No.	Crop	Improved component (nutrient/bioactive compounds)	Mechanism/approach	Remarks	Reference
1	Rice (*Oryza sativa* L.)	Carotenoids	CRISPR/Cas9	Higher carotenoid content was obtained at genomic safe harbour sites in seeds	Dong et al. (2020)
		Vitamin A and iron	Molecular breeding	Increased vitamin A and iron levels enhanced nutrient content of seeds	Lucca et al. (2006)
		Glutenin content	RNAi	Low glutenin content 1 significantly reduced glutenin amount in rice seeds, which is beneficial for kidney patients	Kusaba et al. (2003)
		Alpha-linolenic acid	*Agrobacterium*-mediated transformation	Rice seeds contained more alpha linolenic acid	Anai et al. (2003)
		Protein content	PEG-mediated transformation	Rice seeds were enriched with amino acids	Sindhu et al. (1997)
	Rice (Golden rice 2)	β-Carotene	*Agrobacterium*-mediated transformation	There was a 23-fold increase in total carotenoids and a pre-dominance of β-carotene build-up	Paine et al. (2005)
	Australian rice accession	Vitamin E	Marker assisted breeding	Vitamin E level in rice panicles was increased	Lee et al. (2020)
	Rice (*Oryza sativa* subsp. Indica)	Iron nutrition	Marker assisted selection	Increased iron content prevents iron deficiency	Paul et al. (2012)
	Rice (*Oryza sativa* subsp. japonica)	Folate (vitamin B9)	*Agrobacterium*-mediated transformation	Rice having high folate had improved bio-availability and bio-efficacy	Storozhenko et al. (2007)
		Flavonoids	*Agrobacterium*-mediated transformation	Increased flavonoid content in rice seeds provides plethora of health benefits	Ogo et al. (2013); Shin et al. (2006)
2	Maize (*Zea mays*)	Provitamin A, lysine and tryptophan	Marker assisted stacking	Biofortified maize hybrids would ensure nutritional security	Zunjare et al. (2018)

3	Durum wheat (*Triticum turgidum* L. subsp. *durum*)	Phytic acid	RNAi	Low phytic acid level reduces heavy metal accumulation	Bhati et al. (2016)
	Wheat (*Triticum aestivum*)	Pro-vitamin A	Biolistic transformation	Increased pro-vitamin A content aids in preventing vitamin A deficiency	Wang et al. (2014)
		Grain protein content	Marker assisted backcrossing	Grain protein content was increased in improved cultivar	Vishwakarma et al. (2014)
		Amino acids	Biolistic transformation	Amino acid augmentation increases nutritional and functional quality of wheat seeds	Tamás et al. (2009)
4	Maize	Protein quality	Marker assisted breeding	Improved quality protein maize hybrids prevent protein energy malnutrition	Kaur et al. (2020 b); Hossain et al. (2019); Olsen and Phillips, (2009)
		Pro-vitamin A	Marker assisted backcrossing	Pro-vitamin A augmentation enhanced the nutritional value of quality protein maize	Goswami et al. (2019)
		Vitamin E and tocochromanol	Marker assisted selection	Maize seeds had higher levels of tocochromanol and vitamin E	Diepenbrock et al. (2017)
		Phytic acid	Marker assisted backcrossing	Low phytic acid content provides plethora of health benefits	Tamilkumar et al. (2014)
		Amylose	RNAi	Maize quality improved through high amylose content enrichment	Guan et al. (2012)
		Iron	Biofortification	Increased iron level help in preventing anaemia	Hoekenga et al. (2011)
		Iron (ferritin) and phytase	Biolistic transformation	Iron availability and absorption were increased significantly	Drakakaki et al. (2005)

(continued)

Table 3.2 (continued)

S. No.	Crop	Improved component (nutrient/bioactive compounds)	Mechanism/approach	Remarks	Reference
5	Sorghum (*Sorghum bicolor* L.)	Lysine	Molecular breeding	Enhanced lysine level improved nutritional quality of sorghum grains	Elkonin et al. (2021); Fantaye (2018)
		Pro-vitamin A	*Agrobacterium*-mediated transformation	Pro-vitamin A carotenoids were significantly enhanced in sorghum grains	Lipkie et al. (2013)
6	Barley (*Hordeum vulgare*)	Lysine	*Agrobacterium*-mediated transformation	High lysine content in seeds and leaves of maize hybrid was achieved	Ohnoutkova et al. (2012)

3.8.1 Genetically Modified Crops

A genetically modified organism or GMO, is a living organism into which a gene has been inserted, expressed and inherited to next generation. Genetic engineering/ genetic transformation have been crucial in crop improvement by adding advantageous foreign genes or inhibiting the expression of endogenous genes in crop plants. Genetically modified (GM) crops are agricultural plants whose genomes have been modified using specific gene of interest to improve existing traits or to incorporate a new trait that do not occur naturally in the crop plants. Transgenic plants are those that have specific regions of foreign nucleic acid or gene sequence inserted into their genomes *via* transformation process (Griffiths et al., 2005). Plant breeders use a wide range of hybridization techniques, including intraspecific and interspecific crosses, mutagenesis and so on, to create new varieties. The former results in linkage drag caused by undesirable genes, while the latter results in random outcomes. Genetic transformation/transgenesis, on the other hand, allows researchers to access a broad gene pool because genes from any plant, animal, microorganism or even genes synthesized in the laboratory can be utilized to transform plants using recombinant DNA techniques.

Transgenic crops were grown in 29 nations during 2019, with Africa growing twice as many as the rest of the world (from three to six) (Pallett, 2021). USA, Brazil, Argentina, Canada and India are the top five countries growing transgenic crops, with high transgenic crop acceptance rates in these countries. Despite biosafety and environmental issues, transgenic technology has been a preferred strategy for the quick production of better crop plants with several beneficial features (ISAAA, 2017).

3.8.1.1 Genetically Modified Wheat and Maize

Enhancing folate concentration in agricultural crops by metabolic engineering is recognized as an essential method with significant promise for addressing folate malnutrition. The two gene co-expression strategy resulted in genetically altered wheat and maize with increased folate levels. The study involves cloning of two *GmGCHI* (*GTP cyclohydrolase I*) genes (*Gm8gGCHI* and *Gm3gGCHI*) and one *GmADCS* (*aminodeoxychorismate synthase*) gene extracted from soybean, which is responsible for generating folate precursors pterin and p-aminobenzoate. It was discovered that *Gm8gGCHI* boosted pterin and folate production more than *Gm3gGCHI* in transgenic *Arabidopsis* plants. Then, in maize and wheat, the two major staple crops, researchers co-expressed *Gm8gGCHI* and *GmADCS via* endosperm-specific promoters to increase folate metabolic flux. As a result, transgenic wheat and maize grains so developed were reported to have folate levels that were 2.3 and 4.2 times higher, respectively, than their wild-type cultivars (Liang et al., 2019).

3.8.2 RNA Interference

RNAi-based technique has been implemented, and the results show that it has a lot of potential in a variety of cereal production and improvement domains (Mezzetti et al., 2020). RNAi is a fascinating phenomenon wherein small double-stranded RNA (dsRNA) blocks the expression of particular genes by inducing the destruction of specific mRNA target sequences in the cytoplasm (Napoli et al., 1990).

When it comes to cereal crop nutrition, researchers have discovered that by reducing the expression of maize zein storage proteins (a family of abundant proteins within maize seed low in lysine content), seed-specific RNAi techniques have proven successful in generating dominant high lysine corn (Houmard et al., 2007). In wheat, the starch-branching enzyme was down-regulated using RNAi, leading to high amylose wheat, which has a lot of promise to benefit human health (Sestili et al., 2010). Moreover, glutenin content in rice (hard to digest by renal patients) has been reduced *via* RNAi-mediated gene silencing. *GluB* hairpin RNA was used to create LGC-1 (low glutenin content 1) rice variety that has a low glutenin concentration (Kusaba et al., 2003).

3.8.3 Molecular Breeding

Molecular breeding has revolutionised the process of cereal grain improvement. Molecular plant breeding can be described as the application of genetic modification at DNA level to enhance plant traits of interest. It improves crop breeding efficiency by selecting and stacking favourable alleles at target loci, and hence steadily increasing genetic benefits (Moose & Mumm, 2008). In recent years, the potential for implementing molecular breeding for higher nutritional quality in crops, such as maize have expanded dramatically. Significant progress has also been made, particularly in marker assisted selection for producing quality protein maize cultivars and identification of genes/QTLs governing several quality attributes in maize (Babu and Prasanna, 2014). The two molecular plant breeding technologies that have the potential to improve agricultural crops are: marker assisted selection and marker assisted backcrossing.

3.8.3.1 Marker Assisted Selection

Molecular assisted selection, MAS entails the selection of specific alleles for traits that are influenced by a small number of loci. Marker assisted selection together with QTL analysis and genetic transformation has emerged as the most useful methods for rice molecular breeding that have been used to identify new germplasm and elite rice cultivars (Ding et al., 2004). MAS is a strategy that employs molecular markers closely related to a target gene as a tag for quick indirect selection of the target gene. Rice breeders and geneticists have made tremendous progress in finding QTLs linked to critical agronomic qualities, such as nutrition aspect, grain production efficiency, development and growth, disease and insect resistance and abiotic stress endurance (Chen et al., 2000).

3.8.3.2 Marker Assisted Backcrossing

Marker assisted backcrossing; MABC is a popular approach for breeding that has been used in a variety of studies for finding genomic regions of interest. It is the process of transferring a small number of loci, including transgenes, from one genetic background to another. It is a precise and accurate strategy for introgressing a single locus controlling a trait of interest while maintaining the recurrent parent's key features (Hospital, 2001). MABC works by integrating a specific gene from agronomically unsuitable donor parent into an exclusive breeding line i.e. recurrent parent. The anticipated product is an enhanced line that contains only the specific gene from the donor parent, as well as other genes from recipient parent line that is present throughout the genome. MABC has been discovered to be frequently employed in plant breeding programmes to generate novel varieties or lines in cereal crops, particularly rice (Tanksley et al., 1989).

3.8.4 Genome Editing

Transgenesis has been utilized in plant breeding since 1980s, and it has greatly increased the precision of breeding programs. However, as the gene is randomly integrated in the plant genome during transgenesis, the specific position of the transferred gene cannot be determined. This demands the development of a large number of transgenic events and their analysis for selecting elite events with desired phenotypes and no off-target consequences. A precise and efficient technology known as gene/genome editing (GE) has been developed and is now widely being employed (Yin et al., 2017).

The three major cereal crops *viz.* wheat, rice and maize account for more than 42% of the calories consumed by the world's population. Maintaining a consistent supply of these commodities while enhancing their nutritional value and mitigating climate change is difficult and necessitates the use of a variety of novel agriculture breeding technologies. In this regard, genome editing technologies provide unrivalled prospects for crop improvement with remarkable precision and speed. Genome editing is a revolutionary technique with wide-ranging implications, including crop enhancement in agriculture (Gadal et al., 2019).

DNA insertion, modification, replacement or deletion at pre-determined places in the plant genome is possible using gene editing techniques. Targeted GE causes site-specific double-strand breaks, which are then repaired by the host cell's repair machinery. Genome editing is a breakthrough technology, the bulk of gene-edited crops that are now available have only simple point mutations and no transgene integration. Point mutations could also occur naturally or through conventional mutagenesis. A novel technology known as prime editing system, which incorporates single-strand breaks, has just been developed, adding a new level to breeding precision (Grohmann et al., 2019).

The different GE tools available are:

3.8.4.1 Zinc Finger Nucleases

Zinc finger nucleases, ZFNs are artificial restriction enzymes, which can be tailored to cleave almost any lengthy stretch of double-stranded DNA sequence. ZFNs have been employed in model and crop plant species to promote replacement of native DNA sequences with foreign DNA molecules, and to mediate the integration of targeted transgene into native genome sequence. In cereal crops, ZFNs have been employed to improve phytase content in maize, and are being used as a potential tool for crop improvement (Shukla et al., 2009).

3.8.4.2 Transcription Activator-Like Nucleases

In many different agricultural crops, TALENs have emerged as an important genome editing tool. TALENs boost the effectiveness of genomic alteration by introducing site-specific chromosomal double-strand breaks. Due to the sheer modular structure of the TALE core repeat domains, researchers easily alter DNA recognition specificity and target virtually any DNA sequence (Joung & Sander, 2013).

Rice is among the most important cereal crops, accounting for 20% of global nutritional intake. The fragrance industry was the first to benefit after the introduction of TALENS. The *BADH2* gene, which encodes for betaine aldehyde dehydrogenase, was disrupted using the TALEN technology to produce a competing molecule, γ-aminobutyric acid (*GABA*), instead of dominant taste component, 2-acetyl-1-pyrroline (*2-AP*), from the same primary substrate i.e. γ-amino butyraldehyde (Shan et al., 2015).

3.8.4.3 Clustered Regularly Interspaced Short Palindromic Repeats

CRISPR/Cas technique holds enormous potential for effective genome editing in a variety of agricultural crops (Kaul et al., 2020). Cereals that contain a large amount of amylose are a valuable source of resistant starch. Resistant starch-rich foods are nutrient-dense and lower the risk of a variety of diseases (Jiang et al., 2010). In rice, the CRISPR-Cas9 technology has been utilised to direct mutagenesis of *SBEI* and *SBEIIb* genes encoding starch branching enzymes, which regulate amylose concentration. Despite the fact that there was no discernible difference between *SBEI* edited and wild type rice plants, the *SBEII* edited rice plants had much higher amylose and resistant starch levels as compared to wild type rice plants (Sun et al., 2017).

3.8.5 *Biofortification*

Biofortification, or the practice of breeding nutrients into food crops, is a relatively cost-effective, long-term and sustainable way of increasing micronutrient delivery. Although, biofortified staple foods cannot provide the same daily doses of minerals and vitamins as supplements or commercially fortified meals, they can help individuals meet their daily micronutrient requirements. (Bouis et al., 2011).

Traditional plant breeding can be used to accomplish biofortification, in which parent lines with high vitamin or mineral levels are crossed to produce plants with the necessary nutrient and agronomic properties. When a nutrient does not occur naturally in a crop or when significant amounts of accessible micronutrients cannot be properly bred into a crop, in such cases transgenic techniques are advantageous. However, after obtaining a transgenic line, several years of traditional breeding are required to ensure that the transgenes are incorporated into farmer-preferred cultivars and are inherited in a stable manner. While transgenic breeding can occasionally provide micronutrient advantages not available to conventional breeders, many nations lack the legal structures necessary to allow the distribution and marketing of these cultivars (Saltzman et al., 2013).

3.8.5.1 Golden Rice: A Breakthrough Improvement

Three million preschool-aged children suffer from vitamin A insufficiency each year. Actions were taken to improve vitamin A content of rice in response to this issue. Golden Rice, known for its golden color due to high-carotene content, was developed by Ingo Potyrus research group where two genes from unrelated organisms (daffodil and bacteria *Erwinia uredovora*) were used for introgressing the carotenoid biosynthesis pathway within the rice genome (Tang et al., 2009). By using marker assisted backcross breeding, this new trait was subsequently transferred into high yielding local commercial rice cultivars. The most recent golden rice technology, known as GR2, makes use of genes from two independent provitamin A pathways, including the incorporation of a maize phytoene synthesis gene for the equivalent daffodil gene used in GR1 rice. As a result, GR2 golden rice yielded up to 35 μg β-carotene/g of dry (Shumskaya & Wurtzel, 2013).

3.9 Scenario on Use of New Breeding Techniques in India

India began cultivating Bt cotton in 2002 and is now a leading cotton exporter in the world. More than 96% of the total acreage of cotton-planting land is under Bt cotton (Barwale et al., 2004). Through gene editing, government organisations and agricultural colleges have effectively improved rice, peanut and banana. Nagaraj et al. (2019) employed CRISPR technique to generate thermo-sensitive genic male sterility in rice, which is necessary for the creation of hybrid rice seed. CRISPR-Cas9 has also been used to create rice that is drought and salt tolerant (Farhat et al., 2019; Kumar et al., 2020). Kaur et al. (2020a) used the similar approach to boost β-carotene production in bananas. Through genome editing, the Junagadh Agricultural University in Rajasthan generated a groundnut with low linoleic acid and high oleic acid. Appropriate rules for genome-edited foods have now been drafted in India. Risk assessment should be commensurate to the degree of genetic changes, according to the standards (under development).

3.10 Safety Issues Related to Genetically Modified Foods

The main threat believed to be caused by genetically modified (GM) foods is that their consumption can be harmful to the human beings. Risks of GM foods to human health are related with toxicity, allergenicity or antibiotic resistance of the foods. To talk about first category, it is not the transferred gene that is toxic to human health; rather toxicity risk is related to nature/effects of product synthesized by the transferred gene or changes in composition and metabolism of organisms/plants resulting from gene transfer. For example, some transgenes encode for enzymes that change biochemical pathways, leading to decrease or increase in synthesis of certain biochemical compounds. These metabolic changes can cause an increase in toxin concentrations.

Majority of the risks related to toxicity can be analysed through use of scientific methods. The synthesis of new products in transgenic crops as a result of gene transfer from a source that has not been consumed as food, can cause allergenic effects. It should be noticed that there exists no evidence of transgenic crop products posing more risks than normal food products in triggering allergic reactions. Moreover, the transgenic crops are thoroughly tested for allergenicity before commercial release. For instance, soybean plants genetically modified by introgression of Brazil nut gene to increase methionine content were not released for cultivation after finding that the product of the transgene could cause allergy (ISAAA, 2002).

Sometimes, GM crops are engineered with antibiotic resistant genes, e.g. *nptII* that provides resistance against kanamycin. The researchers make use of this trait to identify the transformed cells carrying desirable gene. It is thought that these genes can move from GM crops/foods into microorganisms residing in the gut of human beings and lead to increase in antibiotic resistance. However, extensive research conducted in this direction has revealed that the possibility of moving such genes into microorganisms is very rare (ISAAA, 2002). Moreover, steps are being taken to avoid this risk by excluding this gene from the gene cassette used for carrying out genetic transformation.

Another issue is risks of GMOs (having herbicide or insect resistance) to environment that include production of 'super weeds' through cross breeding between transgenic plants and non-transgenic related crop species, weedy relatives, development of resistance in insect-pests populations leading to evolution of 'super bugs', impact on biodiversity and non-targeted insects. Resistance in insects can come if selection pressure is strong, i.e. if transgenic crops are grown on a large commercial level, this could promote evolution of resistant insect-pests, which would nullify the efficacy of transgenic crops in a few years. It can also happen that if insect-resistant transgenic plants cause mortality of one particular insect, it will reduce competition, and minor insects would become major. For e.g. outbreak of pink boll worm attack in Punjab due to large scale cultivation of Bt cotton harbouring transgene against *Helicoverpa armigera*. Additionally, it could lead the insect population to migrate to another plant population that was once secure. Similarly, if herbicides are sprayed more regularly due to new transgenic crop cultivars, the weeds would develop resistance against the herbicide tolerated by transgenic crop. This would lead to increased dosage of herbicide or use of another herbicide on plants. However, this risk can be analysed by conducting appropriate experiments prior going for any commercial release. The herbicide-tolerant transgenic crops have been reported to be no more invasive in fields than their non-transgenic control plants (Dale et al., 2002).

It is also feared that residues from insect-resistant or herbicide-resistant plants could harm the useful microorganisms residing in the surrounding soil. Scientist has conducted experiments and provided evidence that the insecticidal Bt proteins are degraded in the soil and do not affect the soil microorganisms (Anonymous, 2006).

There is another issue regarding eating foreign DNA as part of GM food. As DNA is present in all plants, animals and microorganisms, so human beings eat DNA with each meal. This DNA is broken down into molecules during digestion, and small amount is either absorbed in blood or excreted. Till now, there has been no proof of additional risk on human health from consuming DNA of GM foods.

Several other issues regarding GM foods are believed to be as follows:

1. The manufacturers don't provide information on the label that foods have been developed through genetic manipulation, mainly because they believe that doing so will affect their trade negatively, which is a bad practice. According to The Food Safety and Standards Bill 2005, "no person shall manufacture, process, export, import or sell genetically modified articles of food, organic foods, functional foods, nutraceuticals, health supplements etc. except in accordance with the regulations made under this Act" (Anonymous, 2006).
2. Many religious societies feel that GM foods are unnatural, and are against the production of these foods.
3. Some people have ethical issues regarding transfer of genes from animals into plants.
4. It is feared that with increase in production and commercial approval of GM foods, the developing countries would have more dependence on industrialized countries as it is quite probable that in future GM food production would be controlled by them.

3.11 Issues of GM Foods in Context to India

Due to concerns about public health and biodiversity, the Indian government halted field testing of several crops in 2010. The negative public reaction resulted in putting on hold the release of Bt brinjal. Similarly, the Union Ministry of Environment and Forest refused commercial release of transgenic mustard hybrid DMH-11 developed by Delhi University's Centre for Genetic Manipulation of Crop Plants in 2018, citing the need for more field trials to determine its impact on soil biodiversity, beeswax and honey bees. However, an agri-tech body during 2020 appealed Central government to allow for the same, as it emphasized that the studies conducted over past 23 years reported no adverse effects of GM foods on human health.

References

Abdel-Aal, E. S. M., Young, J. C., & Rabalski, I. (2006). Anthocyanin composition in black, blue, pink, purple, and red cereal grains. *Journal of Agricultural and Food Chemistry, 54*(13), 4696–4704.

Adom, K. K., & Liu, R. H. (2002). Antioxidant activity of grains. *Journal of Agricultural and Food Chemistry, 50*(21), 6182–6187.

Adom, K. K., Sorrells, M. E., & Liu, R. H. (2003). Phytochemical profiles and antioxidant activity of wheat varieties. *Journal of Agricultural and Food Chemistry, 51*, 7825–7834.

Anai, T., Koga, M., Tanaka, H., Kinoshita, T., Rahman, S. M., & Takagi, Y. (2003). Improvement of rice (*Oryza sativa* L.) seed oil quality through introduction of a soybean microsomal omega-3 fatty acid desaturase gene. *Plant Cell Reports, 21*(10), 988–992.

Anonymous. (2006). Training workshop on biosafety for state agricultural universities, organized by Ministry of Environment & Forests, Govt. of India and Biotech Consortium India Limited, New Delhi from June 2–3 at PAU, Ludhiana.

Bao, J. S., Corke, H. & Sun, M. (2006). Nucleotide diversity in starch synthase IIa and validation of single nucleotide polymorphisms in relation to starch gelatinization temperature and other physicochemical properties in rice (Oryza sativa L.). *Theoretical and Applied Genetics, 113*(7), pp.1171–1183.

Babu, R. and Prasanna, B. M. (2014). Molecular breeding for quality protein maize (QPM). In Genomics of plant genetic resources (pp. 489–505). Springer, Dordrecht.

Baghurst, P. A., Baghurst, K. I., & Record, S. J. (1996). Dietary fibre, non-starch polysaccharides and resistant starch: A review. *Food Australia, 48*(3), S3–S35.

Barwale, R. B., Gadwal, V. R., Zehr, U., & Zehr, B. (2004). Prospects for Bt cotton technology in India. *AgBioForum, 7*(1&2), 23–26.

Bhati, K. K., Alok, A., Kumar, A., Kaur, J., Tiwari, S., & Pandey, A. K. (2016). Silencing of ABCC13 transporter in wheat reveals its involvement in grain development, phytic acid accumulation and lateral root formation. *Journal of Experimental Botany, 67*(14), 4379–4389.

Bhatty, R. S. (1993). *Non malting uses of barley* (pp. 355–417). Chemistry and Technology.

Bhutta, Z. A., Salam, R. A., & Das, J. K. (2013). Meeting the challenges of micronutrient malnutrition in the developing world. *British Medical Bulletin, 106*(1), 7–17.

Bouis, H. E., Hotz, C., McClafferty, B., Meenakshi, J. V., & Pfeiffer, W. H. (2011). Biofortification: A new tool to reduce micronutrient malnutrition. *Food and Nutrition Bulletin, 32*, S31–S40.

Chen, S., Lin, X. H., Xu, C. G., & Zhang, Q. (2000). Improvement of bacterial blight resistance of 'Minghui 63', an elite restorer line of hybrid rice, by molecular marker-assisted selection. *Crop Science, 40*(1), 239–244.

Clifford, M. (2001). Appendix 1: A nomenclature for phenols with special reference to tea. *Critical Reviews in Food Science and Nutrition, 41*(5), 393.

Cole, C. T. (2003). Genetic variation in rare and common plants. *Annual Review of Ecology, Evolution, and Systematics, 34*, 213–237.

Collins, F. W. (1989). Oat phenolics: Avenanthramides, novel substituted N-cinnamoylanthranilate alkaloids from oat groats and hulls. *Journal of Agricultural and Food Chemistry, 37*(1), 60–66.

Colonna, P., & Buleon, A. (1992). New insights on starch structure and properties. In *9. International cereal and bread congress*.

Cordain, L. (1999). Cereal grains: humanity's double-edged sword. *World Review of Nutrition and Dietetics, 84*, 19–19.

Cristea, S., Freyvert, Y., Santiago, Y., Holmes, M. C., Urnov, F. D., Gregory, P. D., & Cost, G. J. (2013). In vivo cleavage of transgene donors promotes nuclease-mediated targeted integration. *Biotechnology and Bioengineering, 110*(3), 871–880.

Croteau, R., Kutchan, T. M., & Lewis, N. G. (2000). Natural products (secondary metabolites). *Biochemistry and Molecular Biology of Plants, 24*, 1250–1319.

Dale, P. J., Clarke, B., & Fontes, E. M. (2002). Potential for the environmental impact of transgenic crops. *Nature Biotechnology, 20*(6), 567–574.

Dewanto, V., Wu, X., & Liu, R. H. (2002). Processed sweet corn has higher antioxidant activity. *Journal of Agricultural and Food Chemistry, 50*(17), 4959–4964.

Diepenbrock, C. H., Kandianis, C. B., Lipka, A. E., Magallanes-Lundback, M., Vaillancourt, B., Góngora-Castillo, E., Wallace, J. G., Cepela, J., Mesberg, A., Bradbury, P. J., & Ilut, D. C. (2017). Novel loci underlie natural variation in vitamin E levels in maize grain. *The Plant Cell, 29*, 2374–2392.

Ding, X. L., Jiang, L., Liu, S. J., Wang, C. M., Chen, L. M., Cheng, Z. B., et al. (2004). QTL analysis for rice stripe disease resistance gene using recombinant inbred lines (RILs) derived from crossing of Kinmaze and DV85. *Yi chuan xue bao=. Acta Genetica Sinica, 31*(3), 287–292.

Dong, O. X., Yu, S., Jain, R., Zhang, N., Duong, P. Q., Butler, C., et al. (2020). Marker-free carotenoid-enriched rice generated through targeted gene insertion using CRISPR-Cas9. *Nature Communications, 11*(1), 1–10.

Drakakaki, G., Marcel, S., Glahn, R. P., Lund, E. K., Pariagh, S., Fischer, R., et al. (2005). Endosperm-specific co-expression of recombinant soybean ferritin and Aspergillus phytase in maize results in significant increases in the levels of bioavailable iron. *Plant Molecular Biology, 59*(6), 869–880.

Elkonin, L. A., Panin, V. M., Kenzhegulov, O. A., & Sarsenova, S. K. (2021). RNAi-mutants of Sorghum bicolor (L.) Moench with improved digestibility of seed storage proteins. In *Grain and seed proteins functionality*. IntechOpen.

Fantaye, B. M. (2018). Genetic improvement of lysine content in sorghum: A review. *Journal Advanced Plant Science, 1*(3), 307.

FAO, I., WFP. (2013). *The state of food insecurity in the world 2013*. Executive Summary.

Farhat, S., Jain, N., Singh, N., Sreevathsa, R., Dash, P. K., Rai, R., et al. (2019). CRISPR-Cas9 directed genome engineering for enhancing salt stress tolerance in rice. In *Seminars in cell & developmental biology* (Vol. 96, pp. 91–99). Academic Press.

Gadal, N., Shrestha, J., Poudel, M. N., & Pokharel, B. (2019). A review on production status and growing environments of rice in Nepal and in the world. *Archives of Agriculture and Environmental Science, 4*(1), 83–87.

Goesaert, H., Brijs, K., Veraverbeke, W. S., Courtin, C. M., Gebruers, K., & Delcour, J. A. (2005). Wheat flour constituents: How they impact bread quality, and how to impact their functionality. *Trends in Food Science & Technology, 16*(1–3), 12–30.

Goldberg, G. (Ed.). (2003). *Plants: Diet and health. The Report of the British Nutrition Foundation Task Force*. Blackwell.

Goswami, R., Zunjare, R. U., Khan, S., Baveja, A., Muthusamy, V., & Hossain, F. (2019). Marker-assisted introgression of rare allele of β-carotene hydroxylase (crtRB1) gene into elite quality protein maize inbred for combining high lysine, tryptophan and provitamin A in maize. *Plant Breeding, 138*(2), 174–183.

Griffiths, J. F., Griffiths, A. J., Wessler, S. R., Lewontin, R. C., Gelbart, W. M., Suzuki, D. T., & Miller, J. H. (2005). *An introduction to genetic analysis*. Macmillan.

Grohmann, L., Keilwagen, J., Duensing, N., Dagand, E., Hartung, F., Wilhelm, R., et al. (2019). Detection and identification of genome editing in plants: Challenges and opportunities. *Frontiers in Plant Science, 10*, 236.

Guan, S., Ma, Y., Liu, H., Liu, S., Liu, G., Zhao, L., & Wang, P. (2012). Increasing the amylose content of maize through silencing of sbe2a genes. *African Journal of Biotechnology, 11*(30), 7628–7637.

Hano, C. F., Dinkova-Kostova, A. T., Davin, L. B., Cort, J. R. & Lewis, N. G. (2021). Lignans: Insights into their biosynthesis, metabolic engineering, analytical methods and health benefits. *Frontiers in Plant Science, 11*, 2142.

Hahn, D. H., Faubion, J. M., & Rooney, L. W. (1983). Sorghum phenolic acids, their high performance liquid chromatography separation and their relation to fungal resistance. *Cereal Chemistry, 60*(4), 255–259.

Hidalgo, A., Brandolini, A., Pompei, C., & Piscozzi, R. (2006). Carotenoids and tocols of einkorn wheat (Triticum monococcum ssp. monococcum L.). *Journal of Cereal Science, 44*(2), 182–193.

Hizukuri, S., Takeda, Y., Yasuda, M., & Suzuki, A. (1981). Multi-branched nature of amylose and the action of debranching enzymes. *Carbohydrate Research, 94*(2), 205–213.

Hoekenga, O. A., Lung'aho, M. G., Tako, E., Kochian, L. V. and Glahn, R. P. (2011). Iron biofortification of maize grain. *Plant Genetic Resources, 9*(2), 327–329.

Holme, I. B., Gregersen, P. L., & Brinch-Pedersen, H. (2019). Induced genetic variation in crop plants by random or targeted mutagenesis: Convergence and differences. *Frontiers in Plant Science, 10*, 1468.

Holtekjølen, A. K., Kinitz, C., & Knutsen, S. H. (2006). Flavanol and bound phenolic acid contents in different barley varieties. *Journal of Agricultural and Food Chemistry, 54*(6), 2253–2260.

Hooper, L., & Cassidy, A. (2006). A review of the health care potential of bioactive compounds. *Journal of the Science of Food and Agriculture, 86*(12), 1805–1813.

Hospital, F. (2001). Size of donor chromosome segments around introgressed loci and reduction of linkage drag in marker-assisted backcross programs. *Genetics, 158*(3), 1363–1379.

Hossain, F., Sarika, K., Muthusamy, V., Zunjare, R. U., & Gupta, H. S. (2019). Quality protein maize for nutritional security. In *Quality breeding in field crops* (pp. 217–237). Springer.

Houmard, N. M., Mainville, J. L., Bonin, C. P., Huang, S., Luethy, M .H. & Malvar, T. M. (2007). High-lysine corn generated by endosperm-specific suppression of lysine catabolism using RNAi. *Plant Biotechnology Journal, 5*(5), 605–614.

ISAAA. (2002). Pocket Ks, global knowledge center on crop biotechnology (http://www.isaaa.org/), International Service for the Acquisition of Agri-biotech Applications, IRRI, Philippines.

ISAAA. (2006). https://www.isaaa.org/kc/inforesources/publications/biotechinagriculture/Conventional_Plant_Breeding.htm.

ISAAA. (2017). Global status of commercialized biotech/GM crops in 2017: Biotech crop adoption surges as economic benefits accumulate in 22 years. *ISAAA brief, 53*, 25–26.

ISAAA. (2021). *Breaking barriers with breeding: A primer on new breeding innovations for food security. Brief No. 56*. ISAAA.

Jiang, H., Lio, J., Blanco, M., Campbell, M., & Jane, J. L. (2010). Resistant-starch formation in high-amylose maize starch during kernel development. *Journal of Agricultural and Food Chemistry, 58*(13), 8043–8047.

Joung, J. K., & Sander, J. D. (2013). TALENs: A widely applicable technology for targeted genome editing. *Nature Reviews Molecular Cell Biology, 14*(1), 49–55.

Kaul, T., Sony, S. K., Verma, R., Motelb, K. F. A., Prakash, A. T., Eswaran, M., et al. (2020). Revisiting CRISPR/Cas-mediated crop improvement: Special focus on nutrition. *Journal of Biosciences, 45*(1), 1–37.

Kaur, N., Alok, A., Kumar, P., Kaur, N., Awasthi, P., Chaturvedi, S., et al. (2020a). CRISPR/Cas9 directed editing of lycopene epsilon-cyclase modulates metabolic flux for β-carotene biosynthesis in banana fruit. *Metabolic Engineering, 59*, 76–86.

Kaur, R., Kaur, G., Vikal, Y., Gill, G. K., Sharma, S., Singh, J., et al. (2020b). Genetic enhancement of essential amino acids for nutritional enrichment of maize protein quality through marker assisted selection. *Physiology and Molecular Biology of Plants, 26*(11), 2243–2254.

Kim, J. K., Lee, S. Y., Chu, S. M., Lim, S. H., Suh, S. C., Lee, Y. T., et al. (2010). Variation and correlation analysis of flavonoids and carotenoids in Korean pigmented rice (*Oryza sativa* L.) cultivars. *Journal of Agricultural and Food Chemistry, 58*(24), 12804–12809.

Knievel, D. C., Abdel-Aal, E. S., Rabalski, I., Nakamura, T., & Hucl, P. (2009). Grain color development and the inheritance of high anthocyanin blue aleurone and purple pericarp in spring wheat (*Triticum aestivum* L.). *Journal of Cereal Science, 50*(1), 113–120.

Koehler, P., & Wieser, H. (2013). Chemistry of cereal grains. In *Handbook on sourdough biotechnology* (pp. 11–45). Springer.

Kulp, K., & Ponte, J. G. (2000). *Handbook of cereal science and technology* (2nd ed.). Marcel Dekker.

Kumar, V. S., Verma, R. K., Yadav, S. K., Yadav, P., Watts, A., Rao, M. V., & Chinnusamy, V. (2020). CRISPR-Cas9 mediated genome editing of drought and salt tolerance (OsDST) gene in indica mega rice cultivar MTU1010. *Physiology and Molecular Biology of Plants, 26*(6), 1099–1110.

Kusaba, M., Miyahara, K., Iida, S., Fukuoka, H., Takano, T., Sassa, H., et al. (2003). Low glutelin content1: A dominant mutation that suppresses the glutelin multigene family via RNA silencing in rice. *The Plant Cell, 15*(6), 1455–1467.

Lazaridou, A., Biliaderis, C. G., & Izydorczyk, M. S. (2007). Cereal beta-glucans: Structures, physical properties, and physiological functions. *Functional food carbohydrates*, 1–72. Taylor and Francis, CRC Press.

Lee, J. S., Kwak, J., & Hay, F. R. (2020). Genetic markers associated with seed longevity and vitamin E in diverse Aus rice varieties. *Seed Science Research, 30*(2), 133–141.

Liang, Q., Wang, K., Liu, X., Riaz, B., Jiang, L., Wan, X., et al. (2019). Improved folate accumulation in genetically modified maize and wheat. *Journal of Experimental Botany, 70*(5), 1539–1551.

Lipkie, T. E., De Moura, F. F., Zhao, Z. Y., Albertsen, M. C., Che, P., Glassman, K., & Ferruzzi, M. G. (2013). Bioaccessibility of carotenoids from transgenic provitamin A biofortified sorghum. *Journal of Agricultural and Food Chemistry, 61*(24), 5764–5771.

Lucca, P., Poletti, S., & Sautter, C. (2006). Genetic engineering approaches to enrich rice with iron and vitamin A. *Physiologia Plantarum, 126*(3), 291–303.

Lyons, G., Stangoulis, J., & Graham, R. (2003). High-selenium wheat: Biofortification for better health. *Nutrition Research Reviews, 16*(1), 45–60.

Macrae, R., Robinson, R.K. & Sadler, M.J. (1993). Encyclopaedia of food science, food technology and nutrition. Taylor and Francis, CRC Press.

Maga, J. A., & Katz, I. (1978). Simple phenol and phenolic compounds in food flavor. *Critical Reviews in Food Science & Nutrition, 10*(4), 323–372.

Masel, J. (2011). Genetic drift. *Current Biology, 21*(20), R837–R838.

Mazur, W., & Adlercreutz, H. (1998). Natural and anthropogenic environmental oestrogens: The scientific basis for risk assessment. *Pure Applied Chemistry, 70*, 1759–1776.

Mezzetti, B., Smagghe, G., Arpaia, S., Christiaens, O., Dietz-Pfeilstetter, A., Jones, H., Kostov, K., Sabbadini, S., Opsahl-Sorteberg, H. G., Ventura, V., & Taning, C. N. T. (2020). RNAi: What is its position in agriculture? *Journal of Pest Science, 93*, 1125–1130.

Miller, H. E., Rigelhof, F., Marquart, L., Prakash, A., & Kanter, M. (2000). Antioxidant content of whole grain breakfast cereals, fruits and vegetables. *Journal of the American College of Nutrition, 19*, 312S–319S.

Miras-Moreno, B., Sabater-Jara, A. B., Pedreno, M. A., & Almagro, L. (2016). Bioactivity of phytosterols and their production in plant in vitro cultures. *Journal of Agricultural and Food Chemistry, 64*(38), 7049–7058.

Moose, S. P., & Mumm, R. H. (2008). Molecular plant breeding as the foundation for 21st century crop improvement. *Plant Physiology, 147*(3), 969–977.

Muñoz-Amatriaín, M., Cuesta-Marcos, A., Hayes, P. M., & Muehlbauer, G. J. (2014). Barley genetic variation: Implications for crop improvement. *Briefings in Functional Genomics, 13*(4), 341–350.

Nagaraj, H., Kumar, K. K., Sudhakar, D., Amudha, K., Meenakshi, S. P., & Kokiladevi, E. (2019). Genome engineering for thermo-sensitive genic male sterilty (TGMS) in rice using CRISPR/Cas9 editing system (Cyamopsis tetragonoloba L.). *Madras Agricultural Journal, 106*, 150–154.

Napoli, C., Lemieux, C., & Jorgensen, R. (1990). Introduction of a chimeric chalcone synthase gene into petunia results in reversible co-suppression of homologous genes in trans. *The Plant Cell, 2*(4), 279–289.

Nielsen, M. M. L., & Hansen, A. (2008). Rapid high-performance liquid chromatography determination of tocopherols and tocotrienols in cereals. *Cereal Chemistry, 85*(2), 248–251.

Ogo, Y., Ozawa, K., Ishimaru, T., Murayama, T., & Takaiwa, F. (2013). Transgenic rice seed synthesizing diverse flavonoids at high levels: A new platform for flavonoid production with associated health benefits. *Plant Biotechnology Journal, 11*(6), 734–746.

Ohnoutkova, L., Zitka, O., Mrizova, K., Vaskova, J., Galuszka, P., Cernei, N., et al. (2012). Electrophoretic and chromatographic evaluation of transgenic barley expressing a bacterial dihydrodipicolinate synthase. *Electrophoresis, 33*(15), 2365–2373.

Olsen, M. S., & Phillips, R. L. (2009). Molecular genetic improvement of protein quality in maize. In *Impacts of Agriculture on Human Health and Nutrition-Volume II* (p. 60) Taylor and Francis, CRC Press.

Osawaru, M. E., Ogwu, M. C., & Aiwansoba, R. O. (2015). Hierarchical approaches to the analysis of genetic diversity in plants: A systematic overview. *University of Mauritius Research Journal, 21*, 1–33.

Paine, J. A., Shipton, C. A., Chaggar, S., Howells, R. M., Kennedy, M. J., Vernon, G., Wright, S. Y., Hinchliffe, E., Adams, J. L., Silverstone, A. L., & Drake, R. (2005). Improving the nutritional value of Golden Rice through increased pro-vitamin A content. *Nature Biotechnology, 23*(4), 482–487.

Pallett, K. (2021). The Current Status of GM and GE Crops. *Outlooks on Pest Management, 32*(2), 50–52.

Palozza, P., & Krinsky, N. I. (1992). Antioxidant effects of carotenoids in vivo and in vitro: An overview. *Methods in Enzymology, 213*, 403–420.

Panfili, G., Fratianni, A., & Irano, M. (2004). Improved normal-phase high-performance liquid chromatography procedure for the determination of carotenoids in cereals. *Journal of Agricultural and Food Chemistry, 52*(21), 6373–6377.

Paul, S., Ali, N., Gayen, D., Datta, S. K., & Datta, K. (2012). Molecular breeding of Osfer2 gene to increase iron nutrition in rice grain. *GM Crops & Food, 3*(4), 310–316.

Peterson, D. M. (1995). Oat tocols: Concentration and stability in oat products and distribution within the kernel. *Cereal Chemistry, 72*(1), 21–24.

Poole, N., Donovan, J. & Erenstein, O. (2021). Agri-nutrition research: revisiting the contribution of maize and wheat to human nutrition and health. *Food Policy, 100*, 101976.

Russell, J., Dawson, I. K., Flavell, A. J., Steffenson, B., Weltzien, E., Booth, A., et al. (2011). Analysis of> 1000 single nucleotide polymorphisms in geographically matched samples of landrace and wild barley indicates secondary contact and chromosome-level differences in diversity around domestication genes. *New Phytologist, 191*(2), 564–578.

Saikia, D., & Deka, S. C. (2011). Cereals: From staple food to nutraceuticals. *International Food Research Journal, 18*(1).

Saltzman, A., Birol, E., Bouis, H. E., Boy, E., De Moura, F. F., Islam, Y., & Pfeiffer, W. H. (2013). Biofortification: Progress toward a more nourishing future. *Global Food Security, 2*(1), 9–17.

Samanta, A., Das, G., & Das, S. K. (2011). Roles of flavonoids in plants. *Carbon, 100*(6), 12–35.

Sestili, F., Janni, M., Doherty, A., Botticella, E., D'Ovidio, R., Masci, S., Jones, H.D. & Lafiandra, D., (2010). Increasing the amylose content of durum wheat through silencing of the SBEIIa genes. *BMC Plant Biology, 10*(1), 1–12.

Seitz, L. M. (1989). Stanol and sterol esters of ferulic and p-coumaric acids in wheat, corn, rye, and triticale. *Journal of Agricultural and Food Chemistry, 37*(3), 662–667.
Shahidi, F., Janitha, P. K., & Wanasundara, P. D. (1992). Phenolic antioxidants. *Critical Reviews in Food Science & Nutrition, 32*(1), 67–103.
Shalem, O., Sanjana, N. E., Hartenian, E., Shi, X., Scott, D. A., Mikkelsen, T. S., et al. (2014). Genome-scale CRISPR-Cas9 knockout screening in human cells. *Science, 343*(6166), 84–87.
Shan, Q., Zhang, Y., Chen, K., Zhang, K., & Gao, C. (2015). Creation of fragrant rice by targeted knockout of the Os BADH 2 gene using TALEN technology. *Plant Biotechnology Journal, 13*(6), 791–800.
Shibanuma, K., Takeda, Y., Hizukuri, S., & Shibata, S. (1994). Molecular structures of some wheat starches. *Carbohydrate Polymers, 25*(2), 111–116.
Shin, Y. M., Park, H. J., Yim, S. D., Baek, N. I., Lee, C. H., An, G., & Woo, Y. M. (2006). Transgenic rice lines expressing maize C1 and R-S regulatory genes produce various flavonoids in the endosperm. *Plant Biotechnology Journal, 4*(3), 303–315.
Shukla, V. K., Doyon, Y., Miller, J. C., DeKelver, R. C., Moehle, E. A., Worden, S. E., et al. (2009). Precise genome modification in the crop species Zea mays using zinc-finger nucleases. *Nature, 459*(7245), 437–441.
Shumskaya, M., & Wurtzel, E. T. (2013). The carotenoid biosynthetic pathway: Thinking in all dimensions. *Plant Science, 208*, 58–63.
Sindhu, A. S., Zheng, Z., & Murai, N. (1997). The pea seed storage protein legumin was synthesized, processed, and accumulated stably in transgenic rice endosperm. *Plant Science, 130*(2), 189–196.
Slarkin, M. (1985). Gene flow in natural populations. *Annual Review of Ecology and Systematics, 16*(1), 393–430.
Smeds, A. I., Eklund, P. C., Sjöholm, R. E., Willför, S. M., Nishibe, S., Deyama, T., & Holmbom, B. R. (2007). Quantification of a broad spectrum of lignans in cereals, oilseeds, and nuts. *Journal of Agricultural and Food Chemistry, 55*(4), 1337–1346.
Southgate, D. A. T. (1993). Cereals and cereal products. In J. S. Garrow & W. P. T. James (Eds.), *Human nutrition and dietetics* (9th ed., pp. 273–288). Churchill Livingstone.
Stevenson, J. R., Villoria, N., Byerlee, D., Kelley, T. and Maredia, M., (2013). Green Revolution research saved an estimated 18 to 27 million hectares from being brought into agricultural production. *Proceedings of the National Academy of Sciences, 110*(21), 8363–8368.
Storozhenko, S., De Brouwer, V., Volckaert, M., Navarrete, O., Blancquaert, D., Zhang, G. F., et al. (2007). Folate fortification of rice by metabolic engineering. *Nature Biotechnology, 25*(11), 1277–1279.
Sun, Y., Jiao, G., Liu, Z., Zhang, X., Li, J., Guo, X., et al. (2017). Generation of high-amylose rice through CRISPR/Cas9-mediated targeted mutagenesis of starch branching enzymes. *Frontiers in Plant Science, 8*, 298.
Tamás, C., Kisgyörgy, B. N., Rakszegi, M., Wilkinson, M. D., Yang, M. S., Láng, L., Tamás, L., & Bedő, Z. (2009). Transgenic approach to improve wheat (*Triticum aestivum* L.) nutritional quality. *Plant Cell Reports, 28*(7), 1085–1094.
Tamilkumar, P., Senthil, N., Sureshkumar, S., Thangavelu, A. U., Nagarajan, P., Vellaikumar, S., et al. (2014). Introgression of low phytic acid locus ('lpa2-2') into an elite Maize ('Zea mays' L.) inbred through marker assisted backcross breeding. *Australian Journal of Crop Science, 8*(8), 1224–1231.
Tang, G., Qin, J., Dolnikowski, G. G., Russell, R. M., & Grusak, M. A. (2009). Golden Rice is an effective source of vitamin A. *The American Journal of Clinical Nutrition, 89*(6), 1776–1783.
Tanksley, S. D., Young, N. D., Paterson, A. H., & Bonierbale, M. W. (1989). RFLP mapping in plant breeding: New tools for an old science. *Bio/Technology, 7*(3), 257–264.
Topping, D. L., Segal, I., Regina, A., Conlon, M. A., Bajka, B. H., Toden, S., Clarke, J. M., Morell, M. K., & Bird, A. R. (2010). *Resistant starch and human health* (pp. 311–321). Wageningen Academic Publishers.

Van Hung, P., Maeda, T., & Morita, N. (2006). Waxy and high-amylose wheat starches and flours-characteristics, functionality and application. *Trends in Food Science & Technology, 17*(8), 448–456.

Vasil, I. K. (1995). Cellular and molecular genetic improvement of cereals. In *Current issues in plant molecular and cellular biology* (pp. 5–18). Springer.

Vishwakarma, M. K., Mishra, V. K., Gupta, P. K., Yadav, P. S., Kumar, H., & Joshi, A. K. (2014). Introgression of the high grain protein gene Gpc-B1 in an elite wheat variety of Indo-Gangetic Plains through marker assisted backcross breeding. *Current Plant Biology, 1*, 60–67.

Wang, C., Zeng, J., Li, Y., Hu, W., Chen, L., Miao, Y., et al. (2014). Enrichment of provitamin A content in wheat (Triticum aestivum L.) by introduction of the bacterial carotenoid biosynthetic genes CrtB and CrtI. *Journal of Experimental Botany, 65*(9), 2545–2556.

Yamauchi, R., & Matsushita, S. (1979). Light-induced lipid peroxidation in isolated chloroplasts and role of α-tocopherol. *Agricultural and Biological Chemistry, 43*(10), 2157–2161.

Yang, T., Duan, C. L., Zeng, Y. W., Du, J., Yang, S. M., Pu, X. Y., & Yang, S. C. (2013). HPLC analysis of flavonoids compounds of purple, normal barley grain. In *Advanced materials research* (Vol. 634, pp. 1486–1490). Trans Tech Publications Ltd.

Yin, K., Gao, C. and Qiu, J. L. (2017). Progress and prospects in plant genome editing. *Nature plants, 3*(8), 1–6.

Zeeman, S. C., Kossmann, J., & Smith, A. M. (2010). Starch: Its metabolism, evolution, and biotechnological modification in plants. *Annual Review of Plant Biology, 61*, 209–234.

Zunjare, R. U., Hossain, F., Muthusamy, V., Baveja, A., Chauhan, H. S., Bhat, J. S., et al. (2018). Development of biofortified maize hybrids through marker-assisted stacking of β-carotene hydroxylase, lycopene-ε-cyclase and opaque2 genes. *Frontiers in Plant Science, 9*, 178.

Chapter 4
Functional Cereals for Gluten Intolerance

Amardeep Singh Virdi and Narpinder Singh

4.1 Introduction

Modern food diets primarily rely upon the extensive usage of the meal or refined flour of bread wheat. Gluten intake induces inflammatory response in some consumer's and damage villus of the small intestine andleads to the flattening of jejunal mucosa (Shan et al., 2002). In adults, diarrhea or constipation, aphthous ulcers, sore tongue and mouth, dyspepsia, abdominal pain, fatigue, infertility, bloating (weight-loss), neuropsychiatric symptoms, bone pain (osteoporosis), weakness (myopathy, neuropathy) are the primary symptoms, which appear after the intake of gluten. Whereas, infants may show diarrhea, abdominal distension, failure to thrive, anorexia, vomiting, psychomotor impairment, etc., upon the consumption of gluten products. Therefore, such ailments appeared after the consumption of gluten is known as gluten intolerance (CD) or coeliac disease (CD) . The presence of a specific category of gliadins and glutenins in wheat and prolamins (alcohol soluble proteins) from rye and barley are responsible for such immunogenic reactions (Farrell & Kelly, 2002; Fasano & Catassi, 2001; Murray, 1999; Vader et al., 2002). Analysis of the immune epitope database (IEDB) revealed the presence of 190 T-cell stimulatory epitopes for celiac disease in wheat. Among these, 94, 74, and 12 epitopes, respectively, linked with CD are encoded by α-, γ-, and ω-gliadins of wheat. Whereas, 8 and 2 epitopes, respectively, encoded by the low molecular weight and high molecular weight genes in wheat for CD are reported (Comino et al., 2013).

A. S. Virdi (✉)
Department of Food Science and Technology, Amity University Punjab,
Mohali, Punjab, India

N. Singh
Department of Food Science and Technology, Guru Nanak Dev University,
Amritsar, Punjab, India
e-mail: asvirdi@pb.amity.edu

S. Punia Bangar, A. Kumar Siroha (eds.), *Functional Cereals and Cereal Foods*,
https://doi.org/10.1007/978-3-031-05611-6_4

The first approach to combat CD is to exclude wheat flour from the diet and rely upon the composite flours from pseudocereals, pulses, etc., in routine diets. The second approach is the enzymatic treatment of wheat flour or gluten to hydrolyse all/immunogenic proteins or the genetic engineering and plant breeding approaches to silence the immunogenic proteins from wheat and other cereals. A large population in India consumes vegetarian diets, which are based upon the usage of wheat, and rice, the alternatives of wheat with similar viscoelastic properties are limited and also fail to mimic the texture and sensory attributes of wheat products. Therefore, the evaluation of the physico-chemical properties and development of new formulations based on gluten-free composite flour for making high-quality gluten-free products are gaining high popularity among researchers globally. The functional properties, nutritional values, dietary fiber content, and glycemic index of these products are also different from the products of flour from bread wheat. Starches from maize, rice, tubers, and plantains, in combination with dairy products, gums and hydrocolloids, etc., are widely used to produce quality gluten-free products. However, the poor-quality crumb texture and volume of gluten-free bread are major problems (Gujral et al., 2003; Virdi & Singh, 2020). Quinoa, amaranthus and buckwheat are pseudocereals, and are rich source of iron and fibre. Quinoa contains high levels of riboflavin and buckwheat flour is rich in niacin (US Department of Agriculture, 2000), therefore, pseudocereals may serve as suitable diet for CD patients. The functional properties of starch and proteins derived from sorghum and pseudocereals, and possible modifications to improve the functionality of gluten-free products are discussed.

4.2 Sorghum

4.2.1 Starch Characteristics of Sorghum

Sorghum bicolor (L.) Moench is commonly known as great millet, miloa, jowar/jowari and durra. Sorghum is a grass species which belongs to the family Poaceae, a highly thermo tolerant crop, considered as model crop for studying drought and heat stress. Sorghum is the fifth most important crop after wheat, rice, maize and barley, which is grown in arid and semi-arid agro climatic zones of the world (https://www.fao.org/in-action/inpho/crop-compendium/cereals-grains/en/). Khan et al. (2013) reported starch content between 55.6% and 70.0% for the dry grain weight of sorghum cultivars. Sorghum with amylose content of ~25% is considered as normal (*WxWxWx*), whereas, waxy sorghum (*wxwxwx*) is deficient in amylose and composed of 100% amylopectin, while hetero-waxy (*WxWxwx* or *Wxwxwx*) type of sorghum contains intermediate amount (~15%) of amylose content (Sang et al., 2008). Amylose content of 18.2–28.8% for Indian wheat was reported by Singh et al. (2010a), which is close to sorghum starches. Morphological studies revealed the presence of irregular-polyhedral and spherical granules of starches in sorghum (Sandhu et al., 2021). The presence of pores on sorghum starches was also observed (Benmoussa et al., 2006; Huber & Bemiller, 2000; Singh et al., 2010b). Round, doughnut-shaped and polygonal morphology also observed for sorghum starches (Benmoussa et al., 2006). The presence of radial, tube-like channels in the sorghum starch granules were also reported by Huber and BeMiller (2000). The presence of pores, channels may facilitate water access to the granule interior. Amylose content between 11.2% and 28.5% for starches from Indian sorghum was observed against 23.7, 14.0 and 0%, respectively for normal, heterowaxy and waxy sorghum starches (Sang et al., 2008) (Table 4.1). Therefore, normal and heterowaxy sorghum starches consist of high levels of amylose content. The variable solubility behaviour of starches from different sorghum cultivars was also reported. Swelling index ranged between 19.0% and 5.0%, while swelling power (SP) ranged between 6.2 and 15.3 g/g for Indian sorghum starches (Singh et al., 2010a). The solubility of 5% and SP of 8.79 g/g for Nigerian sorghum starches was also reported by Olayinka et al. (2008). The solubility between 17.4% and 22.5%, and SP between 13.8 and 15.2 g/g, for US sorghum starches was also observed (Subrahmanyam & Hoseney, 1995). The swelling power of starches from various Indian wheat varieties was ranged between 13.1 g/g and 24.9 g/g (Singh et al., 2010a). These findings thus demonstrated a wide variation in the SP and solubility of sorghum starches from different varieties, which may be affected by the genetic composition of sorghum cultivars, and by the environmental factors. SP and solubility indicate the strength of interaction between the starch chains found between the crystalline and amorphous domains (Singh et al., 2010a; Punia et al., 2020; Punia Bangar et al., 2021a, b). Starches with higher crystallinity and amylose content have poor SP than starches with higher amylopectin. Higher SP for starches with a greater proportion of short chain amylopectin was observed against starches with a large proportion of long chain amylopectin molecules (Singh et al., 2010a). The formation of a stronger crystalline network in starch granules may be attributed to a higher proportion of the long chains of amylopectin. Therefore, higher SP indicates a weaker interaction among the amorphous and crystalline regions through starch chains. The transition temperatures T_o, T_p, and T_c between 66.1°C and 73.12°C, 70.1°C and 77.79°C, and 75.0°C to 81.24°C, respectively, while the enthalpy of gelatinization (ΔH_{gel}) between 9.26 and 13.5 J/g for Indian sorghum starches was reported by Singh et al. (2010a) (Table 4.1). Seven US sorghum starches showed ΔH_{gel} between 2.84 and 3.39 J/g, while ΔH_{gel} of 7.45 J/g for ten Zimbabwean sorghum starches (Beta et al., 2001) and ΔH_{gel} of 13.7 J/g for Nigerian sorghum starch by Gaffa et al. (2004) was reported. Similarly, T_o, T_p, and T_c of 67.9 °C, 70.7 °C, and 75.7 °C, respectively for normal sorghum starch had been observed (Sang et al., 2008). An average T_p of 67.4 °C for ten Zimbabwean sorghum starches (Beta et al., 2001), and Tc of 90 °C for the Nigerian sorghum starches was also reported (Gaffa et al., 2004). Starches from Indian wheat were exhibited T_o, T_p, T_c and ΔH_{gel} of 55.6–57.3 °C, 60.6–62.1 °C, 65.3–67.5 °C, and 8.0–10.2 J/g, respectively (Singh et al., 2010a) (Table 4.1). Difference between T_c and T_o is known as the difference in gelatinization ranges, (R). The crystalline domain of a starch granule is composed of small crystallites,

Table 4.1 Thermal Properties (onset, peak and end of gelatinization) of flour from different cereals and pseudo cereals

Source	Amylose (%)	T_o (°C)	T_p (°C)	T_c (°C)	ΔH (J/g)	Reference
Indian wheat	18.2–28.8	55.6–57.3	60.6–62.1	65.3–67.5	8.0–10.2	Singh et al., 2010a
Indian sorghum starch	11.2–28.5	66.1–73.12	70.1–77.79	75.0–81.24	9.26–13.5	Singh et al., 2010b
Sorghum starch (normal)	23.7	67.9	70.7	75.7	–	Sang et al., 2008
Sorghum starch (waxy)	0	67.7	73.0	82.1	14.7	Sang et al., 2008
Sorghum starch (hetero waxy)	14.0	69.6	72.8	78.6	13.7	Sang et al., 2008
Amaranths (*A. hypochondriacus*)	5.8[Δ]	63.20–70.01	68.88–72.88	74.47–76.95	8.50–13.94	Singh et al., 2014; [Δ]Gamel et al., 2005
Amaranths (*A. caudatus*)	4.4[e]	60.46–63.28	65.05–67.05	70.93–74.40	11.55–14.38	[e]Okuno and Sakaguchi, 1981 Gamel et al., 2005
Buckwheat[a]	15.95[×]	59[×]	66[×]	72[×]	–	[×]Hager et al., 2012
Buckwheat (1:2 moisture content)	16–18	59.5 to 64.1	63.7–68.4	81.7–85.8	14.5–15	Yoshimoto et al., 2004
Common buckwheat (1:2)	25.6–28.6, 34.5–34.5 g	58.6–60.2	61.5–64.3	70–73	14–15.3	Lu & Baik, 2015
Common buckwheat (1:4 moisture content)		61.2	66.1	75.2	9.0	Li et al., 2014
Chenopodium Quinoa	4.62	52	58	64	–	Hager et al., 2012
Chenopodium Quinoa		53.9	60.6	66.0	10.3	Srichuwong et al., 2017
[a]*Chenopodium Quinoa*	8.4	57.4	66.0	72.7	8.4	
Pearl millet (1:2 moisture content)	21–25	66.2–67.2	69.7–71.4	86.3–91	14.3–14.7	Gaffa et al., 2004
Foxtail millets	16.9–17.5	55	57.5	62	–	Wankhede et al., 1979
Finger millet	38.6	62.5	69	74	–	Malleshi et al., 1986
Maize	17.5–22.1	64.0–68.9	68.9–72.1	73.2–76.8	8.1–11.2	Sandhu & Singh, 2005

[a]Flour; T_o: Onset temperature; T_p: Peak temperature; T_c: End temperature; ΔH: Enthalpy of gelatinization

and the marginal differences in the crystal strength were attributed to variation in R (Banks & Greenwood, 1975).

The enthalpy of retrogradation (ΔH_{ret}) of gelatinized starches indicates tendency to retrograde or recrystalize upon cooling of starch paste after gelatinization. While the ratio of ΔH_{gel} and ΔH_{ret} is defined as percentage retrogradation (%R). Therefore, a higher value of ΔH_{ret} indicates the lower tendency of starches to retrograde and vice-versa. ΔH_{ret} is an indication of the unravelling and melting of double helices formed during storage and influenced by the amylopectin unit chain length distribution (Shi & Seib, 1992). The unravelling and melting of the double-helical regions during the gelatinization of starches are affected by the chain length distribution of amylopectin (Shi & Seib, 1992). Therefore, the breakdown of starch granules during heating, i.e., gelatinization upon heating, and reannealing of starch granules upon cooling relies on the structural arrangements of starch chains within the crystalline and amorphous domains of non-gelatinized starch granules (Perera & Hoover, 1999). T_o, T_p, and T_c between 46.2 and 52.6 °C, 54.18 and 58.61 °C and 61.4 to 65.9 °C, respectively for retrograded Indian sorghum starch pastes stored in a refrigerator were observed. ΔH_{ret} between 1.11 J/g and 4.31 J/g for retrograded Indian sorghum starch pastes were reported (Singh et al., 2010a). These findings thus demonstrated lower transition temperatures and ΔH_{ret} of stored starch pastes than the transition temperatures of gelatinization and ΔH_{gel} fresh starch dispersions. This implies that some of the Indian sorghum cultivars have starches with higher enthalpy of retrogradation (ΔH_{ret}) thereby lower syneresis values and therefore, can be used for making gluten-free products with long storage-shelf life. These findings thus revealed that sorghum starch has higher gelatinization temperature (68–78 °C), than the gelatinization temperature of starches from maize (62–72 °C) and barley(51–60 °C) along with a higher degree of retrogradation (Collar, 2017; Hoseney, 1994). The low number of short chain amylopectin in sorghum may attribute to higher gelatinization temperature and higher degree of retardation than remained cereals (Ai, 2013). Studies have shown that higher gelatinization temperature may have adverse effect on the quality and sensory of baked products (Taylor & Dewar, 2001). However, the texture and sensory of gluten-free chapatti, pan cake, and other food products may not be affected by high gelatinization temperature of sorghum starches.

The pasting properties of starches are crucial for the final texture, sensory, and consumer acceptability. Therefore, the pasting temperature, the breakdown-, and final viscosity of the sorghum starches were evaluated by a rapid visco-analyzer. The peak viscosity (PV) and hot paste viscosity (HPV) ranged from 2541 to 4698 cP, and 919 to 2629 cP, respectively for Indian sorghum starches. The breakdown (BDV) and final viscosity (FV) varied from 911 to 2645 cP and 2314 to 4743 cP, respectively, while the setback viscosity from 1067 cP to 2114 cP for Indian sorghum starches was observed (Singh et al., 2010a). Zimbabwean sorghum cultivars exhibited average PV, HPV, CPV, BD, and SB of 3984 cP, 1392 cP, 2928 cP, 2592 cP, and 1536 cP, respectively (Beta et al., 2001). Whereas, the PV, BD, and SB of 2004 cP, 144 cP, and 1476 cP, respectively for Nigerian sorghum starch was reported by Gaffa et al. (2004). The pasting temperature ranged from 75.2 °C to 80.9 °C for Indian sorghum cultivars and 69 to 70.3 °C for Zimbabwean sorghum varieties (Beta et al., 2001). On the contrary, starches from Nigerian sorghum cultivars were exhibited the PT of 82.6 °C (Gaffa et al., 2004). PT between 82.3 and 89.6 °C for starches from various Indian wheat varieties was observed. Starches with higher PT and higher amylose content demonstrated lower peak, trough, breakdown, setback, and final viscosity (Singh et al., 2010a). The amylose and amylopectin content, agroclimatic conditions, and genetic composition of sorghum influence the pasting properties greatly. Ratnavathi and P. J. (2014) and Khoddami et al. (2021) concluded that sorghum flour with higher hot peak paste viscosity, setback viscosity, water uptake, and low gelatinization temperature are highly suitable for flat breads such as chapatti, whereas sorghum flour with high gelatinization temperature and low peak paste viscosity may be highly suitable for the preparation of stiff porridge such as Indian Sankhati and African tô. The starch digestibility of 33–48% for sorghum starches against 53–58% for corn starches by Sikabbubba (1989) was evaluated. The digestibility of starches from floury and corneous sorghum grains was also different; with starch from the former type of grain revealed higher digestibility than the later one. The lower size particles of the floury grain of sorghum may be digested rapidly by starch solubilizing enzymes *in vitro* and may be attributed to a higher digestibility. A lower starch digestibility of normal sorghum than the waxy was also reported by Hibberd et al. (1982). Since most of the corn starch is utilized by industries for the manufacturing of breakfast cereals, snacks, etc., the availability of corn starch in India is limited. Majority of gluten-free products rely upon corn, rice and potato starches; therefore, sorghum starch may be a good alternative of corn starches. The functionality of sorghum starches is also equivalent to corn starches. Starch from sorghum can be produced by wet milling technology, which is available for corn starch production. These findings thus imply that starches from normal/corneous sorghum starches have a better alternative to corn starches.

4.2.2 Composition and Functionality of Sorghum Proteins

The protein content in sorghum ranged from 80 to 84% of the total grain nitrogen, whereas the germ and pericarp of sorghum contained protein content between 9.4 to 16% and 3.0 to 6.5%, respectively (Serna-Saldivar & Rooney, 1995; Taylor & Schüssler, 1986). Majority of seed storage proteins stored in protein bodies are made from the surrounding layers of lipids. Higher content of α-kafirins, and minor stock of β-, and γ-kafirins, inside the protein bodies (0.3–1.5 μm) of the sorghum endosperm was observed, whereas, higher proportion of β-, and γ-kafirins in the peripheral region of protein bodies was found. Higher cysteine content in β-, and γ-kafirins attributed to crosslinking with each other, which led to the formation of a shell around α-kafirins inside the protein bodies. The minor proportion of glutelin, globulins and albumins in the protein matrix of the sorghum endosperm were also reported. The glutelin, globulin and albumin content of 33.4%, 7.0% and 5.6%, respectively was reported in the matrix of sorghum endosperm by Virupaksha and Sastry (1968). The α-kafirinsshowed two subunits namely α1-, and α2 kafirin with the molecular weight of 23,000 Dalton (Da) and 25,000 Da, respectively. The expression of 19 kafirin encoding genes in sorghum was reported which may encode different subunits of α-kafirins proteins (Xu & Messing, 2008). A gene encoding a methionine richkafirin, known as δ-kafirins, with molecular weight of 16,000 kDa was also reported from sorghum, which accounts for only 1% of total grain proteins. In 92.9 g/kg (N x 5.81) crude protein, the average kafirin content of 48.2 g/kg for 33 Australian sorghum lines/cultivars was observed (Selle et al., 2020), which was 51.9% of total grain nitrogen, therefore, kafirin represents a major proportion of the protein content in sorghum. A substantially higher leucine content of 62.7% in the Australian sorghum kafirin was also observed (Selle et al., 2020). Studies have shown that very high leucine content-based diets may not be suitable for the good performance of the broiler chickens (Selle et al., 2020). Amino acid composition analysis revealed lower lysine and threonine content in the grains of sorghum, as observed for maize (Table 4.2). Since kafirins are homolog of zein proteins of maize, lower lysine and threonine content in sorghum may also be associated with the higher abundance of these endospermic proteins of sorghum. The prolamins also have some degree of viscoelastic properties which rely on the purification methods and proportion of the individual subunits in the purified fraction of each prolamin. The purified zein and kafinrin proteins with higher proportion of α-prolamins showed best viscoelastic properties and the presence of cysteine residues in some of the prolamin subunits showed deleterious effect on the viscoelastic properties of prolamins (Oom et al., 2008; Schober et al., 2011).Therefore, the viscoelastic properties of maize and sorghum proteins can be modified by different processing methods, and could be an alternative source of wheat gluten to enhance the quality of bread upto some extent.

Wet cooking of sorghum resulted in the di-sulfide crosslinking of kafirins, which led to their poor solubility and digestibility. Thus, the digestibility of sorghum decreases upon wet cooking. A higher increase in the proportion of antiparallel β-sheets and decrease in the α-helices may have attributed to the poor solubility of sorghum kafirins (Duodu et al., 2001). Popping and dry-roasting of sorghum grains also did not affect the digestibility of sorghum significantly; however, the addition of reducing agents during cooking enhanced the digestibility of cooked flour (Correia et al., 2010; Hamaker et al., 1987; Parker et al., 1999). These findings thus revealed that sorghum proteins can be used to enhance the viscoelastic properties gluten-free composite dough systems and starches obtained from sorghum can be a very good source of slow digestible starches.

4.3 Roles of Pseudocereals in Health and Nutrition

Pseudocereals, which include amaranth, buckwheat, chenopods and millets etc. provide better nutrition than most major crops and are multipurpose crops. Since these are gluten-free and have superior nutritional attributes, however, the utilization of pseudocereals in making processed food require detailed analysis of structural and functional properties of starch and proteins. The pasting and thermal profile of starches, functional properties of foam of flour, starch and proteins from different pseudocereals appeared to be differential and are discussed here in brief to better understand the utilization of flour, starch and proteins in the designing of different types of gluten-free products.

Table 4.2 Amino acid composition of important cereals, pseudocereals and pulses

Source Amino acid	Jowar	Maize	Rice, raw, milled	Wheat flour, refined	Wheat, whole	Amaranth seed, black	Amaranth seed, pale brown	Quinoa	Tartary buckwheat (B-121)	Common buckwheat (Tomotake et al. (2006)	Bengal gram, whole	Cowpea, white	Lentil whole, brown	Peas, dry	Rajmah, red (Kidney bean)
Histidine	2.07 ± 0.20	2.70 ± 0.21	2.45 ± 0.30	1.95 ± 0.23	2.65 ± 0.31	1.86	1.98 ± 0.50	2.98	294.45 ± 37.90	2.52	2.51 ± 0.18	3.25	2.07 ± 0.14	2.34 ± 0.09	2.70 ± 0.30
Isoleucine	3.45 ± 0.63	3.67 ± 0.22	4.29 ± 0.23	3.19 ± 0.27	3.83 ± 0.20	2.82	2.85 ± 0.04	3.75	75.95 ± 7.24	3.12	4.34 ± 0.23	4.4	4.17 ± 0.05	3.87 ± 0.46	4.23 ± 0.38
Luecine	12.03 ± 1.51	12.24 ± 0.57	8.09 ± 0.40	6.22 ± 0.46	6.81 ± 0.33	4.83	4.94 ± 0.17	6.08	988.87 ± 28.17	5.94	7.40 ± 0.31	7.96	7.36 ± 0.34	7.02 ± 0.38	7.78 ± 0.71
Lysine	2.31 ± 0.40	2.64 ± 0.18	3.70 ± 0.39	2.05 ± 0.18	3.13 ± 0.26	5.45	5.50 ± 0.35	5.55	315.72 ± 45.76	5.68	6.59 ± 0.25	7.14	6.78 ± 0.51	7.12 ± 0.51	6.71 ± 0.81
Methionine	1.52 ± 0.50	2.10 ± 0.17	2.60 ± 0.34	1.64 ± 0.20	1.75 ± 0.21	1.86	1.95 ± 0.12	2.24	403.42 ± 14.96	2.3	1.16 ± 0.16	1.53	0.84 ± 0.03	0.68 ± 0.19	0.88 ± 0.40
Cystine	1.06 ± 0.30	1.55 ± 0.14	1.84 ± 0.18	2.03 ± 0.27	2.35 ± 0.23	1.6	1.51 ± 0.15	1.85	55.09 ± 2.64		1.27 ± 0.09	0.6	1.18 ± 0.04	0.82 ± 0.15	0.70 ± 0.18
Phenyla lanine	5.10 ± 0.50	5.14 ± 0.29	5.36 ± 0.43	4.29 ± 0.28	4.75 ± 0.38	3.98	4.75 ± 0.41	4.35		4.3	6.26 ± 0.70	5.63	4.61 ± 0.68	4.76 ± 0.23	5.90 ± 0.56
Threonine	2.96 ± 0.17	3.23 ± 0.29	3.28 ± 0.27	2.34 ± 0.08	3.01 ± 0.17	3.02	2.99 ± 0.21	3.01	770.66 ± 35.84	3.5	3.55 ± 0.31	4.1	3.35 ± 0.05	3.65 ± 0.15	4.18 ± 0.65
Tryptophan	1.03 ± 0.21	0.57 ± 0.12	1.27 ± 0.14	1.04 ± 0.16	1.40 ± 0.10	1.5	1.69 ± 0.10	1.25		2	0.95 ± 0.07	0.92	0.76 ± 0.04	0.86 ± 0.19	1.05 ± 0.27
Valine	4.51 ± 0.71	5.41 ± 0.71	6.06 ± 0.02	4.01 ± 0.44	5.11 ± 0.05	4.34	4.30 ± 0.27	4.55	243.89 ± 22.90	4.26	4.58 ± 0.51	5.31	4.85 ± 0.06	4.67 ± 0.66	5.07 ± 0.71
Alanine	9.19 ± 1.12	7.73 ± 0.46	5.51 ± 0.40	2.98 ± 0.37	3.64 ± 0.21	4.26	3.83 ± 0.64	4.35	214.64 ± 16.74	3.89	4.67 ± 0.56	5.06	5.75 ± 0.61	4.51 ± 0.44	4.57 ± 0.53
Arginine	3.96 ± 0.43	4.20 ± 0.24	7.72 ± 0.55	3.49 ± 0.28	5.13 ± 0.33	7.77	7.21 ± 0.91	7.85	1475.60 ± 45.76	11.16	8.59 ± 0.58	7.44	7.44 ± 2.06	8.09 ± 0.30	6.10 ± 0.85
Aspartic acid	7.09 ± 0.86	6.55 ± 0.59	8.73 ± 0.80	4.63 ± 0.39	5.44 ± 0.33	12.57	12.70 ± 2.25	8.4	89.26 ± 4.74	9.54	11.78 ± 1.60	11.01	12.48 ± 1.70	11.34 ± 0.72	10.50 ± 0.85
Glutamic acid	21.54 ± 2.81	19.39 ± 0.70	18.92 ± 1.76	31.57 ± 1.80	27.06 ± 1.76	16.12	17.39 ± 1.68	13.75	1372.94 ± 45.80	19.38	17.27 ± 1.08	18.2	17.25 ± 1.68	17.52 ± 0.81	16.01 ± 2.17
Glycine	3.08 ± 0.25	3.27 ± 0.15	4.18 ± 0.16	3.21 ± 0.20	4.19 ± 0.23	8.5	8.28 ± 0.35	4.8	467.6 ± 26.35	5.66	3.95 ± 0.16	4.09	4.78 ± 0.08	4.19 ± 0.16	3.78 ± 0.49
Proline	6.99 ± 0.92	7.88 ± 0.71	4.31 ± 0.78	9.23 ± 0.64	10.25 ± 1.49	3.76	3.83 ± 0.45	5.67	329.1 ± 20.76	7.93	3.74 ± 0.19	4.05	5.01 ± 1.49	3.73 ± 0.19	3.40 ± 0.58
Serine	4.02 ± 0.43	4.58 ± 0.44	4.95 ± 0.21	4.77 ± 0.39	4.80 ± 0.14	7.79	7.27 ± 0.46	4.56	426.16 ± 6.64	4.61	5.10 ± 0.65	4.8	5.51 ± 0.21	4.83 ± 0.38	5.74 ± 0.90
Tyrosine	3.61 ± 0.25	3.71 ± 0.18	4.36 ± 0.41	2.62 ± 0.15	3.12 ± 0.31	2.85	3.10 ± 0.34	1.98	346.33 ± 14.38	3.03	2.88 ± 0.15	3.25	2.40 ± 0.77	3.25 ± 0.19	3.12 ± 0.37

Jowar: *Sorghum vulgare*; Maize, dry: *Zea mays*; Rice, raw, milled: *Oryza sativa*; Wheat flour, refined: *Triticum aestivum;* Wheat, whole: *Triticum aestivum*; Amaranth seed, black: *Amaranthus cruentus*; Amaranth seed, pale brown: Amaranthus cruentus; Quinoa: *Chenopodium quinoa*; Tartary buckwheat (B-121); Common buckwheat: *F. esculentum* (Tomotake et al. (2006); Bengal gram, whole: *Cicer arietinum*; Cowpea, white: *Vigna catjang*; Lentil whole, brown: *Lens culinaris*; Peas, dry: *Pisum sativum*; Rajmah red/Kidney bean: *Phaseolus vulgaris*

Source: Longvah, T., Ananthan, R., Bhaskarachary, K., & Venkaiah, K. (2017). Indian food composition tables, National Institute of nutrition, Indian Council of Medical Research Department of Health Research, Ministry of Health and Family Welfare, government of India, Jamai Osmania (PO), Hyderabad – 500,007, Telangana, India
Phone: +91 4,027,197,334, fax: +91 4,027,000,339, email: nin@ap.nic.in

4.3.1 Amaranth

4.3.1.1 Starch Characteristics of Amaranth

Amaranth, buckwheat chenopods and millets are considered as main pseudocereals. The nutritional quality of amaranth is superior from other pseudocereals. Amaranth grains consist of 73.7–77.0% carbohydrates, 12.5–15.5% proteins, 7.1–8.0% lipids, 3.0–3.5% mineral content, and 19.5–49.3% dietary fiber contents (Pedersen et al., 1990). Major proportion of carbohydrates in amaranth is starch and accounts for 62–65%. Amaranth starch composed of amylose and amylopectin. Where amylose is a linear polymer of glucose, amylopectin is highly branched in nature, made from a linear chain of (1 → 4)-linked α-D-glucose and short chains of (1 → 6)-α-D-glucose-linked branches. Amylopectin content for amaranth ranges between 90 and 98% of the total starch, with 1700 amylopectin/molecules and exhibits smooth polymodal chain length distribution. The degree of polymerization (DP) for amaranth amylopectin also ranges between11–12 (Singh et al., 2014; Wilhelm et al., 2002).The size of amaranth starch granules ranges between 0.5 and 2.5 μm, are polygonal in shape and show unimodal size distribution. Though amaranth starch granules are similar to rice, granules of starches from other cereals are larger in size. The pasting (T_o) and gelatinization temperature (T_p) for amaranth starches ranges from 69 °C to 72 °C and 60 °C to 77 °C, respectively (Zhu, 2017). However, the amylose content, genotype, crystallinity and the presence and/or the absence of amylose-lipid complexes may affect the pasting behaviour of amaranth starches. Amaranth starches contain a minor proportion of starch bound lipids of 0.16–0.28% which also affects their pasting and functionality (Hoover et al., 1998). The small size granules of amaranth starches with low amylose content attributed to a lower breakdown viscosity and a more stable paste upon gelatinization. The small size, granular structure, low amylose content and high tendency to loosen crystallinity results in fast digestibility thereby considered as high glycaemic food. The rapidly digestible starch (RDS) of 30.7% and predicted glycemic-index of 87.2 for the raw seeds of amaranth had been reported earlier. Food with higher content of rapidly digestible starch is considered as high glycaemic food (Capriles et al., 2008). Glycemic index represents the levels of carbohydrate in food in response to postprandial glucose levels after the consumption of food (Jenkins, 2007). Therefore, amaranth may be adversely affecting the postprandial glucose of consumers suffering from diabetic and cardiovascular disease. However, amaranth proteins are having very high nutritional profile mimicking to the nutritional profile of milk. Amaranth grains are rich in minerals like phosphorous, iron, potassium, zinc, calcium, and vitamins such as vitamin B-complexes, vitamin E along with polyphenols such as flavonoids, caffeic acid, p-hydroxybenzoic acid, and ferulic acid. *Amaranth hypochondriacus* from Mexican Highlands showed high content of rutin (4.0–10.2 mg/g flour) and nicotiflorin (7.2–4.8 mg/g flour) (Barba de la Rosa et al., 2009).

4.3.1.2 Composition and Functionality of Amaranthus Proteins

Pulse proteins are deficient in sulfur rich amino acids i.e., cysteine and methionine, while cereal proteins are poor in lysine and tryptophan amino acids (Table 4.2). Conversely, amaranth proteins are rich in essential amino acids such as cysteine and methionine amino acids (Table 4.2). Grains, endosperm, and germ contain protein content of 11% to 17%, 35% and 65%, respectively, whereas average protein content of 15% and 85%, respectively for germ and the endosperm of other cereals has been observed, hence, amaranth is a suitable alternative source of wheat gluten (Singh et al., 2019). As per the Osbornes' (Osborne, 1924) solvent-specific solubility-based classification of seed storage proteins, amaranth grain revealed the accumulation of 51% albumins, 16% globulins, 24% glutelins, as a major fraction, whereas, 1.4% to 2.0% alcohol-soluble prolamine, as minor protein fraction of grain proteins had been reported (Gorinstein et al., 1991; Martínez et al., 1997). Cereals such as maize and wheat, on the contrary, contain alcohol-soluble prolamins as the major storage proteins (Gorinstein et al., 1991). Conversely, a major proportion of grain storage proteins in leguminous crops are salt-soluble globulins. A higher proportion of lysine and valine amino acid residues in the albumin and globulin fraction of the amaranth grain proteins were observed. A higher proportion of leucine and histidine amino acid residue in the glutenin subunits proteins of amaranth were also noticed. The nutritional quality indicators, such as protein digestibility, lysine availability, protein efficiency ratio, etc., are also fairly good for amaranth proteins thus imply that amaranth grain are good source of cereal and pulse proteins (Paredes-Lopez, 2020). Amaranth whole-meal flour showed average protein digestibility of 74.2%, which is significantly improved after thermal processing of grain, like popping, roasting etc., (Bejosano & Corke, 1998). The presence of anti-nutritional substances may have affected the protein digestibility of amaranth whole-meal flour. Whole meal of thermally processed grains of amaranth demonstrated a superior protein digestibility, which may be attributed to the thermal inactivation of anti-nutritional substances. The protein digestibility corrected amino acid score of 0.40 and 0.57, respectively, for wheat and oats was observed against 0.64 for amaranth (Bejosano & Corke, 1998). Therefore, the protein digestibility corrected amino acid score of amaranth whole meal flour was also superior. Apart from these, the functionality of proteins is also important for the development of new product formulations. Gluten-free muffins prepared from amaranth protein isolates demonstrated superior texture (volume, height, springiness, cohesiveness) and sensory attributes (crust, color, appearance and overall acceptance) when compared with muffins made from gluten fortified batter (Shevkani & Singh, 2014).

The presence of angiotensin-converting enzymes inhibitor peptides was reported for the peptides derived from 11S globulin sub-fraction of amaranth. Amaranth glutelin showed antihypertensive and anticarcinogenic activities, which attributed by the presence of the lunasin-like polypeptide in glutelin (Barrio & Añón, 2010; Sabbione et al., 2016). The antioxidant activity of amaranth proteins also enhanced after the gastrointestinal digestion also affects the fatty acid metabolism in liver thus confirmed the hypotriglyceridaemic effect in rat (Escudero et al., 2006) (24). The colonic epithelial cells show a decrease in the expression of *CCL20* gene at transcript levels in the presence of the amaranth peptide (SSEDIKE), therefore, amaranth peptides also possess anti-inflammatory properties. Bioactive peptides with AWEEREQGSR, YLAGKPQQEH, IYIEQGNGITGM, and TEVWDSNEQ amino acid (aa) residues from 11S globulin protein of *Amaranthus mantegazzianus* revealed antioxidant activity. Furthermore, cationic peptide with HVIKPPSRA and KFNRPETT aa residues and a neutral peptide with aa sequence of GDRFQDQHQ demonstrated in vivo inhibition of Cu^{2+}/H_2O_2-induced oxidation of low-density lipoproteins (García Fillería & Tironi, 2017; Orsini Delgado et al., 2016). Gluten-free muffins prepared from Amaranthus protein isolates showed texture and sensory properties like the muffins prepared from protein isolates from kidney bean and field pea. Similarly, the edible/biodegradable film-forming properties of amaranth proteins were also comparable to pulse proteins.

4.3.2 *Buckwheat*

Fagopyrum esculentum (common buckwheat) and *Fagopyrum tataricum* (tartary buckwheat/bitter buckwheat) are widely cultivated in America, Europe, and Asia. On the contrary, *Fagopyrum dibotrys* or *Fagopyrum acutatum* and *Fagopyrum cymosum,* known as golden or tall buckwheat,are a less cultivated buckwheat with potential significance in Asia (Liu et al., 2006). The consumption of buckwheat increased tremendously because of its disease healing and prevention attributes (Ahmed et al., 2014; Cai et al., 2004; Guo et al., 2013; Li & Howard Zhang, 2001). On dry basis, the grain contains ~70% starch, ~12% protein, ~10% dietary fibres, ~3% lipids, 2.5% ash content (Zhu, 2016; Food Data Central, 2020). Hence, starch is a major proportion of buckwheat grains. A wide variation in amylose content in the starch of different buckwheat accession was reported. Minor content of important minerals (Mg, K), vitamins (B, C, E), flavonoids (rutin, quercetin), D-chiro-inositol, fagopyritols, and polyunsaturated essential fatty acids (linoleic acid) in the grains of buckwheat were also observed (Li & Howard Zhang, 2001; Wijngaard & Arendt, 2006).The presence of these biomolecules led to various health benefits like anti-hypertension, hypocholesterolemic activity, fat storage suppression in body, antioxidant and free radical scavenging, anti-inflammatory, etc., (Ahmed et al., 2014; Li & Howard Zhang, 2001; Wijngaard & Arendt, 2006).However, the higher abundance of tannins, protease inhibitors, and phytic acid affects the starch digestibility.

4.3.2.1 Composition and Functionality of Buckwheat Starches

Amylose content ranges between 23 and 29.1% for starches from 30 genotypes of common buckwheat (Ikeda et al., 1997). The amylose content between 3.8% and 16% for waxy or mutant buckwheat was also observed (Gregori & Kreft, 2012). The morphological analysis shows that starch granules had smooth surface and less spherical shape, A type polymorph, with polygonal structure. The starch granules showed average granule size of ~6–7 μm and ranged between~2 to 15 μm (Qian et al., 1998; Zheng et al., 1997).Therefore, the buckwheat starch granules are smaller in size as compared to other cereals (Gregori & Kreft, 2012; Jane et al., 1994; Liu et al., 2015a, b; Vallons & Arendt, 2009). Degree of polymerization (DP) of 94,900 with a minimum and maximum range between 38,000 and 134,000, for starches from buckwheat was observed. Therefore, common buckwheat starch had much higher DP than amaranth, wheat, quinoa, and proso millet than waxy maize (Praznik et al., 1999). DP with two peak maxima ranged between 1020 and 1380 for buckwheat starches. Amylose in common buckwheat starches showed the distribution of chain length from 3.1 to 4.3, with average chain length between 280 and 380 glucosyl residues. Therefore, the common buckwheat amylose resembled with the starches from wheat and barley. According to crystal types, the starches are grouped into A, B, and C types. A type buckwheat starch granules consist of a higher proportion of DP 6–12 chain-length amylopectin molecules compared to amylopectin in B type starch granules, whereas, a lower proportion of medium (DP 16–24) and long chain (DP 25–60) amylopectin molecules in A type starch granule of buckwheat was observed (Punia et al., 2021; Sanderson et al., 2006). Buckwheat starches revealed the higher chain length (CL) (23–24) glycosyl-residues than cereals. Higher amounts of extra-long chains in buckwheat amylopectin could be due to long CL and lower short-to-long chain ratio of amylopectin chains (Hanashiro et al., 2005; Yoshimoto et al., 2004). The gelatinization properties of the buckwheat starches using differential scanning calorimetry (DSC) were accomplished, which revealed that cultivars type, moisture conditions and scanning rate (°C/min) differentially affect the T_o, T_p and T_c of buckwheat starches. T_o, T_p, T_c and ΔH ranged from 59.5 to 64.1 °C, 63.7 to 68.4 °C, 81.7 to 85.8 °C, and 14.5 to 15 j/g, respectively for starches from different common buckwheat accessions(Yoshimoto et al., 2004) (Table 4.1). It ranged from 58.6 to 60.2 °C, 61.5–64.3 °C, 70–73 °C, and 14–15.3 j/g for common buckwheat starches(Lu & Baik, 2015). On the contrary, at higher moisture (1:4) and thermal scanning rates (10 °C/min), T_o, T_p, T_c and ΔH of 61.2 °C, 66.1 °C, 75.2 °C, and 9.0 j/g, respectively for common buckwheat accession was observed (Li et al., 2014). Similar observations of different DSC parameters for tartary buckwheat were also noticed (Zhu, 2016 and reference therein). The apparent amylose content in summer and autumn harvested buckwheat grains of similar accession was similar, thus implying that the amylose content in buckwheat was not affected by agronomic practices (Hurusawa & Miyashita, 1965). The gelatinization profile of buckwheat starches was not affected by the amylose content, however, the chain length distribution of amylopectin revealed strong corelation to T_o, T_p, T_c and ΔH of buckwheat starches. Lower T_p for starched having a higher proportion of amylopectin with DP 7–11 was noticed, whereas buckwheat starches with a higher proportion of amylopectin of DP 12–17 showed a higher ΔH (Noda et al., 1998). A wider gelatinization temperature attributed structural heterogenicity in buckwheat starches than maize and wheat starches. A lower T_o and T_p, and higher ΔH for buckwheat than rice and maize starches was also reported by Zheng et al.

(1997). On the contrary, higher T_o, T_p, and T_c for common and tartary buckwheat starches than potato starches but lower from maize starches were also reported by Gao et al. (2016). However, lower ΔH for buckwheat starches was also observed Gao et al. (2016). On the other hand, lower T_p and T_c of wheat starches than both type of buckwheat starches was also noticed (Li et al., 1997; Qian et al., 1998). Variation in the amylopectin content, DP and morphology of starch granules etc., may be attributed to differential gelatinization behaviour of starches of the buckwheat. Buckwheat starches showed higher peak viscosity than wheat starches (Acquistucci & Fornal, 1997; Li et al., 1997; Praznik et al., 1999), however, the pasting viscosity of buckwheat starches was lower from the potato and higher from the maize starches (Gao et al., 2016). Gao et al. (2016) revealed higher breakdown viscosity for different buckwheat starches, as compared to starches from potato and maize; higher setback viscosity from maize starches while lower from potato starch. The presence of extra-long chains of amylopectin in rice had related to the lower breakdown viscosity (Han & Hamaker, 2001).

Buckwheat starches exhibited lower syneresis as against wheat and maize (Qian et al., 1998). Syneresis was positively correlated to the amylase and resistant starch content of cooked groats of buckwheat (Lu & Baik, 2015).The gelatinization profile of retrograded starches was also evaluated, and T_o, T_p, T_c, and ΔH ranged from 39.3 to 41.5 °C, 49.2 to 51.2 °C, 59.2 to 60.9 °C, and 4.6 to 5.6 J/g, respectively for gelatinized and retrograded buckwheat starches (Lu & Baik, 2015). Lower storage induced water syneresis for buckwheat starches upon storage 3–10 days at 4 °C, and better stability to syneresis after freeze-thawing of gelatinized buckwheat starches at −12 and 25 °C was also reported (Qian et al., 1998). Therefore, buckwheat starches with lower syneresis may be useful to make processed food with long storage-shelf life. However, the application of buckwheat starch in food industry is limited, which may be because of high cost of production and the availability of raw material. Buckwheat starch supplemented cake demonstrated poor baking performance and lower sensory attributes (Lorenz & Dilsaver, 1982). Higher accumulation of long chain amylopectin molecules, wide variation in gelatinization profiles, and small granule size may have contributed to poor baking performance of buckwheat starches. However, the small granule size of starches may be used to replace fats in water-in-oil emulsions (Singer, 1994). Where the octenyl succinic anhydride modified buckwheat starches showed enhanced hydrophobic properties thus can be used as emulsifiers to stabilize pickering emulsions (Timgren et al., 2011). These findings thus revealed that buckwheat starch can be used to prepare the gluten-free products after improvement in the gelatinization profiles through plant breeding or genetic engineering approaches (Zhu, 2016).

4.3.2.2 Structural and Functional Characteristics of Buckwheat Proteins

Buckwheat grain contains ~12% of total proteins, which are primarily albumins, globulins, prolamins and glutelins, as depicted in other cereals. Gálová et al. (2019) also revealed that the grains of common buckwheat exhibited 45% of albumin and globulins, 15% glutelins, and 3% prolamins. On the contrary, the grains of rye and oats consist of 33% and 26% albumins and globulins thus imply that buckwheat has higher proportion of albumin and globulins. Rye grains have 39% of prolamins and 18% of glutelins content was observed against 15% of prolamins and 45% of glutelins for oats. Thus, the proportion of prolamins and glutelins was lower in buckwheat than rye and oats (Gálová et al., 2019).Large diversity in the amino acid composition of buckwheat grains was also observed (Sytar et al., 2016). Common and tartary buckwheat grains are found deficient in leucine, cysteine as compared to other cereals (Bhinder et al., 2020; Motta et al., 2019; Zhang et al., 2017) (Table 4.2). Also, imbalance in the amino acid composition of buckwheat was remarkable; however, a high biological value with the amino acid score of 100 for buckwheat proteins was distinguishable (Sytar et al., 2016; Sytar et al., 2018). SDS-PAGE analysis revealed the presence of 30 to 50 kilo Dalton (kDa), 24 kDa, 19 kDa, 16 kDa and 10 kDa polypeptides (PP) in common buckwheat (Alonso-Miravalles & O'Mahony, 2018). The identity of 50 kDa PP may was appeared to be 13S legumin like and 8S vicilin-like globulin-type PPs, whereas the small molecular weight PPs of 10 to 15 kDa could be albumins (Alonso-Miravalles & O'Mahony, 2018).Though SDS-PAGE analysis not revealed significant differences between PPs of common and tartary buckwheat (Zaika et al., 2019), 2D-PAGE analysis revealed significant variations in protein bands between both types of buckwheat accessions (Capraro et al., 2018).The *in vitro* digestibility of buckwheat proteins was affected by polyphenols (Chen et al., 2019). Buckwheat PPs of 31 and 45 kDa exhibited resistance to proteolytic cleavage, when subjected to simulated gastric and duodenal digestion for 120 min. Whereas, under similar experimental conditions, a 50 kDa PP of buckwheat remained undigested till 180 min exposure to simulated digestion (Gálová et al., 2019).

Attempts have been made to enhance the digestibility of buckwheat proteins. The digestibility of buckwheat protein isolates was enhanced after ultrasound treatment (Jin et al., 2021), and the digestibility of buckwheat proteins improved up to 1% after microwave treatment at 2450 MHz at 850 watts for 30 min;4% after high pressure treatment at 600 MPa pressure for 30 min at 60 °C temperature and 7% after boiling treatments(Deng et al., 2015). However, the protein digestibility of Tartary buckwheat flour was not affected after hydrothermal treatments (Chen et al., 2019). The effect of extrusion cooking on buckwheat protein digestibility remains unexplored. The hypolipidemic effect of Tartary buckwheat proteins by *in vivo* and *in vitro* was validated (Zhang et al., 2017; Zhou et al., 2018), which may appear due to the presence of quercetin in buckwheat flour and the conversion of cholesterol in bile acids (Zhang et al., 2017). Mora et al. (2019) demonstrated blood pressure lowering property and antihypertensive potential of buckwheat peptides by inhibiting the activity of angiotensin-converting enzyme.Peptides derived from the digestion of buckwheat proteins by trypsin and alcalase (gastrointestinal enzymes) demonstrated the inhibitory activity against dipeptidyl peptidase IV enzyme which thus confirmed the antidiabetic properties. Superior dipeptidyl peptidase inhibitory activity was also demonstrated by peptides derived from the hydrolysis of buckwheat proteins than barley and oat peptides. Dipeptidyl peptidase IV is a homodimeric serine peptidase that control the secretion of insulin and glycemic control in human (Wang et al., 2015). The intravascular thrombosis and cardiovascular diseases (CVD) in humans is caused by the aggregation of platelets, therefore, the inhibition of platelet aggregation can help to overcome CVD in humans. Buckwheat protein hydrolysates inhibited the platelet aggregation in a dose-dependent manner and showed superior platelets aggregation inhibitory activity than barley protein hydrolysates (Yu et al., 2016a). The interaction of peptides with amino acid residues composition of ALPVDVLANAYR, ALPIDVLANAYR, EFLLAGNNKR, GEEFDAFTPK, GEEFGAFTPK, LQAFEPLR, QLAQIPR, QKEFLLAGNNK, and TNPNSMVSHIAGK to cyclooxygenase-1 (COX1) through computation modelling was also predicted (Yu et al., 2016b). These findings thus revealed the role of buckwheat protein hydrolysates in the prevention of CVDs in human. Increase in growth of *Bifidobacterium species*, *Enterococcus* and *Lactobacillus*, *Enterococcus* and *Lactobacillus* and decrease in the *E. coil* cell load in the gut of mice after feeding Tartary buckwheat protein rich diet was observed (Zhou et al., 2018). The presence of polyphenols and resistant carbohydrates in buckwheat proteins may attribute to the enhance the growth of gut microbiota in the gut of mice. Allergic responses such as asthma, allergic rhinitis, atopic dermatitis, anaphylaxis, urticarial and enterocolitis were also reported after the consuming buckwheat products. (Miyazaki et al., 2019; Nagai, 2017; Satoh et al., 2020; Satou et al., 2019). Vicilin-like proteins of 55 and 19 kDa, trypsin inhibitory protein of 9 kDa and 16 kDa, 13S protein of 22 kDa, 13S globulin of 22 kDa and proteins with molecular weight of61, 48 and 45 kDa as major immune-responsive proteins in human were identified (Cho et al., 2014; Satoh et al., 2020; Zheng et al., 2018). Among these proteins, 13S globulin as major allergic protein was recognised (Sano et al., 2014). These findings thus revealed that though buckwheat has good health improving characteristics, the allergic reaction needs to be investigated in detail in populations of diverse origins and ethnicities. Processing of buckwheat flour at ultra-high pressure and hydrolysis with alkaline protease reduced the allergic response of buckwheat proteins upto 100% levels (Lee et al., 2017). Plant breeding approaches can also be undertaken to silent or stop the accumulation of 13S globulin in modern buckwheat cultivars. Therefore, the application of buckwheat proteins in the development of gluten-free products require more detailed scientific investigations.

4.3.3 Quinoa

Quinoa (*Chenopodium quinoa*, Willd.; (2n = 4x = 36; x = 9), spinach and beets are the members of the Chenopodiaceae family and 250 species of Chenopodium genus are found world-wide. People of Andes i.e., Peru and Bolvia, domesticated Chenopodium thousands of years ago because of a rich source of proteins with a balanced composition of essential amino acids (Filho et al., 2017; Jancurová et al., 2009). Presently quinoa is largely cultivated in Argentina, Bolivia, Chile, Colombia,

Ecuador and Peru (FAO, 2012). *Chenopodium quinoa* designated as "pseudo-cereal" or a oleaginous "pseudo-seeds" because of the unique panicle-type inflorescence, protein rich grain composition, high sulfer and lysine content (Filho et al., 2017; Vega-Gálvez et al., 2010; Repo-Carrasco-Valencia and Serna, 2003). Chenopodium can grow in diverse climatic conditions, its presence can be observed from sea level to 4000 m above sea level, temperature from −4 °C to 38 °C and humidity between 40% to 88% (Bojanic, 2011). Quinoa is highly tolerant to drought and salt conditions (Jacobsen, 2003). Therefore, different accessions of quinoa have great diversity in physico-chemical, morphological and nutritional quality of quinoa, which may be attributed to its diverse geographic distribution and agro-climatic conditions. The grain of quinoa may be of white, black, red, yellow, etc., in colour (Bhargava et al., 2006; Ruiz et al., 2014; Vega-Gálvez et al., 2010).

4.3.3.1 Morphology, Structure, and Chemical Properties of Quinoa Starch

Quinoa seeds contain 53.5–69.2% of starch on dry matter basis and is a prime component of the grains. Starch granules are small in size with diameter of 1-3 μm and localized in the perisperm of quinoa seeds (Lorenz, 1990; Ruales & Nair, 1994). Some studies revealed size of granule between 0.4–2.0 μm (Li & Zhu, 2017a, b; Lindeboom et al., 2004). The presence of starch bound proteins causes aggregation in QS granules, and these aggregates can be disaggregated into single granules by proteolytic cleavage of starch-bound proteins (Atwell et al., 1983; Ruales & Nair, 1994). The size of 10–30 μm for spherical or oblong shaped aggregates was observed against 14,000–20,000 for single granule size of QS granules (Ando et al., 2002; Lorenz, 1990; Ruales & Nair, 1994; Srichuwong et al., 2017). Polygonal, angular, and irregular shapes for QS starches was observed (Lindeboom et al., 2004; Li & Zhu, 2017). The degree of crystallinity for QS ranged from 21.5 to 43.0%, therefore, QS starches exhibited a lower degree of crystallinity than starches of normal maize, garden orache and amaranth while it was higher from kañiwa, barley, adzuki bean and barley starches (Qian & Kuhn, 1999; Steffolani et al., 2013; Tang et al., 2002; Wright et al., 2002). Amylose content between 4–10.9% for QS from size exclusion chromatography was observed, whereas, debranched QS exhibited 3.5% to 27.0% amylose content. Hence, the presence of amylopectin in starch may interfere with the QS estimation (Li & Zhu, 2017).The degree of polymerization (DP) of amylopectin plays an important role to determine the functionality of starch. Fluorophore-assisted capillary electrophoresis (FACE) and high-performance anion exchange chromatography with pulsed amperometric detection (HPAEC-PAD) are widely used to analyse the DP of amylopectin. QS contained higher proportion of amylopectin with short chains with DP of <12 and a lower proportion of amylopectin with DP of <13–35 as compared to waxy maize starches (Li & Zhu, 2017; Srichuwong et al., 2017; Inouchi et al., 1999). QS revealed a higher ratio between short and long chains of amylopectin (Bertoft et al., 2008; Li & Zhu, 2017).The average chain length of 16–17 glucosyl residues for amylopectinin different QS by

HPAEC-PAD was analysed, whereas SEC revealed the average chain length distribution between 18–21 glycosyl residues(Li & Zhu, 2017; Tang et al., 2002; Watanabe et al., 2007; Watanabe, 2008). The gelatinization profile of QS starches by DSC was also evaluated. T_p and ΔH ranged between56.2 and 65.0 °C, 10.8 and 14.4 J/g, respectively for starches from different quinoa starches as reported by Li et al. (2016). T_o, T_p, T_c and ΔH of 53.9 °C, 60.6 °C, 66.0 °C, and 10.3 J/g, respectively for quinoa starches was noted by Srichuwong et al. (2017) (Table 4.1). The gelatinization profile (T_o, T_p, T_c and ΔH) of white sorghum, red sorghum, millet, corn, wheat, and amaranth starches was higher than quinoa starches (Srichuwong et al., 2017).These findings thus revealed that QS exhibited ΔH similar to sorghum, garden orach, and kañiwa, whereas, maize, adzuki bean, and amaranth starches hold higher ΔH than QS starches. QS starches demonstrated a higher ΔH as compared to wheat and barley starches (Inouchi et al., 1999; Qian & Kuhn, 1999; Steffolani et al., 2013; Tang et al., 2002; Wright et al., 2002). Variations in amylose and amylopectin content among starches from different botanical sources may be linked to wide differences in ΔH, which is linked to the fine structure of amylopectin (Li & Zhu, 2017a). The swelling power of starches is also correlated to amylose content in starches. The peak viscosity of 2860 cp for quinoa starches was observed, whereas, PV ranged between 2240 and 2400 cp for sorghum starches, 2000 cp for amaranth starches, between 1850 and 1891 cp for corn and millet starches, and 1319 cp for wheat starches was reported by Srichuwong et al. (2017). Pasting temperature of 65.3 °C and 63.5 °C for sweat and bitter quinoa was also observed (Wright et al., 2002). While, the peak-, hold-, breakdown-, final and setback viscosities ranged from 367.3 and 402.5 cp, 357.9 and 369.4, 9.4 and 33.1, 495.9 and 495.1, 138.0 and 125.8 cP, respectively, for sweat and bitter starches (Wright et al., 2002). These findings thus demonstrated that peak viscosity (PV) of quinoa starch was higher than that of most other starches. The swelling power and pasting properties of starches are influenced by the interaction of proteins, lipids and non-starch polysaccharides during gelatinization. High swelling power attributed to lower pasting temperature of amaranth starches. Lower amylose content (1.2%) allows starch granule to swell more which led to rupture of starch granules at lower temperature this leads to lower pasting temperatures. The presence of lipids does not affect the pasting properties of amaranth starches due to lower amylose content in amaranth starch. On the contrary, amylose content ranged between 22.9–25.8% for wheat, millet, corn and sorghum starches, which led to a stronger amylose-lipid interaction, associated with higher pasting temperature, lower peak- and breakdown viscosities. QS starches consist of average 8.2% amylose content, and small granule size, which, therefore, resulted in the intermediate or moderate amylose-lipid interactions during gelatinization. Weak starch-lipid interactions, moderate levels of amylose content in QS may have resulted in higher peak- and lower breakdown viscosities. Starches with higher peak viscosity and lower setback viscosities could be used to prepare the food withhold/maintain the consistency of gel but not to be solidified upon cooling, thus it can be said that quinoa starches can be used to prepare rice puddings, instant creamy deserts, and also used as fat replacer for mayonnaise or water-in-oil emulsions.

4.3.3.2 Composition and Functionality of Quinoa Protein

Protein content ranged between 12% and 23% which is higher from cereal grains, while the grain content in quinoa was lesser from pulses (Abugoch James, 2009). Intriguingly, majority of the protein content localized in the embryo of quinoa. Protein content between 15.6 and 18.7% for six quinoa varieties of Southern Europe was reported by Rodríguez Gómez et al. (2021). Quinoa proteins are primarily composed of 37% of globulin and 35% of albumins, however these are deficient in prolamins or least detected. Lower proportion of prolamins of 0.7% to 7.0% for quinoa seed was reported by Abugoch James (2009). The composition of amino acid for quinoa was also evaluated. Quinoa proteins are rich in lysine (5.1–6.4%) and methionine (0.4–1%) amino acids (Prakash and Pal, 1998) (Table 4.2). Superior content of essential amino acids such as histidine and lysine for quinoa was also observed. Histidine of 28.8 mg/g, and lysine content of 54.2 mg/g for quinoa was reported. These findings thus revealed a high nutritional potential of quinoa along with a source of high-quality proteins. Protein efficiency ratio (PER)) and digestibility of quinoa proteins is comparable to the casein protein of milk, while the PER of washed quinoa than raw quinoa was also superior. The presence of saponins may have affected the PER and digestibility of quinoa proteins (Gross et al., 1989; Ruales & Nair, 1993). The digestibility and functionality of quinoa proteins can be enhanced by heat, hydrothermal, microwave and baking treatments. The role of 7S and 11S globulins was crucial in protein aggregation during heating at different pH and temperature regimes (Van de Vondel et al., 2021). However, heat-induced gelling behaviour of quinoa proteins need more detailed investigations. Quinoa seed storage 11S globulin is composed of hexameric protein comprise of six pairs of acidic and basic subunits, which are connected to each other by disulfide bridges (Brinegar & Goundan, 1993). Similarity analysis of 11S globulin at amino acid levels revealed its high homology with glycinin, therefore, quinoa 11S globulin protein is as designated as chenopodin (Barrett, 2006). Chenopodin is a major seed storage protein of quinoa which is 37% of the total seed storage proteins. The acidic and basic subunits of chenopodin (11S globulin) exhibited molecular weight of 30–40 kDa and 20–25 kDa, respectively, which linked together via disulphide bonds (Abugoch James, 2009; Brinegar & Goundan, 1993). Higher content of asparagine, aspartic acid, arginine, serine, leucine, glycine, glutamine-glutamic acid for chenopodin was also observed (Brinegar & Goundan, 1993). Therefore, the amino acid composition of chenopodin matched the leucine, isoleucine, and phenylalanine, and tyrosine amino acid composition of the standards of FAO protein reference (FAO, 1973). A protein with 8–9 kDa with 35% proportion of the total grain protein in quinoa was also identified by Osborne (1924) method. The 8–9 kDa protein was 2S-type protein and belongs to albumins. The 2S-type albumin was rich in arginine, cysteine, and histidine (Brinegar et al., 1996). These findings thus revealed that the11S and 2S-type protein in Chenopodium are the major seed storage proteins and reservoirs of essential amino acids. The quinoa protein isolates exhibited water holding capacity of 2.8–4.5 mL of water/g of sample, while soy protein isolates exhibited the WHC of 4.3 mL of water/g of sample. Thus, the WHC of quinoa protein isolates was comparable soy protein isolates and can be used to fortify gluten-free products.

4.4 Millets

Abiotic and biotic challenges, global warming and abrupt climatic conditions, shrinkage of arable land by urbanization, rising price and increasing demand of cereals globally, are ongoing challenges for cereal production. However, millets are one of the widely consumed grains in arid and semi-arid regions of Asia (India and China) and Africa (Dhull et al., 2021; Yousaf et al., 2021). Millets can withstand up to 64 °C temperature and 350–400 mm annual rainfall (Chivenge et al., 2015). Millets being a C4 crop have highly efficient photosynthesis system and require only 6–8 weeks for seed maturation which may attribute to high yield and thermo tolerance (Hariprasanna et al., 2014). Also, millets are considered as "poor man food" because of lower price and readily availability for population lives in semi-arid and arid zones (Amadou et al., 2013). Millets are considered as first ancient grains domesticated for human use. Millets are considered as first ancient grains domesticated for human use. Millets are round shape small-seeded grains of the Poaceae family, which are of seven types of namely foxtail millet (*Setariaitalica*), finger millet (*Eleusine coracana*), pearl millet (*Pennisetum glaucum*), proso millet (*Panicum miliaceum*), kodo millet (*Paspalum scrobiculatum*), barnyard millet (*Echinochloa crus galli*) and little millet (*Panicum sumatrense*) (Guenard, 2021; Punia Bangar et al., 2021a, b, c, d; Siroha et al., 2021). The total millet production of 31,019,370 tonnes in 2018 was estimated which secured sixth position among other cereals. The largest proportion of millet was produced by India during 2018 followed by Niger, Sudan, and other countries (FAOSTAT, 2020).

4.4.1 Composition and Functionality of Millet Starches

Millet starch revealed amylose content between 6 and 38.6%, lipid between 0.16 and 2.9%, protein 0.2 and 4.3% and ash content of 0.02 to 1.4% (Zhu, 2014 and references therein). The higher accumulation of glycine, glutamine, and aspartic acid in the granules of the foxtail millet starches was also reported by Wankhede et al. (1979). The presence of neutral lipids (linolenic acid), phospholipids lipids (palmitic acid), and glycolipids in the pearl millet starch granules in free and bound form were reported by Hoover, 1995. Polygonal morphology depicted the majority of millet starches (Zhu, 2014 and references therein). The DP affects the pasting properties of starches and DP ranged from 1060 to 1250 and 9000 to 9100, respectively for amylose and amylopectin from pearl millet starches. Whereas, the molecular weight of pearl millet amylose and amylopectin ranged from 105 to 106 and 107, respectively (Madhusudhan & Tharanathan, 1996; Wankhede et al., 1979). The chain length (CL) distribution of glucosyl moieties influences the molecular weight of amylose and amylopectin of a starch granule. The CL between 260 and 270 glucosyl residues with four chains per amylose molecules was observed. Where amylopectin from pearl millet exhibited CL of 18 and 21 glucosyl residues, the external CL ranged between 12 and 14. Along with that, the internal chain length ranged between 4.8 and 6.3 glucosyl residues was also observed (Annor et al., 2014; Gaffa et al., 2004; Madhusudhan & Tharanathan, 1996). It is proposed that the internal chains of amylopectin constitute the amorphous region of a starch granule, whereas, the crystalline region of a starch granule is formed by the external chains of amylopectin (Pérez & Bertoft, 2010). The swelling power and solubility of millet starches between the temperatures range of ~50–90 °C was observed. Therefore, the swelling power and solubility of millet starches is lower than potato starches. Higher amylose content may also be linked with low swelling power and solubility of millet starches at lower temperature. Higher leaching of amylose in rice also linked to higher gruel solid loss which was associated with leaching of amylose during the gelatinization of rice starches followed by gel formation. Since millet starches are composed of higher amylose content, it is imperative that the pasting and rheological properties of these starches should also be analysed. PV between 345 and 425 RVU, BDV between 183 and 237 RVU, SBV between 142 and 188 RVU, and PT between 72.0 and 78.5 °C for different genotypes foxtail millet had been reported (Liu et al., 2011). The thermal behaviour of millet starches by using DSC was also analysed and very high diversity in thermal properties of various accessions of millet was noticed. Pearl millet starches revealed the T_o, T_p, T_c, and ΔH between 66.2 and 67.2 °C, 69.7–71.4 °C, 86.3–91.0 °C, and 14.3–14.7 J/g, respectively (Gaffa et al., 2004) (Table 4.1). T_p and ΔH of between 62.4 to 75 °C and 8.2–13.5 J/g, respectively, for the different accessions of foxtail millets were observed. On the contrary, proso millet starches exhibited T_p between 65.8 and 80.2 °C, while ΔH was ranged between 6.4–11.4 J/g for proso millet starches (Fujita & Fujiyama, 1993). Similar findings for foxtail millet were also observed by Wankhede et al. (1979) (Table 4.1). The starches from pearl millet exhibited good freeze thaw stability than corn and wheat starches (Hoover, 1995), while pearl millet starch revealed poor freeze thaw stability as compared to maize starch (Yañez et al., 1991). The freeze thaw stability is largely influenced by amylose content and proportion of shorter unit chains of amylopectin in starches, and may be attributed to variation in freeze thaw stability among different millet genotypes (Srichuwong et al., 2012). Therefore, detailed investigation of different genotype of millets to improve the functionality of millet starches is required. The glycaemic index of starches is determined by their susceptibility against starch hydrolysing enzymes. Roopa and Premavalli (2008) observed total starch (TS) between 39.7 and 50.1%, RDS between 8.3 and 11.1%, SDS between 26.5 to 35.3%, RS between 0.8 and 1.0% for the flour of finger millet. Whereas, Annor et al. (2013) noticed a total starch (TS) of 83%, RDS of 11.5%, SDS of 31.3%, and RS of 40.1% for Kodo millet flour. On the contrary, Kodo millet starches revealed TS, RDS, SDS, and RS of 94.2%, 21.8%, 33.2% and 37.5%, respectively (Annor et al., 2013). These findings thus revealed that millet starches exhibited relatively fair content of SDS and RS, which is very useful for persons suffering from many chronic life-style related ailments.

4.4.2 Composition and Functionality of Millet Proteins in Millets

The protein content varied among different millets cultivars. Protein content of 14.5% (8.6–19.4%), 11.7% (6.0–14.0%), 13.4% (6.4–15.9%), and 8.0% (6.9–10.9%), respectively for pearl, foxtail, proso, and finger millet, was observed. These findings thus revealed that among different millets, the pearl millets contain the highest protein content of 14.5% whereas, finger millet exhibited lower protein content 8.0% (Taylor & Taylor, 2017). The large protein-rich germ and smaller endosperm may be attributed to high protein content in pearl millet (Serna-Saldivar & Rooney, 1995). The analysis of proteins revealed the presence of albumins and globulins, prolamins, and glutelins in the grain of different millets. Albumin and globulins of different millets exhibited higher abundance of Glutamine/glutamate, asparagine/aspartic acid, arginine, alanine, and leucine, while higher abundance of Glutamine/glutamate, alanine, leucine, proline, and valine in the prolamins subfractions of different millets was observed. On the contrary, the glutelin fraction of different millets revealed the higher accumulation of glutamine/glutamate, arginine, leucine, asparagine/aspartic acid, proline, alanine, methionine, and phenylalanine amino acids. These findings thus demonstrated that the millets are poor in lysine content (21–37 mg/g protein), as observed for other cereals such as wheat etc. (Table 4.2). However, different millets exhibited higher leucine content (122–135 mg/g protein) which may attribute to the higher accumulation of prolamins in these cultivars (Taylor & Taylor, 2017).

4.5 Conclusions

The utilization of gluten-free cereals like sorghum and rice, pseudocereals such as amaranth, quinoa, millets, and buckwheat are gaining popularity. However, significant variation in the composition and functional properties of starches and proteins in pseudocereals vary significantly. Starches from a few Indian sorghum cultivars revealed higher enthalpy of retrogradation and lower synereses values thus can be utilized as superior alternatives of corn starches. Sorghum starches, therefore, can be used for making gluten-free products with long storage-shelf life. On the contrary, proteins in grain sorghum are poor in lysine and threonine amino acid content. Sorghum kafirin proteins showed a higher proportion of α-prolamins with viscoelastic properties which can be used as an alternative to gluten to enhance the viscoelastic properties of gluten-free dough, thus leading to enhance the texture characteristics of gluten-free food products. The poor solubility of sorghum kafirin after wet cooking needs to be addressed in the future. The small size, granular structure, low amylose content and high tendency to loosen crystallinity, and fast digestibility of amaranth starches attributed to consider it as a high glycaemic food. Therefore, amaranth starches cannot be incorporated into gluten-free products. On the contrary, a higher proportion of lysine and valine amino acid residues in the albumin and globulin fraction of the amaranth grain proteins, while a higher proportion of leucine and histidine amino acids in glutenins of amaranth proteins were observed. Also, the protein digestibility and lysine content for amaranth proteins are also good, therefore, amaranth could serve as a superior alternative to the nutrition-rich source of proteins. Common buckwheat amylose resembled the starches from wheat and barley but have lower synereses values. Therefore, buckwheat starches can be used to improve the shelf-life of gluten-free products. Common and Tartary buckwheat grains are deficient in leucine, cysteine amino acids with imbalance amino acid composition, inferior protein digestibility, and allergic response. Therefore, the nutrition and functional characteristics of buckwheat protein are inferior to cereals. Quinoa starches consist of an average of 8.2% amylose content, and small granule size, which, therefore, have a higher peak- and lower breakdown viscosities. Therefore, quinoa starches can be used for making semi-solid foods like pudding and deserts and also replace fat from mayonnaise or water-in-oil emulsions. Quinoa proteins are rich in lysine (5.1–6.4%), methionine (0.4–1%), and histidine amino acids. Also, quinoa proteins revealed high protein efficiency ratio and superior digestibility, which is comparable to the casein protein of milk, therefore quinoa protein isolates can serve as superior alternatives to wheat gluten and rice-bran proteins, which have poor solubility and foaming properties. Though millet proteins are deficient in lysine content (21–37 mg/g protein), pearl millet starches are rich in amylose, hold higher SDS value, and superior freeze-thaw stability than corn and wheat starches, therefore, millet flour could serve as the base of gluten-free flour. Therefore, improvement in the functional properties of starches and proteins of pseudocereals by using processing technology, genetic engineering, and plant breeding approaches in the future will be in the prime focus.

Acknowledgements ASV gratefully acknowledges the facilities provided by Amity University Punjab, Mohali.

References

Abugoch James, L. E. (2009). Quinoa (*Chenopodium quinoa* Willd.): Composition, chemistry, nutritional, and functional properties. *Advances in Food and Nutrition Research, 58*, 1–31. https://doi.org/10.1016/S1043-4526(09)58001-1

Acquistucci, R., & Fornal, J. (1997). Italian buckwheat (*Fagopyrum esculentum*) starch: Physicochemical and functional characterization and in vitro digestibility. *Nahrung, 41*, 281–284.

Ahmed, A., Khalid, N., Ahmad, A., Abbasi, N. A., Latif, M. S. Z., & Randhawa, M. A. (2014). Phytochemicals and biofunctional properties of buckwheat: A review. *Journal of Agricultural Science, 152*(3), 349–369. https://doi.org/10.1017/S0021859613000166

Ai, Y. (2013). Structures, properties, and digestibility of resistant starch. PhD Thesis., Iowa State University.

Alonso-Miravalles, L., & O'Mahony, J. A. (2018). Composition, protein profile and rheological properties of pseudocereal-based protein-rich ingredients. *Food, 7*(5). https://doi.org/10.3390/foods7050073

Amadou, I., Gounga, M. E., & Le, G. W. (2013). Millets: Nutritional composition, some health benefits and processing - a review. *Emirates Journal of Food and Agriculture, 25*(7), 501–508. https://doi.org/10.9755/ejfa.v25i7.12045

Ando, H., Chen, Y. C., Tang, H., Shimizu, M., Watanabe, K., & Mitsunaga, T. (2002). Food components in fractions of quinoa seed. *Food Science and Technology Research, 8*(1), 80–84. https://doi.org/10.3136/fstr.8.80

Annor, G. A., Marcone, M., Bertoft, E., & Seetharaman, K. (2013). In vitro starch digestibility and expected glycemic index of kodo millet (Paspalum scrobiculatum) as affected by starch-protein-lipid interactions. *Cereal Chemistry, 90*(3), 211–217. https://doi.org/10.1094/CCHEM-06-12 0074-R

Annor, G. A., Marcone, M., Bertoft, E., & Seetharaman, K. (2014). Unit and internal chain profile of millet amylopectin. *Cereal Chemistry, 91*(1), 29–34. https://doi.org/10.1094/CCHEM-08-13-0156-R

Atwell, W. A., Patrick, B. M., Johnson, L. A., & Glass, R. W. (1983). Characterization of quinoa starch. *Cereal Chemistry, 60*, 9–11.

Barba de la Rosa, A. P., Fomsgaard, I. S., Laursen, B., Mortensen, A. G., Olvera-Martínez, L., Silva-Sánchez, C., Mendoza-Herrera, A., González-Castañeda, J., & De León-Rodríguez, A. (2009). Amaranth (*Amaranthus hypochondriacus*) as an alternative crop for sustainable food production: Phenolic acids and flavonoids with potential impact on its nutraceutical quality. *Journal of Cereal Science, 49*(1), 117–121. https://doi.org/10.1016/j.jcs.2008.07.012

Barrett, M. (2006). *Identification, sequencing, expression and evolutionary relationships of the 11s seed storage protein gene in Chenopodium quinoa Willd.* Thesis of Master of Science Department of Plant and Animal Sciences Brigham Young University.

Barrio, D. A., & Añón, M. C. (2010). Potential antitumor properties of a protein isolate obtained from the seeds of *Amaranthus mantegazzianus*. *European Journal of Nutrition, 49*(2), 73–82. https://doi.org/10.1007/s00394-009-0051-9

Bejosano, F. P., & Corke, H. (1998). Protein quality evaluation of Amaranthus wholemeal flours and protein concentrates. *Journal of the Science of Food and Agriculture, 76*(1), 100–106. https://doi.org/10.1002/(sici)1097-0010(199801)76:1<100::aid-jsfa931>3.3.co;2-2

Benmoussa, M., Suhendra, B., Aboubacar, A., & Hamaker, B. R. (2006). Distinctive sorghum starch granule morphologies appear to improve raw starch digestibility. *Starch/Staerke, 58*(2), 92–99. https://doi.org/10.1002/star.200400344

Bertoft, E., Piyachomkwan, K., Chatakanonda, P., & Sriroth, K. (2008). Internal unit chain composition in amylopectins. *Carbohydrate Polymers, 74*(3), 527–543. https://doi.org/10.1016/j.carbpol.2008.04.011

Beta, T., Corke, H., Rooney, L. W., & Taylor, J. R. N. (2001). Starch properties as affected by sorghum grain chemistry. *Journal of the Science of Food and Agriculture, 81*(2), 245–251. https://doi.org/10.1002/1097-0010(20010115)81:2<245::AID-JSFA805>3.0.CO;2-S

Bhargava, A., Shukla, S., & Ohri, D. (2006). Chenopodium quinoa - an Indian perspective. *Industrial Crops and Products, 23*(1), 73–87. https://doi.org/10.1016/j.indcrop.2005.04.002

Bhinder, S., Kaur, A., Singh, B., Yadav, M. P., & Singh, N. (2020). Proximate composition, amino acid profile, pasting and process characteristics of flour from different Tartary buckwheat varieties. *Food Research International, 130*, 108946. https://doi.org/10.1016/j.foodres.2019.108946

Bojanic, A. (2011). Quinoa: An ancient crop to contribute to world food security. *PRO, 58*. http://www.fao.org/docrep/017/aq287e/aq287e.pdf

Brinegar, C., & Goundan, S. (1993). Isolation and characterization of Chenopodin, the 11S seed storage protein of quinoa (Chenopodium quinoa). *Journal of Agricultural and Food Chemistry, 41*(2), 182–185. https://doi.org/10.1021/jf00026a006

Brinegar, C., Sine, B., & Nwokocha, L. (1996). High-cysteine 2S seed storage proteins from quinoa (Chenopodium quinoa). *Journal of Agricultural and Food Chemistry, 44*(7), 1621–1623. https://doi.org/10.1021/jf950830+

Cai, C., & Li. (2004). Buckwheat. In C. W. Wrigley, H. Corke, & C. E. Walker (Eds.), *Encyclopedia of grain science* (pp. 120–128). Elsevier.

Capraro, J., Magni, C., Giogri, A., Duranti, M., & Scarafoni, A. (2018). Comparative 1D and 2D-electrophoretic protein profiles of ancestral and modern buckwheat seeds grown in the Italian alpine region. *Italian Journal of Food Science, 30*, 497–503.

Capriles, V. D., Coelho, K. D., Guerra-Matias, A. C., & Arêas, J. A. G. (2008). Effects of processing methods on amaranth starch digestibility and predicted glycemic index. *Journal of Food Science, 73*(7). https://doi.org/10.1111/j.1750-3841.2008.00869.x

Chen, X. W., Luo, D. Y., Chen, Y. J., Wang, J. M., Guo, J., & Yang, X. Q. (2019). Dry fractionation of surface abrasion for polyphenol-enriched buckwheat protein combined with hydrothermal treatment. *Food Chemistry, 285*, 414–422. https://doi.org/10.1016/j.foodchem.2019.01.182

Chivenge, P., Mabhaudhi, T., Modi, A. T., & Mafongoya, P. (2015). The potential role of neglected and underutilised crop species as future crops under water scarce conditions in sub-Saharan Africa. *International Journal of Environmental Research and Public Health, 12*(6), 5685–5711. https://doi.org/10.3390/ijerph120605685

Cho, J., Lee, J. O., Choi, J., Park, M. R., Shon, D. H., Kim, J., Ahn, K., & Han, Y. (2014). Significance of 40-, 45-, and 48-kDa proteins in the moderate-to-severe clinical symptoms of buckwheat allergy. *Allergy, Asthma and Immunology Research, 7*(1), 37–43. https://doi.org/10.4168/aair.2015.7.1.37

Collar, C. (2017). Significance of heat-moisture treatment conditions on the pasting and gelling behaviour of various starch-rich cereal and pseudocereal flours. *Food Science and Technology International, 23*(7), 623–636. https://doi.org/10.1177/1082013217714671

Comino, I., de Moreno, M. L., Real, A., Rodríguez-Herrera, A., Barro, F., & Sousa, C. (2013). The gluten-free diet: Testing alternative cereals tolerated by celiac patients. *Nutrients, 5*(10), 4250–4268. https://doi.org/10.3390/nu5104250

Correia, I., Nunes, A., Barros, A. S., & Delgadillo, I. (2010). Comparison of the effects induced by different processing methods on sorghum proteins. *Journal of Cereal Science, 51*(1), 146–151. https://doi.org/10.1016/j.jcs.2009.11.005

Deng, Y., Padilla-Zakour, O., Zhao, Y., & Tao, S. (2015). Influences of high hydrostatic pressure, microwave heating, and boiling on chemical compositions, Antinutritional factors, fatty acids, in vitro protein digestibility, and microstructure of buckwheat. *Food and Bioprocess Technology, 8*(11). https://doi.org/10.1007/s11947-015-1578-9

Dhull, S. B., Bangar, S. P., Deswal, R., Dhandhi, P., Kumar, M., Trif, M., & Rusu, A. (2021). Development and characterization of active native and cross-linked pearl millet starch-based film loaded with fenugreek oil. *Food, 10*(12), 3097.

Duodu, K. G., Tang, H., Grant, A., Wellner, N., Belton, P. S., & Taylor, J. R. N. (2001). FTIR and solid state 13C NMR spectroscopy of proteins of wet cooked and popped sorghum and maize. *Journal of Cereal Science, 33*(3), 261–269. https://doi.org/10.1006/jcrs.2000.0352

Escudero, N. L., Zirulnik, F., Gomez, N. N., Mucciarelli, S. I., & Giménez, M. S. (2006). Influence of a protein concentrate from Amaranthus cruentus seeds on lipid metabolism. *Experimental Biology and Medicine, 231*(1), 50–59. https://doi.org/10.1177/153537020623100106

Farrell, J. R., & Kelly, P. C. (2002). Coeliac Sprue. *New England Journal of Medicine, 346*(3), 180–188.

Fasano, A., & Catassi, C. (2001). Current approaches to diagnosis and treatment of celiac disease: An evolving spectrum. *Gastroenterology, 120*(3), 636–651. https://doi.org/10.1053/gast.2001.22123

Filho, A. M. M., Pirozi, M. R., Borges, J. T. D. S., Pinheiro Sant'Ana, H. M., Chaves, J. B. P., & Coimbra, J. S. D. R. (2017). Quinoa: Nutritional, functional, and antinutritional aspects. *Critical Reviews in Food Science and Nutrition, 57*(8), 1618–1630. https://doi.org/10.1080/10408398.2014.1001811

Fujita, S., & Fujiyama, G. (1993). The study of melting temperature and enthalpy of Starch from Rice, barley, wheat, foxtail- and Proso-millets. *Starch - Stärke, 45*(12), 436–441. https://doi.org/10.1002/star.19930451207

Gaffa, T., Yoshimoto, Y., Hanashiro, I., Honda, O., Kawasaki, S., & Takeda, Y. (2004). Physicochemical properties and molecular structures of starches from millet (Pennisetum typhoides) and sorghum (Sorghum bicolor L. Moench) cultivars in Nigeria. *Cereal Chemistry, 81*(2), 255–260. https://doi.org/10.1094/CCHEM.2004.81.2.255

Gálová, Z., Palenčárová, E., Špaleková, A., Rajnincová, D., & Drábeková, J. (2019). Protein profiles of buckwheat, rye and oat during in vitro gastro-duodenal digestion. *Journal of Food and Nutrition Research, 58*(3), 266–274.

Gamel, T. H., Linssen, J. P., Mesallem, A. S., Damir, A. A. & Shekib, L. A. (2005). Effect of seed treatments on the chemical composition and properties of two amaranth species: starch and protein. *Journal of the Science of Food and Agriculture, 85*, 319–327.

Gao, J., Kreft, I., Chao, G., Wang, Y., Liu, X., Wang, L., Wang, P., Gao, X., & Feng, B. (2016). Tartary buckwheat (Fagopyrum tataricumGaertn.) starch, a side product in functional food production, as a potential source of retrograded starch. *Food Chemistry, 190*, 552–558. https://doi.org/10.1016/j.foodchem.2015.05.122

García Fillería, S. F., & Tironi, V. A. (2017). Prevention of in vitro oxidation of low density lipoproteins (LDL) by amaranth peptides released by gastrointestinal digestion. *Journal of Functional Foods, 34*, 197–206. https://doi.org/10.1016/j.jff.2017.04.032

Gorinstein, S., Moshe, R., Greene, L. J., & Arruda, P. (1991). Evaluation of four Amaranthus species through protein electrophoretic patterns and their amino acid composition. *Journal of Agricultural and Food Chemistry, 39*(5), 851–854. https://doi.org/10.1021/jf00005a006

Gregori, M., & Kreft, I. (2012). Breakable starch granules in a low-amylose buckwheat (Fagopyrum esculentum Moench) mutant. *Journal of Food, Agriculture and Environment, 10*(2), 258–262.

Gross, R., Koch, F., Malaga, I., de Miranda, A. F., Schoeneberger, H., & Trugo, L. C. (1989). Chemical composition and protein quality of some local Andean food sources. *Food Chemistry, 34*(1), 25–34. https://doi.org/10.1016/0308-8146(89)90030-7

Guenard, R. (2021). FAO 2020. *INFORM, 32*, 6.

Gujral, H. S., Guardiola, I., Carbonell, J. V., & Rosell, C. M. (2003). Effect of cyclodextrinase on dough rheology and bread quality from rice flour. *Journal of Agricultural and Food Chemistry, 51*(13), 3814–3818. https://doi.org/10.1021/jf034112w

Guo, X. D., Zhang, D. Y., Gao, X. J., Parry, J., Liu, K., Liu, B. L., & Wang, M. (2013). Quercetin and quercetin-3-O-glucuronide are equally effective in ameliorating endothelial insulin resistance through inhibition of reactive oxygen species-associated inflammation. *Molecular Nutrition and Food Research, 57*(6), 1037–1045. https://doi.org/10.1002/mnfr.201200569

Hager, A. S., Wolter, A., Jacob, F., Zannini, E., & Arendt, E. K. (2012). Nutritional properties and ultra-structure of commercial gluten free flours from different botanical sources compared to wheat flours. *Journal of Cereal Science, 56*(2), 239–247. https://doi.org/10.1016/j.jcs.2012.06.005

Hamaker, B. R., Kirleis, A. W., Butler, L. G., Axtell, J. D., & Mertz, E. T. (1987). Improving the in vitro protein digestibility of sorghum with reducing agents. *Proceedings of the National Academy of Sciences, 84*(3), 626–628. https://doi.org/10.1073/pnas.84.3.626

Han, X. Z., & Hamaker, B. R. (2001). Amylopectin fine structure and rice starch paste breakdown. *Journal of Cereal Science, 34*(3), 279–284. https://doi.org/10.1006/jcrs.2001.0374

Hanashiro, I., Matsugasako, J., Egashira, T., & Takeda, Y. (2005). Structural characterization of long unit-chains of amylopectin. *Journal of Applied Glycoscience, 52*(3), 233–237. https://doi.org/10.5458/jag.52.233

Hariprasanna, K., Gomashe, S., Ganapathy, K. N., & Patil, J. V. (2014). *Popular Kheti, 3*(3), 170–175.

Hibberd, C. A., Wagner, D. G., Schemm, R. L., Mitchell, E. D., Weibel, D. E., & Hintz, R. L. (1982). Digestibility characteristics of isolated starch from sorghum and corn grain. *Journal of Animal Science, 55*(6), 1490–1497. https://doi.org/10.2527/jas1982.5561490x

Hoover, R. (1995). Starch retrogradation. *Food Reviews International, 11*(2), 331–346. https://doi.org/10.1080/87559129509541044

Hoover, R., Sinnott, A. W., & Perera, C. (1998). Physicochemical characterization of starches from Amaranthus cruentus grains. *Starch/Staerke, 50*(11–12), 456–463. https://doi.org/10.1002/(SICI)1521-379X(199812)50:11/12<456::AID-STAR456>3.0.CO;2-4

Hoseney, R. (1994). *Principles of cereal science and technology* (2nd ed.). American Association of Cereal Chemistry.

Huber, K. C., & Bemiller, J. N. (2000). Channels of maize and sorghum starch granules. *Carbohydrate Polymers, 41*(3), 269–276. https://doi.org/10.1016/S0144-8617(99)00145-9

Hurusawa, Y., & Miyashita, S. (1965). Studies on the components of buckwheat. Part 4. Properties of the starch of buckwheat (4). *Eiyo To Shokuryo*, 17, 415e419. (In Japanese with English abstract).

Ikeda, K., Kishida, M., Kreft, I., & Yasumoto, K. (1997). Endogenous factors responsible for the textural characteristics of buckwheat products. *Journal of Nutritional Science and Vitaminology, 43*(1), 101–111. https://doi.org/10.3177/jnsv.43.101

Inouchi, N., Nishi, K., Tanaka, S., Asai, M., Kawase, Y., Hata, Y., Konishi, Y., Yue, S., & Fuwa, H. (1999). Characterization of Amaranth and Quinoa starches. *Journal of Applied Glycoscience, 46*(3), 233–240. https://doi.org/10.5458/jag.46.233

Jancurová, M., Minarovičová, L., & Dandár, A. (2009). Quinoa - A review. *Czech Journal of Food Sciences, 27*(2), 71–79. https://doi.org/10.17221/32/2008-cjfs

Jane, J. L., Kasemsuwan, T., Leas, S., Zobel, H., & Robyt, J. F. (1994). Anthology of starch granule morphology by scanning electron microscopy. *Starch - Stärke, 46*(4), 121–129. https://doi.org/10.1002/star.19940460402

Jenkins, A. L. (2007). The glycemic index: Looking back 25 years. *Cereal Foods World, 52*(2), 50–53. https://doi.org/10.1094/CFW-52-1-0050

Jin, J., Okagu, O. D., Yagoub, A. E. G. A., & Udenigwe, C. C. (2021). Effects of sonication on the in vitro digestibility and structural properties of buckwheat protein isolates. *Ultrasonics Sonochemistry, 70*. https://doi.org/10.1016/j.ultsonch.2020.105348

Khan, I., Yousif, A., Johnson, S. K., & Gamlath, S. (2013). Effect of sorghum flour addition on resistant starch content, phenolic profile and antioxidant capacity of durum wheat pasta. *Food Research International, 54*(1), 578–586. https://doi.org/10.1016/j.foodres.2013.07.059

Khoddami, A., Messina, V., Vadabalija Venkata, K., Farahnaky, A., Blanchard, C. L., & Roberts, T. H. (2021). Sorghum in foods: Functionality and potential in innovative products. *Critical Reviews in Food Science and Nutrition*. https://doi.org/10.1080/10408398.2021.1960793

Lee, C., Lee, W., Han, Y., & Oh, S. (2017). Effect of proteolysis with alkaline protease following high hydrostatic pressure treatment on IgE binding of buckwheat protein. *Journal of Food Science, 82*(3), 834–839. https://doi.org/10.1111/1750-3841.13627

Li, S. Q., & Howard Zhang, Q. (2001). Advances in the development of functional foods from buckwheat. *Critical Reviews in Food Science and Nutrition, 41*(6), 451–464. https://doi.org/10.1080/20014091091887

Li, G., & Zhu, F. (2017a). Amylopectin molecular structure in relation to physicochemical properties of quinoa starch. *Carbohydrate Polymers, 164*, 396–402. https://doi.org/10.1016/j.carbpol.2017.02.014

Li, G., & Zhu, F. (2017b). Molecular structure of quinoa starch. *Carbohydrate Polymers, 158*, 124–132.

Li, W., Lin, R., & Corke, H. (1997). Physicochemical properties of common and tartary buckwheat starch. *Cereal Chemistry, 74*(1), 79–82. https://doi.org/10.1094/CCHEM.1997.74.1.79

Li, W., Cao, F., Fan, J., Ouyang, S., Luo, Q., Zheng, J., & Zhang, G. (2014). Physically modified common buckwheat starch and their physicochemical and structural properties. *Food Hydrocolloids, 40*, 237–244. https://doi.org/10.1016/j.foodhyd.2014.03.012

Li, G., Wang, S., & Zhu, F. (2016). Physicochemical properties of quinoa starch. *Carbohydrate Polymers, 137*, 328–338. https://doi.org/10.1016/j.carbpol.2015.10.064

Lindeboom, N., Chang, P. R., & Tyler, R. T. (2004). Analytical, biochemical and physicochemical aspects of starch granule size, with emphasis on small granule starches: A review. *Starch/Stärke, 56*(3–4), 89–99. https://doi.org/10.1002/star.200300218

Liu, G., Li, M., Zhu, Q., Li, Y., & Shui, S. (2006). The research advance on resource plant *Fagopyrum dibotrys. Chinese Agricultural Science Bulletin, 22*, 380–389. (In Chinese with English abstract).

Liu, C., Liu, P., Yan, S., Qing, Z., & Shen, Q. (2011). Relationship of physicochemical, pasting properties of millet starches and the texture properties of cooked millet. *Journal of Texture Studies, 42*(4), 247–253. https://doi.org/10.1111/j.1745-4603.2010.00271.x

Liu, H., Guo, X., Li, W., Wang, X., Lv, M., Peng, Q., & Wang, M. (2015a). Changes in physicochemical properties and in vitro digestibility of common buckwheat starch by heat-moisture treatment and annealing. *Carbohydrate Polymers, 132*, 237–244. https://doi.org/10.1016/j.carbpol.2015.06.071

Liu, H., Lv, M., Peng, Q., Shan, F., & Wang, M. (2015b). Physicochemical and textural properties of tartary buckwheat starch after heat-moisture treatment at different moisture levels. *Starch/Stärke, 67*(3–4), 276–284. https://doi.org/10.1002/star.201400143

Lorenz, K. (1990). Quinoa (Chenopodium quinoa) starch — Physico-chemical properties and functional characteristics. *Starch - Stärke, 42*(3), 81–86. https://doi.org/10.1002/star.19900420302

Lorenz, K., & Dilsaver, W. (1982). Buckwheat (Fagopyrum esculentum) starch – Physico-chemical properties and functional characteristics. *Starch - Stärke, 34*(7), 217–220. https://doi.org/10.1002/star.19820340702

Lu, L., & Baik, B. K. (2015). Starch characteristics influencing resistant starch content of cooked buckwheat groats. *Cereal Chemistry, 92*(1), 65–72. https://doi.org/10.1094/CCHEM-04-14-0062-R

Madhusudhan, B., & Tharanathan, R. N. (1996). Structural studies of linear and branched fractions of chickpea and finger millet starches. *Carbohydrate Research, 284*(1), 101–109. https://doi.org/10.1016/0008-6215(95)00404-1

Malleshi, N. G., Desikachar, H. S. R., & Tharanathan, R. N. (1986). Physico-chemical properties of native and malted finger millet, pearl millet and foxtail millet starches. *Starch - Stärke, 38*(6), 202–205. https://doi.org/10.1002/star.19860380608

Martínez, E. N., Castellani, O. F., & Añón, M. C. (1997). Common molecular features among Amaranth storage proteins. *Journal of Agricultural and Food Chemistry, 45*(10), 3832–3839. https://doi.org/10.1021/jf9700384

Miyazaki, A., Watanabe, S., Ogata, K., Nagatomi, Y., Kokutani, R., Minegishi, Y., Tamehiro, N., Sakai, S., Adachi, R., & Hirao, T. (2019). Real-time PCR detection methods for food allergens (wheat, buckwheat, and peanuts) using reference plasmids. *Journal of Agricultural and Food Chemistry, 67*(19), 5680–5686. https://doi.org/10.1021/acs.jafc.9b01234

Mora, L., Aristoy, M. C., & Toldrá, F. (2019). Bioactive peptides. In L. Melton, F. Shahidi, & P. Varelis (Eds.), *Encyclopedia of food chemistry* (pp. 381–389). Elsevier. https://doi.org/10.1016/B978-0-08-100596-5.22397-4

Motta, C., Castanheira, I., Gonzales, G. B., Delgado, I., Torres, D., Santos, M., & Matos, A. S. (2019). Impact of cooking methods and malting on amino acids content in amaranth, buckwheat and quinoa. *Journal of Food Composition and Analysis, 76*, 58–65. https://doi.org/10.1016/j.jfca.2018.10.001

Murray, J. A. (1999). The widening spectrum of celiac disease. *American Journal of Clinical Nutrition, 69*(3), 354–365. https://doi.org/10.1093/ajcn/69.3.354

Nagai, H. (2017). Development of a method for determination of buckwheat allergens using liquid chromatography with tandem mass spectrometry. *Journal of AOAC International, 100*(4), 1051–1057. https://doi.org/10.5740/jaoacint.16-0243

Noda, T., Takahata, Y., Sato, T., Suda, I., Morishita, T., Ishiguro, K., & Yamakawa, O. (1998). Relationships between chain length distribution of amylopectin and gelatinization properties within the same botanical origin for sweet potato and buckwheat. *Carbohydrate Polymers, 37*(2), 153–158. https://doi.org/10.1016/S0144-8617(98)00047-2

Okuno, K., & Sakaguchi, S. (1981). Glutinous and non-glutinous starches in perisperm of grain amaranths. *Cereal Research Communications, 9*(4), 305–310.

Olayinka, O. O., Adebowale, K. O., & Olu-Owolabi, B. I. (2008). Effect of heat-moisture treatment on physicochemical properties of white sorghum starch. *Food Hydrocolloids, 22*(2), 225–230. https://doi.org/10.1016/j.foodhyd.2006.11.004

Oom, A., Pettersson, A., Taylor, J. R. N., & Stading, M. (2008). Rheological properties of kafirin and zein prolamins. *Journal of Cereal Science, 47*(1), 109–116. https://doi.org/10.1016/j.jcs.2007.02.005

Orsini Delgado, M. C., Nardo, A., Pavlovic, M., Rogniaux, H., Añón, M. C., & Tironi, V. A. (2016). Identification and characterization of antioxidant peptides obtained by gastrointestinal digestion of amaranth proteins. *Food Chemistry, 197*, 1160–1167. https://doi.org/10.1016/j.foodchem.2015.11.092

Osborne, T. E. (1924). *The Vegetable Proteins* (2da ed.). Longmans y Green.

Paredes-Lopez, O. (2020). Biotechnology for the improvement of nutritional quality of food crop plants nutritionally important components in Plants. In *Molecular biotechnology for plant food production* (pp. 569–636). https://doi.org/10.1201/9781482293883-21

Parker, M. L., Grant, A., Rigby, N. M., Belton, P. S., & Taylor, J. R. N. (1999). Effects of popping on the endosperm cell walls of sorghum and maize. *Journal of Cereal Science, 30*(3), 209–216. https://doi.org/10.1006/jcrs.1999.0281

Pedersen, B., Knudsen, K. E. B., & Eggum, B. O. (1990). The nutritive value of amaranth grain (Amaranthus caudatus) - 3. Energy and fibre of raw and processed grain. *Plant Foods for Human Nutrition, 40*(1), 61–71. https://doi.org/10.1007/BF02193780

Perera, C., & Hoover, R. (1999). Influence of hydroxypropylation on retrogradation properties of native, defatted and heat-moisture treated potato starches. *Food Chemistry, 64*(3), 361–375. https://doi.org/10.1016/S0308-8146(98)00130-7

Pérez, S., & Bertoft, E. (2010). The molecular structures of starch components and their contribution to the architecture of starch granules: A comprehensive review. *Starch/Staerke, 62*(8), 389–420. https://doi.org/10.1002/star.201000013

Praznik, W., Mundigler, N., Kogler, A., Pelzl, B., Huber, A., & Wollendorfer, M. (1999). Molecular background of technological properties of selected starches. *Starch - Stärke, 51*(6), 197–211. https://doi.org/10.1002/(sici)1521-379x(199906)51:6<197::aid-star197>3.0.co;2-k

Punia Bangar, S., Ashogbon, A. O., Dhull, S. B., Thirumdas, R., Kumar, M., Hasan, M., … Pathem, S. (2021a). Proso-millet starch: Properties, functionality, and applications. *International Journal of Biological Macromolecules, 190*, 960–968.

Punia Bangar, S., Kumar, M., & Whiteside, W. S. (2021b). Mango seed starch: A sustainable and eco-friendly alternative to increasing industrial requirements. *International Journal of Biological Macromolecules, 183*, 1807–1817.

Punia Bangar, S., Sandhu, K. S., Rusu, A. V., Kaur, P., Purewal, S. S., Kaur, M., … Trif, M. (2021c). Proso-millet-starch-based edible films: An innovative approach for food industries. *Coatings, 11*(10), 1167.

Punia Bangar, S., Siroha, A. K., Nehra, M., Trif, M., Ganwal, V., & Kumar, S. (2021d). Structural and film-forming properties of millet starches: A comparative study. *Coatings, 11*(8), 954.

Punia, S., Sandhu, K. S., Dhull, S. B., Siroha, A. K., Purewal, S. S., Kaur, M., & Kidwai, M. K. (2020). Oat starch: Physico-chemical, morphological, rheological characteristics and its applications-a review. *International Journal of Biological Macromolecules, 154*, 493–498.

Punia, S., Kumar, M., Siroha, A. K., Kennedy, J. F., Dhull, S. B., & Whiteside, W. S. (2021). Pearl millet grain as an emerging source of starch: A review on its structure, physicochemical properties, functionalization, and industrial applications. *Carbohydrate Polymers, 117776.*

Qian, J. Y., & Kuhn, M. (1999). Characterization of Amaranthus cruentus and Chenopodium quinoa starch. *Starch/Staerke, 51*(4), 116–120. https://doi.org/10.1002/(sici)1521-379x(199904)51:4<116::aid-star116>3.3.co;2-i

Qian, J., Rayas-Duarte, P., & Grant, L. (1998). Partial characterization of buckwheat (Fagopyrum esculentum) starch. *Cereal Chemistry, 75*(3), 365–373. https://doi.org/10.1094/CCHEM.1998.75.3.365

Ratnavathi, C. V., & P. J. (2014). Sorghum utilization as food. *Journal of Nutrition & Food Sciences, 04*(01). https://doi.org/10.4172/2155-9600.1000247

Repo-Carrasco, R., Espinoza, C., & Jacobsen, S. E. (2003). Nutritional value and use of the andean crops quinoa (*Chenopodium quinoa*) and kañiwa (*Chenopodium pallidicaule*). *Food Reviews International, 19*(1–2), 179–189. https://doi.org/10.1081/FRI-120018884

Rodríguez Gómez, M. J., Matías Prieto, J., Cruz Sobrado, V., & Calvo Magro, P. (2021). Nutritional characterization of six quinoa (Chenopodium quinoa Willd) varieties cultivated in Southern Europe. *Journal of Food Composition and Analysis, 99.* https://doi.org/10.1016/j.jfca.2021.103876

Roopa, S., & Premavalli, K. S. (2008). Effect of processing on starch fractions in different varieties of finger millet. *Food Chemistry, 106*(3), 875–882. https://doi.org/10.1016/j.foodchem.2006.08.035

Ruales, J., & Nair, B. M. (1993). Content of fat, vitamins and minerals in quinoa (Chenopodium quinoa, Willd) seeds. *Food Chemistry, 48*(2), 131–136. https://doi.org/10.1016/0308-8146(93)90047-J

Ruales, J., & Nair, B. M. (1994). Properties of starch and dietary fibre in raw and processed quinoa (Chenopodium quinoa, Willd) seeds. *Plant Foods for Human Nutrition, 45*(3), 223–246. https://doi.org/10.1007/BF01094092

Ruiz, K. B., Biondi, S., Oses, R., Acuña-Rodríguez, I. S., Antognoni, F., Martinez-Mosqueira, E. A., Coulibaly, A., Canahua-Murillo, A., Pinto, M., Zurita-Silva, A., Bazile, D., Jacobsen, S. E., & Molina-Montenegro, M. A. (2014). Quinoa biodiversity and sustainability for food security under climate change. A review. *Agronomy for Sustainable Development, 34*(2), 349–359. https://doi.org/10.1007/s13593-013-0195-0

Sabbione, A. C., Nardo, A. E., Añón, M. C., & Scilingo, A. (2016). Amaranth peptides with antithrombotic activity released by simulated gastrointestinal digestion. *Journal of Functional Foods, 20*, 204–214. https://doi.org/10.1016/j.jff.2015.10.015

Sanderson, J. S., Daniels, R. D., Donald, A. M., Blennow, A., & Engelsen, S. B. (2006). Exploratory SAXS and HPAEC-PAD studies of starches from diverse plant genotypes. *Carbohydrate Polymers, 64*(3), 433–443. https://doi.org/10.1016/j.carbpol.2005.12.026

Sandhu, K. S., & Singh, N. (2005). Relationships between selected properties of starches from different corn lines. *International Journal of Food Properties, 8*(3), 481–491. https://doi.org/10.1080/10942910500267711

Sandhu, K. S., Siroha, A. K., Punia, S., Sangwan, L., Nehra, M., & Purewal, S. S. (2021). Effect of degree of cross linking on physicochemical, rheological and morphological properties of sorghum starch. *Carbohydrate Polymer Technologies and Applications, 2*, 100073.

Sang, Y., Bean, S., Seib, P. A., Pedersen, J., & Shi, Y. C. (2008). Structure and functional properties of sorghum starches differing in amylose content. *Journal of Agricultural and Food Chemistry, 56*(15), 6680–6685. https://doi.org/10.1021/jf800577x

Sano, M., Nakagawa, M., Oishi, A., Yasui, Y., & Katsube-Tanaka, T. (2014). Diversification of 13S globulins, allergenic seed storage proteins, of common buckwheat. *Food Chemistry, 155*, 192–198. https://doi.org/10.1016/j.foodchem.2014.01.047

Satoh, R., Jensen-Jarolim, E., & Teshima, R. (2020). Understanding buckwheat allergies for the management of allergic reactions in humans and animals. *Breeding Science, 70*(1), 85–92. https://doi.org/10.1270/jsbbs.19051

Satou, T., Horino, S., Nihei, M., & Miura, K. (2019). Food protein-induced enterocolitis syndrome caused by buckwheat. *Pediatrics International, 61*(10), 1058–1059. https://doi.org/10.1111/ped.13952

Schober, T. J., Bean, S. R., Tilley, M., Smith, B. M., & Ioerger, B. P. (2011). Impact of different isolation procedures on the functionality of zein and kafirin. *Journal of Cereal Science, 54*(2), 241–249. https://doi.org/10.1016/j.jcs.2011.06.007

Selle, P. H., McInerney, B. V., McQuade, L. R., Khoddami, A., Chrystal, P. V., Hughes, R. J., & Liu, S. Y. (2020). Composition and characterisation of kafirin, the dominant protein fraction in grain sorghum. *Animal Production Science, 60*(9), 1163–1172. https://doi.org/10.1071/AN19393

Serna-Saldivar, S. O., & Rooney, L. W. (1995). Structure and chemistry of sorghum and millets. In D. A. V. Dendy (Ed.), *Sorghum and millets: Chemistry and technology* (pp. 69–124). American Association of Cereal Chemists.

Shevkani, K., & Singh, N. (2014). Influence of kidney bean, field pea and amaranth protein isolates on the characteristics of starch-based gluten-free muffins. *International Journal of Food Science and Technology, 49*(10), 2237–2244. https://doi.org/10.1111/ijfs.12537

Shi, Y. C., & Seib, P. A. (1992). The structure of four waxy starches related to gelatinization and retrogradation. *Carbohydrate Research, 227*(C), 131–145. https://doi.org/10.1016/0008-6215(92)85066-9

Sikabbubba, R. (1989). *The effect of alcohol soluble proteins on the digestibility of sorghum., Etats-unis*. Kansas State University.

Singer, N. S. (1994). Carbohydrate cream substitute. US patent no. 5370894a.

Singh, S., Singh, N., Isono, N., & Noda, T. (2010a). Relationship of granule size distribution and amylopectin structure with pasting, thermal, and retrogradation properties in wheat starch. *Journal of Agricultural and Food Chemistry, 58*(2), 1180–1188. https://doi.org/10.1021/jf902753f

Singh, H., Sodhi, N. S., & Singh, N. (2010b). Characterisation of starches separated from sorghum cultivars grown in India. *Food Chemistry, 119*(1), 95–100. https://doi.org/10.1016/j.foodchem.2009.05.086

Singh, N., Kaur, S., Kaur, A., Isono, N., Ichihashi, Y., Noda, T., & Rana, J. C. (2014). Structural, thermal, and rheological properties of Amaranthus hypochondriacus and Amaranthus caudatus starches. *Starch/Staerke, 66*(5–6), 457–467. https://doi.org/10.1002/star.201300157

Siroha, A. K., Punia Bangar, S., Sandhu, K. S., Trif, M., Kumar, M., & Guleria, P. (2021). Effect of cross-linking modification on structural and film-forming characteristics of pearl millet (Pennisetum glaucum L.) starch. *Coatings, 11*(10), 1163.

Srichuwong, S., Isono, N., Jiang, H., Mishima, T., & Hisamatsu, M. (2012). Freeze-thaw stability of starches from different botanical sources: Correlation with structural features. *Carbohydrate Polymers, 87*(2), 1275–1279. https://doi.org/10.1016/j.carbpol.2011.09.004

Srichuwong, S., Curti, D., Austin, S., King, R., Lamothe, L., & Gloria-Hernandez, H. (2017). Physicochemical properties and starch digestibility of whole grain sorghums, millet, quinoa and amaranth flours, as affected by starch and non-starch constituents. *Food Chemistry, 233*, 1–10. https://doi.org/10.1016/j.foodchem.2017.04.019

Steffolani, M. E., León, A. E., & Pérez, G. T. (2013). Study of the physicochemical and functional characterization of quinoa and kañiwa starches. *Starch/Staerke, 65*(11–12), 976–983. https://doi.org/10.1002/star.201200286

Subrahmanyam, S., & Hoseney, R. (1995). Shear thinning properties of sorghum starch. *Cereal Chemistry, 72*(1), 7–10.

Sytar, O., Brestic, M., Zivcak, M., & Phan Tran, L.-S. (2016). The contribution of buckwheat genetic resources to health and dietary diversity. *Current Genomics, 17*(3), 193–206. https://doi.org/10.2174/1389202917666160202215425

Sytar, O., Chrenková, M., Ferencová, J., Polačiková, M., Rajský, M., & Brestič, M. (2018). Nutrient capacity of amino acids from buckwheat seeds and sprouts. *Journal of Food and Nutrition Research, 57*(1), 38–47.

Tang, H., Watanabe, K., & Mitsunaga, T. (2002). Characterization of storage starches from quinoa, barley and adzuki seeds. *Carbohydrate Polymers, 49*(1), 13–22. https://doi.org/10.1016/S0144-8617(01)00292-2

Taylor, J. R. N., & Dewar, J. (2001). Developments in sorghum food technologies. *Advances in Food and Nutrition Research, 43*, 217–264. https://doi.org/10.1016/S1043-4526(01)43006-3

Taylor, J. R. N., & Schüssler, L. (1986). The protein compositions of the different anatomical parts of sorghum grain. *Journal of Cereal Science, 4*(4), 361–369. https://doi.org/10.1016/S0733-5210(86)80040-6

Taylor, J. R. N., & Taylor, J. (2017). Proteins from sorghum and millets. In *Sustainable protein sources* (pp. 79–104). https://doi.org/10.1016/B978-0-12-802778-3.00005-6

Timgren, A., Rayner, M., Sjöö, M., & Dejmek, P. (2011). Starch particles for food based Pickering emulsions. *Procedia Food Science, 1*, 95–103. https://doi.org/10.1016/j.profoo.2011.09.016

Vader, W., Kooy, Y., Van Veelen, P., De Ru, A., Harris, D., Benckhuijsen, W., Pea, S., Mearin, L., Drijfhout, J. W., & Koning, F. (2002). The gluten response in children with celiac disease is directed toward multiple gliadin and glutenin peptides. *Gastroenterology, 122*(7), 1729–1737. https://doi.org/10.1053/gast.2002.33606

Vallons, K. J. R., & Arendt, E. K. (2009). Effects of high pressure and temperature on buckwheat starch characteristics. *European Food Research and Technology, 230*(2), 343–351. https://doi.org/10.1007/s00217-009-1172-3

Van de Vondel, J., Lambrecht, M. A., Housmans, J. A. J., Rousseau, F., Schymkowitz, J., & Delcour, J. A. (2021). Impact of hydrothermal treatment on denaturation and aggregation of water-extractable quinoa (Chenopodium quinoa Willd.) protein. *Food Hydrocolloids, 115*. https://doi.org/10.1016/j.foodhyd.2021.106611

Vega-Gálvez, A., Miranda, M., Vergara, J., Uribe, E., Puente, L., & Martínez, E. A. (2010). Nutrition facts and functional potential of quinoa (*Chenopodium quinoa*willd.), an ancient Andean grain: A review. *Journal of the Science of Food and Agriculture, 90*(15), 2541–2547. https://doi.org/10.1002/jsfa.4158

Virupaksha, T. K., & Sastry, L. V. S. (1968). Studies on the protein content and amino acid composition of some varieties of grain sorghum. *Journal of Agricultural and Food Chemistry, 16*(2), 199–203. https://doi.org/10.1021/jf60156a022

Virdi, A. S., & Singh, N. (2020). Textural characteristics of Indian foods. In *Textural characteristics of world foods* (pp. 197–222). https://doi.org/10.1002/9781119430902.ch14

Wang, F., Yu, G., Zhang, Y., Zhang, B., & Fan, J. (2015). Dipeptidyl peptidase IV inhibitory peptides derived from oat (Avena sativa L.), buckwheat (Fagopyrum esculentum), and Highland barley (Hordeum vulgare trifurcatum (L.) Trofim) proteins. *Journal of Agricultural and Food Chemistry, 63*(43), 9543–9549. https://doi.org/10.1021/acs.jafc.5b04016

Wankhede, D. B., Shehnaz, A., & Raghavendra Rao, M. R. (1979). Preparation and physicochemical properties of starches and their fractions from finger millet (Eleusine coracana) and foxtail millet (Setariaitalica). *Starch - Stärke, 31*(5), 153–159. https://doi.org/10.1002/star.19790310505

Watanabe, K. (2008). Characterization of quinoa starch and protein. *Foods Food Ingredients, 213*(6), 1.

Watanabe, K., Peng, N. L., Tang, H., & Mitsunaga, T. (2007). Molecular structural characteristics of quinoa starch. *Food Science and Technology Research, 13*(1), 73–76. https://doi.org/10.3136/fstr.13.73

Wijngaard, H. H., & Arendt, E. K. (2006). Buckwheat. *Cereal Chemistry, 83*, 391–401. https://doi.org/10.1094/CC-83-0391

Wilhelm, E., Aberle, T., Burchard, W., & Landers, R. (2002). Peculiarities of aqueous amaranth starch suspensions. *Biomacromolecules, 3*(1), 17–26. https://doi.org/10.1021/bm000138+

Wright, K. H., Huber, K. C., Fairbanks, D. J., & Huber, C. S. (2002). Isolation and characterization of Atriplex hortensis and sweet Chenopodium quinoa starches. *Cereal Chemistry, 79*(5), 715–719. https://doi.org/10.1094/CCHEM.2002.79.5.715

Xu, J. H., & Messing, J. (2008). Organization of the prolamin gene family provides insight into the evolution of the maize genome and gene duplications in grass species. *Proceedings of the National Academy of Sciences of the United States of America, 105*(38), 14330–14335. https://doi.org/10.1073/pnas.0807026105

Yañez, G. A., Walker, C. E., & Nelson, L. A. (1991). Some chemical and physical properties of proso millet (Panicum milliaceum) starch. *Journal of Cereal Science, 13*(3), 299–305. https://doi.org/10.1016/S0733-5210(09)80008-8

Yoshimoto, Y., Egashira, T., Hanashiro, I., Ohinata, H., Takase, Y., & Takeda, Y. (2004). Molecular structure and some physicochemical properties of buckwheat starches. *Cereal Chemistry, 81*(4), 515–520. https://doi.org/10.1094/CCHEM.2004.81.4.515

Yousaf, L., Hou, D., Liaqat, H., & Shen, Q. (2021). Millet: A review of its nutritional and functional changes during processing. *Food Research International, 142*. https://doi.org/10.1016/j.foodres.2021.110197

Yu, Y. J., Amorim, M., Marques, C., Calhau, C., & Pintado, M. (2016a). Effects of whey peptide extract on the growth of probiotics and gut microbiota. *Journal of Functional Foods, 21*, 507–516. https://doi.org/10.1016/j.jff.2015.10.035

Yu, G., Wang, F., Zhang, B., & Fan, J. (2016b). In vitro inhibition of platelet aggregation by peptides derived from oat (Avena sativa L.), highland barley (Hordeum vulgare Linn. Var. nudum Hook. F.), and buckwheat (Fagopyrum esculentum Moench) proteins. *Food Chemistry, 194*, 577–586. https://doi.org/10.1016/j.foodchem.2015.08.058

Zaika, Y., Kozub, N., Sozinov, I., Bidnyk, G., & Karazhbey, P. (2019). Polymorphism of buckwheat seed storage proteins in cultivar groups, differing by their morphotype. *Agricultural Science and Practice, 6*(1), 10–17. https://doi.org/10.15407/agrisp6.01.010

Zhang, C., Zhang, R., Li, Y. M., Liang, N., Zhao, Y., Zhu, H., He, Z., Liu, J., Hao, W., Jiao, R., Ma, K. Y., & Chen, Z. Y. (2017). Cholesterol-lowering activity of tartary buckwheat protein. *Journal of Agricultural and Food Chemistry, 65*(9), 1900–1906. https://doi.org/10.1021/acs.jafc.7b00066

Zheng, G. H., Sosulski, F. W., & Tyler, R. T. (1997). Wet-milling, composition and functional properties of starch and protein isolated from buckwheat groats. *Food Research International, 30*(7), 493–502. https://doi.org/10.1016/S0963-9969(98)00021-0

Zheng, B., Zhang, H., Wang, L., Guo, Y., & Chen, P. (2018). Characterization of 16-kDa major allergen with α-amylase inhibitor domain in tartary buckwheat seeds. *Molecular Immunology, 94*, 121–130. https://doi.org/10.1016/j.molimm.2017.12.024

Zhou, X. L., Yan, B. B., Xiao, Y., Zhou, Y. M., & Liu, T. Y. (2018). Tartary buckwheat protein prevented dyslipidemia in high-fat diet-fed mice associated with gut microbiota changes. *Food and Chemical Toxicology, 119*, 296–301. https://doi.org/10.1016/j.fct.2018.02.052

Zhu, F. (2014). Structure, physicochemical properties, and uses of millet starch. *Food Research International, 64*, 200–211. https://doi.org/10.1016/j.foodres.2014.06.026

Zhu, F. (2016). Buckwheat starch: Structures, properties, and applications. *Trends in Food Science and Technology, 49*, 121–135. https://doi.org/10.1016/j.tifs.2015.12.002

Zhu, F. (2017). Structures, physicochemical properties, and applications of amaranth starch. *Critical Reviews in Food Science and Nutrition, 57*(2), 313–325. https://doi.org/10.1080/10408398.2013.862784

Chapter 5
Functionality of Resistant and Slowly Digesting Starch in Cereals

Aderonke Ibidunni Olagunju and Olufunmilayo Sade Omoba

5.1 Introduction

Cereals belong to the Gramineae family, they are an important class of food as well a principal component of human's daily diet for centuries (Sandhu et al., 2016, 2017). Cereals are used as staple food in most countries and provide above 50% of the caloric intake of man; it also serves as a valuable ingredient in many food products including breakfast cereals, baked goods, pasta, beverages etc. (Carcea, 2020). This class of food represents an important source of nutrients especially carbohydrates, proteins, fibers, specific micronutrients (B-vitamins, vitamin E, minerals). Cereals of economic importance include wheat, rice, maize, oat, rye, barley, sorghum and millet (Awika, 2011; Punia et al., 2017).

Cereals are in form of edible seeds or grains; whole grains are made up of intact, ground or processed kernel of the grain plant and comprises the bran, germ and endosperm while the inedible parts (hull and husk) have been removed (Whole EU Grain, 2021). Whole grains are often a store house of several nutrients (starch, protein, fibre, vitamins and minerals) whereas, refined flours subjected to fractionation (milling) lose a valuable fraction of these components (the nutrition source).

Cereals are classified as carbohydrate-rich foods with an approximate carbohydrate content of 75% (McKevith, 2004; Sandhu & Punia, 2017) and carbohydrates are the major dietary components of cereals. In addition, the quality and type of carbohydrate plays an indisputable role in modulating postprandial glycemia in human nutrition (Devitt et al., 2013; Nutrition Science Corner, 2015). The major component of carbohydrate is starch, accounting for about 85% and comprising two polysaccharides: (i) amylose (ii) amylopectin (Punia, 2020a, b; Punia et al., 2020).

A. I. Olagunju (✉) · O. S. Omoba
Department of Food Science and Technology, Federal University of Technology, Akure, Ondo State, Nigeria
e-mail: aiolagunju@futa.edu.ng

S. Punia Bangar, A. Kumar Siroha (eds.), *Functional Cereals and Cereal Foods*,
https://doi.org/10.1007/978-3-031-05611-6_5

The starch polymers occur in varying ratios in different food materials. Amylose fraction is repeated glucose molecules linked by α-1,4 glycosidic bond whereas, amylopectin represents repeated units of glucose molecules joined by α-1,4 bond branched at α-1,6 (Punia Bangar et al., 2021a). Amylose is said to be difficult to disperse as it forms secondary structures thereby resists digestion thus, cereals with high amount of amylose are highly resistant to enzymatic digestion (Hu et al., 2004).

Cereals are not edible in its raw state, hence, require some form of processing ranging from milling to fermentation, heat processing (cooking), microwave treatment, autoclaving, extrusion etc. It is worthy to note that the food processing involving heat and moisture promote starch digestibility (Anguita et al., 2006).

Several factors influence the amount of starch fraction prevalent in processed food compared to those of the raw product, they include amylose/amylopectin ratio, botanical source of starch and food processing type (Alsaffar, 2011). These processes usually promote the digestibility of starch which is being hydrolyzed into simple sugars for optimum absorption in the body. Starch digestion involves enzymic (salivary α-amylase and pancreatic amylase) action on starch to hydrolyze it into glucose. The type of processing a food material is subjected to also determines the rate of starch digestion which invariably determines the glycemic effects. Singh et al. (2010) highlighted the role different processing technique plays on starch digestibility. For instance, cooking increases starch hydrolysis owing to starch gelatinization occurring during heating (in the presence of water; at high temperature) which promotes enzymatic attack and consequently breakdown of starch to glucose. This is corroborated by the findings by Kaur et al. (2018), cooking of starch slurry caused an increase in the rapidly digesting starch (RDS) of the cooked starch. Extrusion cooking increases starch hydrolysis and digestibility; shearing and kneading taking place during extrusion bring about loss of starch granule structural strength thereby, promoting proneness to enzymatic action. Dehulling, soaking, germination also increases starch digestibility because these processing methods reduce antinutrients (phytate, tannin) and polyphenols. These two broad components naturally impede activities of α-amylase. However, reduced tannin and phytate contents originate a sizeable void in the interior of the starch granules apparently fostering proness to enzyme action and thus, increasing starch breakdown.

Starch is grouped into three principal categories contingent to the rate and extent of enzymatic digestion (degree of digestibility) within the small bowel of the gastrointestinal tract. The groups are (i) rapidly digesting starch portion (RDS), (ii) slowly digesting starch portion (SDS), (iii) resistant starch fraction (RS); its relative proportions determine the postprandial glucose response of food (Englyst et al., 1992; Singh et al., 2013). Cereal grains are high in RS concentrations however, milling of whole grains into flour promotes amylolytic hydrolysis thus cereal flours contain greater fraction of RDS and SDS than RS (Tabibloghmany & Ehsandoost, 2014). It is worthy to note that cereal grains contain as high as five times RS concentration than cereal flour (Bednar et al., 2001).

The different classes of starch can be produced by subjecting cereal grain, or extracted starch to different modification processes namely physical, chemical, enzymatic, genetic modification etc. These modifications bring about physicochemical and techno functional changes to food.

5.2 Categories of Starch in Relation to Digestion

Carbohydrates undergo hydrolysis and assimilation in the human body at different rates; the rate is the utmost predictor of the nutritional properties of this important food class. The extent of starch digestibility in different foods differ to a great extent and is guided by elements like composition, starch source, structure of starch granule, type of processing and amylose/amylopectin ratio (Aller et al., 2011).

The presence of high amylose fraction than the amylopectin counterpart impedes starch digestibility. High amylose is negatively correlated to rapid digestion of starch but on the other hand positively correlated to slow or resisted digestion. Albeit, food processing involves moisture, heat and mechanical action which allows gelatinization of starch and thus, modify the starch digestibility of the processed products (Mishra & Monro, 2012). The different variables, food processing techniques applied to food products especially cereal products is responsible for the different starch digestibility profiles reported for different products. The major determinants of starch susceptiveness to enzymatic break down is the extent of starch gelatinization, this factor also determines the functional properties for food processing (Wang & Copeland, 2013).

Starch hydrolysis into simpler units such as glucose is an important process to aid proper assimilation and utilization by/into the body. Starch is hydrolyzed when amylolytic enzymes precisely salivary and pancreatic α-amylase act on the food material. Starch breakdown starts from the oral cavity (mouth) and it involves the salivary α-amylase acting on the starch-containing food during chewing or mastication, the food material moves to the stomach, which is a more acidic environment, hence inactivates the activities of the amylase enzyme (Singh et al., 2013).

5.2.1 Rapidly Digesting Starch (RDS)

RDS represents the starch fraction which is digested within the first twenty (20) minutes of enzymatic hydrolysis to release glucose. It refers to starch which rapidly digests in the upper small intestine, RDS ensues fast release of blood glucose. At the instance diets containing high quantities of RDS is consumed, the result is an instant upsurge in the levels of blood glucose shortly after ingesting and assimilating a carbohydrate-rich meal. Considering the contributions of RDS to elevated blood glucose levels, RDS therefore represents a leading cause of hyperglycemia. Moreso, it induces a sequence of health complexities, for example diabetes and other cardiovascular diseases (CVD).

5.2.2 Slowly Digesting Starch (SDS)

SDS offers slow and maintained digestion of starch, the period of enzymatic hydrolysis is within 20–120 min. The starch is digested into glucose in the small intestine at a slower progression as opposed to RDS, the glucose is slowly released into the blood, thereby, inducing low glycemic response. SDS is advantageous as it helps in the control and prevention of hyperglycemia-related diseases.

5.2.3 Resistant Starch (RS)

RS represents type of starch undigested following 120 min of enzymatic hydrolysis. RS escapes digestion in the small intestine and proceeds to the colon wherein the starch gets fermented by intestinal bacteria to produce short chain fatty acids (SCFAs). Thus, they are considered as fractions resistant to digestion. The resistant starch is further grouped into five employing the various reasons for enzyme resistance namely; structural characteristics (compactness of starch structure, starch granule configuration, starch gelatinization & cooling, chemical modification) and other intrinsic factors (amylose-lipid complexes, existence of α-amylase inhibitors) which one way or the other makes starch to resist enzymatic hydrolysis (Haralampu, 2000; Nugent, 2005; Lunn & Buttriss, 2007; Fuentes-Zaragoza et al., 2011).

- RS1: represents starch resistant to gastrointestinal tract conditions due to physical inaccessibility to hydrolytic enzymes owing to its entrapment within the food matrix. The presence of seed hulls in incompletely processed cereal is accountable for this property; interestingly, RS1 remains stable to majority of cooking techniques (Sajilata et al., 2006) e.g. RS in partially milled grains and seeds.
- RS2: is the type of resistant starch that remains compact inside the starch matrix thus, prevent enzyme penetration and chain breaking action. They are ungelatinized starches, native, uncooked granules; stable to most cooking operations and resist digestion owing to their compact structure e.g. RS in high amylose starch.
- RS3: refers to the RS formed when starch is subjected to gelatinization and cooling to allow for retrogradation, the high content of retrograded amylose confers on the starch a valid possibility of strong re-association. RS3 is known as retrograded starch.
- RS4: refers to chemically modified starch subjected to either esterification, acetylation, etherification, phosphorylation or cross linking. The modifications successfully block access to hydrolytic enzymes thereby, preventing starch digestion.
- RS5: refers to resistant starch formed as a result of amylose-lipid complex.

From Table 5.1 below, granular (R2) and retrograded (R3) forms from high amylose starches especially from maize are widely employed as RS ingredients in food manufacturing sector. This submission corroborates reports from other studies. Also worthy of mention is the fact that starch from cereals and legumes have higher RS

Table 5.1 Highlight of the characteristics of resistant starch and source

Type of RS	Characteristics	Conditions reducing resistance	Cereal source
RS1	This type is not physically accessible to amylolytic enzymes because the starch is entrapped in a non-digestible matrix	Milling, masticating	Whole grain, incompletely milled grains
RS2	Indigestible starch with crystalline; digestible upon gelatinization	Food processing, cooking	Starches having a high amylose content e.g. high amylose maize starch
RS3	Gelatinized and cooled starch (retrograded)	Food processing	Corn flakes, food products with prolonged and/or repeated moist heat treatment
RS4	Modified starches subjected to chemical modifications (starch acetates/esters/ethers, cross linked)	Poor susceptibility to invitro digestibility	Synthetic starch
RS5	Consist of at least two different components that form starch complex compounds e.g. amylose-lipid complex, maltodextrin-resistant	Amylose-lipid complex	Whole grain

Adapted from Fuentos-Zaragoza et al. (2010)

content than roots and tuber starches (Remya & Jyothi, 2015). In providing health benefits, present-day dietary guidelines are put in place, these guidelines promote the consumption of slowly digesting carbohydrates while starchy foods possess low RS averaging 14% RS (EFSA, 2011).

5.3 Fraction of Starch in Different Cereals

5.3.1 Wheat

Wheat (*Triticum aestivum* L.) is an important cereal crop in many parts of the world. It represents one of the major dietary sources of starch, accounting for approximately 60–70% of the whole wheat grain (Shewry, 2009). The wheat grain is utilized as whole meat or milled (removal of bran) into white flour, admittedly, the milling degree has dire impact on cereal starch digestibility (Alsaffar, 2011). There are several varieties of wheat developed from different genotypes. Recently, a study reported the rapidly, slowly and resistant starch contents of wholemeal wheat flour (from Australian amylose wheat) to be 34.4, 3.4 and 16.9% while the white flour contained 38.4, 9.2 and 6.6% respectively (Štêrbová et al., 2016). The RDS of the whole meal and white flour were not significantly different while the RS content of whole meal flour was almost three times that of refined flour.

5.3.2 Rice

Rice (*Oryza sativa* L.) is an important staple, above 50% of the present world population consumes rice (Polesi et al., 2017), it remains a principal source of carbohydrate in human diet mainly composed of starch (80–90%), a small fraction of protein (8–9%) and minute quantity of dietary fiber (Alhambra et al., 2019). Rice is an important part of human nutrition with starch as its major component. The RS content of rice varies depending on the genetic background or the variety of rice being evaluated. Shu et al. (2009) in their study reported a RS content of 3.3 to 11.7% for seven different rice mutants. Just recently, Tuaño et al. (2021), published on cooked brown rice having more RS content (0.24–1.61%) than cooked milled rice (0.15–0.99%) although, the values reported is lower than for other rice varieties. Whole grains such as brown rice have higher RS content than refined, polished or milled rice (Tuaño et al., 2021; Huang et al., 2021). The lower starch digestion behavior in rice is partially attributed to the high quantity of amylose in the grain moreso, starch digestion rate in cooked rice is a factor of the amylose content (critical component) and non-starch components namely fiber, protein, lipids, polyphenols (Huang et al., 2021).

Different processing techniques have been reported to promote RS content and decrease the RDS content. For instance, parboiling as a form of processing increased the RS content of three different varieties of rice (Darandakumbura et al., 2013). Another processing effect on rice starch fraction was published by Polesi et al. (2017), irradiation of cooked polished white rice at 1–5 KGy exhibited a declining trend of RDS amidst increase in SDS and RS contents.

5.3.3 Maize

Maize (*Zea mays* L.) is adjudged third most relevant cereal asides wheat and rice, representing a major raw material for both food and non-food industry. Maize contains about 70% starch, 7–13% protein and a small fraction of vitamins and minerals (Garg et al., 2020). Wongsagonsup et al. (2008) reported that untreated normal maize and waxy maize comprised a high content (84, 87%) of rapidly digesting starch (RDS) whereas, the SDS contents were 14 and 11% respectively while the RS was 1.8 and 1.7%. However, varied heat moisture treatment lowered the RDS whereas increased the SDS and RS contents to as high as 23.8 and 12.2% respectively. Another study reported a resistant starch content of less than 1.61% in native maize starch (Milašinović-Šeremešić et al., 2012) meanwhile, high-amylose maize starch was reported to have large proportion of resistant starch ranging from 11.5 to 43.2% (Tian & Sun, 2020). Sievert and Pomeranz (1989) reported a 21.3% RS content for an amylo-maize VII hybrid.

5.3.4 Oat

Oat with botanical name *Avena sativa*, is a whole grain food specifically grown for its seed. Essentially, it is a globally reputable, healthy and nutritious food having starch as the most abundant component. It is a very popular starchy cereal used for an array of purposes especially breakfast food and other food products. The whole grain contains about 55–65% starch, 14% protein, 7% lipids and 4% β-glucan (Doehlert et al., 2013; Zhang et al., 2021). Oat starch has approximately 40% of SDS, the slow starch digestion behavior is partially attributed to the high percentage of amylose in the oat starch while other factors include fewer short branch chains and less branching of amylopectin resulting in reduced enzyme accessibility (Xu et al., 2017). Ren et al. (2020) in their study, disclosed that RS accounted for 29.31% of the starch content in oat starch.

Precisely two types of RS identified in oat starch, the first type of RS is RS2, it is reportedly resistant to digestion on account of the natural compactness of the starch granule thus, restricting accessibility to enzymes, the other type of RS referred to as RS5, is owing to resistance to enzymatic breakdown by the amylose-lipid complex (Zhang et al., 2021). Oat comprises of several components which can interfere with starch hydrolysis and the rate of absorbing released glucose into the body. The presence of SDS, RS and other components in oat are contributory factors that determine the glycemic index of the final food product.

In more recent times, research interest has focused on promoting low postprandial glycemic response, in similar fashion, foods with highly indigestible carbohydrate content. In leu of this, SDS and RS have gained more research interest and findings suggest their beneficial health effects.

5.3.5 Resistant Starch Composition of Other Cereals

Barley (*Hordeum vulgare* L.) is an important cereal crop, widespread across human nutrition and animal feeds (Biel et al., 2020). Its production comes in the fourth place after wheat, maize, rice (Biel & Jacyno, 2013). The SDS content of barley was the highest (53.7%) among several cereal starches evaluated, others had contents ranging from 39.5 to 50.7% (Kaur et al., 2018). The same study reported 11.6% RS for raw barley compared to other cereals (wheat, rice, oats, maize, sorghum, millets) which had 9.7, 6.1, 7.1, 8.4, 10.9, 9.23% RS respectively. An earlier study evaluated the RS content of starch from highland barley and reported a high content (53.8 g/kg) than those of oats and buckwheat starch (Shen et al., 2016).

Sorghum (*Sorghum bicolor*) is equally a relevant food crop especially in many developing countries Africa inclusive. It has been tagged number five world's substantially produced cereal (Heba et al., 2022). An earlier study reported a 7.1% SDS and 11.2% RS for native starch from waxy sorghum (Shin et al., 2004). In another study, a wide array of sorghum genotypes (49) was evaluated, and the RS content ranged from 0.3 to 65.6 g/100 g (Teixeira et al., 2015). A more recent study on native sorghum reported 61.27% for SDS and 7.62% for RS while other varieties had lower SDS and RS contents (Li et al., 2021).

5.4 Properties of Slowly Digesting and Resistant Starches

SDS is as important as RS although, RS is a functional dietary component with optimum potential to maintain metabolic and colonic health (Tian & Sun, 2020; Halajzadeh et al., 2020). Consuming foods rich in resistant starch may serve as an important dietary intervention for preventing/managing diabetes. There is an inverse relationship between RS and glycemic index (GI) because RS has low glycemic index and consumption of foods with high RS content results in lower postprandial blood glucose concentrations (Hasjim et al., 2010), moreso, it will promote fat oxidation in place of carbohydrate oxidation. Researchers have reported the potential of low-GI diet to greatly increase satiety at the same time, possess potential to lower plasma insulin response contrary to the implications of high glycemic index diet (Ludwig, 2002; Warren et al., 2003; Wolever, 2003).

Like RS, dietary fiber is the palatable part of plants or carbohydrate fractions, they function to resist absorption in the small intestine whereas, undergoes microbial fermentation in the colon (AACC, 2001). In another light, RS can be recognized as a component of dietary fiber thus, they reduce food consumption and promote satiety.

5.5 Functional Health Benefits of SDS and RS

Carbohydrate-rich foods including cereals constitute the main part of the human diet, nutritional classification with respect to prevailing influence on postprandial glucose response as designated by glycemic index (GI). GI classification of food is a scale ranked from 0 to 100, sectioned into low (0- $\leq$55), medium (56- $\leq$ 70) and high ($\geq$70). GI measures the rate of glucose liberation from the total available carbohydrate in a food sample and consequently absorbed by the body (Dupuis et al., 2014). Carbohydrate-rich food represents category of food having high GI and may induce coronary heart diseases (especially obesity, type 2 diabetes, hyperinsulinemia) owing to the high glycemic response/postprandial glycemia and insulin response linked the regular intake of high GI foods.

SDS and RS has potential health benefits, the former offers a stabilized and sustained blood glucose level while the latter does not contribute to blood glucose (Lehmann & Robin, 2007). In another vein, studies substantiate the fact that the regular intake of low GI diets abates hunger, promotes lower energy intake, control insulin secretion, postprandial glycemia as well reduce triglycerides (Warren et al., 2003; Pereira et al., 2004; Brand-Miller et al., 2009; Ekmekcioglu & Touitou, 2011).

Additionally, low GI diet confers favorable effect in relation to forestalling type 2 diabetes and associated maladies (Blaak et al., 2012; Livesey et al., 2019).

Cereals are important sources of SDS and RS, these starch fractions are used as food additives, dietary ingredient and elicit different health-promoting properties as underlisted.

5.5.1 Glycemic Control

Carbohydrate-rich foods evoke high sugar upon consumption; relative insulin deficiency and or resistance induce elevated blood glucose levels resulting in hyperglycemia also known as diabetes (ADA, 2018). Diabetes mellitus is a global epidemic with estimated prevalence rate of 463 million people and projected population of 700 million by 2045 (IDF, 2017). Glycemic index is a concept that defines the change in postprandial glucose level upon consumption of food (Kaur et al., 2020). High glycemic index food results in high postprandial blood glucose while low glycemic index foods results in low glycemic response. Hence, the need for glycemic control achievable be deliberate production and consumption of low GI food products (Punia et al., 2019; Punia Bangar et al., 2021b).

Studies have revealed a positive out-turn of RS consumption on plasma glucose metabolism in human (Robertson, 2012; Birt et al., 2013). There is little or no breakdown of RS-rich diet in the small intestine therefore, consumption of such food results in low release of glucose (Fuentes-Zaragoza et al., 2010). RS has low GI thus RS diets may attenuate postprandial glucose release thereby minimize the possibility of developing diabetes mellitus and other associated cardiovascular diseases (Fig. 5.1). Moreso, the slow digestion of RS will define glucose release meanwhile, lower glucose response is beneficial for health as it has potential for improved metabolic control in type 2 diabetes mellitus.

RS from a variety of cereals, starch and products have been studied and results show a beneficial impact on glucose control. Grandfeldt et al. (1995) published on treatment with high amylose corn flour RS substantially reduced postprandial plasma glucose and insulin levels. Li et al. (2010) evaluated the GI of gene-modified rice containing 20% RS and result showed the rice had a low GI (44.8) compared to high GI (77.4) in regular rice and the consumption of RS rice meal reduced postprandial glucose by 37%. High amylose VII maize starch was debranched using isoamylase and a RS5 ingredient was obtained and 60% of this ingredient was used in bread, study findings showed that the postprandial plasma glucose and insulin levels reduction was up to 50% (Hasjim et al., 2010). RS from maize and wheat grains, produced significantly low blood glucose response (Brites et al., 2011). Recently, Vinoy et al. (2020) also summited that consumption of low GI cereal product comprising high amount of slowly digesting starch content caused reduction in postprandial glucose response of subjects. Another study on RS-supplemented muffin (30 g RS) showed that the formulated muffin had a low GI (48), the muffin induced significantly low postprandial glucose in subjects (Ma & Lee, 2021).

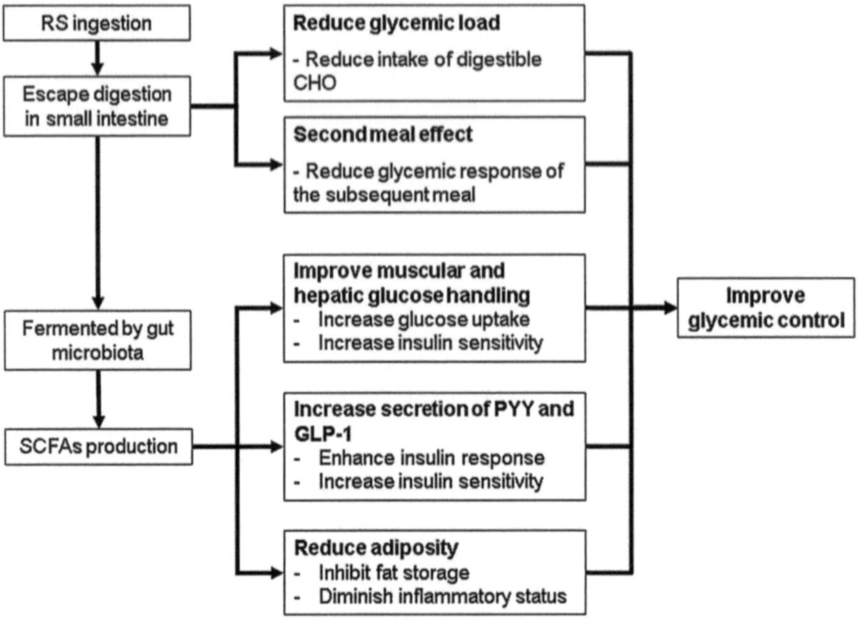

Fig. 5.1 Possible mechanism of the effects of RS consumption on glycemic control. (Adapted from Wong & Louise, 2016). *RS*: resistant starch, *CHO* carbohydrate, *SCFAs* short chain fatty acids, *PYY* peptide *YY*, *GLP-1* glucagon-like peptide-1

Overall, high dose of RS produces higher glucose lowering benefits than low RS dose products (Wong & Louie, 2016). A health claim regarding the potential of high-amylose maize resistant starch to reduce the risk of type 2 diabetes has been recently authorized by the Food and Drug Administration (FDA, 2021).

5.5.2 *Prevention of Colon Cancer*

Colon cancer is a common type of cancer and a definite principal cause of cancer-related deaths. There are postulations from research that RS possess specific anti-cancer properties (Amini et al., 2016). RS is not digested in the small intestine and undergoes an onward transportation to the large intestine where bacterial fermentation of food by selected probiotics takes place; gases are released, likewise potentially beneficial short-chain fatty acids (SCFAs) and organic acids.

Butyrate is a crucial SCFA, a preferred fuel for probiotics, a main nutrient for colonocytes with specific usefulness in minimizing the chances of colonic diseases (colon cancer) by inhibiting the multiplication of cancer cells (i.e. suppresses cell proliferation) and promoting apoptosis (Cummings et al., 1996; Li, 2018). In a similar fashion, butyrate functions to provide energy for epithelial cells and prevents the in vitro malignant transformation of epithelial cells (Raigond et al., 2019).

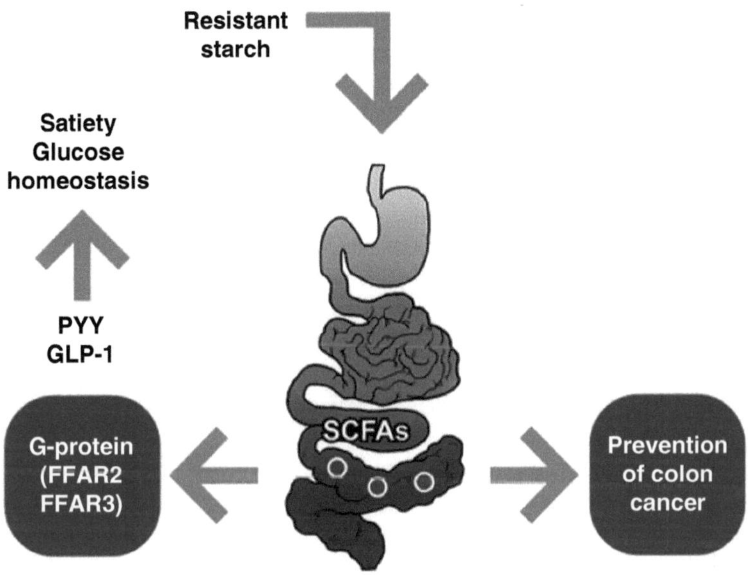

Fig. 5.2 Beneficial effect of resistant starch in prevention of colon cancer. (Source: Magallanes-Cruz et al., 2017). *FFAR2* free fatty acid receptor 2, *FFAR3* free fatty acid receptor 3, *GLP-1* glucagon-like peptide 1, *PYY* peptide YY, *SCFAs* short chain fatty acids

Leu et al. (2009) investigated the end result of maize starch with predominant amylose content on rats' colon health, the study reported that RS diets remarkably boosted SCFA while caecal pH was drastically reduced. Also, RS diets intercepted mucosal atrophy in the rats, likewise butyrate levels showed positive correlation to apoptotic response. Low pH alters microbial composition and favours the proliferation of bacteria that produces butyrate (Walker et al., 2011). Figure 5.2 schematically portrays the physiological and metabolic impacts of fermentation of RS on human health, the SCFAs produced during fermentation binds G-protein coupled receptors (free fatty acid receptor 2 and 3 as seen in Fig. 5.2). The end action is satiety thus promoting decreased appetite and food intake; this is induced by the secretion of leptin and peptide YY (PYY) by the adipocytes and enteroendocrine cells upon receiving signals from the ileal and colonic enteroendocrine L cells (Blaut & Klaus, 2012).

5.5.3 Hypocholesterolemic Effect

Cholesterol,triglycerides and lipoproteins represent vital constituents of the lipid fraction of any human body (Cox & Garcia-Palmieri, 1990). Lipid metabolism entails total lipids, total cholesterol, low density lipoprotein (LDL), high density lipoprotein (HDL), intermediate density lipoprotein (IDL), triglycerides etc. Lipid metabolism disorder is a definite complication of type 2 diabetes mellitus as lipolysis process is hampered; this is characterized by increased cholesterol, triglycerides and LDL levels but decreased HDL levels (Sone et al., 2016; Wahjuningsih et al., 2018). RS affects lipid metabolism as RS provides substrate that produces SCFAs (mainly propionate and butyrate) which forestalls cholesterol biosynthesis (Weitkunat et al., 2015). The hypolipidemic effect of RS is evidenced by the reduced cholesterol level after treatment with RS containing diet.

The consumption of RS-rich meals (produced from high-amylo maize, 16 g RS) was reported to induce reduction in plasma triglycerides (Giacco et al., 1998). Remarkable reduction in serum triglyceride of hamsters fed with oat RS containing extruded snack was earlier reported (Chang et al., 2002). Another study by Park et al. (2004) studied the effect of high RS corn starch (24 g/day) and regular corn starch on hypolipidemic effects, results showed a remarkable downward trend on total serum cholesterol.

Wahjuningsih et al. (2018) in their study treated diabetic rats with sago analog rice rich in RS, results showed that after feeding the rats with the modified rice for 4 weeks, the rats showed decreased concentration of total cholesterol and LDL but increased concentrations of HDL.

5.5.4 Satiety and Weight Control

Overweight and obesity are public health conditions of epidemic proportions linked to development of chronic diseases such as diabetes mellitus, cardiovascular disease, representing the major causes of global deaths (Hruby & Hu, 2015). Weight loss and ability to sustain an appropriate weight has potential to ameliorate the ravaging co-morbidities; weight management strategy is dependent on negative energy balance involving reduced energy intake than expended energy (Higgins, 2014).

Studies regarding consumption of RS-containing foods have shown its potential to promote weight loss due to indigestibility of RS resulting in reduced postprandial glycemia/insulinemia, increased fat oxidation, release of gut satiety peptides, possible increase in satiety (Higgins, 2014).

Two doses (20 g, 30 g) of whole-grain high-amylose corn flour were used as ingredient alongside viscous fiber and type 1 & 2 resistant starch to produce breakfast meal, then fed to human subjects and significant reductions in acute satiety response and energy intake was observed (Harrold et al., 2014).

Furthermore, human study by Emilien et al. (2017) using healthy adults, they were fed with muffin produced from wheat flour or 40% substituted RS; results showed that the meal caused reduced energy intake thus, suggested it may aid weight management.

5.5.5 *Absorption of Minerals*

RS has a beneficial effect on the intestinal absorption of minerals particularly calcium, this is achieved by an enhanced mineral solubility taking place in the cecum and or large intestine. Human and animal studies have been carried out to evaluate the effect of RS-rich diets on absorption of minerals (calcium, magnesium, zinc, iron, copper), results showed significant absorption of all the minerals in rats but a maximal absorption of calcium in human (Younes et al., 1995; Coudray et al., 1997).

Summary of potential benefits of starch fractions

1. Beneficial effect on the digestive tract functionality
2. Promotes the growth of beneficial microbes of colon colon thus, preventing colon cancer
3. Glycemic control and reduced risk of diabetes
4. Lowering total cholesterol level (Ashwar et al., 2021).

5.5.6 *Technological Benefits of Resistant Starch*

Resistant starch is similar to and a part of dietary fiber, in that it resists digestion in the small intestine like the later thus, share in the physiological and functional properties of dietary fiber.

1. High gelatinization temperature: increased gelatinization temperature and enthalpy is observed in RS and this is associated with reorganization of starch chain and newly formed structures resulting from retrogradation. Increased endothermic transition has been associated with structural differences.
2. Good extrusion and film-forming qualities
3. Lower water-holding properties than traditional fiber products
4. Promote formation of low-bulk high-fiber products
5. Improved texture, appearance, and mouth feel compared with traditional high-fiber products
6. Lowering the calorific value of foods; RS has a lower calorie (8 kJ/g) content compared with fully digestible starch (15 kJ/g) (Rochfort & Panozzo, 2007), moreover, suitable for complementing reduced fat and low sugar food formulations.

5.6 Methods of Improving Slowly Digesting Starch and Resistant Starch Content in Cereal Starch and Products

Modification of starch entails the structural modification of starch molecules namely double helical structures and crystallites resulting in reduction of enzymatic hydrolysis. It has potential to influence some techno-functional properties of the starch material, for instance starch solubility, digestibility, palatability and nutritional value. Starch can be modified using several techniques ranging from physical, chemical, enzymatic to genetic methods and each of these processes induce improvement of SDS and RS contents of food materials.

5.6.1 Physical Modification

It is also known as hydrothermal modification which entails the application of heat (with excess or restricted level of moisture) to starchy products in a bid to control granule swelling, amylose leaching, viscosity and susceptibility to digestive enzymes. The commonly used hydrothermal treatments are annealing and heat-moisture treatment (HMT).

Annealing is a hydrothermal modification of starch involving heat treatment (below gelatinization temperature, above glass transition temperature) in the presence of water and for a specified period (Yao et al., 2018). Heat-moisture treatment is also a hydrothermal modification in which starch granules are exposed to heat treatment (above gelatinization temperature, below glass transition temperature) under low moisture conditions for a specified period (Gunaratne & Hoover, 2002). The modification technique promotes increased resistance to amylolytic enzymes.

Hydrothermal treatments positively improve the physicochemical and functional properties of the starches; these include increased water and oil absorption capacities, improved thermal stability, gel hardness, decreased rate of retrogradation, decreased solubility and swelling power (Adebowale et al., 2005; Hormdok & Noomhorm, 2007; Chung et al., 2009; Zavareze & Dias, 2011; Ali & Hasnain, 2016). The observed physicochemical changes were attributed to several factors including: increased interaction between amylose and amylopectin, internal rearrangement of starch granules, reduced amylose leaching, increased cross linking between starch chains in the amylose portion and a host of other factors (Zavareze & Dias, 2011).

The modification techniques have potential to simultaneously disorder and reorder starch molecules to create a densely packed starch fraction without destroying the granule structure. In this state, most of the rapidly digesting starch fractions can be converted to slowly digesting and resistant starch (Ding et al., 2021). Studies have evaluated the role the physical modification technique plays, a few is hereafter underlisted.

HMT treatment of corn starch increased the SDS and RS contents of the modified starch by 2.5% and 7.7% respectively (Chung et al., 2009). The author attributed the increase in the starch fractions to interactions formed during hydrothermal treatment partly hindering access by hydrolyzing enzymes to the starch chains. Furthermore, Silva et al. (2017) observed elevated levels of SDS and RS contents in rice flour after subjecting to heat-moisture treatment (HMT) of 16%, 60 min and 18%, 60 min respectively.

5.6.2 Chemical Modification

This involves the use of chemical reactants through oxidation, esterification, hydroxypropylation, etherification and cross-linking to introduce functional groups to the starch molecules which clearly promotes structural changes and alters the physicochemical properties of the starch and produces starch with low digestibility (Masina et al., 2017; Ding et al., 2021; Punia Bangar et al., 2021c).

Cross-linking of cereal (rice, wheat, corn, oat) starches has been carried out using sodium trimetaphosphate (STMP), sodium tripolyphosphate (STPP), phosphoryl chloride among others and Type 4 RS is produced. Of all the chemical modification approaches, citric acid esterification is the most preferred because it is a mild acid treatment which is considered safe and cheap (Sánchez-Rivera et al., 2017a, b). It is a very effective approach in producing chemically modified starch (RS4).

Chemical modification inhibits digestion of starch by creating a steric barrier at the site of enzyme action, that is, the introduced functional group blocks the starch chain from amylolytic enzymes. Generally, chemically modified starch produces RS type 4, however, the type of chemical to be employed is dependent on the required characteristics and functional properties ranging from use as stabilizer, emulsifier, texturizer etc. (Nurmilah & Subroto, 2021).

Recent studies have reported on chemical modification of starches in this case, cereal starch, and the relative influence on RS contents as well promotion of functional properties have been documented. Acid treated rice reportedly has higher RS (30–39%) than native or heat-moisture treated rice starches (Hung et al., 2016). Similarly, cross-linked rice starches showed higher RS than the native counterpart (Ashwar et al., 2016, 2018). In another study by Ashwar et al. (2017), acetylation of rice starch increased the RS4 content of the starch but decreased the thermal properties, pasting parameters and syneresis of the modified rice starch.

5.6.3 Heat Treatment

This involves the application of heat such as cooking above gelatinization temperature thereafter, drying. This process significantly has effect on susceptibility to hydrolysis by enzymes (Sajillata et al., 2006).

5.6.4 Genetic Modification

This involves strategies developed to generate new cultivars of crops using extensive breeding to produce crops with desired functionality (Ashwar et al., 2018). Certain factors affecting starch functionality are controlled genetically, these factors include starch content, starch structure, cell component interaction and starch granule morphology. Genetic modification is designed to suppress starch synthase (ss3a, ss4b, ss3a/ss2b) and starch debranching enzymes (SBEI, SBEIIb) by applying grain composition engineering to manipulate amylose/amylopectin ratio thereby produce high amylose genotypes with elevated (above 30%) amylose content and increased RS content (Jukanti et al., 2020). High amylose cereal is the target since higher proportion of amylose fraction in a cereal, its starch or by product slows digestibility. High amylose cereal varieties can be achieved by two approaches namely (a) interbreeding mutants with genes promoting increased amylose content and (b) inhibition of the expression of starch-branching enzymes (Dupuis et al., 2014). Several genetic strategies in relation to different cereals (rice, durum wheat, barley) was studied by some researchers (Itoh et al., 2003; Li et al., 2008; Sparla et al., 2014; Botticella et al., 2012; Botticella et al., 2016).

5.6.5 *Enzymatic Modification*

This entails using enzymatic treatments (i.e. debranching enzymes such as pullulanase, isoamylase) on cereal or starch to act on α-(1,6)-glycosidic bonds in order to increase the apparent amylose. The process entails debranching to hydrolyze the amylopectin branch chains thus, increasing the amylose content with formation of tightly packed RS crystals. In addition, realignment of amylose strands forming double helices which is made stable by hydrogen bonds leading to formation of RS type 3 occurs (Li et al., 2011).

Table 5.2 below highlights effects of some modification techniques on the starch fraction in some cereals.

5.7 Applications of Cereal-Based RS in Food Products

The introduction and utilization of ingredients rich in RS is gaining popularity and of utmost interest in research and food industry. Several research works have been carried out in the areas of producing functional starch used as ingredient in an array of food products.

Annealed white sorghum starch (AnnRS) has been reported as a novel RS ingredient suitable for use in baked goods and confections such as biscuit (Cervini et al., 2021a). The RS and dietary fibre contents of the biscuits increased with increase in proportion of AnnRS. In addition, the inclusion of RS ingredient to gluten-free flour did not significantly impair the physical attributes of the biscuit although, some sensory attributes were negatively influenced.

Another study reported the practical application of annealed sorghum starch in pasta (Cervini et al., 2021b); the findings summarize that the incorporation of annealed starch up to 15% effectively increased the RS of the pasta product. A

Table 5.2 Modification and its effects on slowly digesting and resistant starch content in some cereals

Form of modification	Effect	Cereal	References
Physical	Increased SDS and RS contents (heat-moisture treatment)	Rice	Silva et al. (2017), Wang et al. (2018)
Chemical	Increased content of SDS and RS (citric acid treatment, acetylation, oxidation, hydroxypropylation, cross-linking)	Corn	Shin et al. (2007), Chung et al. (2008)
	Increased levels of RS (acid treatment, crosslink-esterification)	Rice	Hung et al. (2016), Ye et al. (2019)
	High RS content (cross-linking, phosphorylation)	Rice	Ashwar et al. (2017), Ashwar et al. (2018)
	A six- to tenfold increase in RS4 content (acetylation)	Rice	Ashwar et al. (2017)
Genetic	High RS2 content of 11.87–13.69% (*N*-methyl-*N*-nitrosourea mutagenesis)	Rice	Yoon et al. (2013)
	High RS3 content (cross breeding, introduction of *SSII*a and/or *GBSSI* mutant)	Rice	Zhang et al. (2011), Itoh et al. (2017)
Enzymatic	Increased RS content	Rice	Ye et al. (2018)
	Increased SDS content (pullulanase debranching)	Maize	Miao et al. (2009)
	Increased RS content from 1.8% in native flour to 24.3% (α-amylase and pullulanase treatment)	Maize	Khan et al. (2020)

previous study used commercial type II RS from Hi-maize to produce gluten-free pasta, they reported that addition of RS enhancedthe quality of the pasta product as it relates to increased firmness and reduced stickiness of cooked pasta (Foschia et al., 2017).

The inclusion of RS ingredient in bread formulas have also been explored. High amylose starches, especially maize based, the popular commercial RS ingredients in bread-making include RS2 (granular) and RS3 (retrograded) types. Barros et al. (2018) replaced a fraction of wheat flour with resistant starch and evaluated the rheology and quality of white bread. From the outcome of the study, the crumb of the RS breads were firmer with lower retrogradation rate and had parameters that indicated up to 15% RS inclusion produced very good quality breads.

5.8 Conclusion

Cereals are a crucial source of resistant starch and the different types of RS (RS1 to RS5) is a factor of the nature of the material, type of processing, type of modification amongst others. RS has an array of nutritional, health and techno-functional properties including positive impact to modulate glycemic response, improved colon health, hypocholesterolemic effect, superior functional properties thus making it a suitable ingredient in food formulations to fulfill the current research and consumer demand for functional ingredients with health benefits.

References

AACC. (2001). The definition of dietary fiber. Association of American Cereal Chemists report. *Cereal Food World, 46*, 112–126.

Adebowale, K. O., Afolabi, T. A., & Olu-Owolabi, B. I. (2005). Hydrothermal treatments of finger millet (Eleusine coracana) starch. *Food Hydrocolloids, 19*(6), 974–983.

Alhambra, C. M., Dhital, S., Sreenivasulu, N., & Butardo, V. M. (2019). Quantifying grain digestibility of starch fractions in milled rice. In N. Sreenivasulu (Ed.), *Rice grain quality. Methods in molecular biology* (Vol. 1892). Humana Press. https://doi.org/10.1007/978-1-4939-8914-0_13

Ali, T. M., & Hasnain, A. (2016). Effect of annealing on morphological and functional properties of Pakistani white sorghum (*Sorghum bicolor*) starch. *Journal of Basic & Applied Sciences, 12*, 358–364.

Aller, E. E., Abete, I., Astrup, A., Martinez, A., & Baak, M. A. (2011). Starches, sugars and obesity. *Nutrients, 3*(3), 341–369.

Alsaffar, A. A. (2011). Effect of food processing on the resistant starch content of cereals and cereal products-A review. *International Journal of Food Science & Technology, 46*(3), 455–462.

American Diabetes Association. (2018). Standards of medical care in diabetes – 2018. *Diabetes Care, 41*.

Amini, A., Khalili, L., Keshtiban, A. K., & Homayouni, A. (2016). Resistant starch as a bioactive compound in colorectal cancer prevention. In R. R. Watson & V. R. Preedy (Eds.), *Probiotics, prebiotics, and synbiotics: Bioactive foods in health promotion* (pp. 773–780). Academic.

Anguita, M., Gasa, J., Martín-Orue, S. M., & Pérez, J. F. (2006). Study of the effect of technological processes on starch hydrolysis, non-starch polysaccharides solubilization and physicochemical properties of different ingredients using a two-step in vitro system. *Animal Feed Science & Technology, 129*(1–2), 99–115.

Ashwar, B. A., Gani, A., Shah, A., & Masoodi, F. A. (2016). Production of RS4 from rice by acetylation: Physico-chemical, thermal, and structural characterization. *Starch/Stärke, 69*, 1600052. https://doi.org/10.1002/star.201600052

Ashwar, B. A., Gani, A., Shah, A., & Masoodi, F. A. (2017). Physicochemical properties, *in-vitro* digestibility and structural elucidation of RS4 from rice starch. *International Journal of Biological Macromolecules, 105*, 471–477.

Ashwar, B. A., Gani, A., Shah, A., & Masoodi, F. A. (2018). Production of RS4 from rice starch and its utilization as an encapsulating agent for targeted delivery of probiotics. *Food Chemistry, 239*, 287–294.

Ashwar, B. A., Gani, A., Shah, A., Ahmad, M., Gani, A., Jhan, F., & Noor, N. (2021). Resistant starch and slowly digestible starch. In A. Gani & B. A. Ashwar (Eds.), *Food biopolymers: Structural, functional and nutraceutical properties*. https://doi.org/10.1007/978-3-030-2706-2

Awika, J. M. (2011). Major cereal grains production and use around the world. *ACS Symposium Series, 1089*, 1–13.

Barros, J. H., Telis, V. R., Taboga, S., & Franco, C. M. (2018). Resistant starch: Effect on rheology, quality, and staling rate of white wheat bread. *Journal of Food Science & Technology, 55*(11), 4578–4588.

Bednar, G. E., Patil, A. R., Murray, S. M., Grieshop, C. M., Merchen, N. R., & Fahey, G. C. (2001). Starch and fiber fractions in selected food and feed ingredients affect their small intestinal digestibility and fermentability and their large bowel fermentability in vitro in a canine model. *The Journal of Nutrition, 131*, 276–286.

Biel, W., & Jacyno, E. (2013). Chemical composition and nutritive value of spring hulled barley varieties. *Bulgarian Journal of Agricultural Science, 19*(4), 721–727.

Biel, W., Kazimierska, K., & Bashutska, U. (2020). Nutritional value of wheat, triticale, barley and oat grains. *Acta Scientiarum Polnorum Zootechnica, 19*(2), 19–28.

Birt, D. F., Boylston, T., Hendrich, S., Jane, J.-L., et al. (2013). Resistant starch: Promise for improving human health. *Advances in Nutrition, 4*(6), 587–601.

Blaak, E. E., Antoine, J. M., Benton, D., Björck, I., Bozzetto, L., Brouns, F., et al. (2012). Impact of postprandial glycaemia on health and prevention of disease. *Obesity Review, 13*(10), 923–984.

Blaut, M., & Klaus, S. (2012). Intestinal microbiota and obesity. *Handbook of Experimental Pharmacology, 209*, 251–273.

Botticella, E., Sestili, F., & Lafiandra, D. (2012). Characterization of SBEIIa homoeologous genes in bread wheat. *Molecular Genetics and Genomics, 287*(6), 515–524.

Botticella, E., Sestili, F., Ferrazzano, G., Mantovani, P., Cammerata, A., D'Egidio, M. G., et al. (2016). The impact of the SSIIa null mutations on grain traits and composition in durum wheat. *Breeding Science, 66*(4), 572–579.

Brand-Miller, J., McMillan-Price, J., Steinbec, K., & Caterson, I. (2009). Dietary glycemic index: Health implications. *Journal of the American College of Nutrition, 28*, 446–449.

Brites, C. M., Trigo, M. J., Carrapico, B., et al. (2011). Maize and resistant starch enriched breads reduce postprandial glycemic responses in rats. *Nutrition Research, 31*(4), 302–308.

Carcea, M. (2020). Nutritional value of grain-based foods. *Food, 9*(4). https://doi.org/10.3390/foods9040504

Cervini, M., Frustace, A., Duserm Garrido, G., Rocchetti, G., & Giuberti, G. (2021a). Nutritional, physical and sensory characteristics of gluten-free biscuits incorporated with a novel resistant starch ingredient. *Heliyon, 7*(3), e06562.

Cervini, M., Gruppi, A., Bassani, A., Spigno, G., & Giuberti, G. (2021b). Potential application of resistant starch sorghum in gluten-free pasta: Nutritional, structural and sensory evaluations. *Food, 10*(5), 908. https://doi.org/10.3390/foods10050908

Chang, Y. K., Martinez-Flores, H. E., Martinez-Bustos, F., & Sgarbieri, V. C. (2002). Effect of extruded products made with cassava starch blended with oat fiber and resistant starch on the hypocholesterolemic properties as evaluated in hamsters. *Nutraceuticals & Food, 7*, 133–138.

Chung, H.-J., Shin, D.-H., & Lim, S.-T. (2008). *In vitro* starch digestibility and estimated glycemic index of chemically modified corn starches. *Food Research International, 41*, 579–585.

Chung, H. J., Liu, Q., & Hoover, R. (2009). Impact of annealing and heat-moisture treatment on rapidly digestible, slowly digestible and resistant starch levels in native and gelatinized corn, pea and lentil starches. *Carbohydrate Polymers, 75*(3), 436–447.

Coudray, C., Bellanger, J., Castigilia-Delavaud, C., Rémésy, C., Vermorel, M., & Rayssignuier, N. Y. (1997). Effect of soluble or partly soluble dietary fibres supplementation on absorption and balance of calcium, magnesium, iron and zinc in healthy young men. *European Journal of Clinical Nutrition, 51*, 375–380.

Cox, R. A., & García-Palmieri, M. R. (1990). Chapter 31: Cholesterol, triglycerides, and associated lipoproteins. In H. K. Walker, W. D. Hall, & J. W. Hurst (Eds.), *Clinical methods: The history, physical, and laboratory examinations* (3rd ed.). Butterworths. Available from: https://www.ncbi.nlm.nih.gov/books/NBK351/

Cummings, J. H., Beatty, E. R., Kingman, S. M., Bingham, S. A., & Englyst, H. N. (1996). Digestion and physiological properties of resistant starch in the human large bowel. *British Journal of Nutrition, 75*(5), 733–747.

Darandakumbura, H. D., Wijesinghe, D. G., & Prasantha, B. D. (2013). Effect of processing conditions and cooking methods on resistant starch, dietary fiber and glycemic index of rice. *Tropical Agricultural Research, 24*(2), 163–174.

Devitt, A. A., Williams, J. A., Choe, Y. S., Hustead, D. S., & Mustad, V. A. (2013). Glycemic responses to glycemia-targeted specialized-nutrition beverages with varying carbohydrates compared to a standard nutrition beverage in adults with type 2 diabetes. *Advances in Bioscience and Biotechnology, 4*(09), 1–10.

Ding, Y., Wang, M., Shen, Y., Shu, X., Xu, D., & Song, W. (2021). Physicochemical properties of resistant starch and its enhancement approaches in rice. *Rice Science, 28*(1), 31–42.

Doehlert, D. C., Simsek, S., Thavarajah, D., Thavarajah, P., & Ohm, J. B. (2013). Detailed composition analyses of diverse oat genotype kernels grown in different environments in North Dakota. *Cereal Chemistry, 90*(6), 572–578.

Dupuis, J. H., Liu, Q., & Yada, R. Y. (2014). Methodologies for increasing the resistant starch content of food starches: A review. *Comprehensive Reviews in Food Science and Food Safety, 13*(6), 1219–1234.

EFSA, European Food Safety Authority. (2011). Scientific opinion on the substantiation of health claims related to resistant starch and reduction of post-prandial glycaemic responses (ID 681), "digestive health benefits" (ID 682) and "favours a normal colon metabolism" (ID 783) pursuant to article 13(1) of regulation (EC) no 1924/2006. EFSA panel on dietetic products, nutrition and allergies (NDA). *The EFSA Journal, 9*, 2024–2041.

Ekmekcioglu, C., & Touitou, Y. (2011). Chronobiological aspects of food intake and metabolism and their relevance on energy balance and weight regulation. *Obesity Reviews, 12*(1), 14–25.

Emilien, C. H., Hsu, W. H., & Hollis, J. H. (2017). Effect of resistant wheat starch on subjective appetite and food intake in healthy adults. *Nutrition, 43–44*, 69–74.

Englyst, H. N., Kingman, S. M., & Cummings, J. H. (1992). Classification and measurement of nutritionally important starch fractions. *European Journal of Clinical Nutrition, 46*(suppl 2), 33–50.

Food and Drug Administration. (2021). RE: Petition for a health claim for high-amylose maize starch (containing type-2 resistant starch) and reduced risk type 2 diabetes mellitus; Docket Number FDA2015-Q-2352. Letter dated December 12, 2016. Available at: http://wayback.archiveit.org/7993/20171115122025/https://www.fda.gov/downloads/Food/IngredientsPackagingLabeling/LabelingNutrition/UCM546963.pdf. Accessed December 2021.

Foschia, M., Beraldo, P., & Peressini, D. (2017). Evaluation of the physicochemical properties of gluten-free pasta enriched with resistant starch. *Journal of the Science of Food & Agriculture, 97*(2), 572–577.

Fuentes-Zaragoza, E., Riquelme-Navarrete, M., Sánchez-Zapata, E., & Pérez-Alvarez, J. (2010). Resistant starch as functional ingredient: A review. *Food Research International, 43*(4), 931–942.

Fuentes-Zaragoza, E., Sanchez-Zapata, E., Sendra, E., Sayas, E., Navarro, C., & Fernandez-Lopez, J. (2011). Resistant starch as prebiotic: A review. *Starch/Stärke, 63*(7), 406–415.

Garg, N. K., Dahuja, A., Singh, A., & Chaudhary, D. P. (2020). Understanding the starch digestibility characteristics of Indian maize hybrids. *Indian Journal of Experimental Biology, 58*, 738–744.

Giacco, R., Clemente, G., Brighenti, F., Mancini, M., D'Avanzo, A., Coppolu, S., et al. (1998). Metabolic effects of resistant starch in patients with type 2 diabetes. *Diabetes Nutrition Metabolism, 11*(6), 330–335.

Granfeldt, Y., Drews, A., & Bjorck, I. (1995). Arepas made from high amylose corn flour produce favorably low glucose and insulin responses in healthy humans. *Journal of Nutrition, 125*(3), 459–465.

Gunaratne, A., & Hoover, R. (2002). Effect of heat-moisture treatment on the structure and physicochemical properties of tuber and root starches. *Carbohydrate Polymers, 49*(4), 425–437.

Halajzadeh, J., Milajerdi, A., Reiner, Ž., Amirani, E., Kolahdooz, F., Barekat, M., Mirzaei, H., Mirhashemi, S. M., & Asemi, Z. (2020). Effects of resistant starch on glycemic control, serum lipoproteins and systemic inflammation in patients with metabolic syndrome and related disorders: A systematic review and meta-analysis of randomized controlled clinical trials. *Critical Reviews in Food Science & Nutrition, 60*(18), 3172–3184.

Haralampu, S. G. (2000). Resistant starch – A review of the physical properties and biological impact of RS3. *Carbohydrate Polymers, 41*(3), 285–292.

Harrold, J., Breslin, L., Walsh, J., Halford, J., & Pelkman, C. (2014). Satiety effects of a whole-grain fibre composite ingredient: Reduced food intake and appetite ratings. *Food & Function, 5*(10), 2574–2581.

Hasjim, J., Lee, S.-O., Hendrich, S., Setiawan, S., Ai, Y., & Jane, J.-l. (2010). Characterization of a novel resistant-starch and its effects on postprandial plasma-glucose and insulin responses. *Cereal Chemistry, 87*(4), 257–262.

Heba, I. M., Eman M. F., Abdul Basit, K., Rafiq L. Mahmoud R. S. (2022). Sorghum: Nutritional Factors, Bioactive Compounds, Pharmaceutical and Application in Food Systems: A Review. Phyton: International Journal of Experimental Botany, https://doi.org/10.32604/phyton.2022.020642

Higgins, J. A. (2014). Resistant starch and energy balance: Impact on weight loss and maintenance. *Critical Reviews in Food Science and Technology, 54*(9), 1158–1166.

Hormdok, R., & Noomhorm, A. (2007). Hydrothermal treatments of rice starch for improvement of rice noodle quality. *LWT-Food Science and Technology, 40*(10), 1723–1731.

Hruby, A., & Hu, F. B. (2015). The epidemiology of obesity: A big picture. *PharmacoEconomics, 33*(7), 673–689.

Hu, P., Zhao, H., Duan, Z., Linlin, Z., & Wu, D. (2004). Starch digestibility and the estimated glycemic score of different types of rice differing in amylose contents. *Journal of Cereal Science, 40*(3), 231–237.

Huang, M., Li, X., Hu, L., Xiao, Z., Chen, J., & Cao, F. (2021). Comparing texture and digestion properties between white and brown rice of Indica cultivars preferred by Chinese consumers. *Scientific Reports, 11*(1). https://doi.org/10.1038/s41598-021-98681-7

Hung, P. V., Vien, N. L., & Phi, N. T. L. (2016). Resistant starch improvement of rice starches under a combination of acid and heat-moisture treatments. *Food Chemistry, 191*, 67–73.

International Diabetes Federation. (2017). *IDF diabetes Atlas* (8th ed.) International Diabetes Federation. [Cited 2019 Jun 23]. Available from: https://www.diabetesatlas.org

Itoh, K., Ozaki, H., Okada, K., Hori, H., Takeda, Y., & Mitsui, T. (2003). Introduction of Wx transgene into rice wx mutants leads to both high-and low-amylose rice. *Plant Cell Physiology, 44*(5), 473–480.

Itoh, Y., Crofts, N., Abe, M., Hosaka, Y., & Fujita, N. (2017). Characterization of the endosperm starch and the pleiotropic effects of biosynthetic enzymes on their properties in novel mutant rice lines with high resistant starch and amylose content. *Plant Science, 258*, 52–60.

Jukanti, A. K., Pautong, P. A., Liu, Q., & Sreenivasulu, N. (2020). Low glycemic index rice-a desired trait in starchy staples. *Trends in Food Science & Technology, 106*, 132–149.

Kaur, H., Gill, B. S., & Karwasra, B. L. (2018). In vitro digestibility, pasting, and structural properties of starches from different cereals. *International Journal of Food Properties, 21*(1), 70–85.

Kaur, B., Koh, M., Ponnalagu, S., & Henry, C. J. (2020). Postprandial blood glucose response: Does the glycaemic index (GI) value matter even in the low GI range? *Nutrition and Diabetes, 10*(15). https://doi.org/10.1038/s41387-020-0118-5

Khan, A., Siddiqui, S., Rahman, U. U., Ali, H., Saba, M., Azhar, F. A., et al. (2020). Physicochemical properties of enzymatically prepared resistant starch from maize flour and its use in cookies formulation. *International Journal of Food Properties, 23*(1), 549–569.

Lehmann, U., & Robin, F. (2007). Slowly digestible starch – Its structure and health implications: A review. *Trends in Food Science & Technology, 18*(7), 346–355.

Leu, R. K., Hu, Y., Brown, I. L., & Young, G. P. (2009). Effect of high amylose maize starches on colonic fermentation and apoptotic response to DNA-damage in the colon of rats. *Nutrition & Metabolism, 6*(11). https://doi.org/10.1186/1743-7075-6-11

Li, X. (2018). Resistant starch and its applications. In Z. Jin (Ed.), *Functional starch and applications in food* (pp. 63–90). Springer.

Li, L., Jiang, H., Campbell, M., Blanco, M., & Jane, J. (2008). Characterization of maize amylose-extender (ae) mutant starches. Part I: Relationship between resistant starch contents and molecular structures. *Carbohydrate Polymers, 74*(3), 396–404.

Li, M., Piao, J.-H., Tian, Y., Li, W.-D., Li, K.-J., & Yang, X.-G. (2010). Postprandial glycaemic and insulinaemic responses to GM-resistant starch-enriched rice and the production of fermentation-related H2 in healthy Chinese adults. *British Journal of Nutrition, 103*(7), 1029–1034.

Li, A., Gao, Q., & Ward, R. (2011). Physicochemical properties and in vitro digestibility of resistant starch from mung bean (*Phaseolus radiatus*) starch. *Starch/Stärke, 63*(3), 171–178.

Li, R., Dai, L., Peng, H., Jiang, P., Liu, N., Zhang, D., Wang, C., & Li, Z. (2021). Effects of microwave treatment on sorghum grains: Effects on the physicochemical properties and in vitro digestibility of starch. *Journal of Food Process Engineering, 44*(10). https://doi.org/10.1111/jfpe.13804

Livesey, G., Taylor, R., Livesey, H. F., Buyken, A. E., Jenkins, D. J. A., Augustin, L. S. A., et al. (2019). Dietary glycemic index and load and the risk of type 2 diabetes: A systematic review and updated meta-analyses of prospective cohort studies. *Nutrients, 11*(6), 1280. https://doi.org/10.3390/nu11061280

Ludwig, D. S. (2002). The glycemic index: Physiological mechanisms relating to obesity, diabetes, and cardiovascular disease. *JAMA, 287*(18), 2414–2423.

Lunn, J., & Buttriss, J. L. (2007). Carbohydrates and dietary fbre. *Nutrition Bulletin, 32*(1), 21–64.

Ma, T., & Lee, C.-D. (2021). Effect of high dose resistant starch on human glycemic response. *Journal of Nutritional Medicine and Diet Care, 7*(1). https://doi.org/10.23937/2572-3278/1510048

Magallanes-Cruz, P. A., Flores-Silva, P. C., & Bello-Perez, L. A. (2017). Starch structure influences its digestibility: A Review. *Journal of Food Science, 82*(9), 2016–2023.

Masina, N., Choonara, Y. E., Kumar, P., du Toit, L. C., Govender, M., Indermun, S., & Pillay, V. (2017). A review of the chemical modification techniques of starch. *Carbohydrate Polymers, 157*, 1226–1236.

McKevith, B. (2004). Nutritional aspects of cereals. *Nutrition Bulletin, 29*, 111–124.

Miao, M., Jiang, B., & Zhang, T. (2009). Effect of pullulanase debranching and recrystallization on structure and digestibility of waxy maize starch. *Carbohydrate Polymers, 76*, 214–221.

Milašinović- Šeremešić, M. S., Radosavljević, M. M., & Dokić, L. P. (2012). Starch properties of various ZP maize genotypes. *Acta Periodica Technologica, 43*, 61–68.

Mishra, S., & Monro, J. (2012). Wholeness and primary and secondary food structure effects on in vitro digestion patterns determine nutritionally distinct carbohydrate fractions in cereal foods. *Food Chemistry, 135*(3), 1968–1974.

Nugent, A. P. (2005). Health properties of resistant starch. *Nutrition Bulletin, 30*, 27–54.

Nurmilah, S., & Subrooto, E. (2021). Chemical modification of starch for the production of resistant starch type-4 (RS4): A review. *International Journal of Engineering Trends & Technology, 69*(7), 45–50.

Nutrition Science Corner. (2015). Mondelez International R&D. In *Slowly digestible starch*. https://www.mondelezinternationalnutritionscience.com/

Park, O. J., Kang, N. E., Chang, M. J., & Kim, W. K. (2004). Resistant starch supplementation influences blood lipid concentrations and glucose control in overweight subjects. *Journal of Nutritional Science and Vitaminology, 50*(2), 93–99.

Pereira, M. A., Swain, J., Goldfine, A. B., Rifai, N., & Ludwig, D. S. (2004). Effects of a low-glycemic load diet on resting energy expenditure and heart disease risk factors during weight loss. *JAMA, 292*(20), 2482–2490.

Polesi, L. F., da Matta Junior, M. D., Sarmento, S. B., & Canniatti-Brazaca, S. G. (2017). Starch digestibility and physicochemical and cooking properties of irradiated rice grains. *Rice Science, 24*(1), 48–55.

Punia, S. (2020a). Barley starch: Structure, properties and in vitro digestibility-A review. *International Journal of Biological Macromolecules, 155*, 868–875.

Punia, S. (2020b). Barley starch modifications: Physical, chemical and enzymatic-A review. *International Journal of Biological Macromolecules, 144*, 578–585.

Punia Bangar, S., Ashogbon, A. O., Dhull, S. B., Thirumdas, R., Kumar, M., Hasan, M., ... Pathem, S. (2021a). Proso-millet starch: Properties, functionality, and applications. *International Journal of Biological Macromolecules, 190*, 960–968.

Punia Bangar, S., Singh Sandhu, K., Trif, M., Rusu, A., IOANA, P., & Kumar, M. (2021b). Increasing yield and enrichment of different health components of barley flour using twin-screw extrusion technology to support nutritionally balanced diets. *Frontiers in Nutrition*, 1259.

Punia Bangar, S., Kumar, M., & Whiteside, W. S. (2021c). Mango seed starch: A sustainable and eco-friendly alternative to increasing industrial requirements. *International Journal of Biological Macromolecules*.

Punia, S., Sandhu, K. S., & Sharma, S. (2017). Comparative studies of color, pasting and antioxidant properties of wheat cultivars as affected by toasting and roasting. *Nutrafoods, 16*, 95–1021.

Punia, S., Siroha, A. K., Sandhu, K. S., & Kaur, M. (2019). Rheological behavior of wheat starch and barley resistant starch (type IV) blends and their starch noodles making potential. *International Journal of Biological Macromolecules, 130*, 595–604.

Punia, S., Sandhu, K. S., Dhull, S. B., Siroha, A. K., Purewal, S. S., Kaur, M., & Kidwai, M. K. (2020). Oat starch: Physico-chemical, morphological, rheological characteristics and its applications-A review. *International Journal of Biological Macromolecules, 154*, 493–498.

Raigond, P., Dutt, S., & Singh, B. (2019). Resistant starch in food. In J.-M. Merillon & K. G. Ramawat (Eds.), *Bioactive molecules in food* (Reference series in phytochemistry) (pp. 815–846).

Remya, R., & Jyothi, A. N. (2015). A comparative study on the resistant starch content from different botanical sources in relation to their physicochemical properties. *Journal of Root Crops, 41*(1), 37–47.

Ren, N., Ma, Z., Xu, J., & Hu, X. (2020). Insights into the supramolecular structure and techno-functional properties of starch isolated from oat rice kernels subjected to different processing treatments. *Food Chemistry, 317*, 126464. https://doi.org/10.1016/j.foodchem.2020.126464

Robertson, M. D. (2012). Dietary-resistant starch and glucose metabolism. *Current Opinion in Clinical Nutrition & Metabolic Care, 15*(4), 362–367.

Rochfort, S., & Panozzo, J. (2007). Phytochemicals for health, the role of pulses. *Journal of Agricultural & Food Chemistry, 55*, 7981–7994.

Sajilata, M. G., Singhal, R. S., & Kulkarni, P. R. (2006). Resistant starch – A review. *Comprehensive Reviews in Food Science & Food Safety, 5*, 1–17.

Sánchez-Rivera, M. M., Núñez-Santiago, M. D. C., Bello Perez, L. A., Agama-Acevedo, E., & Alvarez-Ramirez, J. (2017a). Citric acid esterification of unripe plantain flour: Physicochemical properties and starch digestibility. *Starch/Stärke, 69*, 1700019.

Sánchez-Rivera, M. M., Núñez-Santiago, M. D. C., Bello-Perez, L. A., Agama- Acevedo, E., & Alvarez-Ramirez, J. (2017b). Citric acid esterification of unripe plantain flour: Physicochemical properties and starch digestibility. *Starch/Stärke, 69*. https://doi.org/10.1002/star.201700019

Sandhu, K. S., & Punia, S. (2017). Enhancement of bioactive compounds in barley cultivars by solid substrate fermentation. *Journal of Food Measurement and Characterization, 11*(3), 1355–1361.

Sandhu, K. S., Punia, S., & Kaur, M. (2016). Effect of duration of solid state fermentation by aspergillus awamorinakazawa on antioxidant properties of wheat cultivars. *LWT-Food Science and Technology, 71*, 323–328.

Sandhu, K. S., Punia, S., & Kaur, M. (2017). Fermentation in cereals: A tool to enhance bioactive compounds. In S. K. Gahlawat & J. S. Duhan (Eds.), *Plant biotechnology: Recent advancements and developments* (pp. 157–170). Springer.

Shen, R.-L., Zhang, W.-J., & Dong, J.-L. (2016). Preparation, structural characteristics and digestibility of resistant starches from highland barley, oats and buckwheat starches. *Journal of Food and Nutrition Research, 55*(4), 303–312.

Shewry, P. R. (2009). Wheat. *Journal of Experimental Botany, 60*(6), 1537–1553.

Shin, S. I., Choi, H. J., Chung, K. M., Hamaker, B. R., Park, K. H., & Moon, T. W. (2004). Slowly digestible starch from debranched waxy sorghum starch: Preparation and properties. *Cereal Chemistry, 81*, 404–408.

Shin, S. I., Lee, C. J., Kim, D. I., Lee, H. A., Cheong, J. J., et al. (2007). Formation, characterization, and glucose response in mice to rice starch with low digestibility produced by citric acid treatment. *Journal of Cereal Science, 45*(1), 24–33.

Shu, X., Jia, L., Ye, H., Li, C., & Wu, D. (2009). Slow digestion properties of rice different in resistant starch. *Journal of Agricultural and Food Chemistry, 57*(16), 7552–7559.

Silva, W. M., Biduski, B., Lima, K. O., Pinto, V. Z., Hoffmann, J. F., Vanier, N. L., & Dias, A. R. (2017). Starch digestibility and molecular weight distribution of proteins in rice grains subjected to heat-moisture treatment. *Food Chemistry, 219*, 260–267.

Singh, J., Dartois, A., & Kaur, L. (2010). Starch digestibility in food matrix: A review. *Trends in Food Science & Technology, 21*(4), 168–180.

Singh, J., Kaur, L., & Singh, H. (2013). Food microstructure and starch digestion. In J. Henry (Ed.), *Advances in food and nutrition research* (Vol. 70, pp. 137–179). Academic.

Sivert, D., & Pomeranz, Y. (1989). Enzyme-resistant starch. I. Characterization and evaluation by enzymatic, thermo analytical, and microscopic methods. *Cereal Chemistry, 66*, 342–347.

Sone, H., Nakagami, T., Nishimura, R., & Tajima, N. (2016). Comparison of lipid parameters to predict cardiovascular events in Japanese mild-to-moderate hypercholestemic patients with and without type 2 diabetes: Subanalysis of the MEGA study. *Diabetes Research and Clinical Practice, 113*, 14–22.

Sparla, F., Falini, G., Botticella, E., Pirone, C., Talamè, V., Bovina, R., et al. (2014). New starch phenotypes produced by TILLING in barley. *PLoS One, 9*(10). https://doi.org/10.1371/journal.pone.0107779

Štěrbová, L., Bradová, J., Sedláček, T., Holasová, M., Fiedlerová, V., Dvořáček, V., & Smrčková, P. (2016). Influence of technological processing of wheat grain on starch digestibility and resistant starch content. *Starch/Stärke, 68*(7–8), 593–602.

Tabibloghmany, F. S., & Ehsandoost, E. (2014). Investigation of nutritional and functional properties of resistant starch in food industry: A review. *International Journal of Recent Research and Review, 3*(1), 27–44.

Teixeira, D. C., Rocha, M. C., Amorim, A. C. P., Soares, T. O., Monteiro, M. A. M., et al. (2015). Resistant starch content among several sorghum (Sorghum bicolor) genotypes and the effect of heat treatment on resistant starch retention in two genotypes. *Food Chemistry, 197*(Pt A), 291–296.

The Nutrition Source. Whole Grains: Harvard T.H. Chan School of Public Health. Available at https://www.hsph.harvard.edu/nutritionsource/what-should-you-eat/whole-grains.

Tian, S., & Sun, Y. (2020). Influencing factor of resistant starch formation and application in cereal products: A review. *International Journal of Biological Macromolecules, 149*, 424–431.

Tuaño, A. P., Barcellano, E. C., & Rodriguez, M. S. (2021). Resistant starch levels and in vitro starch digestibility of selected cooked Philippine brown and milled Rices varying in apparent amylose content and glycemic index. *Food Chemistry: Molecular Sciences, 2*. https://doi.org/10.1016/j.fochms.2021.100010

Vinoy, S., Anbert, R., & Chapelot, D. (2020). A cereal product high in slowly digestible starch increases subsequent feelings of satiety and decreases glucose and insulin responses. *Journal of Human Nutrition & Food Science, 8*(1), 1132.

Wahjuningsih, S. B., Haslina, H., & Marsono, M. (2018). Hypolipidaemic effects of high resistant starch sago and red bean flour-based analog rice on diabetic rats. *Materia Socio Medica, 30*(4), 232–239.

Walker, A. W., Ince, J., Duncan, S. H., Webster, L. M., Holtrop, G., Ze, X., Brown, D., Stares, M. D., et al. (2011). Dominant and diet-responsive groups of bacteria within the human colonic microbiota. *The ISME Journal, 5*, 220–230.

Wang, S., & Copeland, L. (2013). Molecular disassembly of starch granule during gelatinization and its effect on starch digestibility: A review. *Food & Function, 4*, 1564–1580.

Wang, H., Liu, Y., Chen, L., Li, X., Wang, J., & Xie, F. (2018). Insights into the multi-scale structure and digestibility of heat-moisture treated rice starch. *Food Chemistry, 242*, 323–329.

Warren, J. M., Henry, C. J. K., & Simonite, V. (2003). Low glycemic index breakfasts and reduced food intake in preadolescent children. *Pediatrics, 112*(5), 414–419.
Weitkunat, K., Schumann, S., Petzke, K. J., Blaut, M., Loh, G., & Klaus, S. (2015). Effects of dietary inulin on bacterial growth, short-chain fatty acid production and hepatic lipid metabolism in gnotobiotic mice. *Journal of Nutritional Biochemistry, 26*(9), 929–937.
WholeEUGrain. (2021). *Whole grain: Definition, evidence base review, sustainability aspects and considerations for a dietary guideline* (pp. 1–97). Ed. Lourenço S.M. WholeEUGrain: Copenhagen.
Wolever, T. M. S. (2003). Carbohydrate and the regulation of blood glucose and metabolism. *Nutrition Reviews, 61*(5 Pt 2), 40S–48S.
Wong, T. H., & Louie, J. C. (2016). The relationship between resistant starch and glycemic control: A review on current evidence and possible mechanisms. *Starch/Stärke, 68*, 1–9.
Wongsagonsup, R., Varavinit, S., & BeMiller, J. N. (2008). Increasing slowly digestible starch content of normal and waxy maize starch and properties of starch products. *Cereal Chemistry, 85*(6), 738–745.
Xu, J., Kuang, Q., Wang, K., Zhou, S., Wang, S., Liu, X., & Wang, S. (2017). Insights into molecular structure and digestion rate of oat starch. *Food Chemistry, 220*, 25–30.
Yao, T., Sui, Z., & Janaswamy, S. (2018). Annealing. In *Physical modifications of starch* (pp. 37–49). https://doi.org/10.1007/978-981-13-0725-6_3.
Ye, J. P., Liu, C. M., Luo, S. J., Hu, X. T., & McClements, D. J. (2018). Modification of the digestibility of extruded rice by enzyme (β-amylolysis): An *in vitro* study. *Food Research International, 111*, 590–596.
Ye, J. P., Luo, S. J., Huang, A., Chen, J., Liu, C. M., & McClements, D. J. (2019). Synthesis and characterization of citric acid esterified rice starch by reactive extrusion: A new method of producing resistant starch. *Food Hydrocolloids, 92*, 135–142.
Yoon, M. R., Lee, J. S., Lee, J. H., Kwak, J., Areum, C., & Kim, B. K. (2013). Content and characteristics of resistant starch in high amylose mutant rice varieties derived from Ilpum. *Korean Journal of Breeding Science, 45*(4), 324–331.
Younes, H., Levrat, M. A., Demige, C., & Remesy, C. (1995). Resistant starch is more effective than cholestyramine as a lipid lowering agent in the rat. *Lipids, 30*, 84 853.
Zavareze, E. R., & Dias, A. R. (2011). Impact of heat-moisture treatment and annealing in starches: A review. *Carbohydrate Polymers, 83*(2), 317–328.
Zhang, N., Sun, J., Xiong, H. J., Xu, K. S., Shu, X. L., & Wu, D. X. (2011). Breeding and characteristics of high resistant starch *Indica* rice for diabetes. *China Rice, 17*(6), 63–65.
Zhang, K., Dong, R., Hu, X., Ren, C., & Li, W. (2021). Oat-based foods: Chemical constituents, glycemic index and the effect of processing. *Food, 10*(6). https://doi.org/10.3390/foods10061304

Chapter 6
The Functionality of β-Glucans and Fibers in Cereals

Mehnaza Manzoor and Sneh Punia Bangar

6.1 Introduction

β-glucan dietary fibers are biologically active non-starch polysaccharides that have become the focus of interest over the years, particularly as health-promoting ingredients and pharmaceuticals because of their physiological and biological characteristics such as blood lipid and blood glucose level reduction, lowering serum cholesterol levels, immunomodulation effect, antidiabetic properties, antitumor activity, obesity prevention and prebiotic effect. As a kind of dietary fiber, *β*-glucans could be found in a variety of natural sources such as wheat, oat, barley, and microorganisms such as bacteria, yeast and algae where they form cell walls and contribute to cell wall integrity, cell exchanges and provide a barrier against environmental stresses and pathogens (Du et al., 2019; Zhu et al., 2015). Because of the difference in *β* glucan source, different structural features emerge, which are major determinants of physical attributes including gelation, viscosity, and water solubility, as well as their physiological activity in the gastrointestinal system.

Plant-derived *β*-glucans, particularly those from the Poaceae family, which includes cereals, are linear homopolymers of glucose linked by *β*-(1/3; 1/4) glucosidic links. The order of β-(1/3) and β-(1/4)-linkages in the chain is not fully random. About 90% of the glucose units are grouped in blocks of two or three consecutive *β*-(1/4)-linked units separated by a single *β*-(1/3)-linkage, forming the

M. Manzoor (✉)
Division of Food Science and Technology, Sher-e-Kashmir University of Agriculture Science and Technology, Jammu, Jammu and Kashmir, India

Department of Food Technology and Nutrition, School of Agriculture, Lovely Professional University, Jalandhar, Punjab, India

S. P. Bangar
Department of Food, Nutrition and Packaging Sciences, Clemson University, Clemson, SC, USA

S. Punia Bangar, A. Kumar Siroha (eds.), *Functional Cereals and Cereal Foods*,
https://doi.org/10.1007/978-3-031-05611-6_6

cellotriosyl and cellotetraosyl units, which are the two building blocks of cereal -glucans. The (1/3) *β*-linkages behave as kinks in the (1/4)-linkages, disrupting their regularity. Longer blocks of up to 14 *β*- (1,4)-linked glucopyranose units make up the remaining less than 10% of the polymer chain. In cereals, cellotriosyl makes up 58–72% of cellotriosyl, while cellotetraosyl makes up 20–34%. Lichenan is a β-glucan derived from *Cetraria islandica* that has a structure similar to cereal β-glucan but is made up of cellotriosyl (78%), cellotetraosyl (4%), and cellulose-like segments (18%) (Lazaridou & Biliaderis, 2007). On the other hand, Microbial-derived- glucans have branching 1, 3 glycosidic linkages with a minority of 1,6 glycosidic bonds (Du et al., 2015).

In general, β-glucans are classified as soluble or insoluble, depending on the degree of polymerization (DP). β-glucans with higher DPs (more than 100) are frequently water-insoluble (Du et al., 2015). Water-insoluble and soluble *β*-glucans may differ in their biological and physical properties. However, both soluble and insoluble, β-glucans are used in almost every facet of daily life. The potential use of *β*-glucans as hydrocolloids in the food industry relies mainly on their rheological behavior i.e. their gelling capacity and ability to increase the viscosity of aqueous solutions. They've been used as viscosity modifiers, emulsifiers, stabilizers, fat replacers, texturizers, and film-forming agents in the food industry to help end products have good functional qualities (Manzoor et al., 2020). β-glucans, for example, operate as viscosity modifiers in beverages to change solution consistency. They can also be used as additives in bread, milk, and yogurt to exhibit cholesterol-lowering and energy-saving effects (Sharafbafi et al., 2014). As a result, these natural polysaccharides have the potential to be used in the food, nutraceutical, and pharmaceutical industries for their functional and nutraceutical qualities.

The nutraceutical potential of *β*-glucans has been linked to improved blood glucose and insulin management, lower blood cholesterol, particularly LDL and total serum cholesterol levels, and modulation of the immunological response to bacterial, viral, and fungal diseases. These beneficial health effects have been attributed to the solubility of *β*-glucans in water and their ability to generate highly viscous solutions. The Food and Drug Administration (FDA) and other prominent institutions throughout the world recognize the importance of *β*-glucan as dietary fiber. The controlled consumption of this dietary fiber is believed to aid in the management of various lifestyle disorders, including hypocholesterolemia, obesity, type-2 diabetes, cancer, tumors, and some skin diseases. LDL and total cholesterol levels are reduced by 0.25 and 0.30 mmol/L, respectively, when β-glucan is consumed (Whitehead et al., 2014). The FDA of the United States (FDA, 2008) has advised a daily intake of 3.0 g of β-glucan that can lower cholesterol levels by 23%. Grains that contain β-glucan are classified as functional foods. "Functional foods influence the organism by increasing the functions of various organs, decreasing the danger of falling ill due to their preventive activity, and also by optimizing the weight and altering the mental state," according to Havrlentova et al. (2011). This justifies the daily introduction of new varieties of β-glucan and other fiber-rich foods to the market.

6.2 Sources of β-Glucans

The key natural sources of β-glucans are plant sources such as cereals (barley, wheat, and oats) and microorganisms (yeasts, fungi, prebiotic bacteria, etc.). In wheat, β-glucans are mostly present in the aleurone cell walls of wheat bran, although they can also be found in other cereal grains. In oats, they are concentrated in the sub-aleurone layer, while they are typically present in the endosperm of barley and rye. It has been investigated that cereal β-glucans are water-soluble, while other β-glucan sources are water-insoluble. Although both types are reported to exhibit health benefits, water-soluble β-glucans have additional health advantages due to their high viscosity in the aqueous medium. The ability to form a viscous solution depends on the concentration and the molecular weight of β-glucans. This chemical variation defines its industrial application and health benefits.

6.2.1 *Cereal Grains*

Certain cereal grains, such as oat and barley, are high in β-glucans, whereas grains like wheat, rye, and sorghum have very low levels. Commercial β-glucans from cereal sources are rich in dietary fibers having more than 50% concentration. The percentage of β-glucan content among different cereals like oats, barley varies with their sources. The highest concentration is found in barley (3–11%) followed by oats (3–7%), rye (1–2%), and wheat (<1%) (Gani et al., 2012). Despite having higher β-glucan content than oat, barley is less commonly used as a dietary fiber source in food products, owing to its appearance and flavor (Limberger-Bayer et al., 2014). Oats, on the other hand, have a pleasant flavor, a longer shelf life, a higher nutritional profile with little to no allergenicity (Sterna et al., 2016). They are used in several meals, including morning cereals, bread, beverages, and even infant feeds. Apart from the commonly known sources, some other storehouse of β-glucans includes beans, lentils (*Lens culinaris*), millet (*Panicum miliaceum*), corn/ maize (*Zea mays*), and canary seeds (*Tropaeolum peregrinum).*

Structurally, cereal β-glucan possesses β-(1-4) linkage as the primary chain, which connects with β-(1-3) linkage at random places, giving the chain a helical appearance (Tosh et al., 2004). The presence of β-(1-3) linkage makes them more soluble and the "β" configuration is not digestible by enzymes in the human gastrointestinal tract, allowing for the classification as soluble dietary fiber (Mejía et al., 2020). However, the structure of cereal β- glucan has not been standardized because the number of cellotriosyl and cellotetraosyl units, connected by a single β-(1-3) linkage and dominating in the chain can vary greatly, while oligomeric glucopyranosyl units of β-(1-4) connected by β-(1-3) D-glucose can occur in some places (Vasanthan & Temelli, 2008). Therefore, molar ratio, defined as the ratio between cellotriosyl/cellotetraosyl units (DP3/DP4), after depolymerization of the β-glucan molecule, varies with each cereal such as for wheat (3.0–4.5), barley (1.8–3.5), rye

(1.9–3.0), and oat (1.5–2.3) (Lazaridou & Biliaderis, 2007). This factor determines the physicochemical and functional properties of extracted β-glucan. Cereal β-glucans have a molecular weight that varies depending on the grain's origin, cultivar, and isolation processes, and has been reported in the range of 0.4–2.5 10^6 (Irakli et al., 2004). Oat and barley β-glucans have a much higher molecular weight than other grain β-glucans, ranging from 200 to 300 kDa (Mikkelsen et al., 2013). Also, the molar mass of β-glucans has a major impact on the health-promoting character of the cereals when consumed. According to Whitehead et al. (2014), a molar mass > 210kD in β-glucans had excellent qualities and contribute to better lipid profile, lower cholesterol levels, and preserving obesity range in breakfast food matrices (muesli, multi-grain flakes). A daily intake of 3 g of β-glucans/day with a molar mass of 100kD or above significantly improved health against chronic cardiovascular and lifestyle disorder-related diseases (Mejía et al., 2020).

6.2.2 Microbial Sources of β-Glucans

In parallel to cereal β-glucans, microbial sources are considered as good sources of dietary fibers including β-glucan such as those obtained from, yeast, fungi, bacteria. Microbial extracted β-glucan can be a linear molecule with a repeating glucose unit linked together by (1–3) links, or it can be made up of glucose units with (1–3), (1–6) glycosidic linkages. These can sometimes be found in cyclic structures with branching at the 1–2 position, resulting in the chemical structure of (1–3, 1–2)-β-glucans. The β-glucan dietary fiber can be produced by both prokaryotes and eukaryotes. Curdlan extracted from microbial source is a linear β-glucan molecule that has gelling and rheological qualities and finds applications in a variety of food products (McIntosh et al., 2005). Curdlan is produced by bacterial species like *Agrobacterium*, *Rhizobium*, *Cellulomonas*, *Streptococcus mutans,* and *Alcaligenes faecalis* and is recognized by JECFA/ FAO and WHO as inert dietary fiber as well as a food additive for use in freezable tofu noodles, ham, sausages, and cakes in countries like Korea, Taiwan, and Japan. Besides, curdlan has also been considered as a meat analog for fish tofu, shrimp ball, and crab meat (Philips et al., 1983). Curdlan fiber being non-toxic, biocompatible, and biodegradable is of Generally Regarded as safe (GRAS) status. *Pseudomonas* sp. QL212, a new source of low molecular weight polysaccharides from *Pseudomonas* species isolated from soil samples, has been identified as a potential source of curdlan. Because of its low molecular weight, it has a higher solubility. Apart from curdlan, the β -glucan extracted from spent brewer's yeast had high apparent viscosity, water-holding, oil-binding, and emulsion stabilizing capacities as studied by Thammakiti et al. (2004). Laminarin is a β- (1,3) linked glucan having storage house in algae and lichens. Source of β-(1,3)- (1,6) glucans are also found in *Lentinula edodes* and are named as lentinan (Synytsya & Novak, 2014). β-glucans from microbial sources hold a huge market share and dominate the biopolymer market in comparison to other biopolymers from microbial sources like pullulan, gellan, curdlan, gum, dextran, laminarin, cellulose, lentinan, xanthan, and bacterial alginate. However industrial-scale production of these microbial β-glucans has not been done as there is a scanty study on the application due to the smaller level of extraction.

6.3 β-Glucan as a Source of Dietary Fiber

In recent years the utilization of β-glucans as useful dietary fiber has sparked a lot of interest in their usage in food systems. Dietary fiber is a word that refers to a set of compounds found in plant matter that resist endogenous mammalian enzymes. Dietary fiber has been officially defined by the American Association of Cereal Chemists' Dietary Fiber Technical Committee (DeVries, 2003). The potential health benefits of dietary fibers include reduced bowel transit time, better gut health, prevention against constipation, and colorectal cancer, lowering of blood cholesterol, and regulation of blood glucose levels for diabetes management (Brennan & Cleary, 2005). Dietary fiber research has looked into the benefits of soluble and insoluble fractions as pure fiber or in naturally fiber-rich whole meals. According to studies by Jenkins et al. (1978) high fiber foods have been associated with the alteration of glycaemic response. In particular, foods high in soluble dietary fiber, have been shown to reduce hyperglycemia and hyperinsulinemia in relation to diabetes control (Brennan & Tudorica, 2003) and the reduction of risk factors for degenerative diseases, such as obesity (Burley et al., 1987), hyperlipidemia (Yang et al., 2003), hypertension (Anderson, 1990), cardiovascular diseases (Keogh et al., 2003) and cancer (Sier et al., 2004).

Many studies have been conducted to determine the mechanisms by which dietary fiber and β-glucans exert their benefits. The potential reduction in glycaemic response after ingestion of dietary fiber has led to theories involving: the amount and quality of fiber; an increased intrinsic viscosity of the food in combination with fluids and thus the gastrointestinal environment; food material physical integrity and incomplete starch gelatinization (Brennan & Cleary, 2005). Cereal fiber's cholesterol-lowering potential is thought to be due to effects in the upper gastrointestinal tract. These, in turn, could be linked to cereal fiber's propensity to create a gel-like network and change gastrointestinal viscosity (Reimer et al., 2000). An increase in viscosity in the upper gut portion may lead to slow digestion and absorption and delay starch degradation. This may be due to the weak action of pancreatic amylase on starches. Some other factors, including nonstarch polysaccharides, might also be involved during this mechanism to define the physiological functionality of β-glucan well (Regand et al., 2011). So, whatever the source of β-glucan, it may provide beneficial effect through its high viscosity or gel formation properties and delay the absorption of certain nutrients, including glucose and other sugars, thereby slowly passing the glucose into the bloodstream and lessening the requirement of insulin entering into cells.

The availability of a variety of sources for β-glucans, as well as its health implications, has encouraged industrialists' interest in adding this beneficial element into food items for the development of nutraceutical foods. For diabetes patients, there is a variety of low glycemic index (GI) meals available, many of which are made with β-glucan. Food scientists are working hard to create novel functional foods containing β-glucan fibers that will benefit diabetic, cardiovascular, and cancer patients and act as a prophylactic against these diseases (American Diabetes Association, 1999). As a result, various health organizations, including the FDA and WHO, advocate incorporating β-glucan in one's regular diet. These recommendations have sparked the introduction of β-glucan into a variety of food products and are being used as a stabilizer, viscosity modifier, fat replacer, and thickener.

6.4 Functional Properties of Cereal *β*-Glucan

β-Glucans display all the functional characteristics of viscous and gel-forming food hydrocolloids combined with all the physiological characteristics of dietary fibers. The overall functionalities of *β*-glucans like solubility, hydration, and gelation in the aqueous systems are controlled by structural characterization, such as their distribution of cellulosic oligomers, their linkage pattern, and their molecular weight, as well as by temperature and concentration. The rheological behavior of *β*-glucans is frequently linked to their technical and nutritional value.

6.4.1 Solubility

The solubility of cereal *β*-glucan and other polysaccharides represents the maximum amount of compound that can be diluted in water to form a homogenous solution under controlled pressure and temperature (Mejía et al., 2020). The molecular irregularity of *β*-glucan is reflected in their water solubility properties, although they are classified as soluble fibers (El Khoury et al., 2012). Certain necessary substitution like the presence of COOH or SO_2 increases polysaccharide solubility (Elleuch et al., 2011). The chemical confirmation of a molecule allows for extensive interactions and associations between its chains and water molecules; nevertheless, a high degree of polymerization may reduce the molecule's solubility (El Khoury et al., 2012). Cereal *β*-glucans with cellulose-like segments may contribute to the stiffness of the molecules in solution (Varum & Smidsrod, 1988); *β*-glucans with blocks of adjacent *β*-(1-4) linkages may have a tendency for interchain aggregation (and thus lower solubility) due to strong hydrogen bonds along with the cellulose-like portions. The *β*-(1-3) linkages disrupt the regularity of the *β* -(1-4) linkage sequence, resulting in a more soluble and flexible molecule. It has been proposed that the non-ordered conformation of the polysaccharide is caused by the irregular spacing of (1–3) linked *β*-glucosyl residues in the *β*-glucan chain, and therefore the chains are unable to align closely over entended regions, keeping the polysaccharide in solution (Woodward et al., 1988). On the other hand, it has been suggested that in mixed linked (1-3),(1-4) β-glucans, helical segments made up of at least three consecutive cellotriosyl residues may form a conformationally stable motif (ordered domain) (Tvaroska et al., 1983) and that the β-(1-3) linkages may be involved in the ordered conformation of barley β-glucans. As a result, a higher amount of consecutive cellotriosyl units could impose some conformational regularity on the β-glucan chain, resulting in a higher degree of organization of these polymers in solution (and hence poorer solubility) (Izydorczyk et al., 2003). According to these two aggregation mechanisms, higher amounts of cellotriosyl fragments and a higher ratio of (1–4):(1–3) linkages could explain the solubility differences between cereal β-glucan fractions obtained using different aqueous or alkali extraction conditions and/or different ammonium sulphate concentrations.

6.4.2 *Rheological Properties*

The rheological property of a material is described as its response to deformation and flow behaviors. Cereal β-glucan falls rheologically into the category of viscoelastic fluids behaving similarly to random coil-type polysaccharides and their rheological properties may change depending on their molecular characteristics (size, structure), the concentration of dietary fiber, and their solubility. Most cereal β-glucans behave as a Newtonian solution at low concentrations, but they tend to form gels at high concentrations, and their gelling properties are influenced by molecular weights structural characteristics, and concentration of polysaccharides, (Lazaridou et al., 2003; Lazaridou & Biliaderis, 2004). At concentrations, more than 1% β-glucan solution exhibits pseudoplastic behavior, which further rises with concentration (Burkus & Temeli, 2000). At increasing concentrations, β-glucan with a low molecular weight forms soft gels, whereas β-glucan with a high molecular weight forms pseudoplastic and viscous solutions (Doublier & Wood, 1995). On the other hand, β-glucan with short-chain molecules and low molecular weight showed higher mobility. Since these short chains with low molecular weight β-glucan structures diffuse more readily, they have a greater possibility of forming junctions with neighboring chains (Doublier & Wood, 1995). This reaction mechanism indicated that there is an inverse relationship between gelation time and molecular weight of the polysaccharide (Lazaridou et al., 2003). Along with the concentration of β-glucan and molecular weight, viscosity properties are also influenced by tri/tetra ratios, cellulose-like fragments, molecular weight distribution, and molecular size of cereal β-glucan. Furthermore, they can alter some other physiological responses when they are intended for use in cereal-based products (Vaikousi et al., 2004).

The rheological behavior of β-glucan solutions is generally explained based on Power Law. A pseudoplastic behavior of high viscosity β-glucan gums has already been proven using this model. According to Burkus and Temelli (2005), these samples exhibit a high coefficient of consistency (K) and a low ($n < 1$) flow behavior index. Dongowski et al. (2005) evaluated the rheological behavior of β-glucan solutions prepared from extruded meal and bran using oscillatory and rheological measurements. The Power Law parameters like storage moduli (G'') and loss moduli (G″) of β-glucan preparations were found to rise continuously with increasing frequency, demonstrating dominant viscous behavior. It was observed that attraction forces between molecules in freshly produced barley β-glucan solutions were weaker, however, after an induction period, *G′* and *G″* increase with time and aqueous solutions/dispersions of β-glucan may adopt gel-like properties.

In vitro digestion can also be used to determine the viscosity of β-glucan solutions. In this regard, Pentikäinen et al. (2014) calculated the in vitro viscosity of various β-glucan containing breakfast items. The findings revealed that regardless of food form (liquid or solid) concentration of incorporated β-glucans controls the viscosity in gastrointestinal conditions. In a recent study, Mäkelä et al. (2020) used an in vitro gastrointestinal simulation to predict the status of β-glucans from diverse oat products The authors claimed that viscosity was quite important and varied substantially between product categories and functionality of β-glucan was highly dependent on product type. The variation could likely be explained by the difference in ingredients and process.

Food formulation also has an impact on the gelation ability and rheological properties of cereal *β*-glucan gels (Irakli et al., 2004; Lazaridou & Biliaderis, 2007). Incorporation of various sugars (sucrose, fructose, glucose, ribose, and xylose) into aqueous dispersions of barley *β*-glucan (6%) at a concentration of 30% increase gelation time when cured at room temperature (Irakli et al., 2004). Similarly, adding sugars (sucrose, fructose, glucose, and xylose) at concentrations ranging from 5–30% (w/w) to aqueous dispersions of *β*-glucan (3% w/w) with or without skimmed milk (12% w/w) and subjecting them to repeated freezing-thawing caused a significant delay in the transition to a gel state. The resulting cryo-gels were weaker, and xylose and fructose had stronger inhibitory effects than glucose and sucrose. Sorbitol, on the other hand, seems to encourage network creation (Lazaridou & Biliaderis, 2007). While polyols had a substantial influence when added to high molecular weight -glucan preparations (210×10^3), for low molecular weight *β*-glucan preparations (e.g. 140 and 70×10^3) the effect seemed to be independent on type of polyol.

6.4.3 Thermal Properties

Calorimetry analysis is used to explain the phase transition and thermal stability of *β*-glucan molecules by calculating the enthalpy changes (ΔH) and denaturation temperature (T_d). Td and ΔH are closely related to thermal denaturation. The former assesses the energy required to overcome non-covalent contacts during the denaturation process, whereas the latter describes the thermal stability of proteins. The greater the Td and the lower the ΔH, the better the thermal stability. The DSC thermal scans of cereal *β*-glucan hydrogels generated by curing at temperatures above 0 °C and freeze-thaw cycling revealed relatively large endothermic gel to sol transitions at 55–80 °C (Lazaridou & Biliaderis, 2004; Lazaridou et al., 2003; Vaikousi et al., 2004). The rate of endotherm development (short-range molecular order) during gel curing at room temperature Lazaridou et al., 2003), as well as the apparent melting enthalpy values (plateau ΔH) of the gels, increase with decreasing molecular size and increasing DP3:DP4 ratio (Lazaridou et al., 2003; Vaikousi et al., 2004). Cryogel ΔH values also rise as the DP3:DP4 ratio rises and the molecular weight of cereal *β*-glucans decreases.

Furthermore, as assessed by dynamic rheometry and DSC, the melting temperatures of the gel networks investigated increased with the molecular weight and number of cellotriosyl units in the polysaccharide chains (Lazaridou & Biliaderis, 2004; Lazaridou et al., 2003; Vaikousi et al., 2004). Although greater molecular weight *β*-glucan gels at a slower rate, the gel network structure appears to contain structural features (micro aggregates) that are better organized and/or involve interchain interactions over longer chain lengths. It's also worth noting that cryo-gels made from repeated freeze-thaw cycles with low (3%) initial polysaccharide concentrations (Lazaridou & Biliaderis, 2004) are comparable to gel networks formulated at room temperature with much higher polysaccharide concentrations (10% w/v) (Lazaridou & Biliaderis, 2004). DSC thermal scans of these two systems show that the apparent melting enthalpy (H) values for both forms of gels are similar, however, a comparison of the broadness of the endotherm peak reveals that cryo-gelation gives a more cooperative gel to sol transition in general. Similarly, after curing 3% *β*-glucan dispersions at 5 °C for 135 days, which is long enough for G′ to reach a 'pseudo-plateau' value, the strength (G′ value) of the resultant gel is lower than that of cryo-gels made from dispersions with an equivalent initial polysaccharide concentration after 14 freeze-thaw cycles (Lazaridou et al., 2007). The higher concentration (cryoconcentration) of the polysaccharide in the unfrozen bulk phase, which promotes associative contacts among the polysaccharide chains, is responsible for the reinforcement of the *β*-glucan network architectures caused by cryo-structurization (physical cross-linking).

6.5 Physiological Activities of β-Glucans

β-Glucan as a dietary fiber is often recommended as a great nutraceutical ingredient. Several researchers have reported the potential health implications of β-glucan from various sources resulting from its high viscosity or gel formation properties. Typical (1 → 3) (1 → 4)-*β*-glucans from cereals have received the most attention due to their significant protective role in preventing or delaying the onset of chronic diseases and disorders such as coronary heart disease (Truswell, 2002), colon dysfunction (Sudha et al., 2007), lowering blood glucose and LDL cholesterol, controlling obesity, type-2 diabetes, fighting against cancer and cardiovascular disease (CVD).

6.5.1 Improving Gut Microbiota

The potential and benevolent physiological activities of β -glucans have been largely contributed by their ability to increase the viscosity of the gut content causing modification in the absorption rates pattern of nutrients and bile acids. In this respect, the importance of molecular weight (MW) of β -glucan on its nutritional and health-related benefits are well known. The higher the MW of β-glucan, the higher the viscosity and health advantages (Queenan et al., 2007).

In aqueous media, water soluble-G creates a highly viscous solution, which is the basis of its biological activity in the human gastrointestinal system (Cui & Wang, 2009). Unlike water-insoluble β -glucan, which does not break down, soluble β -glucan is digested in the large intestine by the gut bacteria. It is metabolized here to make short-chain fatty acids such as acetic acid, propionic acid, and butyric acid which are having hypocholesterolemic and anti-carcinogenic characteristics (Alminger & Eklund-Jonsson, 2008; Drzikova et al., 2005a, b) and result in gas production (H_2, CO_2, and CH_4). Insoluble β -glucan, on the other hand, which is resistant to the fermentation process in the large intestine, absorbs a huge amount of water, softening undigested food and making the excretion process easier.

In vitro investigations of various β -glucan sources found that it can help bacteria like *Lactobacillus acidophilus, Lactobacillus casei*, and *Bifidobacterium animalis lactis* grow faster (Su et al., 2013). β -glucan from barley and oats were reported to boost *B. animalis lactis* growth and viability in yogurt (Vasiljevic et al., 2007). Another in vitro investigation carried out by Hughes and Rowland (2001) discovered that fermenting oat and barley β -glucan inoculated in human faeces results in changes in SCFA synthesis and bacterial mass (*Clostridium histolyticum* and *Bacteroides-Prevotella*).

6.5.2 Hypocholesterolemic Effect

Viscous fibers associated with cereal β-glucan are responsible for beneficial physiological responses in human, animal, and animal-alternative *in vitro* models (Cheryl et al., 2006). These responses are triggered, controlled, and often altered primarily by β-glucan along or by the presence of other non-starch polysaccharides which may also influence these changes since these types of dietary fiber tend to increase luminal viscosity in the gastrointestinal tract contents, that is a key factor responsible for reduced absorption of carbohydrates, cholesterol and absorption and reabsorption of bile acid (Newman & Newman, 1991). Whole grains of oat and barley, as well as oat bran, have been shown to lower LDL cholesterol and total serum cholesterol in humans and animals, with the effects attributable to the presence of β-glucans (Maier et al., 2000). The effect of β-glucan on lowering the LDL cholesterol and the glycemic index was also investigated by Duss and Nyberg (2004). Davidson et al. (1991) evaluated the effect of oat bran and oatmeal (same quantity) on reduction of LDL-cholesterol, where oat bran was found to have a greater capability over oatmeal to reduce LDL-cholesterol levels. This could imply that the oat bran being a rich source of β-glucan, was more effective in the regulation of glucose and insulin responses compared to oat β-glucan (Hallfrisch et al., 2003; Granfeldt et al., 2008). The current leading explanation for cholesterol-lowering actions in the blood is that binding of viscous β-glucan in the small intestine causes reduced reabsorption and increased excretion of bile salts (Gunness et al., 2016). The cholesterol-lowering mechanism and binding of bile acid have been obediently studied where it was noticed that β-glucan containing extrudates from oat have an ability to bind bile acids. These replenish the deficiency of bile acid wherein more cholesterol from the body is consumed for the synthesis of bile acid thus lowering the serum cholesterol level in the body (Drzikova et al., 2005a, b). β-glucan is also found effective in reducing the risk of chronic diseases including cardiovascular disease by the combined effect of reducing LDL and increasing HDL (Shimizu et al., 2008). In a review of Schmidt, 2022 a range of dietary fiber including β-glucan have been reported to exert beneficial physiological influences that they exert on the human body. According to Kerckhoffs et al. (2003), these physiological responses could be triggered with a recommended daily intake of 3 g soluble fiber that may lower the total cholesterol levels by 0.41 $mmolL^{-1}$ in hypercholesterolemic persons and 0.13 $mmolL^{-1}$ in normal cholesterolemic persons (Kerckhoffs et al., 2003). On a similar work by Behall et al. (1997), it was reported that on ingestion of 2.1 g of β-glucan daily, the total cholesterol reduction could be achieved by 9.5%, whereas according to Jenkins et al., 2002, it was indicated that 4 units decline in glycemic index can be achieved by taking 1 g of β-glucan per 50 g of carbohydrates on an RDA basis. These reports have been in similarity with the regulations of the FDA which has also recommended daily consumption of 3 g β-glucan to achieve such health benefits (FDA, 1997a, b).

6.5.3 Immunomodulation Effect of β-Glucan

Several in vitro, animal and human studies have demonstrated the well know immunomodulatory activities of β-glucan molecules which include the development of resistance against infection caused by bacteria, fungi, and viruses (Bohn & BeMiller, 1995). The immunomodulatory activity of β-glucan dietary fibers is related to macrophage activation (NO production, reactive oxygen species generation, TNF-*α production,* phagocytosis), activation of the reticuloendothelial system, natural killer (NK) cells, and production of higher amounts of antibodies. Of these, macrophages are the best-known targets of β-glucan. β-glucan through binding with macrophage, natural killer cells, and neutrophils produces innate and acquired immune response. The identification of β-glucan by these cells has been attributed to numerous receptors such as CR3, lactosylceramide, TLR, scavenger receptors, and Dectin-1 (Battle et al., 1998). These receptors identify β-glucan and stimulate an effective immune response in monocyte/macrophage and neutrophils. Dectin-1 recognizes both soluble and particulate β 1 → 3 and/or β 1 → 6 linked glucans. The outcome of Dectin-1 activation leads to the release of cytokines like interleukin (IL) – 12, IL-10, IL-6, tumor necrosis factor (TNF)-α, which in turn activates both humoral and cell-mediated immunity (Kubala et al., 2003). Both *in vitro* and *in vivo* revealed that the immunostimulating activity depends on the structure, molecular weight, and branching of β-glucan. Maity et al. (2015) reported the immunological activity of β-glucan extracted from *Entoloma lividoalbum* by stimulating the activity of macrophages, splenocytes, and thymocytes. In an earlier study, Wang et al. (2014) reported the immunostimulatory activity of β-glucan obtained from *Ganoderma lucidum* by decreasing the nitric oxide (NO) production, *via* blocking NF-κB and inhibiting the phosphorylation of JNK MAPK signal pathways. Another study by Nandi et al. (2014) reported the effectiveness of mushroom β-glucan in stimulating macrophage activation via nitric oxide (NO) production as well as splenocytes and thymocytes proliferation. Furthermore (Davis et al., 2004) demonstrated that rats fed with oat-β-glucan based diet showed reduced morbidity and mortality along with an increase in macrophage antiviral resistance. No cytotoxicity was observed for NK cells with the ingestion of oat β-glucan (Davis et al., 2004).

6.5.4 Antidiabetic Activity

Cereal-glucans can also help prevent type-2 diabetes by regulating postprandial blood glucose and insulin levels. A study by Ekström, et al. (2017) has reported that even low amounts of oat-glucan are effective in lowering insulinemia and glycemia. Although some studies reported a positive relationship between β-glucan dose and a drop in plasma glucose (Cavallero et al., 2002), most evidence suggests that the influence of viscosity of β-glucan is a key determinant in lowering glycemic, insulinemic, and LDL-cholesterol levels (Bai et al., 2019). Another putative mechanism for cereal β-glucan in controlling postprandial blood glucose and insulin levels includes the interaction of β-glucan with intestinal mucus (Mackie et al., 2016). However, evidence on blood glucose response has revealed that coil overlap is responsible for the essential function of the aforementioned processes in the lowering of postprandial glycemic response by β-glucan (Rieder et al., 2017). Evidence suggests that oat-glucan particles reside in a network-like natural structure as part of the cell wall, which could encapsulate proteins and starch to form a complex matrix within the cell wall. This matrix might make enzymes less accessible, resulting in less starch breakdown and a reduced postprandial glycemic response (Zhang et al., 2017). Highly viscous oat β-glucan may inhibit glucose uptake in normal absorptive gut epithelial cells IEC-6 by altering the expression of intestinal glucose transport protein 1 (SGLT1) and transporter 2 (GLUT2) (Abbasi et al., 2016).

A diet consisting of a high amount of barley β-glucan has been shown to prevent insulin resistance which is beneficial for diabetic patients (Hlebowicz et al., 2008). The β-glucan enriched diet enhanced the release of hepatic insulin signaling by reducing serine phosphorylation of insulin receptors (Choi et al., 2010). Beck et al. (2009) observed a sharp reduction in insulin response and increment of postprandial cholecystokinin levels after intake of β-glucan in overweight human subjects. Similarly, Granfeldt et al. (2008) suggested a minimum intake of 4 g oat β-glucans to reflect a significant decrease in glucose and insulin responses in healthy subjects creating a favorable condition for diabetic patients. These theories have been advocated by several researchers to be re-evaluated for the concentration, the food vectors, and the tolerability of β-glucan products to improve the metabolic profile in type 2 diabetic subjects for long-term health goals (Cugnet-Anceau et al., 2010).

6.5.5 Anticancer Activity

Cereal β-glucans have been widely employed in tumor therapies due to their properties to influence immune response. Though there are many anti-cancer therapies available, like chemotherapy and anti-cancer drugs that are known to have cytotoxicity on cancer cells, simultaneously they are toxic to normal cells and harmful to the immune system (Yang et al., 2003). Thus, researchers in pursuit of identifying an alternate strategy have found that cereal β-glucan is effective against various cancers and immune disorders (Goodridge et al., 2009). It is known that various reactive oxygen species (ROS) have been linked to a variety of chronic disorders, including Alzheimer's disease, diabetes, rheumatoid arthritis, inflammation, malignancies, and aging processes. Cereal β-glucans have been shown to have anticancer, free radical scavenging properties, and trapping reactive oxygen species by donating hydrogen thus inhibiting ROS-induced chain reaction (Sener et al., 2007; Maity et al., 2015). Zhang et al. (2018) reported the immunomodulatory activity of β-glucan based nanoparticles on the Tumor Micro-environment (TME). The study reported that β-glucan and similar nanoparticles are capable of manipulating the TME; however, the action mechanism of β-glucan may vary with its structure, molecular weight, solubility, and mode of administration. The low molecular weight glucan from Avena sativa has been reported to exhibit strong antitumor activity induced by increased caspase-12 expression in Me45 and A431 cancer cell lines (Choromanska et al., 2015). (1-3, 1-4)- β-D-glucan isolated from oat, was able to induce proapoptotic of human melanoma HTB-140 cells via induction of caspase-3/7 activation and the presence of phosphatidylserine on the external surface of cellular membranes (Parzonko et al., 2015).

Oral administration of β-glucans is known to improve the anti-tumor effects of monoclonal antibodies (mAB) (Cheung, et al., 2002; Takeda & Okumura, 2004) besides stimulating the proliferation and activation of peripheral blood monocytes in patients with advanced breast cancer (Demir, et al., 2007). Furthermore, β-glucan has anti-cytotoxic, anti-mutagenic, and anti-tumorigenic properties (Mantovani et al., 2008).

6.5.6 Antiobesity Effect

Obesity has been linked to a high fat intake combined with a low fiber intake. Plant β-glucans, as a dietary fiber, play a vital role in the prevention of obesity and serum biochemical markers linked to obesity, fatty liver, and adipocyte growth in those who eat a high-fat diet. Consumption of barley β-glucan has been demonstrated to significantly reduce visceral fat in a safe manner (Aoe et al., 2017). The interaction mechanisms of dietary fiber in preventing obesity include (a) increasing the viscosity of intestinal content, which delays the gastric emptying rate and result in extended satiety after the meal (Drzikova et al., 2005a, b); (b) dietary fibers affect intestinal microflora and subsequent products in the colon. The increasing acetate and butyrates produced during fermentation play a direct role in obesity prevention (Lin et al., 2012) (c) increase in short-chain fatty acids in the large intestine which promotes the release of the anorectic gut hormones PYY and GLP-1, leading to reduced food intake (Lin et al., 2012). *β*-glucan from *Prowashonupana* barley rich in *β*-glucan reduced intestinal fat deposition (IFD) in a *C. elegans* model system, which appeared to be primarily mediated by *sir-2.1*, *daf-16*, and *daf-16/daf-2* gene regulation(Gao et al., 2015).

6.6 Potential Application of β-Glucan in Food Formulations

β-Glucan dietary fibers have been found to have great potential for being used as a functional ingredient in bread, biscuits, breakfast cereals, dairy products, fat replacers, and salad dressings. It is the promising functional properties of *β*-glucans such as thickening, stabilizing, emulsification, and gelling ability apart from various health and nutritional benefits that determine its suitability for incorporation in various food products. Barley *β*-glucan is particularly well suited for such applications, being capable of imparting a smooth mouthfeel to beverage products, and also makes the beverage an excellent source of soluble dietary fiber. These properties in turn favor its application as traditional beverage thickeners for replacement of gum Arabic, alginates, pectin, xanthan gum, and carboxymethyl-cellulose (Giese, 1992).

Enrichment of flour with arabinoxylan and *β*-glucan for breadmaking (Trogh et al., 2004; Ahmad et al., 2008) was found to have a remarkable improvement in bread loaf volume and also increased firmness of the bread crumb (Lazaridou et al., 2007) and with increased soluble fiber content (Trogh et al., 2004). The addition of *β*-glucan from barley and oat sources was recently reported by Kalinga and Mishra (2009) with promising effects on the rheological and physical properties of low-fat cake batter. Symons and Brennan (2004) studied the incorporation of barley β-glucan fractions (at 2.5 and 5%) into wheat bread for evaluation of in vitro digestion process. The results demonstrated a satisfactory reduction in starch degradation and sugar release that was proportional to the amount of β-glucan incorporated into the bread. In another study, Cavallero et al. (2002) enriched wheat bread with different concentrations of barley β-glucan rich fractions. It was found that an increase in the concentration of β-glucan level in the bread (notably the increased soluble β-glucan level) was responsible for the reducing glycaemic index through its effect on digesta viscosity and glucose absorption. The ability of high fiber diets rich in β-glucans to influence the rate of starch degradation and hence the glycaemic index of foods has obvious benefits with regard to various lifestyle diseases such as coronary heart diseases, obesity, and diabetes. Also in pasta products, *β*-glucans were found effective in lowering glycemic index, making it effective against metabolic diseases. Jenkins et al. (2002) observed a reduced glycemic index (GI = 43) in type-2 diabetic patients fed with β-glucan (6.5%) enriched breakfast bar. Similarly, a reduced GI was reported in β-glucan enriched pasta products (Yokoyama et al., 1997). Brennan et al., 2013 proposed that β-glucan rich barley and mushroom fractions can be employed in the creation of extruded ready-to-eat snacks. The study reported a substantial reduction in the *in vitro* glycemic response of extruded products (up to 25% compared to control products) illustrating the possibility of utilizing these fractions to reduce energy intake and help modulate overall glycemic response.

Another possible application of β-glucans in the food industry is in the production of low-fat dairy-based products, low-fat ice creams, yogurts, and beverages. The incorporation of β-glucan in these products can improve the mouthfeel and sensory properties that resemble those of full-fat products. β-glucans when used in combination with other soluble dietary fibers was found to improve the rheological characteristics and gelling properties of low-fat cheese curds (Tudorica et al., 2004). Moreover, the addition of β-glucan solutions to milk modifies curd formation via increasing curd yield and reducing curd cutting time. These effects appear to be related to the gelling capacity of β-glucans and their ability to form a highly structured and elastic casein–protein–glucan matrix (Tudorica et al., 2004). Additionally, Sharafbafi et al. (2014) introduced high molecular weight oat β-glucan into milk to produce low calorie and cholesterol-lowering dairy products and studied the phase behavior, rheological properties, and microstructure of this dairy product. The findings revealed that the flow behavior of mixtures with concentrations above the binodal curve was influenced not only by the existence of β-glucan chains but also by their production. Furthermore, Rinaldi et al. (2015) observed that enrichment of yogurts with β-glucan and pectin resulted in faster proteolysis, reduction in the release of large peptides, and a higher proportion of free amino acids than in yogurts with starch or without β-glucan. Lumaga et al. (2012) revealed that a sucrose-sweetened beverage with 3 g barley-glucans can help people control their food intake even up to 24 h by modulating gastrointestinal hormone response and even reduce their 24-hour energy intake when compared to control beverage without dietary fiber.

6.7 β-Glucan in Drug Delivery Applications

β-glucan based polysaccharides are currently under investigation for drug delivery purposes. It is a naturally occurring non-digestible polysaccharide that can resist the degradation caused by endogenous mammalian enzymes. They have attracted increasing attention to serving as potential matric for the drug delivery system creating a faster drug release with enhanced efficiency. It is due to their versatile structural modification and desirable biocompatibility via polyhydroxy functional groups for loading drugs as well as their targeted recognition by immune cells (Liu et al., 2018). Soto and Ostroff (2012) investigated the drug delivery function of β-glucan for macrophage targeted drug delivery, encased in spherical shaped glucan particles derived from Saccharomyces cerevisiae, where it was discovered that glucan particles can enhance the delivery of multiple nanoparticles to the macrophage through a single application with improved drug efficacy. In a similar vein to the efficacy of -glucan as a drug delivery agent, Chang et al. (2006) observed that sulfated β-glucan becomes more soluble in water as the number of ions combined increases, boosting their medicinal action. To fulfill the applicability of the drug delivery method, many derivatives such as sulfated-glucan, carboxymethyl-glucan, and cyclic-glucan have been developed. The most important feature of β-glucan is its ability to bond with macrophages. As a result, in the event of food contamination resulting in pathogen attack, β-glucan serves as a site of recognition for macrophages. Verma and Gu (2012) investigated macrophage-targeted drug delivery of drugs encapsulated in β-Glucan, which was advantageous because of the high viscosity of β-Glucan, which allows for gradual and controlled drug release, improving pharmacological action.

6.8 Conclusion

A well-balanced diet rich in dietary fibers will help reduce the mortality related to various lifestyle diseases. The β-glucan polysaccharide can be recommended either in the form of β-G fortified food or as a dietary supplement in preventing lifestyle diseases. Cereal β-glucans are gaining considerable attention owing to their distinctive solubility in aqueous solutions. Additionally, cereal β-glucan shows great potential as a thickener or stabilizer in food products Promising results of various in vitro, animal and human studies suggest various biological activities such as immunomodulatory, antidiabetic, antiobesity, anticancer, hypocholesterolemic of β-glucan molecules. It can thus be concluded that cereal β-glucans offer potential rheological and nutraceutical advantages that can be explored in food and pharmaceutical applications. However, more research is needed on biomedical applications included targeted drug delivery.

References

Abbasi, N. N., Purslow, P. P., Tosh, S. M., & Bakovic, M. (2016). Oat β-glucan depresses SGLT1- and GLUT2-mediated glucose transport in intestinal epithelial cells (IEC-6). *Nutrition Research, 36*(6), 541–552.

Ahmad, A., Anjum, F. M., Zahoor, T., Chatha, Z. A., & Nawaz, H. (2008). Effect of barley β-glucan on sensory characteristics of bread. *Pakistan Journal of Agricultural Sciences, 45*, 88–94.

Alminger, M., & Eklund-Jonsson, C. (2008). Whole-grain cereal products based on a high-fibre barley or oat genotype lower post-prandial glucose and insulin responses in healthy humans. *European Journal of Nutrition, 47*(6), 294.

American Diabetes Association. (1999). Consensus development conference on diabetic foot wound care: 7-8 April 1999, Boston, Massachusetts. American Diabetes Association. *Diabetes Care, 22*(8), 1354–1360.

Anderson, J. W. (1990). Dietary fiber and human health. *HortScience, 25*(12), 1488–1495.

Aoe, S., Ichinose, Y., Kohyama, N., Komae, K., Takahashi, A., Abe, D., … Yanagisawa, T. (2017). Effects of high β-glucan barley on visceral fat obesity in Japanese individuals: A randomized, double-blind study. *Nutrition, 42*, 1–6.

Bai, J., Ren, Y., Li, Y., Fan, M., Qian, H., Wang, L., … Rao, Z. (2019). Physiological functionalities and mechanisms of β-glucans. *Trends in Food Science & Technology, 88*, 57–66.

Battle, J., Ha, T., Li, C., Della Beffa, V., Rice, P., Kalbfleisch, J., … Williams, D. (1998). Ligand binding to the (1→ 3)-β-D-glucan receptor stimulates NFκB activation, but not apoptosis in U937 cells. *Biochemical and Biophysical Research Communications, 249*(2), 499–504.

Beck, E. J., Tosh, S. M., Batterham, M. J., Tapsell, L. C., & Huang, X. F. (2009). Oat β glucan increases postprandial cholecystokinin levels, decreases insulin response and extends subjective satiety in overweight subjects. *Molecular Nutrition & Food Research, 53*(10), 1343–1351.

Behall, K. M., Scholfield, D. J., & Hallfrisch, J. (1997). Effect of beta-glucan level in oat fiber extracts on blood lipids in men and women. *Journal of the American College of Nutrition, 16*(1), 46–51.

Bohn, J. A., & BeMiller, J. N. (1995). (1→ 3) β d Glucans as biological response modifiers: A review of structure-functional activity relationships. *Carbohydrate Polymers, 28*(1), 3–14.

Brennan, C. S., & Cleary, L. J. (2005). The potential use of cereal (1→ 3, 1→ 4)-β-D-glucans as functional food ingredients. *Journal of Cereal Science, 42*(1), 1–13.

Brennan, C. S., & Tudorica, C. M. (2003). The role of complex carbohydrates and non-starch polysaccharides in the regulation of postprandial glucose and insulin responses in cereal foods. *Journal of Nutraceuticals, Functional & Medical Foods, 4*(2), 49–55.

Brennan, M. A., Derbyshire, E., Tiwari, B. K., & Brennan, C. S. (2013). Integration of β-glucan fibre rich fractions from barley and mushrooms to form healthy extruded snacks. *Plant Foods for Human Nutrition, 68*(1), 78–82.

Burkus, Z., & Temelli, F. (2000). Stabilization of emulsions and foams using barley β-glucan. *Food Research International, 33*(1), 27–33.

Burkus, Z., & Temelli, F. (2005). Rheological properties of barley β-glucan. *Carbohydrate Polymers, 59*(4), 459–465.

Burley, V. J., Leeds, A. R., & Blundell, J. E. (1987). The effect of high and low-fibre breakfasts on hunger, satiety and food intake in a subsequent meal. *International Journal of Obesity, 11*, 87–93.

Cavallero, A., Empilli, S., Brighenti, F., & Stanca, A. M. (2002). High (1→ 3, 1→ 4)-β-glucan barley fractions in bread making and their effects on human glycemic response. *Journal of Cereal Science, 36*(1), 59–66.

Chang, Y. J., Lee, S., Yoo, M. A., & Lee, H. G. (2006). Structural and biological characterization of sulfated-derivatized oat β-glucan. *Journal of Agricultural and Food Chemistry, 54*(11), 3815–3818.

Cheryl, L., Dikeman, A., & Fahey, G. C. (2006). Viscosity as related to dietary fiber: A review. *Critical Reviews in Food Science and Nutrition, 46*, 649–663.

Cheung, N. K. V., Modak, S., Vickers, A., & Knuckles, B. (2002). Orally administered β-glucans enhance anti-tumor effects of monoclonal antibodies. *Cancer Immunology, Immunotherapy, 51*(10), 557–564.

Choi, J. S., Kim, H., Jung, M. H., Hong, S., & Song, J. (2010). Consumption of barley β-glucan ameliorates fatty liver and insulin resistance in mice fed a high-fat diet. *Molecular Nutrition & Food Research, 54*(7), 1004–1013.

Choromanska, A., Kulbacka, J., Rembialkowska, N., Pilat, J., Oledzki, R., Harasym, J., & Saczko, J. (2015). Anticancer properties of low molecular weight oat beta-glucan–an in vitro study. *International Journal of Biological Macromolecules, 80*, 23–28.

Cugnet-Anceau, C., Nazare, J. A., Biorklund, M., Le Coquil, E., Sassolas, A., Sothier, M., … Moulin, P. (2010). A controlled study of consumption of β-glucan-enriched soups for 2 months by type 2 diabetic free-living subjects. *British Journal of Nutrition, 103*(3), 422–428.

Cui, S. W., & Wang, Q. (2009). Cell wall polysaccharides in cereals: Chemical structures and functional properties. *Structural Chemistry, 20*(2), 291–297.

Davidson, M. H., Dugan, L. D., Burns, J. H., Bova, J., Story, K., & Drennan, K. B. (1991). The hypocholesterolemic effects of β-glucan in oatmeal and oat bran: A dose-controlled study. *JAMA, 265*(14), 1833–1839.

Davis, J. M., Murphy, E. A., Brown, A. S., Carmichael, M. D., Ghaffar, A., & Mayer, E. P. (2004). Effects of moderate exercise and oat β-glucan on innate immune function and susceptibility to respiratory infection. *American Journal of Physiology-Regulatory, Integrative and Comparative Physiology, 286*(2), R366–R372.

Demir, G., Klein, H. O., Mandel-Molinas, N., & Tuzuner, N. (2007). Beta glucan induces proliferation and activation of monocytes in peripheral blood of patients with advanced breast cancer. *International Immunopharmacology, 7*(1), 113–116.

DeVries, J. W. (2003). On defining dietary fibre. *Proceedings of the Nutrition Society, 62*(1), 37–43.

Dongowski, G., Drzikova, B., Senge, B., Blochwitz, R., Gebhardt, E., & Habel, A. (2005). Rheological behaviour of β-glucan preparations from oat products. *Food Chemistry, 93*(2), 279–291.

Doublier, J. L., & Wood, P. J. (1995). Rheological properties of aqueous solutions of (1--> 3)(1--> 4)-bD-glucan from oats (Avena sativa L.). *Cereal Chemistry, 72*(4), 335–340.

Drzikova, B., Dongowski, G., & Gebhardt, E. (2005a). Dietary fibre-rich oat-based products affect serum lipids, microbiota, formation of short-chain fatty acids and steroids in rats. *British Journal of Nutrition, 94*(6), 1012–1025.

Drzikova, B., Dongowski, G., Gebhardt, E., & Habel, A. (2005b). The composition of dietary fibre-rich extrudates from oat affects bile acid binding and fermentation in vitro. *Food Chemistry, 90*(1–2), 181–192.

Du, B., Lin, C., Bian, Z., & Xu, B. (2015). An insight into anti-inflammatory effects of fungal beta-glucans. *Trends in Food Science & Technology, 41*(1), 49–59.

Du, B., Meenu, M., Liu, H., & Xu, B. (2019). A concise review on the molecular structure and function relationship of β-glucan. *International Journal of Molecular Sciences, 20*(16), 4032.

Duss, R., & Nyberg, L. (2004). Oat soluble fibers (beta-glucans) as a source for healthy snack and breakfast foods. *Cereal Foods World, 49*(6), 320–325.

Ekström, L. M., Bok, E. A. H., Sjöö, M. E., & Östman, E. M. (2017). Oat β-glucan containing bread increases the glycaemic profile. *Journal of Functional Foods, 32*, 106–111.

El Khoury, D., Cuda, C., Luhovyy, B. L., & Anderson, G. H. (2012). Beta glucan: Health benefits in obesity and metabolic syndrome. *Journal of Nutrition and Metabolism, 2012*.

Elleuch, M., Bedigian, D., Roiseux, O., Besbes, S., Blecker, C., & Attia, H. (2011). Dietary fibre and fibre-rich by-products of food processing: Characterisation, technological functionality and commercial applications: A review. *Food Chemistry, 124*(2), 411–421.

FDA. (1997a). Final rule for food labelling: Health claims: Oats and coronary heart disease. *Federal Regulations, 62*, 3584–3681.

FDA. (1997b). 21 CFR part 101. Food labeling: Health claims; soluble dietary fiber from certain foods and coronary heart disease. *Federal Register, 62*(15), 3584–3601.

Food and Drug Administration, HHS. (2008). Food labeling: health claims; soluble fiber from certain foods and risk of coronary heart disease. Interim final rule. *Federal Register, 73*(37), 9938–9947.
Gani, A., Wani, S. M., Masoodi, F. A., & Hameed, G. (2012). Whole-grain cereal bioactive compounds and their health benefits: A review. *Journal of Food Processing and Technology, 3*(3), 146–156.
Gao, C., King, M. L., Fitzpatrick, Z. L., Wei, W., King, J. F., Wang, M., … Zheng, J. (2015). Prowashonupana barley dietary fibre reduces body fat and increases insulin sensitivity in Caenorhabditis elegans model. *Journal of Functional Foods, 18*, 564–574.
Giese, J. (1992). Developing low-fat meat products. *Food Technology.*
Goodridge, H. S., Wolf, A. J., & Underhill, D. M. (2009). β-Glucan recognition by the innate immune system. *Immunological Reviews, 230*(1), 38–50.
Granfeldt, Y., Nyberg, L., & Björck, I. (2008). Muesli with 4 g oat β-glucans lowers glucose and insulin responses after a bread meal in healthy subjects. *European Journal of Clinical Nutrition, 62*(5), 600–607.
Gunness, P., Michiels, J., Vanhaecke, L., De Smet, S., Kravchuk, O., Van de Meene, A., & Gidley, M. J. (2016). Reduction in circulating bile acid and restricted diffusion across the intestinal epithelium are associated with a decrease in blood cholesterol in the presence of oat β-glucan. *The FASEB Journal, 30*(12), 4227–4238.
Hallfrisch, J., Scholfield, D. J., & Behall, K. M. (2003). Physiological responses of men and women to barley and oat extracts (nu-trimX). II. Comparison of glucose and insulin responses. *Cereal Chemistry, 80*(1), 80–83.
Havrlentova, M., Petrulakova, Z., Burgarova, A., Gago, F., Hlinkova, A., & Šturdík, E. (2011). β-Glucans and their significance for the preparation of functional foods-a review. *Czech Journal of Food Sciences, 29*(1), 1–14.
Hlebowicz, J., Darwiche, G., Björgell, O., & Almér, L. O. (2008). Effect of muesli with 4 g oat β-glucan on postprandial blood glucose, gastric emptying and satiety in healthy subjects: A randomized crossover trial. *Journal of the American College of Nutrition, 27*(4), 470–475.
Hughes, R., & Rowland, I. R. (2001). Stimulation of apoptosis by two prebiotic chicory fructans in the rat colon. *Carcinogenesis, 22*(1), 43–47.
Irakli, M., Biliaderis, C. G., Izydorczyk, M. S., & Papadoyannis, I. N. (2004). Isolation, structural features and rheological properties of water-extractable β-glucans from different Greek barley cultivars. *Journal of the Science of Food and Agriculture, 84*(10), 1170–1178.
Izydorczyk, M. S., Dexter, J. E., Desjardins, R. G., Rossnagel, B. G., Lagasse, S. L., & Hatcher, D. W. (2003). Roller milling of Canadian hull-less barley: Optimization of roller milling conditions and composition of mill streams. *Cereal Chemistry, 80*(6), 637–644.
Jenkins, D. J., Wolever, T. M., Leeds, A. R., Gassull, M. A., Haisman, P., Dilawari, J., … Alberti, K. G. (1978). Dietary fibres, fibre analogues, and glucose tolerance: Importance of viscosity. *Br Med Journal, 1*(6124), 1392–1394.
Jenkins, A. L., Jenkins, D. J. A., Zdravkovic, U., Würsch, P., & Vuksan, V. (2002). Depression of the glycemic index by high levels of β-glucan fiber in two functional foods tested in type 2 diabetes. *European Journal of Clinical Nutrition, 56*(7), 622–628.
Kalinga, D., & Mishra, V. K. (2009). Rheological and physical properties of low fat cakes produced by addition of cereal β-glucan concentrates. *Journal of Food Processing and Preservation, 33*(3), 384–400.
Keogh, G. F., Cooper, G. J., Mulvey, T. B., McArdle, B. H., Coles, G. D., Monro, J. A., & Poppitt, S. D. (2003). Randomized controlled crossover study of the effect of a highly β-glucan–enriched barley on cardiovascular disease risk factors in mildly hypercholesterolemic men. *The American Journal of Clinical Nutrition, 78*(4), 711–718.
Kerckhoffs, D. A., Hornstra, G., & Mensink, R. P. (2003). Cholesterol-lowering effect of β-glucan from oat bran in mildly hypercholesterolemic subjects may decrease when β-glucan is incorporated into bread and cookies. *The American Journal of Clinical Nutrition, 78*(2), 221–227.

Kubala, L., Ruzickova, J., Nickova, K., Sandula, J., Ciz, M., & Lojek, A. (2003). The effect of (1→ 3)-β-D-glucans, carboxymethylglucan and schizophyllan on human leukocytes in vitro. *Carbohydrate Research, 338*(24), 2835–2840.

Lazaridou, A., & Biliaderis, C. G. (2004). Cryogelation of cereal β-glucans: Structure and molecular size effects. *Food Hydrocolloids, 18*(6), 933–947.

Lazaridou, A., & Biliaderis, C. G. (2007). Molecular aspects of cereal β-glucan functionality: Physical properties, technological applications and physiological effects. *Journal of Cereal Science, 46*(2), 101–118.

Lazaridou, A., Biliaderis, C. G., & Izydorczyk, M. S. (2003). Molecular size effects on rheological properties of oat β-glucans in solution and gels. *Food Hydrocolloids, 17*(5), 693–712.

Lazaridou, A., Duta, D., Papageorgiou, M., Belc, N., & Biliaderis, C. G. (2007). Effects of hydrocolloids on dough rheology and bread quality parameters in gluten-free formulations. *Journal of Food Engineering, 79*(3), 1033–1047.

Limberger-Bayer, V. M., de Francisco, A., Chan, A., Oro, T., Ogliari, P. J., & Barreto, P. L. (2014). Barley β-glucans extraction and partial characterization. *Food Chemistry, 154*, 84–89.

Lin, H. V., Frassetto, A., Kowalik, E. J., Jr., Nawrocki, A. R., Lu, M. M., Kosinski, J. R., ... Marsh, D. J. (2012). Butyrate and propionate protect against diet-induced obesity and regulate gut hormones via free fatty acid receptor 3-independent mechanisms. *PLoS One, 7*(4), e35240.

Liu, B., Yang, T., Luo, Y., Zeng, L., Shi, L., Wei, C., ... Luo, F. (2018). Oat β-glucan inhibits adipogenesis and hepatic steatosis in high fat diet-induced hyperlipidemic mice via AMPK signaling. *Journal of Functional Foods, 41*, 72–82.

Lumaga, R. B., Azzali, D., Fogliano, V., Scalfi, L., & Vitaglione, P. (2012). Sugar and dietary fibre composition influence, by different hormonal response, the satiating capacity of a fruit-based and a β-glucan-enriched beverage. *Food & Function, 3*(1), 67–75.

Mackie, A., Rigby, N., Harvey, P., & Bajka, B. (2016). Increasing dietary oat fibre decreases the permeability of intestinal mucus. *Journal of Functional Foods, 26*, 418–427.

Maier, S. M., Turner, N. D., & Lupton, J. R. (2000). Serum lipids in hypercholesterolemic men and women consuming oat bran and amaranth products. *Cereal Chemistry, 77*(3), 297–302.

Maity, P., Sen, I. K., Maji, P. K., Paloi, S., Devi, K. S. P., Acharya, K., ... Islam, S. S. (2015). Structural, immunological, and antioxidant studies of β-glucan from edible mushroom Entoloma lividoalbum. *Carbohydrate Polymers, 123*, 350–358.

Mäkelä, N., Brinck, O., & Sontag-Strohm, T. (2020). Viscosity of β-glucan from oat products at the intestinal phase of the gastrointestinal model. *Food Hydrocolloids, 100*, 105422.

Mantovani, A., Allavena, P., Sica, A., & Balkwill, F. (2008). Cancer-related inflammation. *Nature, 454*(7203), 436–444.

Manzoor, M., Ahmad, M., Bandral, J. D., Gani, A., Singh, J., & Shams, R. (2020). Food hydrocolloids: Functional, nutraceutical and novel applications for delivery of bioactive compounds. *International Journal of Biological Macromolecules, 165*, 554–567.

McIntosh, M., Stone, B. A., & Stanisich, V. A. (2005). Curdlan and other bacterial (1→ 3)-β-D-glucans. *Applied Microbiology and Biotechnology, 68*(2), 163–173.

Mejía, S. M. V., de Francisco, A., & Bohrer, B. (2020). A comprehensive review on cereal β-glucan: Extraction, characterization, causes of degradation, and food application. *Critical Reviews in Food Science and Nutrition, 60*(21), 3693–3704.

Mikkelsen, M. S., Jespersen, B. M., Larsen, F. H., Blennow, A., & Engelsen, S. B. (2013). Molecular structure of large-scale extracted β-glucan from barley and oat: Identification of a significantly changed block structure in a high β-glucan barley mutant. *Food Chemistry, 136*(1), 130–138.

Nandi, A. K., Samanta, S., Maity, S., Sen, I. K., Khatua, S., Devi, K. S. P., ... Islam, S. S. (2014). Antioxidant and immunostimulant β-glucan from edible mushroom Russula albonigra (Krombh.) Fr. *Carbohydrate Polymers, 99*, 774–782.

Newman, R. K., & Newman, C. W. (1991). Barley as a food grain. *Cereal Foods World, 36*, 800.

Parzonko, A., Makarewicz-Wujec, M., Jaszewska, E., Harasym, J., & Kozłowska-Wojciechowska, M. (2015). Pro-apoptotic properties of (1, 3)(1, 4)-β-D-glucan from Avena sativa on human melanoma HTB-140 cells in vitro. *International Journal of Biological Macromolecules, 72*, 757–763.

Pentikäinen, S., Karhunen, L., Flander, L., Katina, K., Meynier, A., Aymard, P., … Poutanen, K. (2014). Enrichment of biscuits and juice with oat β-glucan enhances postprandial satiety. *Appetite, 75*, 150–156.

Phillips, K. R., Pik, J., Lawford, H. G., Lavers, B., Kligerman, A., & Lawford, G. R. (1983). Production of curdlan-type polysaccharide by Alcaligenes faecalis in batch and continuous culture. *Canadian Journal of Microbiology, 29*(10), 1331–1338.

Queenan, K. M., Stewart, M. L., Smith, K. N., Thomas, W., Fulcher, R. G., & Slavin, J. L. (2007). Concentrated oat β-glucan, a fermentable fiber, lowers serum cholesterol in hypercholesterolemic adults in a randomized controlled trial. *Nutrition Journal, 6*(1), 1–8.

Regand, A., Chowdhury, Z., Tosh, S. M., Wolever, T. M., & Wood, P. (2011). The molecular weight, solubility and viscosity of oat beta-glucan affect human glycemic response by modifying starch digestibility. *Food Chemistry, 129*(2), 297–304.

Reimer, R. A., Thomson, A. B., Rajotte, R. V., Basu, T. K., Ooraikul, B., & McBurney, M. I. (2000). Proglucagon messenger ribonucleic acid and intestinal glucose uptake are modulated by fermentable fiber and food intake in diabetic rats. *Nutrition Research, 20*(6), 851–864.

Rieder, A., Knutsen, S. H., & Ballance, S. (2017). In vitro digestion of beta-glucan rich cereal products results in extracts with physicochemical and rheological behavior like pure beta-glucan solutions–a basis for increased understanding of in vivo effects. *Food Hydrocolloids, 67*, 74–84.

Rinaldi, L., Rioux, L. E., Britten, M., & Turgeon, S. L. (2015). In vitro bioaccessibility of peptides and amino acids from yogurt made with starch, pectin, or β-glucan. *International Dairy Journal, 46*, 39–45.

Schmidt, M. (2022). Cereal beta-glucans: An underutilized health endorsing food ingredient. *Critical Reviews in Food Science and Nutrition, 62*(12), 3281–3300.

Sener, G., Toklu, H. Z., & Cetinel, S. (2007). β-Glucan protects against chronic nicotine-induced oxidative damage in rat kidney and bladder. *Environmental Toxicology and Pharmacology, 23*(1), 25–32.

Sharafbafi, N., Tosh, S. M., Alexander, M., & Corredig, M. (2014). Phase behaviour, rheological properties, and microstructure of oat β-glucan-milk mixtures. *Food Hydrocolloids, 41*, 274–280.

Shimizu, C., Kihara, M., Aoe, S., Araki, S., Ito, K., Hayashi, K., … Ikegami, S. (2008). Effect of high β-glucan barley on serum cholesterol concentrations and visceral fat area in Japanese men – A randomized, double-blinded, placebo-controlled trial. *Plant Foods for Human Nutrition, 63*(1), 21–25.

Sier, C. F., Gelderman, K. A., Prins, F. A., & Gorter, A. (2004). Beta-glucan enhanced killing of renal cell carcinoma micrometastases by monoclonal antibody G250 directed complement activation. *International Journal of Cancer, 109*(6), 900–908.

Soto, E., & Ostroff, G. (2012). Glucan particles as carriers of nanoparticles for macrophage-targeted delivery. In *Nanomaterials for biomedicine* (pp. 57–79). American Chemical Society.

Sterna, V., Zute, S., & Brunava, L. (2016). Oat grain composition and its nutrition benefice. *Agriculture and Agricultural Science Procedia, 8*, 252–256.

Su, Y., Miao, B., Wang, H., Wang, C., & Zhang, S. (2013). Splenic abscess caused by streptococcus gallolyticus subsp. pasteurianus as presentation of a pancreatic cancer. *Journal of Clinical Microbiology, 51*(12), 4249–4251.

Sudha, M. L., Vetrimani, R., & Leelavathi, K. (2007). Influence of fibre from different cereals on the rheological characteristics of wheat flour dough and on biscuit quality. *Food Chemistry, 100*(4), 1365–1370.

Symons, L. J., & Brennan, C. S. (2004). The influence of (1→ 3)(1→ 4)-β-D-Glucan-rich fractions from barley on the physicochemical properties and in vitro reducing sugar release of white wheat breads. *Journal of Food Science, 69*(6), C463–C467.

Synytsya, A., & Novak, M. (2014). Structural analysis of glucans. *Annals of Translational Medicine, 2*(2).

Takeda, K., & Okumura, K. (2004). CAM and NK cells. *Evidence-based Complementary and Alternative Medicine, 1*(1), 17–27.

Thammakiti, S., Suphantharika, M., Phaesuwan, T., & Verduyn, C. (2004). Preparation of spent brewer's yeast β-glucans for potential applications in the food industry. *International Journal of Food Science & Technology, 39*(1), 21–29.

Tosh, S. M., Wood, P. J., Wang, Q., & Weisz, J. (2004). Structural characteristics and rheological properties of partially hydrolyzed oat β-glucan: The effects of molecular weight and hydrolysis method. *Carbohydrate Polymers, 55*(4), 425–436.

Trogh, I., Sørensen, J. F., Courtin, C. M., & Delcour, J. A. (2004). Impact of inhibition sensitivity on endoxylanase functionality in wheat flour breadmaking. *Journal of Agricultural and Food Chemistry, 52*(13), 4296–4302.

Truswell, A. S. (2002). Cereal grains and coronary heart disease. *European Journal of Clinical Nutrition, 56*(1), 1–14.

Tudorica, C. M., Jones, T. E. R., Kuri, V., & Brennan, C. S. (2004). The effects of refined barley β-glucan on the physico-structural properties of low-fat dairy products: Curd yield, microstructure, texture and rheology. *Journal of the Science of Food and Agriculture, 84*(10), 1159–1169.

Tvaroska, I., Ogawa, K., Deslandes, Y., & Marchessault, R. H. (1983). Crystalline conformation and structure of lichenan and barley β-glucan. *Canadian Journal of Chemistry, 61*(7), 1608–1616.

Vaikousi, H., Biliaderis, C. G., & Izydorczyk, M. S. (2004). Solution flow behavior and gelling properties of water-soluble barley (1→ 3, 1→ 4)-β-glucans varying in molecular size. *Journal of Cereal Science, 39*(1), 119–137.

Vårum, K. M., & Smidsrød, O. (1988). Partial chemical and physical characterisation of (1→ 3),(1→ 4)-β-D-glucans from oat (Avena sativa L.) aleurone. *Carbohydrate Polymers, 9*(2), 103–117.

Vasanthan, T., & Temelli, F. (2008). Grain fractionation technologies for cereal beta-glucan concentration. *Food Research International, 41*(9), 876–881.

Vasiljevic, T., Kealy, T., & Mishra, V. K. (2007). Effects of β-glucan addition to a probiotic containing yogurt. *Journal of Food Science, 72*(7), C405–C411.

Verma, M. S., & Gu, F. X. (2012). 1, 3-Beta-Glucans: Drug Delivery and Pharmacology. In *The complex world of polysaccharides*. IntechOpen.

Wang, J., Yuan, Y., & Yue, T. (2014). Immunostimulatory activities of β-D-glucan from Ganoderma lucidum. *Carbohydrate Polymers, 102*, 47–54.

Whitehead, A., Beck, E. J., Tosh, S., & Wolever, T. M. (2014). Cholesterol-lowering effects of oat β-glucan: A meta-analysis of randomized controlled trials. *The American Journal of Clinical Nutrition, 100*(6), 1413–1421.

Woodward, J. R., Phillips, D. R., & Fincher, G. B. (1988). Water-soluble (1→ 3, 1→ 4)-β-d-glucans from barley (Hordeum vulgare) endosperm. IV. Comparison of 40 C and 65 C soluble fractions. *Carbohydrate Polymers, 8*(2), 85–97.

Yang, J. L., Kim, Y. H., Lee, H. S., Lee, M. S., & Moon, Y. K. (2003). Barley β-glucan lowers serum cholesterol based on the up-regulation of cholesterol 7α-hydroxylase activity and mRNA abundance in cholesterol-fed rats. *Journal of Nutritional Science and Vitaminology, 49*(6), 381–387.

Yokoyama, W. H., Hudson, C. A., Knuckles, B. E., Chiu, M. C. M., Sayre, R. N., Turnlund, J. R., & Schneeman, B. O. (1997). Effect of barley β-glucan in durum wheat pasta on human glycemic response. *Cereal Chemistry, 74*(3), 293–296.

Zhang, J., Luo, K., & Zhang, G. (2017). Impact of native form oat β-glucan on starch digestion and postprandial glycemia. *Journal of Cereal Science, 73*, 84–90.

Zhang, M., Kim, J. A., & Huang, A. Y. C. (2018). Optimizing tumor microenvironment for cancer immunotherapy: β-glucan-based nanoparticles. *Frontiers in Immunology, 9*, 341.

Zhu, F., Du, B., Bian, Z., & Xu, B. (2015). Beta-glucans from edible and medicinal mushrooms: Characteristics, physicochemical and biological activities. *Journal of Food Composition and Analysis, 41*, 165–173.

Part II
Functional Cereal Foods

Chapter 7
Prebiotic and Probiotic Potential of Cereals

Kartik Sharma, Ramandeep Kaur, Vikas Kumar, Satish Kumar, Arashdeep Singh, and Neha Gautam

7.1 Introduction

A food that improves human life and cures ailments is comprises of nutrient (prebiotic), which is processed by specialised bacteria (probiotic), and products (postbiotic) that are physiologically active, is one of the most intricate loops. Clinicians, microbiologists, dieticians, nutritionists, food technologists, and farmers are all interested in probiotics, their nutrients–prebiotics, and postbiotics. Probiotics play vital role for preventing and treating a variety of infectious and non-infectious diseases (Salmerón, 2017). Anti-diarrheal, anti-carcinogenic, anti-mutagenic, anti-inflammatory, antibacterial action, immune system activation, improvement in lactose metabolism, and suppression of gastrointestinal infections are all documented benefits. Mainly probiotic meals, which include microorganisms from the lactic acid bacteria family, are generally limited to drinks and cheese. Health-promoting foods are usually dairy-based, consisting of milk and its fermented products. Furthermore, lactose sensitivity and the cholesterol level of dairy products, together with the vegetarian tendencies of a wide range of third-world people, have

K. Sharma
Department of Biotechnology, Council of Scientific and Industrial Research-Institute of Himalayan Bioresource Technology (CSIR-IHBT), Palampur, HP, India

R. Kaur · V. Kumar (✉) · A. Singh
Department of Food Science and Technology, Punjab Agricultural University, Ludhiana, Punjab, India

S. Kumar
College of Horticulture and Forestry, Thunag, Mandi. Dr YS Parmar University of Horticulture and Forestry, Nauni, Solan, HP, India

N. Gautam
Department of Basic Sciences, Dr. Y. S. Parmar University of Horticulture and Forestry, Nauni, Solan, HP, India

S. Punia Bangar, A. Kumar Siroha (eds.), *Functional Cereals and Cereal Foods*,
https://doi.org/10.1007/978-3-031-05611-6_7

pushed for the recent adoption of non-dairy beverages. On the other hand, several studies have focused on the probiotic capability of fermented cereal-based beverages, which are particularly popular in poor nations with low nutritional safety and a high frequency of gut infections (Mohammadi et al., 2021). Probiotic bacteria originate usually from humans or animals, however strains that are identified as probiotics can also be found in fermented non-dairy foods (Agrawal et al., 2000; Aidoo et al., 2006).

In current era, the chronic diseases, especially the non-communicable ones along with inflammatory disorders, usually associated with wrong habit of feeding the foods with poor fiber content in them and with higher content of refined carbohydrates and fats alongside. This resulted in increased number of diseases, making the researchers to focus on the importance of prebiotics which has active mechanism in controlling the process of infectious diseases (Rolim, 2015). Cereals, being one of the major components of diet possess dietary fiber along with their prebiotic potential. The effects of dietary fibers on human health are well documented and widely accepted as well, and prebiotic foods nowadays are getting much importance due to their beneficial effects and high nutritional status. According to solubility in water, dietary fibers are categorized as water soluble or water insoluble. The former one i.e. water soluble or water extractable dietary fibers includes β-glucan, arabinoxylans, fructo-oligosaccharides whereas the later one i.e. water insoluble or water unextractable dietary fibers include resistant starch, cellulose, etc. Water insoluble dietary fibers are generally more resistant to the colonic fermentation and therefore possess more prebiotic effects in contrast to water soluble dietary fibers. However, the potential of prebiotic depends upon the physicochemical properties and is affected by chemical structure, degree of polymerization and many more (Abdi & Joye, 2021).

Lactic acid fermentation of cereals is a widely adopted technique that is used to make a variety of foods throughout Asia and Africa, including porridge and beverages. Cereal grains, primarily sorghum and maize are soaked for 0.5–2 days in clean water. Soaking softens the grains, making it simpler to smash or wet-mill them into a slurry, from which hulls, bran particles, and germs can be sieved off. Mixed fermentations, including lactic acid fermentation, take place during the slurring or doughing stage, which lasts 1–3 days. During fermentation, the pH drops while the acidity rises as lactic and other organic acids accumulate as a result of microbial activity.

Lactobacilli and bifidobacteria have complicated dietary requirements that vary greatly from species to species, including vitamins, nucleic acid derivatives, fatty esters, peptides, amino acids, and carbohydrates (Severson, 1998). Starch, water-soluble or -insoluble dietary fibre components, and numerous free sugars such as arabinose, sucrose, maltose, fructose, xylose, stachyose, glycerol and glucose are the major component of cereals, determine the contents of these components. The ability of a probiotic starter to survive the acidic environment of the final fermented product, as well as the harsh conditions of the gastrointestinal tract, is a critical element in its selection. The metabolites generated by the starter, such as lactic acid and acetic acid, hydrogen peroxide, and bacteriocins, may affect the probiotic bacteria's viability in vitro (Saarela et al., 2000).

7.2 Why Cereal-Based Foods

Cereals, no doubt being a rich source of nutrition offers various challenges also, such as its swelling power upon cooking due to the presence of starch in them, limiting presence of amino acids in them along with limited and lower mineral bioavailability because of presence of number of anti-nutritional factors such as phytic acid, tannins, etc. Besides this, cereals are grown in more than 73% of the total harvested area globally and it contributes to more than 60% of the total production of food in the world. They are the major contributors of the protein, dietary fibers, vitamins, minerals and provide energy in optimum amount that is desired for the human health (Nout, 2009).

Therefore, the cereals or its constituents can possibly be employed for the formulation of functional foods, where the cereals can be used as the fermentable substrate for growing of probiotic microorganisms such as *bifidobacteria* and *lactobacilli*. These can also be used as the major source of dietary fiber for promoting the health benefits to the consumers. Also, due to the presence of specific content of non-digestible carbohydrates in them, they can either serve as prebiotic or as probiotic for increasing their stability (Charalampopoulos et al., 2002).

Various studies demonstrates that whole grain cereals play important role in protecting the body against various diseases such as cancer, cardiovascular disease, diabetes and other related diseases (Venn & Mann, 2004). This might be attributed to the presence of dietary fibers in whole grain cereals and due to presence of minerals in their outer fractions, which help them in fighting against the oxidative stress, carcinogenesis, hyperglycaemia and inflammatory disorders. Cereals with whole grain are carriers of micronutrients which include folates, zinc, selenium, manganese, betaine, sulphur amino acids, lignans, alkyl resorcinols, vitamin E, phenolic acids, iron, copper, carotenoids, choline, phytic acid, etc. (Das et al., 2012).

7.3 History of Cereal Based Beverages

Homo sapiens appear between 2000 and 1,000,000 years ago. The only drinks they drink are water and breast milk. About 11,000 years ago, the other beverages like animal milk, tea, beer, coffee, and afterwards fruit juices and soft drinks enter our diet. A large amount of archaeological evidence indicates that the production and consumption of beer was between 4000 and 3500 BC. At the same time, distilled alcoholic beverages were also developed. The Sumerians and Egyptians are considered to be the earliest winemakers. The domestication of crops has played a key role in beer brewing. Incorporating grains into the human diet is considered an important step in human evolution, because converting grains into staple food requires extraordinary technology and cooking skills (Wolf et al., 2008).

Poor water quality is one of the main factors for mankind to seek new beverages. Therefore, fermentation has been used to make various beverages, mainly for their safety. Fermented alcoholic and non-alcoholic beverages were made in the Middle Ages. Most non-alcoholic cereal beverages are produced in Africa and South America (Altay et al., 2013). Some traditional African non-alcoholic beverages include Bushera from millet or sorghum, Kunan-Zaki from sorghum, Gowé from millet, Mahewu from corn, Obiolor from millet and sorghum and Obiolor from millet and sorghum, Borde from corn, millet, wheat, sorghum and barley. The last two beverages consist of several grains. South American non-alcoholic fermented beverages include Pozol, Agua-agria, Napú, Fubá, Acupe and Champuz. They are mainly made of corn. Kali as well as Ambil are the traditional fermented beverages from country India. The first one is made with millet and the second one is made with rice. Ambil uses yoghurt to ferment. Recently, in the past 200 years, beverage mixes made from malted grains have appeared. They are intended to be consumed with milk to increase its safety and flavor. The first malt powder was developed by William Horlick in the 1870s. By 1882, he had mastered the process of drying milk with malt and wheat to make the product easily soluble in water. The doctor recommended Horlick's food to their patients (Fernandesa et al., 2021).

7.4 Risks Associated with the Consumption of Dairy Products

Many health risks are associated with the consumption of dairy based fermented foods, adhering to which people are now moving more towards cereal based fermented foods rather than dairy based fermented food products. Some of these issues include milk related allergies, lactose intolerance, high content of cholesterol in dairy based products, higher protein content, etc. Few of these are discussed in detail in the chapter below.

7.4.1 Allergy Associated with Milk Proteins

Atopic dermatitis is one of the diseases which is commonly associated with the allergy present in food amongst the children. The occurrence rate of atopic dermatitis in the initial first year lies between 2–3% (Høst, 2002; Jøhnke et al., 2006; Ricci et al., 2006). Various studies demonstrate that the use of some probiotics is beneficial in reducing the atopic dermatitis occurrence in human beings (Reid & Kirjaivanen, 2005). Also, it has been observed that not all of the probiotic preparations are used for the children who suffer from allergies associated with the milk of cow. However, by using the selective strains of probiotics the severity of the sensitivity can be reduced up to some extent in such children (Moro et al., 2006). Moreover, few studies suggests that the supplementation of probiotic has no significant effect on the any of the symptoms related to atopic dermatitis and in such cases, there are chances of occurrence of increased allergen sensitization in the children suffering from atopic dermatitis (Lee et al., 2007; Moneret-Vautrin et al., 2006; Taylor et al., 2007).

7.4.2 *Lactose Intolerance*

Lactose intolerance or mal absorption of lactose is one amongst the most common type of disorder related with mal absorption of carbohydrate. It is recognised when a person is not able to digest the lactose into its individual components i.e. galactose and glucose. This usually happens because of lack of enzyme responsible for digestion of lactose, which is known as lactase (Hauck et al., 2011). During birth, the activity of the enzyme lactase is at its highest, as the children consumes more of the milk as compared to adults and the activity of this enzyme decreases post weaning. In cases where lactose is not sub divided into its constituents, the colonic bacteria metabolizes the lactose and produces gases such as CH_4 (methane) and H_2 (hydrogen) along with the production of short chain fatty acids (Lee, 2014). The person starts facing the problems related to lactose intolerance just after consumption of milk or related products containing lactose after half an hour or 2 hours. The symptoms of lactose intolerance are recognized by cramping, loose stool, flatulence and bloating. In some cases there occurs irritability in the bowel syndrome also. The population of lactose intolerance people is highest among the Asian people, African Americans and Native Americans, however the lowest population of lactose intolerant people are found in Northern European.

7.4.3 *High Cholesterol and Fat Content*

Milk is one the major sources of fats and the content of fat in ilk depend majorly on the source from where it is obtained. In case of cow milk, it has around 4–5% of the fat, while in case of buffalo milk; the fat content ranges from 7% to 8%. The ratio of poly unsaturated fatty acid to saturated fatty acid is 0.05. Consumption of large volume of milk generally enhances the low density lipoproteins cholesterol and levels of total cholesterol content in the blood. The saturated fat is majorly responsible for increasing the levels of plasma cholesterol, this in turn is responsible for increasing the risks associated with coronary heart disease. Such related risks can be reduced up to some extent by reducing the consumption of low density lipoproteins, which can be further be achieved by reducing the saturated fat content in diets. It has been observed that probiotics has potential beneficial effect on customers suffering from the hypercholesterolemia (Ebel et al., 2014).

7.5 Prebiotic Compounds in Cereal Dietary Fibers

7.5.1 Water Soluble Dietary Fibers

7.5.1.1 β-Glucan

It is one of the most important and major cereal water soluble dietary fiber. It is found in cell walls of endosperm in cereal grains. β-glucan (Fig. 7.1) comprises of D-glucopyranosyl units that are joined in a linear fashion to each other with a mixture of β-1-4 and β-1-3 glycosidic linkage. It has wide impact on the gastro intestinal health of the host due to variation in length and branching. Apart from cereal grains β-glucan is also observed in various marine plants, algae and mushrooms, with barley and oats having the highest content of β-glucan in them (Cloetens et al., 2012).

β-glucan is getting attention from industrial point of view because of its unique properties like stabilizing, gelling and thickening. Besides this, it has biological effects due to its prebiotic property, thereby increasing the interests among the researchers to incorporate the dietary fibers in different foods. Also, due to variation in molecular weight of β-glucan, it becomes feasible to access the growth of microbes in colon. For instance, hydrolysates of 37 kDa, 150 kDa and 172 kDa promotes the growth of Prevotella and Bacteriodes species when tested in vitro, whereas the larger hydrolysates above 230 kDa had non-significant effect in enhancing the growth rate of the tested bacteria (Hamaker & Tuncil, 2014). Cancer prevention, anti-inflammatory action, skin protection, antioxidant, immunological modulation, and glycaemia and serum cholesterol reduction are only a few of the functional qualities that offer promise for human health. Any increase in arbinoxylan substitution resulted in bacterial population and short-chain fatty acid decrease. Because wheat has been used by humans for thousands of years, we hypothesised that the human gut microbiota has co-evolved to digest the ratio of dietary fibre

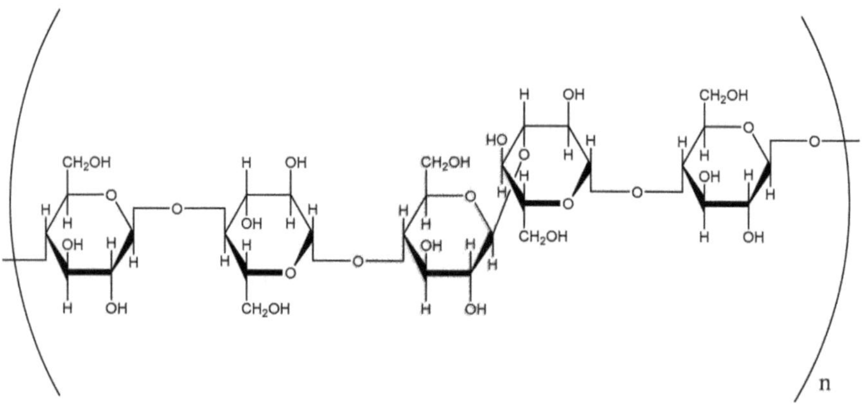

Fig. 7.1 Structure of β-glucan in cereals. (Abdi & Joye, 2021)

polysaccharides present in wheat more efficiently than other polysaccharide ratios. Beneficial bacteria have been demonstrated to preferentially ferment arbinoxylan over -glucan, with the ideal arbinoxylan:glucan ratio found in wheat (3:1) (Shoukat & Sorrentino, 2021).

7.5.1.2 Arabinoxylans

Arabinoxylans are the fragments of non-starch polysaccharides present in cell walls of cereals. It is composed of units of β-1,4 xylopyranose with moieties of arabinofuranose at C_2 position of xylopyranose units. Arabinoxylans are further divided into water extractable arabinoxylans and water unextractable arabinoxylans, on the basis of their solubility. Both types of arabinoxylans possess general structure. The water extractable arabinoxylans act as building blocks for the water unextractable arabinoxylans (Broekaert et al., 2011).

Arabinoxylans, being potential prebiotics are hydrolyzed by various hydrolytic enzymes of bacteria in large intestine. These enzymes include arabinofuranosidase and xylanase. The prebiotic activity of arabinoxylans is associated with its chemical structure. Various studies relating to the structural features of arabinoxylans claims that substitution of ferulic acid play important role in fermentation of arabinoxylans. There must exist a negative relation between fermentability of arabinoxylans and presence of esterified ferulic groups (Hopkins et al., 2003). Gut microbial fermentation can release ferulic acid found in arabinoxylans from various cereal grains. By stimulating the signalling pathways of Kelch-like ECH-associated protein-1 and nuclear factor E2-related factor-2, it can increase antioxidase activity and reduce reactive oxygen species levels. Cereal-derived arabinoxylans have been shown to have an important role in gut microbial population regulation, intestinal barrier function, immunological processes, and glucose and fatty acid metabolism. Activation of G protein-coupled receptors or suppression of histone deacetylase activity to reduce NF-κB signalling are currently the mechanisms of action of arabinoxylans on host health and metabolism, as mediated by short chain fatty acids generated by microbial fermentation. The gut bacteria ferment them to produce free ferulic acid and short-chain fatty acids. Ferulic acid activates the Nrf2 signalling pathway, which has pre-biotic effects on antioxidant capacity and inflammatory responses (Zhang et al., 2021).

Feruloylated arabinoxylan is not degraded by endogenous digestive enzymes in humans or animals, but it can be fermented by intestinal microbiota in the caecum and colon to create short chain fatty acids (Hald et al., 2016). As a result, insoluble ferulic acid is separated from cereal arabinoxylans in the intestine by microbial fermentation, which releases free ferulic acid, which is subsequently absorbed by the host. Bacteroides are the most common bacteria in the proximal colon, and they breakdown low-branched arabinoxylans into feruloylated xylo-oligosaccharide with low polymerization degrees (Pereira et al., 2021). When wheat arabinoxylan and barley -glucan were combined at a 3:1 ratio, the best prebiotic activity was identified, based on concentrations of total short chain fatty acids and increases in total bacteria as well as beneficial Bifdobacterium, Clostridium coccoides, and Eubacterium groups (Harris et al., 2019).

Short chain fatty acids, primarily lactic acid, acetic acid, propionic acid, and butyric acid, are produced by microbial fermentation of cereal-derived arabinoxylans. The production of short chain fatty acids by microbial fermentation is mediated by a set of fundamental processes mediated by the makeup and quantity of gut bacteria (Koh et al., 2016; Louis et al., 2014). Lactic acid is produced largely in the upper gut by microbial fermentation of soluble polysaccharides and indigested oligosaccharides, while acetic, propionic, and butyric acids are produced in the colon and cecum. Short-chain fatty acid synthesis is connected with a specific microbial population found in the host's foregut and hindgut. Lactobacillus is the most common lactic acid-producing bacterium; however, due to changes in the gut environment, such as pH and oxygen concentration, Lactobacillus abundance is reduced in the hindgut. Furthermore, the bacterial families Prevotellaceae, Ruminococcaceae, and Lachnospiraceae are abundant in the hindgut and are the most important producers of short chain fatty acids via microbial fermentation of cereal fibres (Liu et al., 2017). The chemical structure of Arabinoxylan is presented on Fig. 7.2.

7.5.1.3 Fructans

Cereals and their derivatives are the major source of fructans in our daily food. Although eating large amounts of cereal products can cause fructans to have a huge impact on colon health, eating large amounts of fructans (for example, up to 20 grams per day) may also cause bloating and mild flatulence depending up on consumer health and sensitivity. The fructan content of immature grains of most grains is significantly higher than that of mature grains. For instance, in the immature grains of triticale, barley, rye and wheat the reported fructan content is about 23.7–39.0 g/100 g of dry flour. In wheat grains, fructans are reported to accumulate during cell division and expansion stages (Van Loo, 2006).

Both *in vitro* and *in vivo* studies have proved that such fructans have the potential to promote health and have been regarded as prebiotics. Fructans are resistant to

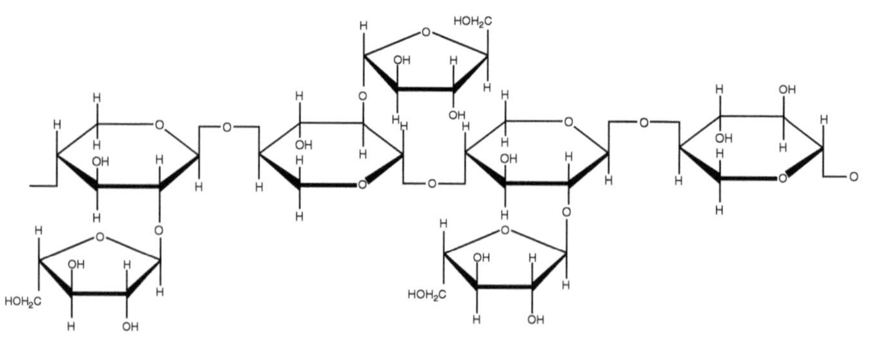

Fig. 7.2 Chemical structure of Arabinoxylan (Sinha et al., 2011)

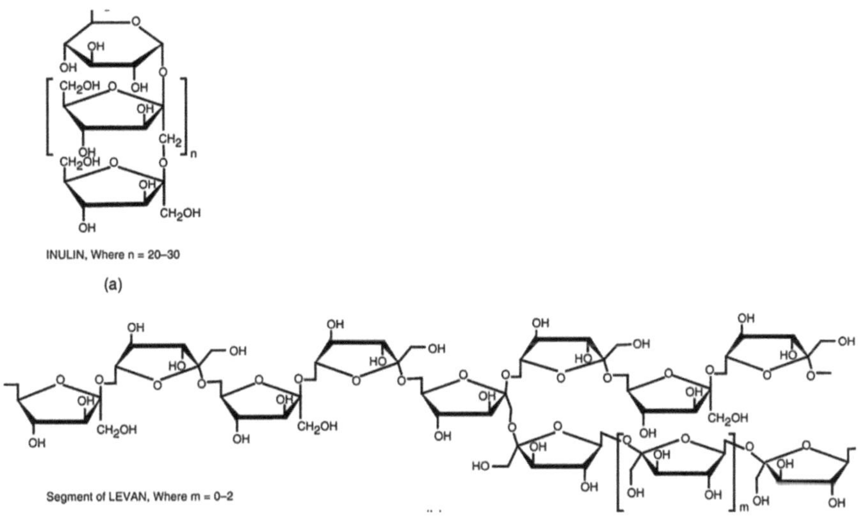

Fig. 7.3 Chemical structure of two fructans: inulin and levan (Izydorczyk et al., 2005)

enzymes that occur in the upper gastrointestinal tract and they are easily hydrolyzed by specific hydrolytic enzymes of certain bacterial species in the lower gastrointestinal tract, so they are fermented there and produce fatty acids (short-chain) or gases. The positive health effects of inulin-like fructans are not limited to their effects on colonic bacteria, because recent evidence suggests that these fructans have a direct immunomodulatory effect on the innate immune system. Fructans arouse Toll-like receptors and are accountable for activating cellular responses, therefore, they seem to trigger a protective effect against oxidation *in vitro* (Pasqualetti et al., 2014). Figure 7.3 represents the chemical structure of Fructans.

7.5.1.4 Fructooligosaccharides

Fructooligosaccgarides are low molecular-weight oligosaccharides of fructose with degree of polymerization ≤10. These are produced by sucrose's transfructosylation and contain 2-,4-β,2–1 linked fructosyl segments. Fructooligosaccharides are found in plant tissues in the form of osmoregulators or source of carbohydrates. Among different plants, the level of fructooligosaccharide is 10 times more in durum wheat. Also, the levels of fructooligosaccharide in immature wheat are comparatively higher as compared to the matured one. Fructooligosaccharides are well known for their potential to exhibit bifidogenic effects, however these effects are influenced by the polymerization length (Yeung et al., 2005). The structure of fructooligosaccharide is presented in Fig. 7.4.

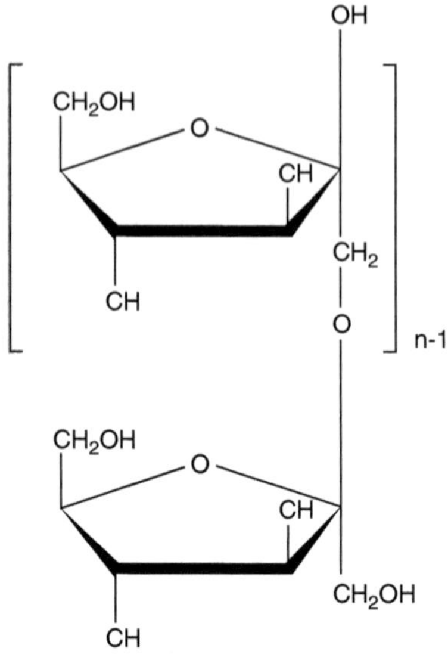

Fig. 7.4 Structure of fructooligosaccharide (Yeung et al., 2005)

7.5.2 *Water Insoluble Dietary Fibers*

7.5.2.1 Resistant Starch

Maltodextrin or resistant starch, is the starch that resists digestion in small intestine by an enzyme called pancreatic amylase. This undigested starch therefore reaches the colon. The behaviour of resistant starch is similar to fermentable fibers, say guar gum, which results in lowering of colonic pH and increasing the faecal matter. It also improves the bowel health, glycaemic control, thereby behaves like compounds known as dietary fibers. Several factors are responsible for the resistant starch that makes them resist to digestion which includes the fragment size of starch, its conformation and structure of the starch granule. Resistant starch is commonly found in cereals, seeds, nuts and vegetables (Fuentes-Zaragoza et al., 2011). These resistant starches are used as encapsulating material for reduced gastric pH. The structural representation of resistant starch is presented in Fig. 7.5.

7.5.2.2 Cellulose

Cellulose, a polysaccharide consists of ≤3000 D-glucose units linked with β-1,4 linkage. Due to the lack of cellulase in the digestive tract, cellulose cannot be ingested in monogastric animals. It is one of the basic components of the cell walls of plant. Cells with higher cell content of cellulose in them possess stronger as well as thicker walls. Cellulose is basically straight chain polymer with no branching and coiling at all (Ummartyotin & Manuspiya, 2015). The structure of cellulose is presented in Fig. 7.6.

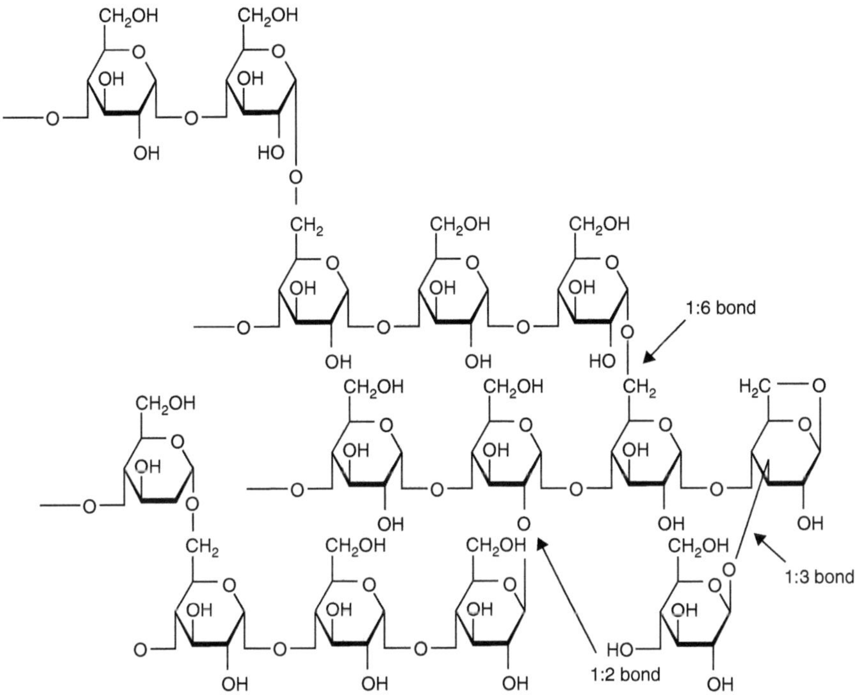

Fig. 7.5 Structure of starch (Slavin et al., 2009)

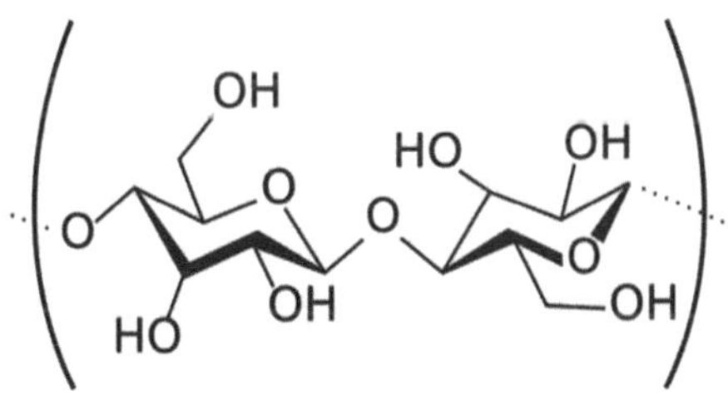

Fig. 7.6 Chemical structure of cellulose (Visanko et al., 2016)

7.6 Effect of Processing on Pre-biotic Potential of Cereals

7.6.1 Fermentation

Processing techniques such as flour pre-processing or wholemeal pre-processing generally improves the prebiotic potential of the cereal crops. It has been observed that certain techniques such as sprouting or pre fermentation also affect the prebiotic characteristics of cereal grains. Fermentation is one amongst the traditional technique used for preservation of food. However, fermentation nowadays is becoming the centre of attraction for the scientists as it has ability to enhance the functional as well as nutritional properties of the product, say, the products made from sour dough after fermentation has positive effects on intestinal health via various mechanisms such as, it changes the fermentability and dietary fiber population, produces exopolysaccharides and other secondary metabolites which ultimately affects the bacteria present in gut. It is found that the fermentation of whole-wheat starter products will enhance the wheat bran arabinoxylan solubility. Hence, the fermentation of starter can affect the prebiotic properties of arabinoxylan by making it easier for the growth of beneficial intestinal bacteria. Using yeast or more specifically, pre-fermenting bran with yeast and lactic acid bacteria may be a simple tool to increase the prebiotic properties of cereal ingredients. The fermentation effect is believed to be caused by changes in enzyme activity and pH during the process, which will change the fermentation mode of dietary fiber. In a study, hydrolase from Trichoderma successfully converted the insoluble dietary fiber in durum wheat into soluble dietary fiber with prebiotic potential. In this study, the enzymatic treatment of insoluble dietary fiber resulted in the production of soluble dietary fiber, which supports the growth of lactobacilli and bifidobacteria in the intestinal model (Lappi et al., 2010; Napolitano et al., 2009; Noori et al., 2017).

7.6.2 Germination or Sprouting

Sprouting/ germination are another ancient traditional grain pretreatment technology, which can be regarded as a green tool to improve the health characteristics of grains. According to various studies, durum wheat varieties germinated in an *in vitro* digestion model significantly enhanced the prebiotic index. Though, this increase depends on the wheat genotype. The influence of germination on prebiotic index is related to alter of nutritional characteristics (Singh et al., 2015). In the study of wheat kernels, it was observed that during the prolonged germination period of 168 hours, total dietary fiber and soluble dietary fiber increased significantly while the content of insoluble dietary fiber was halved. Noori et al. (2017) determined the potential prebiotic activity of germinated rye in an in vitro study. Their results showed that the concentration of germinated rye extract was positively correlated with the vigor of *Lactobacillus acidophilus* and *Bifidobacterium animalis.*

7.6.3 Baking

Bread is a common way to consume cereal grains, however breads are primarily made from wheat. By utilising flours with higher amounts of dietary fibre or adding fibre-rich fractions, the amount of dietary fibre in bread can be increased. Endogenous enzymes prevalent in wheat have the ability to modify the β-glucan polymer structure dramatically (Andersson et al., 2004). Andersson et al. (2008) reported a steady decline in molecular weight of β-glucan as fermentation duration increased within the first several minutes. Endogenous enzymes were also important for enhancing β-glucan polymer extractability by degrading other cell wall polysaccharides in the grain. When compared to native flour, short fermentation times doubled extractability, while extended fermentation times lowered extractability to values below natural extractability. The content of water-extractable β-glucans was unaffected by the level of endogenous enzyme activity. Finally, both the molecular weight and extractability of β-glucan were unaffected by oven baking. As a result, the impact of bread processing on β-glucan related viscous behaviour will be greatly influenced by the fermentation process. According to other study, some barley cultivars are more resistant to these fermentation-related alterations, which open up new possibilities for the production of -glucan enriched breads (Djurle et al., 2018).

Tosh et al. (2008) reported on the self-aggregation of enzymatically formed β-glucan oligomers on muffin preparation with β-glucanase . Extractability of β-glucans was altered to variable degrees by partial depolymerization caused by enzyme addition: extractability increased with decreasing molecular weight and then fell following the self-assembly of gel networks in solutions caused by partially depolymerized β-glucans. At low molecular weight, the lowest extractability values were 17%, whereas at high molecular weight, they were 54%. The baking of treated muffins (30–60 minutes at 180 °C) had no effect on the β-glucans, as previously indicated for bread.

7.6.4 Cooking

Cooking also affected the dietary fibre content of the cereal grains. Foods that have been demonstrated to have high dietary fibre include oat bran, baked porridge, muffins, and morning cereals. The cooking influence of the dietary fibre may also be modulated by the physical format of the ingredient. For example, Beer et al. (1997) found that heating oat bran into porridge had no effect on the extractability of β-glucans, but that cooking rolled oats had a modest increase in extractability.

According to Johansson et al. (2007), adding oat flakes to boiling water and simmering for 10 minutes enhanced the amount of soluble dietary fibre (β-glucans). Similarly, Åman et al. (2004) found that using a large bran particle size and a short fermentation time can reduce -glucan breakdown during bread making. According to Beer et al. (1997), freezing reduced the extractability of β-glucans in muffins but had no effect on molecular weight. The extractability was lowered by 50% after 8 weeks of freezer storage. These alterations were attributable to molecular organisation during storage, which caused water to be repelled from the polysaccharide matrix, according to the authors.

7.6.5 Extrusion

Cooking by extrusion is a popular method of processing dietary fibre rich foods, in addition to muffins and bread making. The high pressure and high shear that will be applied to the product distinguishes extrusion from typical porridge cooking or dough baking. Depending on the settings used, extrusion can affect both molecular weight and extractability. Authors Gaosong and Vasanthan (2000) measured increased extractability that was both cultivar- and process-parameter-dependent when processing two different barley cultivars (waxy and regular) using twin-screw extrusion at different moisture levels (20–50% moisture) and temperatures (90, 100, 120, and 140 °C). At each extrusion temperature examined, solubility of the waxy material declined as moisture content increased. At each temperature level, the regular cultivar demonstrated higher extractability following extrusion and increased extractability with increasing moisture. Extrusion of the waxy cultivar revealed some fragmentation, which was particularly apparent around 120 °C. Extrusion seems to be a problem for the standard cultivar.

Zhang et al. (2011) used oat bran alone and found that extrusion parameters (feed moisture and temperature) had a substantial impact on the soluble dietary fibre content. On hulled barley husks, an improvement in extractability was also recorded (up to 8% at high temperature–low moisture extrusion conditions). As a result, traditional processing of -glucan-rich foods will always result in a change in dietary fibre structure, extractability (cell wall release or aggregation), or both. Even if some general trends emerge, these changes are difficult to forecast since they are highly process-dependent. Furthermore, the physical source of dietary fibre such as bran, flour, or rolled grain, as well as the type of the processed recipe, will influence the structural change (Sharma & Gujral, 2013).

7.7 Cereal-Based Probiotics Products

Probiotics are usually selective live microorganisms present in different foods that provide promising health benefits by stabilizing the gut micro flora upon consumption. They are well known for their potential to exhibit preventive properties against number of non-infective and infective disorders. They play important role in reducing the levels of cholesterol serum, improves lactose metabolism, protein digestibility, stimulates the immune system, absorption of calcium and possess various anti carcinogenic properties, anti-mutagenic properties, anti-diarrhoeal properties as well as aid in curing of bowel disease. Unfortunately, the majority of probiotic foods available in market such as cheese, beverages are usually milk based or confined to its fermented products, comprising of mainly the live microorganisms belonging to the family of lactic-acid bacteria (Enujiugha & Badejo, 2017). However, there are few beverages and probiotics containing products which are cereal based despite of the fact that cereals are widely distributed throughout and has rich nutritive value as well. But due to emerging demand of different tastes among the consumers and also because of vegan consumers, the researchers are now focussing towards development of cereal based probiotic products. Such products are not only boon for the vegan people but also for the people who suffers from allergies associated with milk or milk products, especially for the ones who are intolerant to lactose. Cereal based probiotic products have further advantages over the milk based probiotic products in terms of longer shelf life and nutritional properties (Hassan et al., 2012). Various cereal based probiotic beverage products are enlisted in Table 7.1 along with their probiotic microorganisms. In addition, various labs are still developing more of the cereal based products to exploit the potential of

Table 7.1 Cereal based probiotic beverages

Sr. No.	Product name	Substrate	Microorganism/ specie	Protocol or conditions	Countries	References
1.	Akamu	Millet, Maize, Sorghum	*L. fermentum, L. plantarum, S. lactis*	Steeping of grains at room temperature for 2–4 days Souring Sieving Wet milling	Nigeria	Enujiugha and Badejo (2017)
2.	Pozol	Maize	*Lactobacillus, Leuconostoc, Lactococcus, Pediococcus*	Maize is cooked in 1% w/v of lime solution Washing is done using water Grinding is done to till dough is formed Balls off dough are made wrapping is done in banana leaves Fermentation for upto 4 days at ambient temperature.	Mexico	Prado et al. (2008)
3.	Gowe	Sorghum	*L. fermentum, P. acidilactici, Lb. mucosae, W. confusa*	Blending of non malted and malted Sorghum flour Fermentation to be done where the surrounding has moisture content of 52–87%.	Benin Republic	Vieira-Dalodé et al. (2008)
4.	Boza	Millet, Rye, Maize, wheat	*L. plantarum, L. brevis, Lb. acidophilus, L. mesenteroides, L. fermentum, L. ewffinolactis, L. coprophilus*	Cereals are boiled in water Cooling and sieving Addition of sugar	Bulgaria, Romania, Albania, Turkey	Blandino et al. (2003)

(continued)

Table 7.1 (continued)

Sr. No.	Product name	Substrate	Microorganism/ specie	Protocol or conditions	Countries	References
5.	Bushera	Millet, Sorghum	*L. lactis, Lb. brevis, L. mesenteroides, S. lactis*	Mixing of grains with boiling water Cooled to ambient temperature Germinated flour is added to the mixture Mixture is allowed to ferment for 1–6 days at room temperature.	Uganda	Muyanja et al. (2003)
6.	Symbiotic functional drink	Oats	*Lb. plantarum* B28	Fermentation was carried out for 8 hours with the help of sucrose at 37 °C.	Bulgaria	Angelov et al. (2006)
7.	Mahewu	Millet malt/ Sorghum, corn meal	*Lb. brevis, L. lactis, Lb. bulgaricus*	Maize porridge is mixed with water Addition of cereal flours Allowed to ferment at ambient temperature.	African and Arabian Gulf nations	Prado et al. (2008)
8.	Grainfields	Grains, seeds (maize, alfalfa, oats, rice, mung beans, barley, millets, rye, wheat)	*S. boulardii, Lb. acidophilus, Sc. Cerevisiae, Lb. delbreukii*	Natural organic fermentation	Various countries	Prado et al. (2008)

(continued)

Table 7.1 (continued)

Sr. No.	Product name	Substrate	Microorganism/ specie	Protocol or conditions	Countries	References
9.	Kunun zaki	Sorghum, Millet	*Lb. fermentum, L. lactis, Lb. plantarum*	Steeping of grains or millets Wet milling Addition of spices usually pepper or ginger Wet sieving Addition of sugar Fermentation is carried out for upto 48 hours.	Nigeria	Agarry et al. (2010)
10.	Koko	Millets	*Lb. paraplantarum, Lb. confusa, Lb. fermentum, P. pentosaceus*	Steeping of grains overnight Milling Addition of water and mixing Sieving Fermentation for 2–3 hours Addition of sugar	Kenya	Lei and Jakobsen (2004)
11.	Togwa	Finger millets, Maize	*Lb. fermentum, W. confusa, Lb. plantarum, Lb. cellobiosus, Lb. brevis*	Cooking of cereal flour in water Cooled at 35 °C Addition of culture of togwa along with cereal flour Fermentation is done till pH reaches at 4.0	Africa	Oi and Kitabatake (2003)
12.	Borde	Finger millets, Maize, tef, Barley	*Lb. fermentum, Lb. acidophilus*	25% unmalted and 75% malted cereals are allowed to ferment for 4 upto 4 days in four different phases.	Ethiopia	Abegaz et al. (2002)

(continued)

Table 7.1 (continued)

Sr. No.	Product name	Substrate	Microorganism/ specie	Protocol or conditions	Countries	References
13.	Mageu	Wheat, Maize	*Lb. fermentum, Lb. lactis*	Boiling of Maize in water Addition of wheat flour Allowed to ferment for 24 hours at 35 °C	South Africa	Enujiugha and Badejo (2017)
14.	Kwete	Millets, Maize	*L. lactis, Lb. plantarum, L. brevis*	Soaking of grains for upto 48 hours Allowed to germinate upto 3 days Subjected to sun drying for 1 to 2 days Souring is performed at 24 to 30 °C for 1 day. Allowed to ferment for 24 hours Filtration	Uganda	Muyanja & Namugumya (2009)

probiotics in the field of cereals, for instance, Kedia and others have made probiotic beverages based on malt; different oat based beverages are developed by Angelov and others in their experiment.

7.8 Other Products

Apart from cereal based probiotic beverages, several efforts have been made by researchers to impart the benefits of probiotics in traditional fermented foods so that it gets easily acceptable amongst the consumers, the examples of this includes the traditional food products such as Adai made of legumes and cereals by using lactic acid bacteria (LAB); Atole, made from maize with LAB as the probiotic microorganisms; Ben saalga, made of pearl millet using LAB, etc. The common traditional foods which are popularly consumed as breakfast or dinner in Southern India and in other countries like Sri Lanka i.e. as Dosa, Idli, are made up of Bengal gram, rice, legume or cereals can be inoculated with different probiotic microorganisms such as *L. mesenteroides, LAB, Lb. fermentum, S. cerevisiae* (Agrawal et al., 2000; Aidoo et al., 2006; Farnworth, 2005; Tou et al., 2006). Other traditional fermented foods include Tempeh, Uji, Saurkraut, Ilambazi lokubilisa, etc. which can be turned into probiotic foods after incorporation of live probiotic microorganisms in them. However there are several probiotic products that are not dairy based but are fruits or vegetable based, soy based or cereal based and are available in market as well. These include puddings based on cereals (Helland et al., 2004), Yosa or pudding based on oat bran (Blandino et al., 2003), rice based yogurt (Wongkhalaung & Boonyaratanakornkit, 2000).

Several probiotic products are produced globally based on the cereal matrices. These includes cereal bars which are available by different commercial names viz. Effi foods probiotic care bars, vega one, Good! Greens bars, all in one Meal bars. These are originated in United States of America (USA) by using single probiotic microbe i.e. *B. coagulans*. PROBAR meal bars is another example of baking mix made by using culture of *B. coagulans*. Other products such as Yog active probiotic cereals (origin- Germany, probiotic microbe- *Lb. acidophilus*), probiotic granola bars (commercial name- Udi's Gluten free, origin- USA, organism- *B. coagulans*), oatmeal bar (commercial name- Pop culture probiotic, origin- California, organism- *B. coagulans*), cereal bar (commercial name- Welo Probiotic bar, origin- Canada, organism- *B. coagulans*), Poppers (commercial name- Brad's Broccoli peppers, origin- Pipersville, organism- *B. coagulans*), Muesli (commercial name- Nutrus Slim Muesli, origin- India, organism- *B. coagulans*), Muesli (commercial name- Something to crow about- probiotic Muesli, origin- New Zealand, organism- *B. coagulans*), Burritos (commercial name- Probiotic Burritos, origin- USA, organism- *B. coagulans*) and many more (Dey, 2018).

7.9 Health Benefits of Pre-biotic and Probiotic

Probiotics serves both, the health as well as the nutritional benefits upon their consumption. Enzymatic hydrolysis of bacteria escalates the protein bioavailability, which in turn is beneficial for the consumer if he or she is suffering from the deficiency of production of endogenous protease. This happens because of increased free amino acid production due to enzymatic hydrolysis by bacteria as discussed earlier. Various strains of lactic acid bacteria are also responsible for increasing the content of vitamins, especially B-complex in various fermented foods.

Probiotics exhibit the beneficial effects when consumed in sufficient quantity or concentration, say, as lower as 10^8 colony forming unit per gram of the product and this must will be capable of surviving the harsh conditions present in human stomach such as, low pH of the gastric juice and then it reaches the small intestine and last to the colons. The best effect of probiotics is believed to be achieved when microorganisms colonize themselves in intestinal epithelium. At this time, they are capable of affecting the immune system of intestine by providing antioxidants and anti-mutagens, and also by displacing the enteric pathogens.

` Probiotics also posses anti microbial properties which might be attributed to the production of number of inhibitory compounds such as bacteriocins, hydrogen per oxide and organic acids, and to its competition towards the nutrients. The organic acids such as acetic acids and lactic acid produced by the bacteria are responsible for lowering the pH and thereby become responsible for exhibiting the bacteriostatic effect or sometimes bactericidal effect also. However, the role of probiotic strain in the inhibition of pathogen is limited as the bacteriocins that are present since ages, do possess the inhibitory action against the closely related species like that of *Clostridium* (spore-forming), *Bacillus* (spore-forming) and other species of *Lactobacillus*. The other metabolites especially, the low molecular weight ones which includes acetic acid, hydrogen per oxide, various aroma compounds and lactic acid along with different secondary metabolites are considered to be the most important since they exhibit wide range of inhibitory spectrum against number of harmful organisms such as *Chlostridium, E. coli, Salmonella, Helicobacter,* etc. Also, the lactic acid bacteria from *Lactococcus* genera, *Lactobacillus* genera and *Pediococcus* genera are responsible for producing the di-acetyl that is rarely found in the food fermentations is one of the major contributor of anti-bacterial activity (Enujiugha & Badejo, 2017).

7.10 Effect of Prebiotic Dietary Fibers on Health

7.10.1 Effect on Composition of Hind Gut Bacteria

The genera of these bacteria are the well-known common markers for the health of microbiota that are responsible for targeting the dietary stimulation. Lactobacillus has been shown to down-regulate the inflammation of the gastrointestinal mucosa. Lactobacillus plays a role in helping patients with lactose intolerance to digest the lactose, relieve constipation, improve symptoms of IBS (irritable bowel syndrome), and may help to prevent diarrhea in case of travellers. Bifidobacterium naturally exists in gastrointestinal tract of humans and possess strong affinity for fermenting certain oligosaccharides, thereby establishing them as common marker to have prebiotic ability. It has been observed that Bifidobacterium does not produce any carcinogens in the body. The level of bifidobacteria is negatively correlated with weight gain and obesity. The decrease of bifidobacteria and the decrease of bacterial diversity are related to the increase of IBS and inflammation. Though, various studies reveal the beneficial nature of these bacteria towards health, however, the mechanism of the same is not yet clear (Carlson et al., 2018).

7.10.2 Effects on the Absorption of Mineral

Prebiotics are well known in improving and maintaining the bone structure of elder people and adolescents in the era where the bone fracture and osteoporosis is a common problem. Intake of prebiotics usually increases the calcium absorption along with its bioavailability, thereby reduces the risk of osteoporosis. Distal intestine, the primary site for the absorption of calcium, which is escalated by various chemical changes and increased acid-fermentation of prebiotic dietary fibers with the help of various bacteria.

7.10.3 Effects on Production of Metabolites

Various metabolites, primary as well as secondary metabolites are formed by indirect or direct fermentation of various selective compounds that are correlated with number of beneficial effects on human health. Various SCFA (short chain fatty acids) are produced upon fermentation of amino acids, nutrients, carbohydrates and other different compounds by gut microbiota, which is usually absorbed in the small intestine. These include butyrate, acetate and propionate, which contribute up to 95% of the SCFA that are produced in colon. SCFAs are known to possess number if positive outcomes in humans. It has been observed that inulin type fructans when fermented, increases the levels of hippurate, which is microbial a co-metabolite, usually present in higher concentration in lean people as compared to the obese ones. Its increased levels are known to possess beneficial effect on humans upon fermentation of inulin after consumption.

7.10.4 Protein Fermentation

Protein fermentation from endogenous or undigested protein sources occurs in the absence of fermentable carbohydrates, can lead to the accumulation and formation of harmful metabolites, such as ammonia, sulphides, phenols and amines. Also, at the same time, the concentration of SCFAs decreases and the pH value of the environment increases, thereby providing a favorable environment (distal colon) for protein fermentation to occur efficiently. This in turn leads to the production of branched-chain fatty acids and various indoles and phenols, which is unique to the bacterial metabolism.

7.10.5 Effects on Risks Associated with Allergy

Microbial diversity of gut is responsible for development of inflammatory conditions such as diseases related to allergy. Allergies during the first 5 years of life are generally associated with the decreased level of Lactobacilli and bifidobacteria. Dietary oligosaccharides have immuno-modulating effects against development of rhinoconjuctionvitis and eczema. They also possess various allergy protective effects.

7.11 Future Perspectives

Cereals are one amongst the most suitable substrate that can be used for growing the probiotic strains, which are basically human derived. Irrespective of the number of differences in the performances in between the specie, complexity of substrates, a stable systematic approach is to be implemented which is capable of enhancing the survival and the growth of the probiotic strain. Also, the functionality can be improved by addition of various non digestible components of cereal based matrix which can act as prebiotics. Novel functional ingredients are advantageous for the manufacturing companies as they can add more of value to the products. The development of such novel products requires campaigns so as to make consumers aware of the products and its benefits. Also, consumer needs some additional adaptation time to accept the new product. Therefore, the approach of making fortifying the traditional cereal based food products are of more advantageous as these are easily acceptable among the consumers. Cereals are capable enough to deliver the probiotic lactic acid bacteria to human guts and can act as substrate for the growth of bacteria as well.

7.12 Summary

Cereal-based foods have been consumed since ancient times, and their popularity continues to grow due to their consumer acceptance and health benefits. However, due to advancements in food technology and interest in the production of novel goods with improved health qualities, their design has evolved. Probiotics, prebiotics, and synbiotics have all had a lot of success in the field of functional foods over the previous two decades. Various cereal-based products can now be formulated with synbiotic features that include additional health-improving molecules without altering their physicochemical, rheological, or sensory properties. Food businesses can add additional value to items that consumers are already familiar with by developing new functional ingredients. New food development necessitates greater marketing campaigns, and consumers may require time to adjust to the new product. Nutritional food components enable producers to meet and exceed the demands of today's health-conscious consumer by either inventing new and inventive products or simply reformulating existing ones. Cereals contain possibly pre-biotic compounds whose activity as well as the ability to develop and transfer probiotic lactic acid bacteria to the human gut should be investigated.

References

Abdi, R., & Joye, I. J. (2021). Prebiotic potential of cereal components. *Foods, 10*(10), 2338.

Abegaz, K., Beyene, F., Langsrud, T., & Narvhus, J. A. (2002). Indigenous processing methods and raw materials of borde, an Ethiopian traditional fermented beverage. *Journal of Food Technology in Africa, 7*(2), 59–64.

Agarry, O. O., Nkama, I., & Akoma, O. (2010). Production of Kunun-zaki (A Nigerian fermented cereal beverage) using starter culture. *International Research Journal of Microbiology, 1*(2), 18–25.

Agrawal, R., Rati, E. R., Vijayendra, S. V. N., Varadaraj, M. C., Prasad, M. S., & Nand, K. (2000). Flavour profile of idli batter prepared from defined microbial starter cultures. *World Journal of Microbiology and Biotechnology, 16*(7), 687–690.

Aidoo, K. E., Rob Nout, M. J., & Sarkar, P. K. (2006). Occurrence and function of yeasts in Asian indigenous fermented foods. *FEMS Yeast Research, 6*(1), 30–39.

Altay, F., Karbancıoglu-Güler, F., Daskaya-Dikmen, C., & Heperkan, D. (2013). A review on traditional Turkish fermented non-alcoholic beverages: microbiota, fermentation process and quality characteristics. *International Journal of Food Microbiology, 167*(1), 44–56.

Åman, P., Rimsten, L., & Andersson, R. (2004). Molecular weight distribution of β-glucan in oat-based foods. *Cereal Chemistry, 81*(3), 356–360.

Andersson, A. A., Armö, E., Grangeon, E., Fredriksson, H., Andersson, R., & Åman, P. (2004). Molecular weight and structure units of (1→ 3, 1→ 4)-β-glucans in dough and bread made from hull-less barley milling fractions. *Journal of Cereal Science, 40*(3), 195–204.

Andersson, A. A. M., Rüegg, N., & Åman, P. (2008). Molecular weight distribution and content of water-extractable β-glucan in rye crisp bread. *Journal of Cereal Science, 47*(3), 399–406.

Angelov, A., Gotcheva, V., Kuncheva, R., & Hristozova, T. (2006). Development of a new oat-based probiotic drink. *International Journal of Food Microbiology, 112*(1), 75–80.

Beer, M. U., Wood, P. J., Weisz, J., & Fillion, N. (1997). Effect of cooking and storage on the amount and molecular weight of (1→ 3)(1→ 4)-β-D-glucan extracted from oat products by an in vitro digestion system. *Cereal Chemistry, 74*(6), 705–709.

Blandino, A., Al-Aseeri, M. E., Pandiella, S. S., Cantero, D., & Webb, C. (2003). Cereal-based fermented foods and beverages. *Food Research International, 36*(6), 527–543.

Broekaert, W. F., Courtin, C. M., Verbeke, K., Van de Wiele, T., Verstraete, W., & Delcour, J. A. (2011). Prebiotic and other health-related effects of cereal-derived arabinoxylans, arabinoxylan-oligosaccharides, and xylooligosaccharides. *Critical Reviews in Food Science and Nutrition, 51*(2), 178–194.

Carlson, J. L., Erickson, J. M., Lloyd, B. B., & Slavin, J. L. (2018). Health effects and sources of prebiotic dietary fiber. *Current Developments in Nutrition, 2*(3), nzy005.

Charalampopoulos, D., Wang, R., Pandiella, S. S., & Webb, C. (2002). Application of cereals and cereal components in functional foods: a review. *International Journal of Food Microbiology, 79*(1–2), 131–141.

Cloetens, L., Ulmius, M., Johansson-Persson, A., Åkesson, B., & Önning, G. (2012). Role of dietary beta-glucans in the prevention of the metabolic syndrome. *Nutrition Reviews, 70*(8), 444–458.

Das, A., Raychaudhuri, U., & Chakraborty, R. (2012). Cereal based functional food of Indian subcontinent: a review. *Journal of Food Science and Technology, 49*(6), 665–672.

Dey, G. (2018). Non-dairy probiotic foods: Innovations and market trends. In *Innovations in technologies for fermented food and beverage industries* (pp. 159–173). Springer.

Djurle, S., Andersson, A. A., & Andersson, R. (2018). Effects of baking on dietary fibre, with emphasis on β-glucan and resistant starch, in barley breads. *Journal of Cereal Science, 79*, 449–455.

Ebel, B., Lemetais, G., Beney, L., Cachon, R., Sokol, H., Langella, P., & Gervais, P. (2014). Impact of probiotics on risk factors for cardiovascular diseases. A review. *Critical Reviews in Food Science and Nutrition, 54*(2), 175–189.

Enujiugha, V. N., & Badejo, A. A. (2017). Probiotic potentials of cereal-based beverages. *Critical Reviews in Food Science and Nutrition, 57*(4), 790–804.

Farnworth, E. R. (2005). The Beneficial health effects of fermented foods-potential probiotics around the world. *Journal of Nutraceuticals, Functional & Medical Foods, 4*(3–4), 93–117.

Fernandesa, C. G., Sonawaneb, S. K., & Arya, S. S. (2021). Cereal based functional beverages: a review. *Journal of Microbiology, Biotechnology and Food Sciences, 2021*, 914–919.

Fuentes-Zaragoza, E., Sánchez-Zapata, E., Sendra, E., Sayas, E., Navarro, C., Fernández-López, J., & Pérez-Alvarez, J. A. (2011). Resistant starch as prebiotic: A review. *Starch-Stärke, 63*(7), 406–415.

Gaosong, J., & Vasanthan, T. (2000). Effect of extrusion cooking on the primary structure and water solubility of β-glucans from regular and waxy barley. *Cereal Chemistry, 77*(3), 396–400.

Hald, S., Schioldan, A. G., Moore, M. E., Dige, A., Lærke, H. N., Agnholt, J., … Dahlerup, J. F. (2016). Effects of arabinoxylan and resistant starch on intestinal microbiota and short-chain fatty acids in subjects with metabolic syndrome: a randomised crossover study. *PLoS One, 11*(7), e0159223.

Hamaker, B. R., & Tuncil, Y. E. (2014). A perspective on the complexity of dietary fiber structures and their potential effect on the gut microbiota. *Journal of Molecular Biology, 426*(23), 3838–3850.

Harris, S., Monteagudo-Mera, A., Kosik, O., Charalampopoulos, D., Shewry, P., & Lovegrove, A. (2019). Comparative prebiotic activity of mixtures of cereal grain polysaccharides. *AMB Express, 9*(1), 1–7.

Hassan, A. A., Aly, M. M., & El-Hadidie, S. T. (2012). Production of cereal-based probiotic beverages. *World Applied Sciences Journal, 19*(10), 1367–1380.

Hauck, F. R., Thompson, J. M., Tanabe, K. O., Moon, R. Y., & Vennemann, M. M. (2011). Breastfeeding and reduced risk of sudden infant death syndrome: a meta-analysis. *Pediatrics, 128*(1), 103–110.

Helland, M. H., Wicklund, T., & Narvhus, J. A. (2004). Growth and metabolism of selected strains of probiotic bacteria in milk-and water-based cereal puddings. *International Dairy Journal, 14*(11), 957–965.

Hopkins, M. J., Englyst, H. N., Macfarlane, S., Furrie, E., Macfarlane, G. T., & McBain, A. J. (2003). Degradation of cross-linked and non-cross-linked arabinoxylans by the intestinal microbiota in children. *Applied and Environmental Microbiology, 69*(11), 6354–6360.

Høst, A. (2002). Frequency of cow's milk allergy in childhood. *Annals of Allergy, Asthma & Immunology, 89*(6), 33–37.

Izydorczyk, M., Cui, S. W., & Wang, Q. (2005). *Polysaccharide gums: structures, functional properties, and applications* (pp. 293–299). Chemistry, Physical Properties, and Applications.

Johansson, L., Tuomainen, P., Anttila, H., Rita, H., & Virkki, L. (2007). Effect of processing on the extractability of oat β-glucan. *Food Chemistry, 105*(4), 1439–1445.

Jøhnke, H., Norberg, L. A., Vach, W., Høst, A., & Andersen, K. E. (2006). Patterns of sensitization in infants and its relation to atopic dermatitis. *Pediatric Allergy and Immunology, 17*(8), 591–600.

Koh, A., De Vadder, F., Kovatcheva-Datchary, P., & Bäckhed, F. (2016). From dietary fiber to host physiology: short-chain fatty acids as key bacterial metabolites. *Cell, 165*(6), 1332–1345.

Lappi, J., Selinheimo, E., Schwab, U., Katina, K., Lehtinen, P., Mykkänen, H., et al. (2010). Sourdough fermentation of wholemeal wheat bread increases solubility of arabinoxylan and protein and decreases postprandial glucose and insulin responses. *Journal of Cereal Science, 51*(1), 152–158.

Lee, B. H. (2014). *Fundamentals of food biotechnology*. Wiley.

Lee, T. T. T., Morisset, M., Astier, C., Moneret-Vautrin, D. A., Cordebar, V., Beaudouin, E., et al. (2007). Contamination of probiotic preparations with milk allergens can cause anaphylaxis in children with cow's milk allergy. *Journal of Allergy and Clinical Immunology, 119*(3), 746–747.

Lei, V., & Jakobsen, M. (2004). Microbiological characterization and probiotic potential of koko and koko sour water, African spontaneously fermented millet porridge and drink. *Journal of Applied Microbiology, 96*(2), 384–397.

Liu, P., Zhao, J., Guo, P., Lu, W., Geng, Z., Levesque, C. L., et al. (2017). Dietary corn bran fermented by Bacillus subtilis MA139 decreased gut cellulolytic bacteria and microbiota diversity in finishing pigs. *Frontiers in Cellular and Infection Microbiology, 7*, 526.

Louis, P., Hold, G. L., & Flint, H. J. (2014). The gut microbiota, bacterial metabolites and colorectal cancer. *Nature Reviews Microbiology, 12*(10), 661–672.

Mohammadi, M., Nouri, L., & Mortazavian, A. M. (2021). Development of a functional synbiotic beverage fortified with different cereal sprouts and prebiotics. *Journal of Food Science and Technology, 58*(11), 4185–4193.

Moneret-Vautrin, D. A., Morisset, M., Cordebar, V., Codreanu, F., & Kanny, G. (2006). Probiotics may be unsafe in infants allergic to cow's milk. *Allergy, 61*(4), 507–508.

Moro, G., Arslanoglu, S., Stahl, B., Jelinek, J., Wahn, U., & Boehm, G. (2006). A mixture of prebiotic oligosaccharides reduces the incidence of atopic dermatitis during the first six months of age. *Archives of Disease in Childhood, 91*(10), 814–819.

Muyanja, C. M. B. K., Narvhus, J. A., Treimo, J., & Langsrud, T. (2003). Isolation, characterisation and identification of lactic acid bacteria from bushera: a Ugandan traditional fermented beverage. *International Journal of Food Microbiology, 80*(3), 201–210.

Muyanja, C., & Namugumya, B. S. (2009). Traditional processing, microbiological, physiochemical and sensory characteristics of kwete, a Ugandan fermented maize based beverage. *African Journal of Food, Agriculture, Nutrition and Development, 9*(4), 1–14.

Napolitano, A., Costabile, A., Martin-Pelaez, S., Vitaglione, P., Klinder, A., Gibson, G. R., & Fogliano, V. (2009). Potential prebiotic activity of oligosaccharides obtained by enzymatic conversion of durum wheat insoluble dietary fibre into soluble dietary fibre. *Nutrition, Metabolism and Cardiovascular Diseases, 19*(4), 283–290.

Noori, N., Hamedi, H., Kargozari, M., & Shotorbani, P. M. (2017). Investigation of potential prebiotic activity of rye sprout extract. *Food Bioscience, 19*, 121–127.

Nout, M. R. (2009). Rich nutrition from the poorest–Cereal fermentations in Africa and Asia. *Food Microbiology, 26*(7), 685–692.

Oi, Y., & Kitabatake, N. (2003). Chemical composition of an East African traditional beverage, togwa. *Journal of Agricultural and Food Chemistry, 51*(24), 7024–7028.

Pasqualetti, V., Altomare, A., Guarino, M. P. L., Locato, V., Cocca, S., Cimini, S., et al. (2014). Antioxidant activity of inulin and its role in the prevention of human colonic muscle cell impairment induced by lipopolysaccharide mucosal exposure. *PLoS One, 9*(5), e98031.

Pereira, G. V., Abdel-Hamid, A. M., Dutta, S., D'Alessandro-Gabazza, C. N., Wefers, D., Farris, J. A., et al. (2021). Degradation of complex arabinoxylans by human colonic Bacteroidetes. *Nature Communications, 12*, 459.

Prado, F. C., Parada, J. L., Pandey, A., & Soccol, C. R. (2008). Trends in non-dairy probiotic beverages. *Food Research International, 41*(2), 111–123.

Reid, G., & Kirjaivanen, P. (2005). Taking probiotics during pregnancy. Are they useful therapy for mothers and newborns? *Canadian Family Physician, 51*(11), 1477–1479.

Ricci, G., Patrizi, A., Baldi, E., Menna, G., Tabanelli, M., & Masi, M. (2006). Long-term follow-up of atopic dermatitis: retrospective analysis of related risk factors and association with concomitant allergic diseases. *Journal of the American Academy of Dermatology, 55*(5), 765–771.

Rolim, P. M. (2015). Development of prebiotic food products and health benefits. *Food Science and Technology, 35*(1), 3–10.

Saarela, M., Mogensen, G., Fondén, R., Mättö, J., & Mattila-Sandholm, T. (2000). Probiotic bacteria: safety functional and technological properties. *Journal of Biotechnology, 84*, 197–215.

Salmerón, I. (2017). Fermented cereal beverages: From probiotic, prebiotic and synbiotic towards Nanoscience designed healthy drinks. *Letters in Applied Microbiology, 65*(2), 114–124.

Severson, D. K. (1998). Lactic acid fermentations. In T. W. Nagodawithana & G. Reed (Eds.), *Nutritional Requirements of Commercially Important Microorganisms* (pp. 258–297). Esteekay Associates.

Sharma, P., & Gujral, H. S. (2013). Extrusion of hulled barley affecting β-glucan and properties of extrudates. *Food and Bioprocess Technology, 6*(6), 1374–1389.

Shoukat, M., & Sorrentino, A. (2021). Cereal β-glucan: a promising prebiotic polysaccharide and its impact on the gut health. *International Journal of Food Science & Technology, 56*(5), 2088–2097.

Singh, A. K., Rehal, J., Kaur, A., & Jyot, G. (2015). Enhancement of attributes of cereals by germination and fermentation: a review. *Critical Reviews in Food Science and Nutrition, 55*(11), 1575–1589.

Sinha, A. K., Kumar, V., Makkar, H. P., De Boeck, G., & Becker, K. (2011). Non-starch polysaccharides and their role in fish nutrition–A review. *Food Chemistry, 127*(4), 1409–1426.

Slavin, J. L., Savarino, V., Paredes-Diaz, A., & Fotopoulos, G. (2009). A review of the role of soluble fiber in health with specific reference to wheat dextrin. *Journal of International Medical Research, 37*(1), 1–17.

Taylor, A. L., Dunstan, J. A., & Prescott, S. L. (2007). Probiotic supplementation for the first 6 months of life fails to reduce the risk of atopic dermatitis and increases the risk of allergen sensitization in high-risk children: a randomized controlled trial. *Journal of Allergy and Clinical Immunology, 119*(1), 184–191.

Tosh, S. M., Brummer, Y., Wolever, T. M., & Wood, P. J. (2008). Glycemic response to oat bran muffins treated to vary molecular weight of β-glucan. *Cereal Chemistry, 85*(2), 211–217.

Tou, E. H., Guyot, J. P., Mouquet-Rivier, C., Rochette, I., Counil, E., Traoré, A. S., & Trèche, S. (2006). Study through surveys and fermentation kinetics of the traditional processing of pearl millet (Pennisetum glaucum) into ben-saalga, a fermented gruel from Burkina Faso. *International Journal of Food Microbiology, 106*(1), 52–60.

Ummartyotin, S., & Manuspiya, H. (2015). A critical review on cellulose: from fundamental to an approach on sensor technology. *Renewable and Sustainable Energy Reviews, 41*, 402–412.

Van Loo, J. (2006). Inulin-type fructans as prebiotics. In *Prebiotics: Development & application* (pp. 57–100). Wiley Publisher

Venn, B. J., & Mann, J. I. (2004). Cereal grains, legumes and diabetes. *European Journal of Clinical Nutrition, 58*(11), 1443–1461.

Vieira-Dalodé, G., Hounhouigan, J., Jespersen, L., & Jakobsen, M. (2008). Use of starter cultures of lactic acid bacteria and yeasts as inoculum enrichment for the production of gow, a sour beverage from Benin. *African Journal of Microbiology Research, 2*(7), 179–186.

Visanko, M., Liimatainen, H., Sirvio, J., Hormi, O., & Illikainen, M. (2016, March). Functionalized nanocelluloses and their use in barrier and membrane thin films. In *Abstracts of papers of the American Chemical Society* (Vol. 251). American Chemical Society.

Wolf, A., Bray, G. A., & Popkin, B. M. (2008). A short history of beverages and how our body treats them. *Obesity Reviews, 9*(2), 151–164.

Wongkhalaung, C., & Boonyaratanakornkit, M. (2000). Development of a yogurt-type product from saccharified rice. *Agriculture and Natural Resources, 34*(1), 107–116.

Yeung, C. K., Glahn, R. E., Welch, R. M., & Miller, D. D. (2005). Prebiotics and iron bioavailability—is there a connection? *Journal of Food Science, 70*(5), 88–92.

Zhang, M., Bai, X., & Zhang, Z. (2011). Extrusion process improves the functionality of soluble dietary fiber in oat bran. *Journal of Cereal Science, 54*(1), 98–103.

Zhang, Z., Yang, P., & Zhao, J. (2021). Ferulic acid mediates prebiotic responses of cereal-derived arabinoxylans on host health. *Animal Nutrition.* https://doi.org/10.1016/j.aninu.2021.08.004

Chapter 8
Cereal Based Fermented Foods and Non-alcohol Beverages

Gayatri Mishra and Brajesh Kumar Panda

8.1 Introduction

Fermentation has a very long history of preserving food, perhaps 12,000 years old and is defined as the biochemical process of complex carbohydrate breakdown into simpler sugar by the action of bacteria and yeast. The enzyme produced by the selected microorganisms, either singularly or in combination alter the final product and make it desirable for human consumption. It is an inexpensive and low-energy process. It provides a natural way to destroy undesirable components and improves shelf-life, sensory, nutritional value of the product (Şanlier et al., 2019). Most bacterial fermentation produce lactic acid and yeast fermentation produce alcohol as secondary metabolites. During fermentation, various taste, aroma, and flavor active components are also produced, which are highly valued as they improve the overall acceptability of the final product. Indigenous fermented foods have been prepared for thousands of years and are strongly linked to cultures and traditions of people all over the world. The earliest record of fermentation appears back in 6000 BC, when people used the process in an artisan way to preserve milk, fish, and vegetables without knowing the role of microorganisms involved. Cereal fermentation was started with leavening wheat dough with yeast by ancient Egyptians in 3500 BC and indigenous souring cereal-legume batters were done by ancient Indians later in 500 AD (Nout et al., 2007). Since then, cereal fermented foods, in various forms has become a staple in diets of many population groups.

G. Mishra (✉)
Centre for Applied Research, Innovation and Entrepreneurship, Lethbridge College, Lethbridge, AB, Canada

UniSA STEM, University of South Australia, Adelaide, SA, Australia

B. K. Panda
Centre for Applied Research, Innovation and Entrepreneurship, Lethbridge College, Lethbridge, AB, Canada

S. Punia Bangar, A. Kumar Siroha (eds.), *Functional Cereals and Cereal Foods*,
https://doi.org/10.1007/978-3-031-05611-6_8

Cereals are rich sources of carbohydrates, proteins, vitamins, minerals, and fiber, consumed as a primary source of energy by the people around the world. These are important substrates for fermented foods as they contain enough substrate for growth of microorganisms. However, nutritional, sensorial and functional quality of cereal-based products is inferior. The reason behind the inferiority is due to lack in lysine and presence of antinutritional compounds like phytic acid, tannins, and polyphenols. Several processing techniques, including fermentation has been implemented to ameliorate the acceptability of cereal-based products. Fermentation enhances the digestibility of cereal grains by assimilating nondigestible poly and oligosaccharides and providing optimum pH conditions for enzymatic degradation of antinutritional compounds. Fermentation also improves the nutritional quality by enhancing the production of lysine and B-group vitamins (Blandino et al., 2003). Cereal grains have recently come into focus for production of nonalcoholic fermented beverages due to the presence of natural sugars, antioxidants, vitamins, and other beneficial nutrients. Cereal-based beverages are mostly based on the grain suspension and depending upon the processing step involved, they are categorized either into fermented and non-fermented beverages or alcoholic and nonalcoholic beverages. Nonalcoholic beverages are mostly in the form of fermented functional beverages and stimulants like tea, coffee, cereal-based milk, soft drinks, or energy drinks. Fermented nonalcoholic cereal beverages have unique and specific health benefits depending upon the microbiological strain used for production.

Fermented foods and beverages prepared indigenously are remained as house culinary art or produced in small-scale industries. Therefore, there was no verified information of economical, nutritional, and quality control of traditional fermented foods. Rising awareness towards health benefits of functional foods in last 20 years necessitates commercialization of indigenous fermented foods and beverages to meet global demands. Other than the improved nutritional quality, cereal based fermented food possess gut-health promoting aspects, owing to the presence of edible beneficial microbes, also called probiotics: fermentable sugars (i.e. prebiotics), and group of microbe-derived hydrolytic enzymes, etc. Consequently, advance scientific works are being carried out to understand the manufacturing process and engineering interventions. This review focuses on the production processes of most common cereal-based fermented foods and nonalcoholic beverages highlighting the biochemical properties and commercialization of the final product. This will encourage food industries for large-scale production of the indigenous fermented foods to meet the growing market demand of functional foods.

8.2 Biochemical Process Involved in Cereal Grain Fermentation

Cereal fermentation involves conversion of complex organic compounds into energy through the metabolic process of microorganisms. The pioneer of fermentation process Louis Pasteur described fermentation as "a cascade of chemical events where oxygen is transferred from one part of a sugar molecule to another, resulting in a highly oxidized product (CO_2) and a highly reduced product (alcohol)". He succinctly defined the process of fermentation is "la vie sans l'air," or "life without air" (Alba-Lois & Segal-Kischinevzky, 2010). Furthermore, Traube discovered the involvement of protein-like substance enzymes which catalyze the process of fermentation. Later, Eduard Buchner in 1907 won Nobel prize in Chemistry by demonstrating sucrose fermentation and coined the term "zymase" to delineate the cellular enzyme which catalyzes the process of conversion of sugar to alcohol (Barnett & Lichtenthaler, 2001). The process of fermentation is diverse as different microorganism has different mechanism for the conversion of organic compounds into energy. Fermentation can also be described as a process of microbial cellular respiration or microbial glycolysis. Glycolysis is the sequence of chemical reactions where glucose converts into pyruvate. When oxygen is available, pyruvic acid enters aerobic glucose oxidization process, but lack of oxygen leads the pyruvic acid to follow two different pathways, depending on the type of cell. First is alcoholic fermentation pathway, where Pyruvate converts into alcohol and CO_2, and second pathway is lactic acid fermentation where pyruvate transforms into lactate (Voet & Voet, 2004; Taillefer & Sparling, 2016). Lactic Acid Bacteria (LAB) is a group of gram-positive bacteria that converts carbohydrate into lactic acid at the end of fermentation cascades. Moreover, other common fermenting bacterial species are *Leuconostoc, Streptococcus, Pediococcus, and Micrococcus*. The common fermenting fungi species are *Saccharomyces* responsible for alcoholic fermentation in cereal grains (Blandino et al., 2003).

Natural fermentation in cereals results decrease in the carbohydrate content as well as the nondigestible poly and oligosaccharides, which improve the functional properties of cereal grains. Moreover, fermentation improve the protein quality by significantly increasing lysine content in lysine in cereal-based products. It was also found that fermentation improves the level of lysine, methionine, and tryptophan in corn meal (Nanson & Field, 1984). The pH of fermented food facilitates degradation of phytate which present in grain matrix combined with polyvalent cations such as iron, zinc, calcium, magnesium, and proteins. Thus, phytate degradation leads to availability of more soluble iron, zinc, and calcium in the final product. Fermentation process is known to increase shelf-life, taste, and aroma. During fermentation, several volatile compounds and organic acids are formed which are mostly responsible for improved sensory properties in the final product. Organic acids formed during fermentation of cereal grains are namely, lactic acid, acetic acid, butaric acid, formic acid, succinic acid etc. helps reducing the pH to below 4.0, making it difficult for some spoilage organisms to survive (Stanojević-Nikolić et al., 2016). Similarly, during yeast fermentation various alcohols are formed, viz. ethanol, n-propanol, isobutanol, amyl alcohol, isoamyl alcohol, etc. Aldehydes, ketones, and carbonyl compounds formed during fermentation are the basic flavour compounds in the final product (Karovičová & Kohajdova, 2007).

8.3 Cereal-Based Fermented Foods

8.3.1 Rice-Based Fermented Foods

Rice is one of the most important cereal grain and staple food of South-Asian subcontinent. Major component of rice is starch making it suitable media for the growth of pervasive groups of microorganisms. Rice-based fermentation process involves two types of reaction, viz. acidic or alcoholic fermentation or both consecutively. The fermentation process includes pretreatment of rice grains to loosen the compact structure of starch and simultaneously release of antinutrient components out of the grain matrix. The pretreatments are done by soaking, grinding, and boiling. Sometimes, overnight soaking of rice in some food preparations activates various hydrolytic enzymes, help slack the dense starch by breaking the complex bonds. To complement the nutritional quality, some pulses are usually mixed with the rice to form the batter. During fermentation, formation of CO_2 in the batter makes the final product spongy. *Lactobacillus* and *Bifidobacterium* constitutes most of the LAB group which are commonly used for rice-based fermentation. Following indigenous rice-based fermented products are popular all over the world and produced commercially in various forms. Besides, some of other popular cereal-based fermented products and the microorganisms involved in their fermentation are given on Table 8.1.

8.3.1.1 *Idli, Dosa* and *Uttapam*

Idli is a savory and spongy cake prepared from a thick batter of rice (*Oryza sativum*) and dehulled splitted black gram (*Vigna mungo*) and is very popular traditional fermented food of South Asia. Rice and black grams in a ratio of 3:1 is soaked separately for 6–8 h and then combinedly grind to make a thick batter after draining the soaked water. The batter is then fermented for 10–12 h at room temperature (about 30 °C) and the fermented batter is steamed inside a pan contains concave moulds to prepare starchy and spongy *idli*. It is mostly consumed during breakfast with sambhar, which is basically a lentil soup. Fermentation of batter is mediated by mixed culture of LAB including *Lactobacillus fermenti, Lactobacillus lactis,* Lactobacillus delbrueckii, *Lactobacillus plantarum, Leuconostoc mesenteroides, Streptococcus faecalis* and *Pediococcus cerevisiae.* The major physico-chemical changes such as decrease in pH and increase in the titrable acidity in idli batter is due to lactic acid fermentation. Bernard et al. (2021) observed remarkable drop in pH from 6.28 to 3.72 and increase in acidity from 0.24% to 0.92% during the period of 0–32 h of fermentation. They also concluded *L. mesenteroides* and *S. faecalis* strains are found to be responsible for leavening of the batter and lowering the pH. Increase in acidity of batter helps activity of yeasts by producing optimum pH for their growth. The yeast strains, viz. *Torulopsis holmii*, *Torulopsis candida*, *Trichosporon pullulans*, *Geotrichum candidum*, *Candida fragilola*, *Candida tropicalis*, *Candida kefyr*, *Rhodotorula graminis*, and *Hansenula anomala* have also been isolated from

Table 8.1 Most common indigenous cereal-based fermented foods

Product	Substrates	Microorganisms	Nature of use	Regions of production
Adai	Rice with legumes	*Pediococcus, Streptococcus, Leuconostoc*	Breakfast/ snacks	India
Ambeli	Rice	*Leuconostoc mesenteroids, Lactobacillus fermentum, Streptococcus faecalis*	Breakfast meal	India
Anarshe	Rice	LAB	Breakfast/ snacks	India
Ang-kak	Red rice	*Monascus purpureus*	Dry red powder as food colorant	India
Appam	Rice	LAB	Breakfast	India
Atole	Maize	LAB	Porridge	Southern Mexico
Babru	Rice	LAB (*Lactobacillus plantarum, Lactococcus lactis*), yeasts such as *Saccharomyces cerevisiae, Debaromycis sp.*	Breakfast	India
Banku	Maize	LAB, molds	Dough	Ghana
Bhattejaanr	Rice	*Hansenula anomala, Mucor rouxianus*	Sweet and sour paste	India
Chakuli	Rice/black gram	*LAB, mold*	Breakfast	India
Chee-fan	Soybean, wheat	*Mucor, Aspergillus glaucus*	Cheese like product	China
Dalaki	Millets	*Unknown*	Thick porridge	Nigeria
Enduri Pitha	Rice and black gram	*Lactobacillus fermentum*	Spongy cake	India
Hamanatto	Wheat, soybeans	*Streptococcus, Pediococcus* and *Aspergillus oryzae,*	Flavoring compound	Japan
Ilambazi lokubilisa	Maize	LAB, yeasts, and molds	Porridge	Zimbabwe
Injera	Sorghum, maize or wheat	*Candida guilliermondii*	Bread like product	Ethiopia
Jalebi	Wheat, Bengal gram	*Saccharomyces bayanus*	Confection	India
Jamin-bang	Maize	*Yeasts, bacteria*	Spongy cake	Brazil
Kaanga-Kopuwai	Maize	*Bacteria, yeasts*	Soft, slimy eaten as vegetable	New Zealand
Khanomjeen	Rice	*Lactobacillus, Streptococcus*	Noodle	Thailand
Kichudok	Rice	*Saccharomyces*	Sponge cake	Korea
Kisra	Sorghum	Unknown	Staple food like bread	Sudan

(continued)

Table 8.1 (continued)

Product	Substrates	Microorganisms	Nature of use	Regions of production
Koko	Maize	LAB *(Lactobacillus platarum, L. brevis), yeasts (Saccharomyces cerevisiae, Candida mycoderma), Enterobacter clocae, Acinetobacter*	Porridge	Ghana
Lao-chao	Rice	*Rhizopus oryzae, R. chinensis, Chlamydomucor oryzae, Saccharomycopsis*	Spongy buns	China, Indonesia
Mahewu	Maize	*Streptococcus lactis*	Porridge	South Africa
Mantou	Wheat	*Saccharomyces*	Steamed cake	China
Mutwiwa	Maize	*LAB, bacteria and moulds*	Porridge	Zimbawe
Naan	Wheat	*Saccharomyces cerevisiae, LAB*	Flattened bread	India, Pakistan, Afghanistan, Iran
Puto	Rice	*Leuconostoc mesenteroides, Strepromyces faecalis, yeasts*	Solid paste	Philippines
Soy sauce	Wheat and soybean	*Aspergillus oryzae or A. soyae, Zygosaccharomyces rouxi,* LAB	Liquid seasoning	Japan, China, Taiwan
Selroti	Rice	LAB *(Leuconostoc mesenteroids, Lactobacillus curvatus), Enterococcus faecium, Pediococcus pentosaceus, Saccharomyces cerevisiae, Saccharomyces kluyveri, Debaryomyces hansenii, Pichia burtonii and Zygosaccharomyces rouxii*	Spongy and ring-shaped rice bread	India
Sierra rice	Rice	*Aspergillus flavus, A. candidus, Bacillus subtilis*	Brownish-yellow rice for lunch/dinner	Ecuador
Sour rice	Rice	LAB *(Lactobacillus bulgaricus, Lactobacillus casei), Sacchaeromyces sp.* and *Pediococcus acidilactici, Streptococcus faecalis, Streptococcus thermophilus, Macrobacterium flavum*	Cooked rice as staple food	India
Sez	Rice	Saccharolytic and ethanol producing microbes	Cooked rice with seasonings	India
Taotjo	Wheat/rice and soybean	*Aspergillus oryzae*	Condiment	India
Uji	Maize. Sorghum, millet	*Leuconostoc mesenteriodes, Lactobacillus platarum*	Porridge	Kenia, Uganda, Tanganyika
Vada	Rice and black gram	*Pediococcus, Streptococcus, Leuconostoc*	Fried spongy snacks	India

fermented *idli* batter (Shaikh et al., 2021). LAB and yeasts work synergistically in the fermented *idli* batter and enhance the nutritional value and produce desirable texture and flavor. The aroma and flavour attributes of idli mainly depends on the production of organic compounds arise from both the raw materials and fermentation process. Major flavour compounds arise from rice and black gram are ketones (ethanone, pentanone and butanones) and polyunsaturated fatty acids, respectively (Agrawal et al., 2000). With reference to the nutritional attributes, fermentation enhances the level of soluble solids, essential amino acids, viz. lysine, cystine and metheionine, non-protein nitrogen, soluble vitamins (folate, vitamin A, B1, B2 and B12) content in the batter and reduces antinutrient phytic acid content making *idlis* more functional.

Dosa and uttapam batters are similar except the proportion of raw material used and particle size during grinding. The ratio of rice and black gram is about 4:1 and the soaked rice used are ground more finely to prepare *dosa* and *uttapam* batter. Instead of steaming, the batter is spread and heated with a little oil, over a flat pan. *Dosa* is spread over the heated pan as a thin layer, while *uttapam* is spread as a thick layer with some vegetable toppings to prepare the final product as pancakes. Like *idli* batter, the *dosa* and *uttapam* batters are also fermented by a mix culture of LAB and yeasts increasing the acidity and pH. Minor yeast strains found in *dosa* batter are belonging to *Saccharomyces cerevisiae*, *Debaryomyces hansenii*, and *Trichosporon beigelli*.

With the growing demands for low-calorie functional breakfast and snack foods, active research works have been conducted to replace rice in *idli, dosa* and *uttapam* batters with other grains, viz. brown rice, pearl millets, amaranth, finger millets and buck wheat. Though incorporation of these functional cereals improves the nutritional attributes greatly, it decreased the palatability, texture, and overall acceptability of the product (Rani et al., 2019; Dhillon et al., 2020; Kumari et al., 2020). Therefore, to maintain balance between the nutritional and sensory attributes, researchers are more inclined towards partial replacement of rice with other cereals. Increasing priority towards convenience have resulted in the development of ready-to-cook foods and instant food premixes of the fermented foods. Shelf-life enhancement and instantization are two major processes for commercialization of the fermented batter. Use of combined thermal and electron beam radiation of the fermented batter and packaging of the treated batter in oxygen barrier multi-layered packaging film for increasing shelf-life of *idli* and *dosa* batter are highly studied area in the last decade (Mulmule et al., 2017; Gaikwad et al., 2020). Accelerating the process of fermentation using selected starter culture is another aspect for commercial production of fermented batter (Chelliah et al., 2017).

8.3.1.2 *Dhokla*

Dhokla is a soft and spongy rice cake popularly consumed as breakfast in Western India. It is made from a batter of polished white rice and Bengal gram (*Cicer arietinum* L.). Rice and Bengal gram are soaked separately for 5–10 h, then drained to ground into a fine batter. The batter is then mixed with salt (approximately 1% w/w) and fermented at room temperature (30–32 °C) for 12–14 h. Additional seasonings and chopped green leafy vegetables are often added to the fermented batter and steamed over a flat pan for 10–15 min. The product is then cooled and cut in the shape of diamond and seasoned with heated oil, mustard seeds, and curry leaves (*Murraya koenigii).* Sponginess and volume in the batter is resulted due to the production of CO_2 by Yeasts like *Torulopsis sp., Candida sp., Tpullulans.* LABs, viz. *L. fermentum, Leuconostoc mesenteroids, Pichia silvicola, streptococcus faecalis*, etc. produce lactic acid and acetoin in the batter during fermentation, which impart sour taste and a pleasant flavour (Aidoo et al., 2006; Ray et al., 2016). Like other fermented foods, the biological value and net protein utilization of *dhokla* are significantly improved due to fermentation. Antinutritional compounds like tannins, phytic acid, total biogenic amines, trypsin inhibitor and haemagglutinating activities reduced because of fermentation (Sharma et al., 2018). It can be used as supplementary diet for children suffering from malnutrition and patients with digestive disorders (Desai & Salunkhe, 2018). Industrial manufacturing of *dhokla* depends largely on the scale of manufacturing, location of production, climatic conditions. Due to increased demand towards convenient foods, manufacturers mostly produce instant *dhokla* mix by drying the fermented batter. To improve the functional property of *dhokla*, several researchers have studied the nutritional and organoleptic properties of dhokla after incorporation of soybean, ragi, gareden cress seeds, okra, and pumpkin flour (Lohekar & Arya, 2014; Ray et al., 2017). Incorporation of the functional ingredients improves the nutritional parameters, viz. antioxidant properties, carotene content, folic acid and vitamin C, but reduced the sensory values significantly. Therefore, dhokla can be prepared with incorporating optimum ration of rice, Bengal gram and other functional ingredient to maintain balance between the nutritional and sensory parameters.

8.3.1.3 Miso

Miso is a traditional Japanese seasoning paste produced by fermenting the mixture of rice, soybeans, and salt. Preparation of miso involves two successive stages of fermentation. In the first stage, rice is soaked and steamed for complete conversion of raw starch (β-starch) to gelatinized starch (α-starch) and then inoculated with strains of *Aspergillus oryzae.* The mold rice is called 'koji' serves as a source of enzyme for catalyzing the second stage fermentation. In the meantime, whole soybeans are soaked and cooked to gelatinize the starch. In the second stage, koji, cooked soybeans and salt are ground together to form a thick paste and inoculated with the miso from previous fermentation. The mixture is then kept for fermentation 42–48 h at a temperature 28–32 °C to produce the final product as *miso* (Kusumoto et al., 2021). Based on the proportion of soybean used for preparation, miso can be of three types, viz. white, red, or dark miso. The amount of soybean used is less, equal and high in white, red and dark miso, respectively. Although indigenous *miso* preparation was done in small batches in Japanese households, commercialized production has more recently been superseded to meet global demand for consistent, high quality and safe food. During industrial production, temperature and humidity control and use of commercial strain of microorganism reduced the fermentation time of miso. In addition, pasteurization process, use of suitable packaging material and invention of smart packaging increased the shelf-life of commercially produced miso (Allwood et al., 2021). Barley, legumes, nuts, functional seeds, mushrooms can be used as koji substrates during miso preparation to replace rice (Shurtleff & Aoyagi, 2018).

8.3.2 Wheat-Based Fermented Products

Wheat is an important source of carbohydrate, protein, vitamins, and minerals. It contains gluten protein, which is responsible for most of the viscoelastic properties of wheat flour doughs and is the main factor dictating the use of a wheat for making spongy fermented products. Starch is the major component of wheat flour which helps in the water absorption and fermentation time requirements of bread-making doughs and crumb textural properties of the final products. Following are some of the popular fermented products made from wheat and their commercial aspects of production.

8.3.2.1 Bread

Bread is a popular fermented product produced from wheat dough by baking process. Bread dough is usually leavened by naturally occurring microbes (e.g., sourdough), or yeast (*Saccharomyces cerevisiae*), or high-pressure aeration which creates the gas bubbles that fluffs up during baking. It is believed that ancient Egyptians first fermented wheat dough and baked them in clay ovens during 3500 BC (Özdemir & Altıner, 2021). Preparation steps of bread involves mixing of flour, water and yeast to form a visco-elastic dough, allowing the yeast to fermentation at 24–28 °C for 2–4 h to an aerated mass and then baking the structure inside a high temperature oven at 220 °C for 20–25 min. The fermentation in bread dough is primarily done by bakers' yeast or *Saccharomyces cerevisiae*, which is a top fermenting yeast. Growth of the yeast is optimum at pH range of 4–5 and at temperature of 25–28 °C (Narendranath & Power, 2005). The kneading process of dough introduces oxygen which converts the accumulated sugars to pyruvate by the yeast cells through aerobic glycolytic pathway. Later, the dough started anaerobic fermentation due to production of CO_2 in glucose metabolism (Heitmann et al., 2018). CO_2 production primarily makes the fermented dough into a spongy and elastic structure by entrapped in the gluten matrix. Remixing of the dough facilitates additional CO_2 production and growth of LAB to produce organic acids, which lowers the pH of the dough and give acidic pleasant aroma and flavour during baking of bread.

Since the beginning of modern civilization, bread is manufactured commercially as it is a staple food for Middle East, Central Asia, North Africa, Europe, and in America. Temperature and relative humidity (RH) are two important parameters in industrial bread baking as it affects yeast growth and fermentative capacity. Followed by fermentation, the dough mass is cut into loaf-sized portions and proofing is done for 30–60 min at 35–42 °C and RH of 70–80% prior to baking. The proofed loaves are baked in high temperature ovens in five stages, viz. (a) enzyme inactivation zone (30–60 °C), (b) yeast inactivation zone (55–60 °C), (c) starch gelatinization zone (55 °C to <90 °C); (d) water evaporation (>100 °C); and (v) Maillard browning and aroma formation (>200 °C) (Erkmen & Bozoglu, 2016). Several aroma and flavour-active compounds are formed during baking and the process is called Maillard browning. Major compounds are organic acids, ethyl esters, alcohols, aldehydes, ketones, sulfur-containing compounds, maltol, isomaltol, melanoidin, furan derivatives, lactones, pyrazines, pyridines, and pyrroles (Birch et al., 2014). Sever heating condition produces certain pigments responsible for brown crust colour formation. Although vitamin B and niacin content increases due to bread dough fermentation, baking at high temperature reduces some nutritional quality of bread due to loss of some amino acids and reducing sugars in Maillard browning reactions.

8.3.2.2 Soy Sauce

Soy sauce is a traditional condiment made from wheat and soybean blend, used as a key seasoning in Asian countries. Soy sauce was originated as *chiang-yu* in China over 2500 years ago and popular as *shoyu* in Japan, *kunjang* in Korea, *toyo* in the Philippines and *kicap* in Indonesia and Malaysia (Yokotsuka, 1993). Soy sauce has a salty and unique umami taste due to the phytochemicals present in it, which also helps in reduction of total and low-density lipoprotein cholesterol (Larkin et al., 2009). Manufacturing of soy sauce involves two primary fermentation process, a solid-state fermentation process and a long-term liquid-state fermentation process. Cooked soybeans are mixed with coarse wheat flour keeping initial moisture content of 55% (w/w) and the mixture is allowed to ferment by *Aspergillus* mold (*A. oryzae* or *A. sojae*) for 3 days at 25–35 °C. Aspergillus mold produce enzyme that break down the complex molecules of wheat and soybean, which helps in the development of halophilic bacteria, LABs, bacillus species and yeasts (Zygosaccharomyces rouxii, *T. halophilus* and Candida species). The fermented mixture is called as *koji* is then immersed in a brine solution with 18–20% salt content in the ratio of 1:3 (w/v). The brine solution containing *koji* is called as *moromi*, which is left to ferment for 1 month to 4 years of maturation to finally produce soy sauce (Sassi et al., 2021). The detailed flowchart of soy sauce manufacturing is given in Fig. 8.1. After the long-term maturation, the soy sauce is then filtered, pasteurized, and bottled (Harada et al., 2017; Det-Udom et al., 2019). Raw soy sauce is usually heat pasteurized at 70–80 °C for 30 min or can be flash pasteurized at 115 °C or higher for 3–5 s to increase the shelf life and stability. The characteristics taste and aroma of soy sauce is generated due to the enzymatic activities of yeasts and

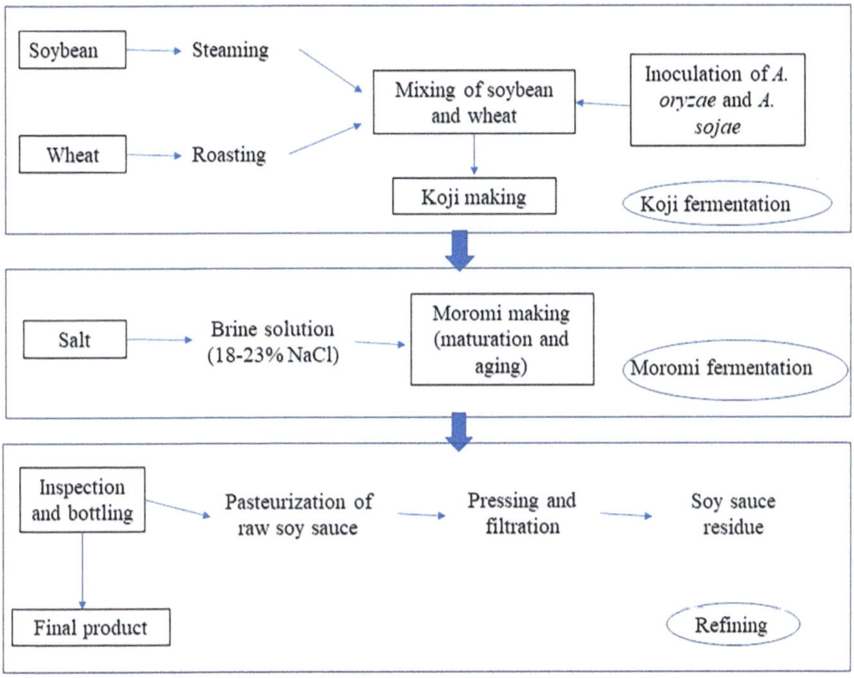

Fig. 8.1 Process flow chart of soy sauce manufacturing involving two stage fermentation

some osmophilic LABs. Major odor active compounds found in soy sauce are Methional, ethyl lactate, furfuryl alcohol, furanone, propanal etc. The pH of soy sauce is generally 4.6–4.8 and typical salt concentration is about 17–19%. High salt concentration in soy sauce may be a concern hypertension, health driven industries are trying to reduce the sodium concentration by maintaining the salt percentage as salt less than 16% can result in the development of putrefactive species during fermentation (Devanthi et al., 2018).

8.3.2.3 *Kishk*

Kishk is a popular dried fermented product of Egypt and middle east made from sour milk and parboiled wheat (bulgar) mixture. The raw materials required for *kishk* preparation are parboiled wheat flour, *laban zeer* (Tamime & O'connor, 1995). Wheat is slowly boiled in a large, covered cooking pans followed by cooling and drying to produce bulgar or *belila*. The dried and hard bulgar is then coarsely ground and mixed with *laban zeer* to form a homogenous paste, called *hamma*. *Laban zeeer* is the remaining by-product obtained from squeezing the sour milk through a skin bag. The *hamma* is then fermented for 24 h and then kneaded by hand and divided into small dough portions and dried in the open air to form the final product, called *kishk* (Morcos et al., 1973). Sometimes, cumin and pepper are added as seasoning to the *laban zeer* for enhancing the flavour. *Kishk* is rich in various nutrients, including many vitamins and growth factors. The microorganisms responsible for fermentation of *kishk* are *Lactobacillus plantarum*, *Lactobacillus casei*, Lactobacillus *brevis*, Lactobacillus delbrueckii subsp. bulgaricus, and *Bacillus subtilis* (Abd El-Ghani et al., 2014). To facilitates consistent industrial production of *kishk*, the modern manufacturing process uses a combination of starter culture containing *Streptococcus thermophilus* and *Lactobacillus bulgaricus* at 45 °C for fermentation, which leads to more rapid acid production, and suppresses the growth of other pathogenic bacteria (Abou-Zeid, 2016). The functionality of kishk can be improved by incorporation of barley, pearl millets and quinoa instead of bulgar in optimized proportions to get highly nutritious product without compromising the sensory quality.

8.3.2.4 Tarhana

Tarhana is a traditional Turkish fermented food prepared from wheat flour, yoghurt and different vegetables, followed by for fermentation from 1 to 7 days (Daglioğlu, 2000). Fermented *tarhana* mixture is usually dried to a moisture content of 6–10% for longer shelf life. The dried *tarhana* mixture is then reconstituted and consumed as soup at lunch and dinner. *Tarhana* is produced by simultaneous LAB and yeast fermentation. During fermentation, complex molecules like carbohydrates, protein and fat are break down to lower molecules and organic acids constituents by microorganisms, which gives acidic taste to the product. LAB involved in tarhana fermentation are *Streptococcus thermophilus*, *Lactococcus lactis*, *Lactococcus diacetylactis*, *Lactobacillus bulgaricus*, *Lactobacillus acidophilus*, *Leuconostoc cremoris*, *Lactobacillus casei,* responsible for lowering the pH to 3.3–5.0. The yeast, *Saccharomyces cerevisiae* gives the final product the body, taste and flavour (Sengun et al., 2009). Traditionally, tarhana was prepared in households by the way that have been learned from their ancestors, but to meet increasing demand for consistent, convenient and safe food, it is started manufactured in the commercial plants. In industries, there are two methods for tarhana production, viz. straight method and sourdough method. In straight method, the ingredients are combined by kneading at 50 rpm for 15 min, spread to a depth of 1–1.5 cm, kept for fermentation at 35 °C for 5 days, dried at 55 °C for 28 h to obtain moisture content less than 10%, ground to particle size <800 μm and finally sieved (Ozdemir et al., 2007). Sourdough method is done in three stages given in Fig. 8.2. Although there are various packaging materials are available for storage of fermented dough to read-to-use in the households, it is difficult to avoid moisture absorption during storage, resulting unwanted growth of microbes like *E. coli O157:H7, S. aureus, S. typhimurium and B. cereus.* Therefore, irradiation of the fermented dough is emerging method adopted by industries to reduce the risk of microbiological encroachment during the storage of *tarhana* (TaŞoğulları & ŞimŞek, 2020).

Tarhana soup is highly flavored and nutritious and can be easily digested. The nutritional importance of tarhana is remarkable due to improvement of basic amino

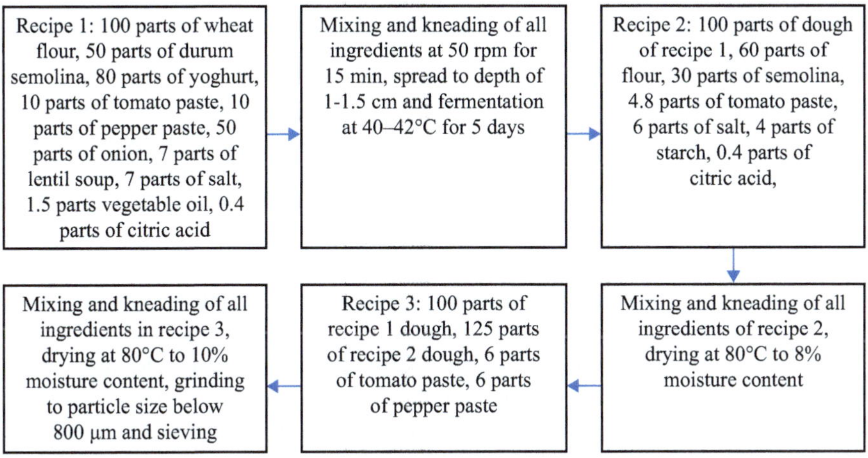

Fig. 8.2 The flow chart of commercial tarhana production using sourdough method. (Daglioğlu, 2000)

acid content of cereal and proteolytic activity by adding the dairy protein and fermentation, respectively. Increase in water soluble vitamins were reported to increase during fermentation, while drying after fermentation had a adverse effect. Tarhana is a good source of total minerals, viz. Ca, Mg, and K with favorable bioavailibities due to reduction of phytic acid, which is an antinutritional compound binds minerals and proteins, altering alters their solubility, functionality, digestibility, and absorption. Several researchers have replaced the wheat flour with other cereals like buck wheat, barley, quinoa, legumes, wheat germ, or wheat bran to raise the nutritional value of tarhana (Erkan et al., 2006; Bilgiçli, 2009; Demir, 2014).

8.3.3 *Maize-Based Fermented Products*

Maize is considered as one of the principal staple foods and has the highest global production among all the cereal grains. Maize kernel consists of three major parts, viz. the endosperm, the germ and the pericarp. The endosperm predominantly consists of starch (70–73%), which makes maize as a good base for fermentation. The fat and protein content of germ are high are around 33% and 19%, respectively makes the fermented product highly nutritious. Fermentation of maize usually consists of four common steps, viz. cleaning, soaking overnight to soften the kernel, grinding and fermentation. The fermentation process depends on temperature, pH, the quantity of inoculum and types of microorganism used. Following are some of the popular maize-based fermented products and their commercial aspects of production.

8.3.3.1 *Ogi*

Ogi is a traditional fermented food popular in Western Africa and Nigeria is a sour starch cake made from fermentation of maize, sorghum or millets flour. *Ogi* is usually consumed as a smooth porridge by boiling the sour cake at the time of breakfast. Considering the health benefits of *ogi*, it is considered as an important weaning food for infants in West Africa. Traditionally *ogi* is prepared by steeping the grain in an earthen pot for 1–3 days, allowing it for spontaneous fermentation at 28 ± 2 °C. After softening of the grain, it is milling into a smooth paste. The paste is then sieved to remove the unwanted bran, germ and hull and the filtrate is allowed to undergo a secondary fermentation for about 24–72 h in order to develop its characteristic sour taste (Adisa & Enujiugha, 2020). LABs, yeast and molds are responsible for fermentation of maize grain to produce *ogi*. *Lactobacillus plantarum* is the dominant bacteria in ogi and other bacteria such as *Corynebacterium, B. suhtilis, Klebsiella oxytoca, Staphylococcus aureus* hydrolyze the corn starch to make the grain soften. In the secondary fermentation, yeasts of the *Saccharomyces* and Candida species also contribute to flavour development (Adegoke & Babalola, 1988). Indigenous production involves unskilled producers who doesn't follow good manufacturing practices and cleanliness, resulting contamination from unwanted microorganisms during fermentation. In addition, to overcome the limitations of labor-intensive and time-consuming traditional process, commercial manufacturing of *ogi* has been started to meet the increasing demand day by day. Flowchart for industrial production of ogi is illustrated in Fig. 8.3. The nutritional value of ogi, specifically aminoacids and is reduced by 20–50% as that of the original cereal grains used because of the loss of aleurone layer and germ during wet milling and filtration. To compensate the losses, lysine and methionine excreting mutants of Lactobacillus and yeasts have been used to fortify ogi by several researchers (Odunfa & Oyewole, 1998).

8.3.3.2 *Kenky*

Kenkey is a popular maize sourdough mostly consumed in Ghana. It is usually eaten in form of dumplings or porridge depending upon the final preparation method. In the first type of *kenkey* production, maize grains are soaked for 1–2 days at room temperature and then wet-milling of the hydrolyzed grain is done after draining the water. The milled maize meal is kept for solid state fermentation to produce stiff sourdough, which can be consumed in form of dumplings. In the second type of *kenkey* preparation, filtration of the chaff from the soaked maize meal is done to give it a smooth texture. Then the meal is kept for overnight fermentation followed by discarding the water to get a wet mash, which is used for porridge making (Ganguly et al., 2021). *Kenkey* fermentation is mostly regulated by obligatively heterofermentative LABs, viz. *Lactobacillus fermentum* and *L. reuteri*. Yeasts and mould flora, viz. *Candida, Saccharomyces, Penicillium, Trichosporon, Kluyveromyces and*

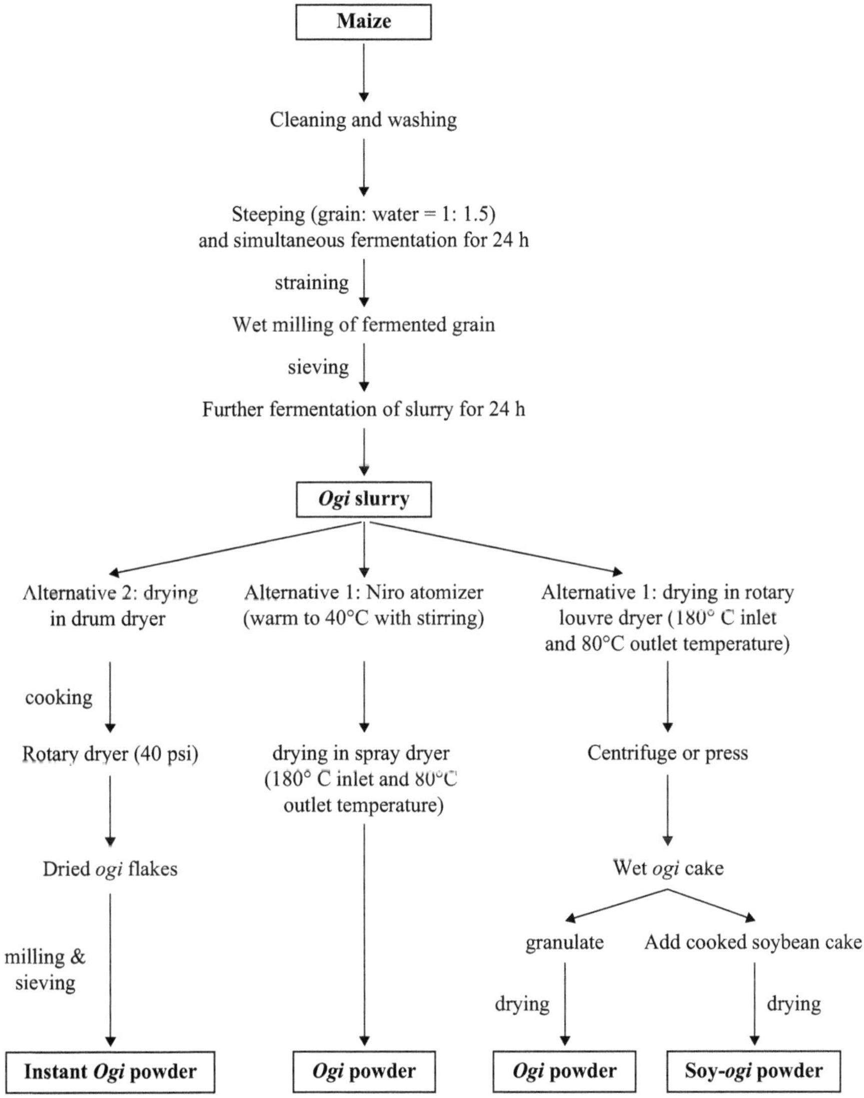

Fig. 8.3 Flowchart of industrial *ogi* production

Debaryomyces were also reported to form flavour compounds in the final product (Ekpa et al., 2019).

Several aroma compounds and organic acids are formed during kenkey production, which are responsible for the pleasant flavour and sour taste of the final product. Aroma compounds such as propanol, acetoine, Butanediol and organic acids such as propanoic acid, butanoic acid, pentanoic acid, hexanoic acid, heptanoic acid, benzanoic acid, a furan, two phenolic compounds and an alkene were isolated from GC/MS analysis of fermented kenkey dough by Annan et al. (2003). Indigenous preparation of *kenkey* sourdough dumplings is labor intensive and time-consuming takes usually 4–6 days. For commercial production, accelerated fermentation, dry-milled maize and pure starter cultures have been used to increase the productivity and consistency of the final product (Nout et al., 1995).

8.4 Millet-Based Fermented Products

8.4.1 Injera

Injera is an Ethiopian traditional pancake made from ancient grain teff. It can also be made from other grains like sorghum, barley, or finger millet. Sometimes people add some rice flour to make *injera* white. Gebrekidan and Belainesh (1981) described *injera* as circular, soft, spongy and resilient pan cake with uniformly spaced honeycomb-like 'eyes' on the top. Teff grains are dehulled and milled into flour and one part of flour is mixed with two parts of water in a clay container called "bohaka" to form a batter. The batter is then mixed with 16% seed culture or "ersho" and fermented for 2–3 days. Ersho is leftover fluid saved from previously fermented batter used as seed culture (Neela & Fanta, 2020). After fermentation, the batter is thinned down slightly and poured on to a lightly greased pan in form of a thin circular cake. The pan temperature during injera baking usually reached up to 200 °C and baking time is about 2–3 min (Adem & Ambie, 2017). Yeasts responsible for fermentation present in the seed culture or ersho are *Pichia fermentans*, *Pichia occidentalis*, *Candida humilis*, *Saccharomyces cerevisiae*, and *Kazachstania bulderi* (Tadesse et al., 2019). Some of the researchers reported the presence of LABs namely *Pediococcus pentosaceus*, *Lactobacillus fermentum*, *Lactococcus piscium*, *Lactococcus plantarum*, *Pediococcus acidilactici*, *Leuconostoc mesenteroides* etc. (Desiye & Abegaz, 2013; Desiye et al., 2017). *Injera* has slightly sour flavour due to fermentation and rich in nutrients, mostly calcium and iron. Fermentation time and viscosity of batter are to important parameter for final quality of *injera.*

8.4.2 Kisra

Kisra is like *injera* and mostly consumed in Sudan and some parts of Arabian Gulf and Iraq (Blandino et al., 2003). The Kisra dough is made from sorghum and pearl millet flour by adding water followed by fermentation. Traditionally fermentation was initiated by adding seed culture from previous batch of dough. The fermented dough is then rolled into thin sheets and then dried for few hours to store it for a year. The *kisra* sheets are then baked as unleavened bread on a hot steel plate

(150–160 °C) for 30–40 s in a process called as "aowasa" (Ibnouf, 2012). *Kisra* is usually consumed with certain types of stew prepared from vegetables and meat. Microorganisms responsible for *kisra* fermentation were studied by some researchers and the main stains isolated are *C. intermedia, C. krusei, Debrayomyces hansenii, Enterococcus faecium, L. amylovorus, L. brevis, L. confusus, L. fermentum, Pichia kudriavzevii* (Adebo, 2020). Nutritional benefits of kisra bread were studied extensively and reported that fermentation increased riboflavin content of the dough and significantly decreased thiamine content, however there was no significant effect on mineral content (Mahgoub et al., 1999).

8.5 Cereal-Based Non-alcoholic Beverages

8.5.1 Cereal-Based Sour Milk

Cereal-based sour milk are prepared by fermentation of cereal-based milk substitutes like rice milk and oat milk. Rice and oat milk are prepared by soaking the grains in water for hydration followed by grinding and filtration. Fermentation is carried by LABs yoghurt like product, which is then diluted and seasoned with cumin seeds and black pepper to produce cereal-based sour milk. Cereal-based milks and yoghurts are now popular in market, however cereal-based non-dairy fermented beverages are still under research for optimization of starter culture, sensory attributes and storability. Beneficial microbes like *Streptococcus thermophiles, Lactobacillus acidophilus, Lactobacillus reuteri and Bifidobacterium bifidum* etc. has been tested for their use production of cereal-based fermented beverages and the final products are proven to be of great nutritional interest, specifically for gut health (Bernat et al., 2015; Atwaa et al., 2020). Table 8.2 summarizes some of the popular cereal-based beverages consumed by various parts of the world.

8.5.2 Boza

Boza is a traditional nonfermented beverage popular in Turkey and Bulgaria. It is a thick homogeneous suspension, sweet to sour in taste, indigenously made from maize, wheat and rice flours in the ratio of 2:1:1, respectively. The grain flours are mixed with five times (w/v) of water and boiled for 1 h with continuous stirring. The slurry is cooled and diluted with 2.5 times with water and sugar (20% w/v). The slurry is kept for fermentation at 15–25 °C for 24 h using the boza from previous batch (Hancioğlu & Karapinar, 1997). Two types of fermentation are involved in boza production, viz. yeast fermentation produces carbon dioxide bubbles and increases the volume and LAB fermentation produces sour taste (Arici &

Table 8.2 Common indigenous cereal-based fermented non-alcoholic beverages

Product	Substrates	Microorganisms	Regions of production
Borde	Maize, sorghum or millet	LAB *(Lactobacillus brevis, Lactobacillus viridescens), Weissella confusa, Pediococcus pentosaceus* and *P. pentosaceus subsp. intermedius*	Ethiopia
Braga	Millet	Lactobacillus plantarum, Lactobacillus fermentum, and Lactobacillus delbrueckii	Romania
Busa	Rice or millet	*Lactobacillus, Saccharomyces*	Syria, Egypt, Turkestan
Darassum	Millet	Unknown	Mongolia
Haria	Rice	LAB, yeast and molds	India
Mangisi	Millet	lactic acid bacteria and yeast and molds	Zimbawe
Munkoyo	Kaffir corn, millet or maize in addition to roots of munkoyo	*Streptococcaceae, Leuconostocaceae, Enterobacteriaceae, Lactabacillales, Bacillaceae* and *Aeromonadaceae*	Africa
Tchang	Millets	Yeast (*Saccharomycopsis fibuligera, Saccharomyces cerevisiae),* Molds *(Mucor cicinelloides, Rhizopus chinensis, R. stolonifer var. lyococcus), Lactobacillus sp.*	India
Tobwa	Maize	LAB	Zimbawe
Uji	Maize, millet, sorghum or cassava flours	LAB *(Lactobacillus plantarum, Lactobacillus paracasei ssp. paracasei, Lactobacillus fermentum, Lactobacillus buchneri), Pediococcus acidilactici*	Sub-Saharan Africa
Yu	Rice	Unknown	India
Yosa	Oat	LAB or *bifidobacteria*	Finland and Scandinavia

Daglioglu, 2002). *Lactobacillus* and *Leuconostoc* genera, *Saccharomyces cerevisiae, Streptococcus spp., Micrococcus spp.* are common microorganisms isolated from traditionally fermented boza. Total titratable acidity by means of lactic acid is about 0.2–0.5% in sweet boza and 0.5–1.0% in sour boza (Gotcheva et al., 2001). Indigenously fermented boza cann't be stored for more than 2 days due to uncontrolled growth of microbes making the product sour and unacceptable. Therefore, industrial boza production uses standard starter culture and controlled environment for consistent product quality. Sugar or saccharine is added before bottling of boza as significant decrease in pH, viscosity, free amino nitrogen content and dry matter during fermentation.

8.5.3 Torani/Kanji

Torani is a traditional non-alcoholic probiotic beverage popular in Eastern states of India (Odisha and West Bengal), consumed for improved gut health and electrolyte balance in the body. *Torani* is made from rice, water and curd taste mostly like sour rice. It is also called as *kanji* in North India, prepared by boiling the *torani* with vegetables and spices (Sahoo et al., 2017). To make *torani*, cooked rice is soaked overnight in a clay pot for fermentation. After fermentation, the soaked water is drained and mixed with some curd and seasoned with heated oil, crackled mustard seeds and curry leaves. The fermented rice is also called as *pakhala* used as a staple food by indigenous people of Odisha (Ray & Swain, 2013). The fermented rice water is a store house of various beneficial microbes, which produces short-chain fatty acids play important role in improving gut health. The production of *pakahala*, *torani* and *kanji* is mostly indigenous; however, in last decade researchers have shown interest in studying the commercial feasibility of *torani*, owing to the health benefits. Though not much scientific literature available regarding the microbiological characterization of *torani*, preliminary research suggested that *pakahla* and *torani* contains several yeasts (*Saccharomyces spp.*) and *lactobacilli*, which bring typical aroma and sourness to the final product.

8.5.4 Kvass

Kvass is a cereal-based nonalcoholic refreshment drink, prepared from fermented barley, rye malt, rye flour. It achieved high commercial success and popularly manufactured in Russia, Poland, Latvia and Lithuania (Basinskiene & Cizcikiene, 2020). Traditionally, *Kvass* is made by cutting stale rye bread into small pieces followed by adding hot water (1:3 w/v) with continuous stirring for 1 h. Sugar, yeast and raisins are then added to the extracted slurry and kept for fermentation at 28 °C for 24 h. Following fermentation, the sediments are filtered to get the kvass. The microorganism responsible for fermentation are LABs and *Saccharomyces cerevisiae,* symbiotically produce the characteristic sour taste and flavor. Industrial kvass is produced in a slightly modified way, where the pre-extruded whole rye flour is hydrolyzed by α-amylase, amyloglucosidase, and β-xylanolytic enzymes at 100 °C for 10 min. After cooling, starter cultures (LAB, yeast) are added and kept for fermentation at 30–37 °C for 48–60 h. The fermented slurry is mixed with sugar and caramelized malt and filtered to get kvass, which is usually bottled and stored at refrigerator temperature (8 °C) (Dlusskaya et al., 2008). As a result of microbial fermentation complex starch is converted to oligosaccharides, simpler sugars viz. maltose, glucose, fructose, xylose and proteins are converted to amino acids. Other metabolites formed are lactic acid, acetic acid, ethanol, minerals, and vitamins (Lidums et al., 2017). The volatile compounds formed during *kvass* production are esters, alcohols (<1%), organic acids, aldehydes and ketones, mostly responsible for the flavor of the final product (Lidums et al., 2015). *Kvass* helps in eliminating digestive disorders and increases the bioavailability of minerals. It doesn't require any heat treatment after fermentation, thus contains viable yeast and LAB cells over 10^7 cfu/mL making the product probiotic (Basinskiene et al., 2016).

8.5.5 Togwa

Togwa is a traditional non-alcoholic beverage consumed as weaning food in rural villages of East Africa. Raw materials used for *togwa* preparation are flour of maize and germinated finger millet flour, which are source of starch and amylase, respectively (Mashau et al., 2021). At times, as a source of starch, sorghum flour and cassava flour are also used. For preparation of *togwa,* suspension of maize flour is heated with stirring up to 80 °C to form a gel paste and cooled to about 50 °C, then the germinated finger millet flour is added to the warm porridge kept at the same temperature for 20 min. After addition of finger millet flour, the gel of maize flour is converted to a viscous liquid, which is then fermented for about 15 h (Kitabatake et al., 2003). After fermentation, the viscous paste is diluted to make a non-alcoholic drink. *Togwa* has opaque and brownish colour due to presence of hull of finger millet and maize and has a sweet and sour taste. Microorganisms responsible for togwa fermentation were reported to be *Lactobacillus species*, *Issatchenkia orientalis* and *Saccharomyes cerevisiae*. Unhygienic production process of *togwa* and poor shelf life led to declination of popularity amongst the Tanzanian people. Use of selective culture, good manufacturing process and proper packaging could improve the industrial value of *togwa* manufacturing.

8.5.6 Mahewu

Mahewu is a Southern African and Zimbabwean non-alcoholic beverage made from maize fermentation (Olusanya et al., 2021). It is prepared by mixing one part of maize meal with nine parts of boiling water and cooking for about 10–15 min. Cooking results maize starch gelatinization and swelling of the starch granules making the amylase leaching and increase in the viscosity of porridge. The gelatinized mixture is then allowed to cool, and wheat flour (5% w/w of maize flour) is added as a starter or inoculum source (Fadahunsi & Soremekun, 2017). The porridge is then kept for spontaneous fermentation for 36 h at room temperature (20–30 °C) with stirring at the beginning. The fermented *Mahewu* can be stored in a cool place for 20–25 days (Simango, 2002). The major microflora involved in the *Mahewu* fermentation is *Lactococcus lactis* subsp. *Lactis,* however other researchers have also tried *Lactobacillus delbruckii, L. bulgaricus* and some yeasts to produce acceptable mahewu at room temperature (Mashau et al., 2021). *Mahewu* is reported to address childhood diarrhea challenges by enhancing the microbial balance in the gut. Moreover, Mahewu is now considered as a important functional food as it helps enhancing the immune system, and bioavailability of nutrients (Olusanya et al., 2021).

8.6 Commercialization Status and Future Prospects

The demand for fermented food has become significant in modern society as it possesses great combination of very good sensory as well as nutritional attributes. Optimization of fermented products for commercial production and scaling up the process is important to obtain consistent quality. As fermentation process is dynamic in nature, the knowledge of involvement of microorganisms and their mode of functioning directed food scientists to use specific strains as starter culture to get desirable characteristics under controlled conditions. In addition, efficient use of novel food process methods, viz. co-culturing, molecular tools, genetic engineering, recombinant DNA technology and accelerated fermentation helps design and development of specific starter cultures which can perform better than the natural one in a controlled environment. For maintaining safety during fermentation and to increase the shelf life of the final product without hampering the nutritional value, large scale food industries are keen to implement advance thermal and non-thermal techniques, viz., ohmic heating, pulse electric field heating, microwave processing, gamma irradiation, UV light disinfection, etc. FDA (Food and Drug Administration, USA) has already approved the use of cobalt-60, cesium-137 and UV light for use in food applications. For safe production of the cereal-based fermented foods, small and medium-scale industries should implement good manufacturing practices (GMP) and hazard analysis critical control point (HACCP) principles.

Fermentation technology needs evolution as manufacturing process of many indigenous cereal-based fermented foods is yet to be optimized at commercial scale. For higher yield, the areas of modern fermenters, automation of the process and sensor development are important future aspects. Biochemical changes in the cereal starch during fermentation using microbial enzymes is another important future prospect for manufacturing of high quality and consistent product. Further research is needed in the area of supplementation or fortification of cereal-based fermented foods with protein isolated from plant sources and feasibility of the process in commercial scale. Moreover, to improve the functionality of cereal-based beverages, use of probiotic microbes and their safe delivery into the colon through nano emulsions are extensively being researched. Future studies should concentrate on developing novel cereal-based fermented foods that are fortified with underutilized millets and indigenous herbs targeting improved health benefits.

References

Abd El-Ghani, S., Bahgaat, W. K., & Fouad, M. T. (2014). The microbiological quality and physicochemical attributes of Egyptian traditional Sa'eedi Kishk. *Journal of Food and Nutrition Sciences, 4*, 13–21.

Abou-Zeid, N. A. (2016). Review of Egyptian cereal-based fermented product (Kishk). *International Journal of Agriculture Innovations and Research, 4*(4), 600–609.

Adebo, O. A. (2020). African sorghum-based fermented foods: Past, current and future prospects. *Nutrients, 12*(4), 1111.

Adegoke, G. O., & Babalola, A. K. (1988). Characteristics of micro-organisms of importance in the fermentation of fufu and ogi—two Nigerian foods. *Journal of Applied Bacteriology, 65*(6), 449–453.

Adem, K. D., & Ambie, D. A. (2017). A review of injera baking technologies in Ethiopia: Challenges and gaps. *Energy for Sustainable Development, 41*, 69–80.

Adisa, A. M., & Enujiugha, V. N. (2020). Microbiology and safety of ogi fermentation: A review. *European Journal of Nutrition & Food Safety, 12*(5), 90–100.

Agrawal, R., Rati, E. R., Vijayendra, S. V. N., Varadaraj, M. C., Prasad, M. S., & Nand, K. (2000). Flavour profile of idli batter prepared from defined microbial starter cultures. *World Journal of Microbiology and Biotechnology, 16*(7), 687–690.

Aidoo, K. E., Rob Nout, M. J., & Sarkar, P. K. (2006). Occurrence and function of yeasts in Asian indigenous fermented foods. *FEMS Yeast Research, 6*(1), 30–39.

Alba-Lois, L., & Segal-Kischinevzky, C. (2010). Yeast fermentation and the making of beer and wine. *Nature Education, 3*(9), 17.

Allwood, J. G., Wakeling, L. T., & Bean, D. C. (2021). Fermentation and the microbial community of Japanese koji and miso: A review. *Journal of Food Science, 86*, 2194–2207.

Annan, N. T., Poll, L., Plahar, W. A., & Jakobsen, M. (2003). Aroma characteristics of spontaneously fermented Ghanaian maize dough for kenkey. *European Food Research and Technology, 217*(1), 53–60.

Arici, M., & Daglioglu, O. (2002). Boza: A lactic acid fermented cereal beverage as a traditional Turkish food. *Food Reviews International, 18*(1), 39–48.

Atwaa, E. H., Hassan, M. A. A., & Ramadan, M. F. (2020). Production of probiotic stirred yoghurt from camel milk and oat milk. *Journal of Food and Dairy Sciences, 11*(9), 259–264.

Barnett, J. A., & Lichtenthaler, F. W. (2001). A history of research on yeast 3: Emil Fischer, Eduard Buchner, and their contemporaries, 1880–1900. *Yeast, 18*, 363–388.

Basinskiene, L., & Cizeikiene, D. (2020). Cereal-based nonalcoholic beverages. In: Galanakis, C.M., *Trends in Non-alcoholic Beverages*,cAcademic Press 63–99.

Basinskiene, L., Juodeikiene, G., Vidmantiene, D., Tenkanen, M., Makaravicius, T., & Bartkiene, E. (2016). Non-Alcoholic beverages from fermented cereals with increased oligosaccharide content. *Food Technology and Biotechnology, 54*(1), 36–44.

Bernard, D., Jeyagowri, N., & Madhujith, T. (2021). Characterization of lactic acid bacteria isolated from idli batter and their susceptibility to antibiotics. *Tropical Agricultural Research, 32*(4), 453–461.

Bernat, N., Cháfer, M., González-Martínez, C., Rodríguez-García, J., & Chiralt, A. (2015). Optimisation of oat milk formulation to obtain fermented derivatives by using probiotic Lactobacillus reuteri microorganisms. *Food Science and Technology International, 21*(2), 145–157.

Bilgiçli, N. (2009). Effect of buckwheat flour on chemical and functional properties of tarhana. *LWT-Food Science and Technology, 42*(2), 514–518.

Birch, A. N., Petersen, M. A., & Hansen, Å. S. (2014). Aroma of wheat bread crumb. *Cereal Chemistry, 91*(2), 105–114.

Blandino, A., Al-Aseeri, M. E., Pandiella, S. S., Cantero, D., & Webb, C. (2003). Cereal-based fermented foods and beverages. *Food Research International, 36*(6), 527–543.

Chelliah, R., Ramakrishnan, S. R., Premkumar, D., & Antony, U. (2017). Accelerated fermentation of idli batter using Eleusine coracana and Pennisetum glaucum. *Journal of Food Science and Technology, 54*(9), 2626–2637.

Daglioğlu, O. (2000). Tarhana as a traditional Turkish fermented cereal food. Its recipe, production and composition. *Food/Nahrung, 44*(2), 85–88.

Demir, M. K. (2014). Use of quinoa flour in the production of gluten-free tarhana. *Food Science and Technology Research, 20*(5), 1087–1092.

Desai, B. B., & Salunkhe, D. K. (2018). Dhokla and khaman. In N. R. Reddy, M. D. Pierson, & D. K. Salunkhe (Eds.), *Legume-based fermented foods* (pp. 162–172). CRC Press.

Desiye, A., & Abegaz, K. (2013). Isolation, characterization and identification of lactic acid bacteria and yeast involved in fermentation of teff (Eragrostis tef) batter. *Advanced Research in Biological Sciences, 1*(3), 36–44.

Desiye, A., Abegaz, K., Negera, E., & Gobena, E. (2017). The microbiology of teff (Eragrostis tef) enjera. *London School of Hygiene and Tropical Medicine research online, 2*(2), 115–120. Accessed from https://www.scischolars.com/journals/index.php/sjr...

Det-Udom, R., Gilbert, C., Liu, L., Prakitchaiwattana, C., Ellis, T., & Ledesma-Amaro, R. (2019). Towards semi-synthetic microbial communities: Enhancing soy sauce fermentation properties in B. subtilis co-cultures. *Microbial Cell Factories, 18*(1), 1–8.

Devanthi, P. V. P., Linforth, R., Onyeaka, H., & Gkatzionis, K. (2018). Effects of co-inoculation and sequential inoculation of Tetragenococcus halophilus and Zygosaccharomyces rouxii on soy sauce fermentation. *Food Chemistry, 240*, 1–8.

Dhillon, B., Sodhi, N. S., Gandotra, S., Kaur, S., & Jaiswal, S. (2020). Physico-chemical and textural (sensorial and electromyographic) evaluation of idlis formulated with brown rice and pearl millet flours. *Journal of Food Measurement and Characterization, 14*(5), 2898–2906.

Dlusskaya, E., Jänsch, A., Schwab, C., & Gänzle, M. G. (2008). Microbial and chemical analysis of a kvass fermentation. *European Food Research and Technology, 227*(1), 261–266.

Ekpa, O., Palacios-Rojas, N., Kruseman, G., Fogliano, V., & Linnemann, A. R. (2019). Sub-Saharan African maize-based foods-processing practices, challenges, and opportunities. *Food Reviews International, 35*(7), 609–639.

Erkan, H., Çelik, S., Bilgi, B., & Köksel, H. (2006). A new approach for the utilization of barley in food products: Barley tarhana. *Food Chemistry, 97*(1), 12–18.

Erkmen, O., & Bozoglu, T. F. (2016). *Fermented Cereal and Grain Products. In: Food microbiology: Principles into practice. John Wiley and Sons Ltd.* 349–373.

Fadahunsi, I. F., & Soremekun, O. O. (2017). Production, nutritional and microbiological evaluation of mahewu a South African traditional fermented porridge. *Journal of Advances in Biology & Biotechnology, 14*(4), 1–10.

Gaikwad, P. S., Yadav, B. K., Anandakumar, S., Loganathan, M., & Shanmugasundaram, S. (2020). Development of high oxygen barrier multi-layered packaging film for shelf life extension of Ready-To-Cook (RTC) idli batter. *International Research Journal of Pure and Applied Chemistry*, *21*, 13–27.

Ganguly, S., Sabikhi, L., & Singh, A. K. (2021). Cereal-based fermented foods for enhanced nutritional attributes and better gut health. *International Journal of Fermented Foods, 10*(1), 1–12.

Gebrekidan, B., & Belainesh, G. H. (1981, October 28–31). Sorghum injera preparations and quality parameters. In: *Proceedings of the international symposium on sorghum grain quality,* ICRISAT Center Patancheru, India, pp. 55–65.

Gotcheva, V., Pandiella, S. S., Angelov, A., Roshkova, Z., & Webb, C. (2001). Monitoring the fermentation of the traditional Bulgarian beverage boza. *International Journal of Food Science & Technology, 36*(2), 129–134.

Hancioğlu, Ö., & Karapinar, M. (1997). Microflora of Boza, a traditional fermented Turkish beverage. *International Journal of Food Microbiology, 35*(3), 271–274.

Harada, R., Yuzuki, M., Ito, K., Shiga, K., Bamba, T., & Fukusaki, E. (2017). Influence of yeast and lactic acid bacterium on the constituent profile of soy sauce during fermentation. *Journal of bioscience and bioengineering, 123*(2), 203–208.

Heitmann, M., Zannini, E., & Arendt, E. (2018). Impact of Saccharomyces cerevisiae metabolites produced during fermentation on bread quality parameters: A review. *Critical Reviews in Food Science and Nutrition, 58*(7), 1152–1164.

Ibnouf, F. O. (2012). The value of women's indigenous knowledge in food processing and preservation for achieving household food security in rural Sudan. *Journal of Food Research, 1*(1), 238.

Karovičová, Z. K. J., & Kohajdova, J. (2007). Fermentation of cereals for specific purpose. *Journal of Food and Nutrition Research, 46*(2), 51–57.

Kitabatake, N., Gimbi, D. M., & Oi, Y. (2003). Traditional non-alcoholic beverage, Togwa, in East Africa, produced from maize flour and germinated finger millet. *International Journal of Food Sciences and Nutrition, 54*(6), 447–455.

Kumari, S., Bhinder, S., Singh, B., Kaur, A., & Singh, N. (2020). Effect of buckwheat incorporation on batter fermentation, rheology, phenolic, amino acid composition and textural properties of idli. *LWT, 122*, 109042.

Kusumoto, K. I., Yamagata, Y., Tazawa, R., Kitagawa, M., Kato, T., Isobe, K., & Kashiwagi, Y. (2021). Japanese traditional miso and koji making. *Journal of Fungi, 7*(7), 579.

Larkin, T. A., Astheimer, L. B., & Price, W. E. (2009). Dietary combination of soy with a probiotic or prebiotic food significantly reduces total and LDL cholesterol in mildly hypercholesterolaemic subjects. *European journal of clinical nutrition, 63*(2), 238–245.

Lidums, I., Karklina, D., Kirse, A., & Sabovics, M. (2017). Nutritional value, vitamins, sugars and aroma volatiles in naturally fermented and dry kvass. In *11th Baltic conference on food science and technology-food science and technology in a changing world* (Vol. 1, pp. 61–65).

Lidums, I., Karklina, D., Sabovics, M., & Kirse, A. (2015). Evaluation of aroma volatiles in naturally fermented kvass and kvass extract. *Research for Rural Development, 1*, 143–149.

Lohekar, A. S., & Arya, A. B. (2014). Development of value added instant dhokla mix. *International Journal of Food and Nutritional Sciences, 3*(4), 78–83.

Mahgoub, S. E., Ahmed, B. M., Ahmed, M. M., & El Nazeer, A. A. (1999). Effect of traditional Sudanese processing of kisra bread andhulu-mur drink on their thiamine, riboflavin and mineral contents. *Food Chemistry, 67*(2), 129–133.

Mashau, M. E., Maliwichi, L. L., & Jideani, A. I. O. (2021). Non-alcoholic fermentation of maize (Zea mays) in Sub-Saharan Africa. *Fermentation, 7*(3), 158.

Morcos, S. R., Hegazi, S. M., & El-Damhougy, S. T. (1973). Fermented foods in common use in Egypt I. The nutritive value of kishk. *Journal of the Science of Food and Agriculture, 24*(10), 1153–1156.

Mulmule, M. D., Shimmy, S. M., Bambole, V., Jamdar, S. N., Rawat, K. P., & Sarma, K. S. S. (2017). Combination of electron beam irradiation and thermal treatment to enhance the shelf-life of traditional Indian fermented food (Idli). *Radiation Physics and Chemistry, 131*, 95–99.

Narendranath, N. V., & Power, R. (2005). Relationship between pH and medium dissolved solids in terms of growth and metabolism of lactobacilli and Saccharomyces cerevisiae during ethanol production. *Applied and Environmental Microbiology, 71*(5), 2239–2243.

Nanson, N. J., & Fields, M. L. (1984). Influence of temperature of fermentation on the nutritive value of lactic acid fermented cornmeal. *Journal of Food Science, 49*(3), 958–959.

Neela, S., & Fanta, S. W. (2020). Injera (An ethnic, traditional staple food of Ethiopia): A review on traditional practice to scientific developments. *Journal of Ethnic Foods, 7*(1), 1–15.

Nout, M. J. R., Kok, B., Vela, E., Nche, P. F., & Rombouts, F. M. (1995). Acceleration of the fermentation of kenkey, an indigenous fermented maize food of Ghana. *Food Research International, 28*(6), 599–604.

Nout, M. R., Sarkar, P. K., & Beuchat, L. R. (2007). Indigenous fermented foods. In *Food microbiology: Fundamentals and frontiers* (3rd ed., pp. 817–835). American Society of Microbiology.

Odunfa, S. A., & Oyewole, O. B. (1998). African fermented foods. In *Microbiology of fermented foods* (pp. 713–752). Springer.

Olusanya, N. R., Kolanisi, U., Van Onselen, A., & Ngobese, N. (2021). Preparation, storage, and utilization of Mahewu (a non-alcoholic maize meal beverage) in Ntambanana, South Africa. *African Journal of Food, Agriculture, Nutrition and Development, 21*(2), 17492–17508.

Özdemir, G., & Altıner, D. D. (2021). Cereal grain-fermented products: Health, nutrition and microbiology. *Engineering Sciences Innovative Approaches*, 146–167.

Ozdemir, S., Gocmen, D., & Yildirim Kumral, A. (2007). A traditional Turkish fermented cereal food: Tarhana. *Food Reviews International, 23*(2), 107–121.

Rani, M., Amane, D., & Ananthanarayan, L. (2019). Impact of partial replacement of rice with other selected cereals on idli batter fermentation and idli characteristics. *Journal of Food Science and Technology, 56*(3), 1192–1201.

Ray, M., Ghosh, K., Singh, S., & Mondal, K. C. (2016). Folk to functional: An explorative overview of rice-based fermented foods and beverages in India. *Journal of Ethnic Foods, 3*(1), 5–18.

Ray, R. C., & Swain, M. R. (2013). Indigenous fermented foods and beverages of Odisha, India: An overview. In *Indigenous fermented foods of South Asia* (pp. 1–6). CRC Press.

Ray, S., Saha, S. K., Raychaudhuri, U., & Chakraborty, R. (2017). Preparation of okra-incorporated dhokla and subsequent analysis of nutrition, antioxidant, color, moisture and sensory profile. *Journal of Food Measurement and Characterization, 11*(2), 639–650.

Sahoo, S., Lenka, C., & Biswal, G. (2017). Knowledge and Awareness about Health Benefits of indigenous fermented foods: A comprehensive study. *Education, 85*(15), 2.

Şanlier, N., Gökcen, B. B., & Sezgin, A. C. (2019). Health benefits of fermented foods. *Critical Reviews in Food Science and Nutrition, 59*(3), 506–527.

Sassi, S., Wan-Mohtar, W. A. A. Q. I., Jamaludin, N. S., & Ilham, Z. (2021). Recent progress and advances in soy sauce production technologies: A review. *Journal of Food Processing and Preservation, 45*(10), e15799.

Sengun, I. Y., Nielsen, D. S., Karapinar, M., & Jakobsen, M. (2009). Identification of lactic acid bacteria isolated from Tarhana, a traditional Turkish fermented food. *International Journal of Food Microbiology, 135*(2), 105–111.

Shaikh, A., Siddique, I., Shaikh, M., & Shetty, S. (2021). Isolation and enumeration of probiotic microorganisms from fermented idli batter. *African Journal of Biological Sciences, 3*(3), 59–63.

Sharma, A., Kumari, S., Nout, M. J., & Sarkar, P. K. (2018). Preparation of antinutrients-reduced dhokla using response surface process optimization. *Journal of Food Science and Technology, 55*(6), 2048–2058.

Shurtleff, W., & Aoyagi, A. (2018). *The book of miso* (2nd ed.). Soyinfo Centre.

Simango, C. (2002). Lactic acid fermentation of sour porridge and mahewu, a non-alcoholic fermented cereal beverage. *Journal of Applied Science in Southern Africa, 8*(2), 89–98.

Stanojević-Nikolić, S., Dimić, G., Mojović, L., Pejin, J., Djukić-Vuković, A., & Kocić-Tanackov, S. (2016). Antimicrobial activity of lactic acid against pathogen and spoilage microorganisms. *Journal of Food Processing and Preservation, 40*(5), 990–998.

Tadesse, B. T., Abera, A. B., Tefera, A. T., Muleta, D., Alemu, Z. T., & Wessel, G. (2019). Molecular characterization of fermenting yeast species from fermented teff dough during preparation of injera using ITS DNA sequence. *International Journal of Food Science, 1291863*(1-7).

Taillefer, M., & Sparling, R. (2016). Glycolysis as the central core of fermentation. In: Hatti-Kaul, R., Mamo, G. and Mattiasson, B. *Anaerobes in Biotechnology*, Springer, 55–77.

Tamime, A. Y., & O'connor, T. P. (1995). Kishk—a dried fermented milk/cereal mixture. *International Dairy Journal, 5*(2), 109–128.

TaŞoğulları, N., & ŞimŞek, Ö. (2020). The effect of irradiation on the quality properties of tarhana. *Applied Sciences, 10*(14), 4749.

Voet, D., & Voet, J. (2004). Biochemistry. In *Biomolecules, mechanisms of enzyme action, and metabolism* (Vol. 1, 3rd ed.). Wiley.

Yokotsuka, T. (1993). Japanese shoyu. In K. H. Steinkraus (Ed.), *Handbook of indigenous fermented foods* (pp. 511–517). Marcel Dekker.

Chapter 9
Functional Cereal-Based Bakery Products, Breakfast Cereals, and Pasta Products

Maria Di Cairano, Roberta Tolve, Nazarena Cela, Lucia Sportiello, Teresa Scarpa, and Fernanda Galgano

9.1 Introduction

Consumers' interest in consuming functional foods is increasing and food industry has been developing new products that are healthier and more diversified (Nicole et al., 2021; Punia, Sandhu, & Kaur, 2020; Punia, Dhull, et al., 2020). Among the different legislations and definitions on functional foods, the common denominator is that they are those foods able to provide beneficial effects beyond the provision of nutrients. They are not medicines, and they should be an integral part of the eating habits of the population. Bakery products and pasta are a staple food in many parts of the world, representing an ideal matrix to be functionalized. This chapter will explore recent trends in the functionalization of bakery goods, breakfast cereals, and pasta through various ingredients from those traditional to more innovative ones (Fig. 9.1). Before going into the topic it follows a general overview of the products covered in the chapter and their potential to be functional foods.

Bread is one of the most ancient food products and the art of baking has been passed from generations (Nehra et al., 2021). Flour, yeast, salt and water are the main ingredients used for bread making. It is a source of carbohydrates, protein, vitamins, mainly B group, and minerals. Bread was originally produced from barley (*Hordeum vulgare* L.) and emmer wheat (*Triticum dicoccum* L.). However, nowadays several other ingredients and innovative preparation methods are used to fulfil consumers' preference or to take account of environmental and health issues (de Pinho Ferreira Guiné & dos Reis Correia, 2013; Miskelly, 2017). Bread is one of the most consumed cereal-based products worldwide, it is convenient as food carrier

M. Di Cairano (✉) · N. Cela · L. Sportiello · T. Scarpa · F. Galgano
School of Agricultural, Forestry, Food and Environmental Sciences, University of Basilicata, Potenza, Italy

R. Tolve
Department of Biotechnology, University of Verona, Verona, Italy

S. Punia Bangar, A. Kumar Siroha (eds.), *Functional Cereals and Cereal Foods*,
https://doi.org/10.1007/978-3-031-05611-6_9

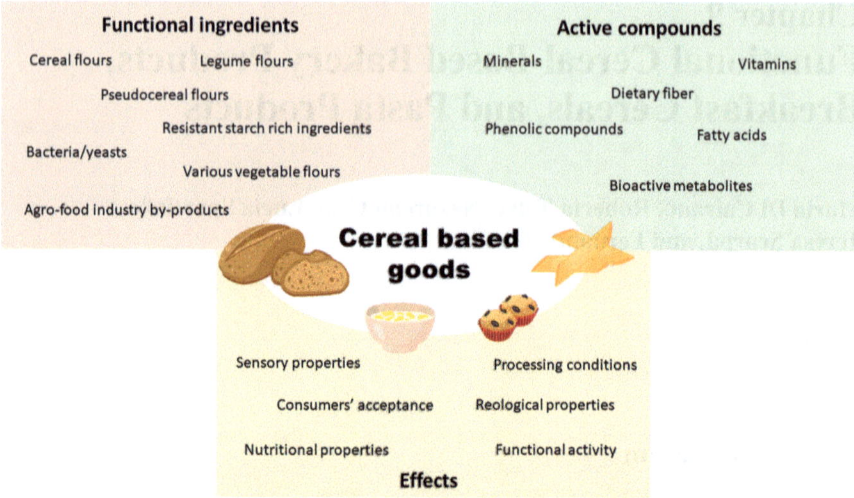

Fig. 9.1 Functional ingredients used in cereal based goods, the active compounds contained and their effects. (Images designed by pch.vector/Freepik)

since its economic, well-known and easy to implement process, due to its large daily intake and its acceptable sensory properties. For this reason, bread is well suited for the incorporation of bioactive compounds in order to improve the nutritional value of the product and to develop a functional bread (Betoret & Rosell, 2020).

Biscuits are cereal based products, baked to a moisture content lower than 5% (Manley, 2000) which allows to have a long shelf life. Biscuit main ingredients are wheat flour (*Triticum aestivum* L.), fats and sugars to which a variety of other ingredients may be added, often including eggs. Dough piece formation can be obtained through different technologies, but biscuit production process is quite simple in outline. Indeed, gluten network development is not as fundamental as in bread, biological leavened products, and pasta (Schober et al., 2003); hence, the addition of new ingredients in replacement of, or in addition to, wheat flour is possible without affecting excessively the technological properties of the dough and the texture of the biscuits. However, the fact remains that it is important to prepare a dough with the ideal characteristics for the dough piece formation technology to be used, otherwise the dough could not be workable and the final quality of the biscuits could result impaired. Biscuits are generally appreciated by most of the population, they are affordable, and have a long shelf life when properly stored. For these reasons, biscuits represent an ideal matrix to be reformulated in the view of producing new functional products.

The enrichment of baked goods other than bread and biscuits with functional ingredients may be an effective way to prompt people's health. Many attempts are being made to improve the nutritional value and functionality of snacks by modifying their nutritive composition. Such effects are very often achieved by increasing the nutrient density in basic recipes (Ainsworth et al., 2007; Sun-Waterhouse et al., 2010). These baked goods can be both sweet and savory, ranging from cakes to muffins, crackers, and breadsticks. All of them are appealing for consumers and eaten quite often even if dietary guidelines suggest to have a moderate consumption of some of them due to their richness in sugars and fats. Since they are widely consumed, they are suitable to be functionalized (Marchetti et al., 2018) also in the view of enhancing their nutritional profile.

Besides baked goods, breakfast cereals represent another vast food industry sector based on cereals. It is in constant evolution and its global market value was 55 billion dollars in 2020 (Wunsch, 2020). These products are defined as processed grains for human consumption and they are mostly made of wheat, corn, rice, barley and oats (Caldwell et al., 2004). Data collected through numerous surveys conducted by the International Breakfast Initiative found that breakfasts containing ready-to-eat cereals, oats, and/or muesli in association with milk exhibited better nutritional profiles and were classified as having higher diet quality scores (Gibney et al., 2018). This has been proven for all age categories, sex groups, and culturally different diet patterns. Regular consumption of ready-to-eat breakfast cereals was also associated with higher diet quality. In addition, it was also shown that those who chose this category of products were less likely to skip breakfast, a parameter that contributed to a better overall nutrient intake (Michels et al., 2016). The chemical composition of breakfast cereals is characterized by a wide presence of phytochemicals, such as phenolic acids, flavones, phytic acid, flavanoids, coumarins, and terpenes (Sidhu et al., 2007). On the other hand, these products have certain limitations from a nutritious point of view, especially linked to the small amount and amino acid profile of their proteins, the swelling of the starch during heating process and the limited bioavailability of their mineral component due to the presence of some anti-nutritional factors (e.g. phytic acid) (Nout, 2009). These limitations can be the starting point for an improvement in the nutritional quality of breakfast cereals and the achievement of functional foods.

Pasta has a primary role in human nutrition, and it is included at the base of the food pyramid since daily consumption is strongly recommended in the Mediterranean diet. Conventional durum wheat (*Triticum turgidum* L. var. *durum*) semolina is the most suitable cereal for high–quality pasta because of its sensorial and cooking properties (Simonato et al., 2019). Pasta made with durum semolina is considered a good source of carbohydrates (74–78% DM) but lacks in proteins (11–15% DM) and is deficient in lysine and threonine as well as most cereal products (Abdel-Aal & Hucl, 2002). Moreover, pasta has low fat and sodium levels and shows large acceptability by the population for its palatability, long shelf life, and low cost. All these characteristics give a reason to consider pasta as an appropriate carrier of different bioactive compounds, increasing nutritional and functional properties (Simonato et al., 2019). In addition, pasta production technology easily allows its functionalization.

9.2 Bread

Each bread making step causes physicochemical, sensory and nutritional changes: raw materials, mixing, fermentation, baking, and storage conditions affect final bread quality. Refining process during milling reduces nutritional value of grains. Indeed, the majority of bioactive compounds are present in the outer layers of the kernels such as arabinoxylans or β-glucans and dietary fiber. In addition, phenolic compounds are not uniformly distributed in wheat kernel but mainly located in its outer layer. During milling process, it is separated from endosperm causing a loss of these bioactive compounds. Therefore, the use of wholegrain is recommended by dietary guidelines (Angelino et al., 2017; Ciudad-Mulero et al., 2020). Also β-glucans, useful in promoting the reduction of blood cholesterol, undergo changes according to mixing and fermentation time of the dough (de Pinho Ferreira Guiné and dos Reis Correia, 2013). Thermal process during baking has a significant impact on nutritional value of bread. For example, a reduction in lysine content and consequently in the protein quality of the final product occurs due to heat destruction (Lindenmeier & Hofmann, 2004). Moreover, high temperatures negatively affect thermolabile compounds with health effect, such as polyphenols and vitamins.

Considering the growing consumers' demand for healthier foods, several studies focused on incorporation of bioactive ingredients in bread, for example legumes (Stamataki et al., 2016), prebiotics (Mollakhalili-Meybodi et al., 2021), fruit and vegetables (Betoret & Rosell, 2020) in order to develop functional bread, taking into account physicochemical and sensory properties of final product.

Dietary fiber consists in carbohydrates such as hemicellulose, inulin, resistant starch and other polysaccharides and oligosaccharides. They have technological influence on bread making, influencing rheological parameters. Their positive health benefits are well-know: dietary fiber is not digested nor absorbed promoting a better gastrointestinal motility, controlling blood cholesterol and glucose content, preventing diabetes, cancers and weight gain (Kurek & Wyrwisz, 2015). Taking into account these aspects, the supplementation of fiber in bread has been explored in the recent years. Rice bran is a good source of fiber, but also of B-vitamins, minerals and polyunsaturated fatty acids. For this reason, it was added in wheat bread to enhance its nutritional profile. As reported by Ameh et al. (2013), the supplementation of increased levels of rice bran in wheat bread resulted in a significant increase in minerals, protein and fiber content. As regards the physical properties, there was a significant reduction of bread weight and volume. Different sensory properties in terms of texture, crumb and crust color, aroma, and overall liking were highlighted. Control sample scored higher than bread supplemented with rice bran, 7.95 and 7.20, respectively. Anyway, the incorporation of rice bran resulted in a product with enhanced nutritional aspect, consequential it may be considered a potential functional bread. Hu et al. (2009) stated that the addition of defatted rice bran in bread enhance hemicellulose and insoluble fiber content. The final product had acceptable sensory properties despite a significant increase in firmness. Punia Bangar et al. (2021) and Punia, Kumar, et al. (2021, Punia, Sandhu, et al. (2021)) reported that fermented barley and rice bran may improve the bioactive profile of various food products. Nehra et al. (2021) evaluated the effect of pearl millet addition on nutritional and sensory properties of bread. It is observed that after the addition of pearl millet flour nutritional and antioxidant properties of bread improved.

Another challenge for the market is the re-use of agro-food by-products in order to reduce food waste. Most of them contains bioactive compounds, thus their incorporation in foodstuffs meets aspects related both to the circular economy and the consumers' requirements for functional foods (Martins et al., 2017). Brewers spent grains (BSG) are the main by-products from brewery industry. They contains dietary fiber, mainly β-glucans and resistant starch, but are also source of essential amino acids, minerals and polyphenols. The addition of 5% of BSG resulted in two-fold increase of the fiber content of the bread and five-fold when 20% of BSG were added in bread formulation. In this way the BSG enriched bread allowed to respect the recommended daily intake of fiber required for healthy nutrition (about 30 g per day) (Fărcaş et al., 2014). Volatile profile of bread containing BSG showed a higher content of volatile compound groups such as alcohols and aldehydes, generally associated to "malty" flavour, compared to 100% wheat bread suggesting that there was a difference in aroma profile between samples. As regards the sensory acceptability, the ideal amount of BSG in order to not adversely affect the sensory acceptability of the enriched product was 10%. Taste and texture were the main sensory attributes influencing the liking of final product: increased levels of BSG corresponded to a reduction in the texture score, the hardness of the crumb was higher, due to the crosslink between fiber and gluten proteins. In addition, incorporation of BSG resulted in a darker color of the crust and crumb of the bread (Fărcaş et al., 2014). Significant changes in rheological properties were also shown by Amoriello et al. (2020), who reported a decreased volume of bread loaves and a higher water absorption and dough strength as BSG % in the formulation increased. The second major by-product of brewery industry is the dry spent yeast, low in fat, carbohydrates and sodium but with an appreciable content in β-glucans. It was added in a homemade bread causing an increase in β-glucans content from 65 to 125 mg in 50 g of bread without affecting sensory attributes. Fruit and vegetables by-products such as banana (Eshak, 2016), cupuassu (*Theobroma grandiflorum*) (Salgado et al., 2011), orange (Stoll et al., 2015) and pineapple (Wu & Shiau, 2015) peels, apple pomace (Pyanikova et al., 2021), sweet potato leaf and stem (Liu et al., 2007), pea and broad bean pods (Belghith Fendri et al., 2016) were frequently used in supplementation of bread to uplift their nutritional value. All of these studies stated that consumption of bread enriched with 5–10% of by-products from fruit and vegetables industry allowed to increase the daily intake of dietary fiber without affecting overall acceptability (Stoll et al., 2015).

Chicory-inulin enriched bread showed altered rheological properties in terms of dough machinability due to the interactions among protein, inulin and starch. Furthermore, loaf volume and crust color were significantly influenced by inulin addition. Five percent inulin was considered the ideal concentration in order to not drastically influence the bread quality. More studies are needed to confirm whether addition level of maximum 5% inulin allows to make bread "functional" since that inulin beneficial effects are due to a daily recommended intake of at least 10–15 g (Sirbu and Arghire, 2017).

Phenolic compounds have a crucial role in preventing inflammation, oxidative stress and they have cardio protective effects (Sandhu & Punia, 2017; Punia, Sandhu, et al., 2019). There is a correlation between consumption of phenols-rich foods and reduction of chronic disease suggesting antioxidant properties (Angelino et al., 2017). As reported by Pathak et al. (2016), incorporation of ripe mango peels showed a significant increase in total phenolic content (from 220 to 756 mgGAE/100 gm) compared to control sample and also for antioxidant capacity based on DPPH and FRAP assay results. For these reasons, ripe mango peel could be considered a potential health-promoting ingredient. However, rheological properties of bread, in terms of loaf height, weight loss percentages and specific volume, were negatively correlated with the amount of ripe mango peels added. On the other hand, an increase in hardness, cohesiveness and springiness was observed in enriched bread. Five percent was considered the optimum amount of ripe mango peels to incorporate in bread, without excessively affecting rheological and sensory properties, and to improve at the same time the antioxidant capacity. The same occurred when grape (*Vitis Vinifera* L.) pomace was added in bread improving its functional properties in terms of total polyphenols content (from 35 mgGAE/100 g in control sample to 89 mgGAE/100 g in bread with 10% of grape pomace) and anti-radical activity. The incorporation significantly influenced color parameters of crust and crumb. Anyway, overall liking was not influenced by 2–5% addition of grape pomace (Hayta et al., 2014). This is in agreement with Walker et al. (2014) who reported significant increase in total phenolic content and DPPH radical scavenging activity in white bread enriched with spearmint (*Mentha spicata*) aqueous extract. The spearmint extract could be a proper functional ingredient but further studies will be useful in optimization of sensory aspects of the final product (Shori et al., 2021).

Seaweeds is related to beneficial health effects due to its composition in bioactive compounds such as polysaccharides, vitamins, phenols. Its addition in bread resulted in product with an increased phenolic compound and antioxidant activity. Color of crumb and crust was highly influenced with enhanced darkness and green coordinates (Amoriello et al., 2021).

Different *in vitro* studies examined the bio accessibility of the phenol compounds present in enriched bread. Incorporation of ground flaxseed hulls in bread increased phenolic content and, as consequence, antioxidant capacity. Flaxseed by-products are good source of lignans, functional compounds with beneficial effect on human health helping in the prevention of cancers, diabetes and cardiovascular disease. Phenols resulted bio accessible according to digestion *in vitro* (Sęczyk et al., 2017; Wirkijowska et al., 2020). Artichoke stem powder is rich in polyphenols, mainly cynaropicrin. *In vitro* simulated gastrointestinal digestion showed a high bio accessibility of polyphenols (82%) from artichoke heads used in the formulation of unconventional breads suggesting this ingredient as a promising functional compound in order to exert an intestinal protective action but also able to modulate glucose metabolism (Colantuono et al., 2018).

Vitamins and minerals in cereal-based products undergo significant changes during bread making. Milling of grains causes a depletion of vitamin B and minerals content. Even if fermentation could increase vitamin B levels, different studies aimed at overcoming the depletion of vitamins and minerals during baking (Batifoulier et al., 2005). The addition of rice bran could be one of the strategies to improve vitamins and minerals content in bread. Ameh et al. (2013) showed a significant increase in vitamin B1 and B2 content in comparison with control bread sample. Both "contribute to normal psychological function" and to "normal energy-yielding metabolism" (European Commission, 2012). Bread supplemented with rice bran contained significant higher levels of iron, potassium, calcium and magnesium than 100% wheat bread showing a successful nutritional profile.

Red bell pepper contains a high amount of carotenoids and the incorporation of its powder in bread was performed with the aim to enhance carotenoids content and make bread a functional product with high antioxidant capacity (Kaur et al., 2020). Also pumpkins by-products are rich in carotenes. Kampuse et al. (2015) showed that the addition of pumpkin pomace resulted in five-fold increase in carotenoids content against an increase of 13 times higher when pumpkin powder was used as supplementation of bread. Moreover, as regards sensory aspects, addition of 50% of pumpkin pomace scored the highest in terms of overall acceptability.

Bread enriched with 5–15% of pomegranate whole fruit bagasse proved to be "high in Cu" (Bhol et al., 2016) showing increased antioxidant properties than control bread since significant amount of copper "contributes to the protection of cells from oxidative stress" (European Commission, 2012). However, this supplementation negatively affected sensory attributes of enriched bread imparting a gritty mouthfeel.

Calcium is a vital nutrient contributing to neurotransmission and to the right formation of bones. This is the reason for the development of calcium-enriched bread. As reported by Agrahar-Murugkar and Dixit-Bajpai (2020), different ingredients were added for this purpose such as malted finger millet, sesame seeds, moringa leaves and cumin seeds. Also Weisstaub et al. (2018) developed a functional bread with resistant starch, garlic and calcium citrate showing an increased calcium bioavailability.

Polyunsaturated fatty acids (PUFA), mainly eicosapentaenoic acid (EPA C20:5) and docosahexaenoic acid (DHA C22:6) have an important role in the prevention of cardiovascular disease, "contributing to the normal function of the heart if daily consumption is above 250 mg of EPA and DHA" (European Commission, 2012). Rice bran oil is unique among edible vegetable oils because of its unique fatty acid composition, phenolic compound (γ-oryzanol, ferulic acid) and vitamin E (tocopherol and tocotrienol) (Punia, Sandhu, & Kaur, 2020; Punia, Dhull, et al., 2020). Fish oil is one of the main dietary source of PUFA and it was used to produce a functional bread (Kolanowski & Laufenberg, 2006). No significant differences emerged in sensory profile of control and enriched breads. The latter scored above six in a nine-point hedonic scale suggesting an acceptable overall liking of final product despite the increase in hardness (Sridhar et al., 2021). γ-Aminobutyric acid (GABA), a non-proteinogenic amino acid considered as bioactive compound due to its health functions such as reduction of hypertension, depression and anxiety, prevention of diabetes, cancer and chronic disease. Few studies reported its incorporation in bread. Unconventional flours, made of buckwheat, amaranth, chickpea, quinoa, were selected for bread making and fermented by *Lactobacillus plantarum* C48 (now reclassified as *Lactiplantibacillus plantarum*) in order to synthetize GABA. Sourdough fermentation allowed to enhance phenolic compounds and antioxidant activity of bread. From sensorial point of view, an enhanced perception of acidity and acid flavour was observed. Nevertheless, final product had an acceptable taste suggesting that bread enriched with GABA could be a suitable functional food (Coda et al., 2010).

Pseudocereals are also used in gluten-free bread both for their compliance with "gluten-free" statement and their nutritional composition. Gluten-free bread made from pseudocereals flour demonstrated higher protein, fat, fibers and minerals (Alvarez-Jubete et al., 2009), total polyphenols content and increased antioxidant activity (Ibrahim et al., 2015).

9.3 Biscuits

New trends involvethe production of functional biscuits, which if inserted into a healthy diet plan, could help to provide physiological benefits other than purely nutritional effects. In recent years, we witnessed to the changes that many producers made to their product portfolio by the addition of healthy biscuits. On the other hand, also the academic world is working on the development and characterization of functional biscuits realized with varied raw materials and new combinations of ingredients.

Biscuits can be "functional" thanks to the flours used, but also to the addition of other functional ingredients into the recipe both from unconventional food ingredients and from by-products of agro-food industry. One of the simplest approach to realize functional biscuits is the use of flours richer in phytochemical and fiber to totally or partially replace wheat flour. Pasqualone et al. (2015) used purple wheat line CItr 14,629 (*Triticum turgidum* ssp. *durum* (Desf.) Husnot) to make functional biscuits. The quality characteristic of biscuits resulted impaired by the use of purple wheat; nevertheless, purple biscuits had higher values of total anthocyanins, phenolic content, and antioxidant activity. For example, purple biscuits had 13.86 mg/kg Cy-3-Glu which was not detected in conventional biscuits. The sensory profile of conventional and purple biscuits was similar, except for friability and color. Nevertheless, no information was reported about the liking of the products. Inyang et al. (2018) studied biscuits made with acha (*Digitaris exilis* (Kippist) Stapf), a typical West African crop, and kidney bean (*Phaseolus vulgaris* L.). The use of the combination of these flours allowed obtaining biscuits with an increased fiber content and a higher amount of some minerals, mainly calcium and magnesium compared to control biscuit made with wheat flour. Sensory analysis showed that the use of acha and kidney beans did not lower overall acceptability of the biscuits. This result should be confirmed with further test since the panel employed to collect this data was quite small.

Always with a view to fulfil the needs of consumers looking for healthy baked goods together with the requirement of a target group of population with specific dietary needs, gluten-free biscuits with underexploited flours have been developed. Di Cairano, Condelli, Caruso, et al. (2021) formulated gluten-free biscuits with pseudocereal, legume and cereal flour other than rice and maize. The aim of the researchers was the reduction of the glycemic index of the products. Nowadays, there is a growing attention to the glycemic index of cereal-based product since a high glycemic index diet could favor the onset of some metabolic diseases, even if this topic is controversial (Livesey et al., 2019; Vega-López et al., 2018). The estimation of the glycemic index in biscuits made with combinations of buckwheat, sorghum, millet lentil and chickpea flours, showed a reduction of this parameter when experimental samples were compared to a wheat based control and two commercial gluten-free biscuits. All the biscuits had a good acceptance score, except for the formulation containing millet that gave a bitter taste to biscuits, lowering its liking score. The use of more nutrient dense flours, could potentially lead to healthier biscuits, as also observed by Pellegrini and Agostoni (2015), Thejasri et al. (2017) and Molinari et al. (2018).

The flours used in the production of functional biscuits could also be obtained by plant matrices other than grains. For example, Dyshlyuk et al. (2017) evaluated the effect of the addition of pumpkin flour to biscuits on hypocholesterolemic, antioxidant, hepatoprotective and prebiotic properties of laboratory animals. Pumpkin contains a range of compounds such as fiber, antioxidants, fatty acids and vitamins that are able to confer it a wide spectrum of biological activity (Yadav et al., 2010). Milled biscuits were added to animal feed in order to evaluate the effect of biscuits on the health parameters mentioned above. The functional properties were actually proved by the study; indeed after 6 weeks there was a decrease in hypocholesterolemic values in serum of laboratory animals and there was a reduction of pathogenic and growth of lactic- and bifidobacteria in the gastrointestinal tract of the studied groups of animals. Nevertheless, there was any insight on technological or sensory properties of the biscuits.

Banana flour thanks to its composition, richness in resistant starch and minerals (Agama-Acevedo et al., 2012) is gaining more and more attention as functional ingredients in different cereal-based products. Mabogo et al. (2021) used unripe banana flour to partially replace wheat flour in biscuits. Experimental biscuits had less crispy/crunchy texture as compared to control biscuits. In addition, the replacement lead to biscuits with a higher content of phenolic compounds and antioxidant activity. A higher amount of phenolic compounds was also recorded by Mahloko et al. (2019) who used banana flour to replace part of wheat flour in biscuits. In this case, banana flour was obtained by the peel and not from the pulp and it was used alone or in combination with prickly pear (*Opuntia ficus-indica* (L.) Mill) peel flour. They also measured crude fiber content, which increased in biscuits with banana and prickly pear peel flours, alone or in combination. Specifically, the use of banana flour, mainly the one obtained from unripe bananas, is able to enrich the content of resistant starch of the finished product. Resistant starch, is the fraction of the starch that is not digested and act as a fiber at gastrointestinal level (Lockyer & Nugent, 2017). García-Solís et al. (2018) used plantain flour, a type of banana, to reduce starch digestibility and increase fiber in gluten-free cookies. Results showed a higher fiber content and a reduced glycemic index for experimental biscuits. Unfortunately, no indication on the liking of these product based on banana flours product were reported. The use of ingredients rich in resistant starch represent a strategy to obtain biscuits with potential health benefits. In this regard, Cervini et al. (2021) used a novel resistant starch ingredient obtained from annealed white sorghum starch to make biscuits. The highest total dietary fiber and resistant starch content was measured in biscuits with the highest replacement level (45%), and consequently the starch hydrolysis index decreased at increasing replacement levels. The use of resistant starch increased hardness of the biscuits. This result is in contrast with data from Pourmohammadi et al. (2019) and Di Cairano et al. (Di Cairano, Caruso, Galgano, et al., 2021) reporting lower hardness values for biscuits with added resistant starch, but this was probably due to a different measurement method. Cervini et al. (2021) found lower acceptability scores due to the addition of resistant starch, but the values were still above the acceptability threshold. Di Cairano et al. (2022) used a commercial RS2-type of resistant starch to partially replace flour in gluten-free biscuits based on buckwheat, sorghum and lentil flours with a view to reduce their glycemic index. In addition, maltitol and inulin were used as sucrose replacers. In spite of what previously reported, the use of resistant starch did not have any effect on *in vitro* glucose release and consequently on glycemic index, as opposed to using sucrose replacers. Thanks to the very composition of the flours, these biscuits also had a higher protein and fiber content. Moreover, this work actually demonstrated the feasibility of producing potentially functional gluten-free biscuits on large scale, since biscuits were made at an industrial plant. Consumer test showed a general appreciation for the biscuits; anyway, the liking of the product was strictly related to eating habits of the consumers.

Sahin et al. (2021) to compensate quality loss due to addition of fiber rich ingredients adopted an innovative strategy to produce functional biscuits. Indeed, they used the lactic acid bacterium strain *Leuconostoc citreum* TR116 to ferment oat and wheat bran prior to their use in biscuit making. Fermentation increased biscuit spreading, influenced biscuit snap force, enhanced crunchiness and color formation, and lowered the predicted glycemic index. There results were mainly related to the acidification of bran and the higher presence of monosaccharides.

Moringa oleifera is getting high attention as food fortificant (Oyeyinka and Oyeyinka, 2018) thanks to its richness in protein, fiber and mineral. Hedhili et al. (2021) added dried moringa leaf to biscuits, obtaining a higher protein and iron content for supplemented biscuits, but the protein digestion was incomplete. The incomplete digestion of the proteins, together with a lower palatability when high substation levels were employed, suggest adding cautiously moringa to food products. Other authors (Giuberti et al., 2021), used moringa leaf powder to improve nutritional quality of gluten-free biscuits obtaining a higher dietary fiber content, even at low replacement level, and an enhanced protein, resistant starch content, and reduced hydrolysis index.

In recent years, environmental sustainability is under the spotlight, explaining why agro-food by-products are widely being studied to become new food ingredients to valorize food "waste" and minimize environmental impact. This trend involves also research about biscuits. For example, Castaldo et al. (2021) added spent coffee ground to biscuits. Spent coffee grounds are rich in melanoidins, chlorogenic acid and caffeine, which are compounds with good antioxidant activity. It was simulated a gastrointestinal digestion to evaluate the phenolic compound release, results showed that the highest bio accessibility is after the colonic stage, with potential advantages for human health. Vázquez-Sánchez et al. (2018) tried to valorize spent coffee grains by extracting antioxidant dietary fiber from spent coffee grounds and adding it to biscuits. They stated that anti diabetic compounds might be released during the digestion with beneficial effects on the regulation of sugar metabolism of diabetic people. Spent coffee grounds from instant coffee were also used to make biscuits without sucrose (Martinez-Saez et al., 2017). A good number of research report the use of coffee by-products in biscuit recipes; hence, wondering about their safety is legit. A first answer was given by Martinez-Saez et al. (2017) who evaluated the food safety of spent coffee ground through microbiological analyses and measuring the amount of acrylamide and hydroxymethylfurfural, reporting values not causing concerns. Garcia-Serna et al. (2014) did not find any trace of acrylamide in the digests of biscuits enriched with coffee silver skin, suggesting that it is not bio accessible in coffee digests. Nevertheless, they also reported that no chlorogenic acid was found in the digests casting doubts on its bioavailability. In this case, coffee silver skin was added to improve the color of sucrose free biscuits and enrich them with dietary fiber; both the goals were achieved, however no clue on the sensory acceptability of the biscuits is reported in the paper.

Fruit peels are a residual of fruit processing which can reach up to 50% of fresh product; they represent a source of valuable compounds, generally polyphenols and dietary fiber, which can be used to enrich biscuits. For example, pomegranate peel (Colantuono et al., 2016), orange peel (Obafaye & Omoba, 2018), prickly pear peel (Bouazizi et al., 2020), passion fruit peel (Weng et al., 2021), banana, carrot and apple (Rahman et al., 2020), watermelon rinds and orange pomace (Ogo et al., 2021) were all used in biscuits recipes. They were useful to enhance fiber and bioactive compounds content. However, when dealing with the employ of food by-products in baked goods, the intention of consumers to buy them should not be taken for granted. An interesting study by Grasso and Asioli (2020) illustrates consumer preferences for upcycled ingredients. Three groups of consumers were identified: traditionalist, price sensitive and environmentalist. The latter, was the group more interested carbon trust label, and had the strongest preference for biscuits made with upcycled ingredients.

9.4 Baked Snacks

Different baked foods both savory and sweet exist beside bread and biscuits. They range from cakes, to muffins, croissants, crackers, breadsticks etc. All of them are appealing for consumers and eaten quite often even if dietary guidelines suggest to have a moderate consumption of some of them due to their richness in sugars and fats. Since they are widely consumed, they are suitable to be functionalized (Marchetti et al., 2018) also in the view of enhancing their nutritional profile. Many attempts are being made to improve the nutritional value and functionality of snacks by modifying their nutritive composition. Such effects are often achieved by increasing the nutrient density in basic recipes (Ainsworth et al., 2007; Sun-Waterhouse et al., 2010). Additionally, a very important aspect of food functionality is its antioxidant capacity since there is much scientific evidence indicating the important role of food antioxidants in the prevention of different types of cancer and coronary heart diseases (Marlett et al., 2001).

Among snack foods, crackers are a versatile food consumed on a frequent basis due to the appealing taste, long shelf life and relatively low cost. Functional crackers are now gaining more and more popularity (Ahmed & Abozed, 2015).

Polat et al. (2020) produced functional crackers with the enrichment of germinated lentil extract (GLE). Legumes are suitable for producing functional foods due to their rich phenolic content, antioxidant activity and fiber, moreover when germination the nutritional value of legumes results enriched (Polat et al., 2020). In particular, lentils (*Lens culinaris* L.) have the highest phenolic content and antioxidant activity compared to other legume species (Singh et al., 2017). They are rich in proteins, carbohydrates, fibers, minerals, and vitamins (Ajila et al., 2008; Amarowicz et al., 2010). Germination is an economical and effective method to improve the quality of legumes. It has been suggested as an effective way to increase total antioxidant activity. However, the antioxidant properties, which are often related to phenolic content, depend on the type of legume and germination conditions (Dueñas et al., 2015). Polat et al. (2020) showed that crackers could be enriched with GLE for increasing the nutritional value of the product. The GLE addition increased the total protein content (from 1.68% to 2.43%), total phenolic content (from 0.78 to 3.33 mg GAE/g) and antioxidant activity (from 0.34 to 0.84 μmol trolox/mg) of the crackers. Sensory analysis proved that 5%-enriched crackers could be accepted by consumers. Crackers enriched with GLE can be considered as a food with better beneficial and functional properties as compared to conventional crackers, which are poor in phenolic compounds. Hence, they would be appealing functional food thanks to the retention of sensory properties.

Functional crackers were also produced and characterized by Nicole et al. (2021). Crackers were based on fermented soybean (tempeh) paste to which wheat flour and soy protein isolate (SPI) were incorporated to enhance the global quality of the crackers.

Fermented soybean is often regarded as an alternative to meat as it is a complete source of protein and contains vitamin B12 (Erkan et al., 2020). Tempeh have many properties such as antioxidant, antimicrobial, anticancer, antihypertensive, antithrombotic and hypocholesterolaemic effects, which are known to be beneficial to human health (Sanjukta & Rai, 2016). SPI is a food additive derived from separating and extracting dehulled and defatted soybean meal. SPI normally contains around 90% protein and it is regarded as a high-quality protein that is nutritionally balanced and is relatively low in cost. In its application, it can be used alone or in a combination with other protein sources such as wheat proteins. Soy proteins are widely used in the industries due to its hydrating capacity, solubility, colloidal stability, gelation, emulsification, foaming and adhesion properties (Martins & Netto, 2006). However, these properties are strongly influenced by processing methods, treatment parameters and the interactions of soy protein with other food components (Zhao et al., 2020). Nicole et al. (2021) incorporated various proportions of wheat flour and SPI into tempeh paste crackers, the results showed that the addition of wheat and SPI affected the physicochemical properties of crackers. The incorporation of SPI was successful in boosting the protein content of crackers and did not affect the color and the moisture content, together with other textural and physicochemical properties. The findings supported the potential of crackers with tempeh paste to be considered as a good high protein plant-based product.

Ahmed and Abozed (2015) investigated the possibility of improving the quality of wheat flour based cracker by supplementing the basic recipe with different amount of *Hibiscus sabdariffa* calyxes residue (HRS) to enhance their dietary fiber and antioxidant content.

Hibiscus sabdariffa L. is one of the most common flower plants grown worldwide and is used to make jellies, jams and beverages. Recently, it has gained importance as a soft drink material in many parts of the world. It is a good source of phytochemicals and has antioxidant compounds activity (Mahadevan & Kamboj, 2009; Chen et al., 2003). Ahmed and Abozed (2015) showed that *Hibiscus sabdariffa* calyxes remaining after drink preparation are characterized by high dietary fiber content, low fat content and considerable proportion of other biologically active compounds, mainly polyphenols. Crackers prepared with HSR exhibited lower protein, fat content and higher content of dietary fiber compared to commercial crackers. At increasing amounts of HSR weight, height, specific volume, moisture, and pH of crackers decreased. Lower moisture as well as pH favors improved shelf life of the crackers. Phenolic content had also a positive contribution on nutritional excellence of the developed cracker. Sensory ratings for crackers containing 1.25% and 2.50% HSR replacement of wheat flour were positive. Specifically, taste, crispness odor and overall acceptability ratings for these crackers were superior compared to commercial cracker. HSR is a potential functional food ingredient high in fiber content and antioxidants activity that may be processed into flour and used in food applications, such as baked goods. Still in the field of savory bakery goods, recently, Rainero et al. (2022) formulated breadsticks with red grape pomace, a by-product obtained by wine industry, and evaluated their physico-chemical and sensory properties. The addition of grape pomace, reduced hardness and fracturability of the breadsticks, antioxidant activity and total phenolic content increased for increasing replacement level of wheat flour with grape pomace. Sensory profile slightly changed due to the addition of the pomace; however, the acceptability was above the acceptance threshold.

Sweet bakery products represent an important part of baked goods. These products are usually considered poorly healthy and they should be consumed sparingly. This is one of the reasons why enhancing their nutritional quality and conferring functional properties could be relevant. Salehi and Aghajanzadeh (2020) studied the effects of the addition of dried fruit and vegetable powder in cake formulations. These powders are good source of different vitamins, natural colorants, minerals, fibers and antioxidants. Their use affected the appearance, physicochemical, textural and sensorial properties of the cakes. The presence of fiber increased the moisture content of the cakes due to its ability to absorb water. In most cases, replacing wheat flour with fruit and vegetable powders reduced the gluten content of the batter and lead to cake with a lower volume and firmer texture. The color of the cakes resulted darker, and it was found a relationship between color and sensory acceptability.

Dyshlyuk et al. (2017) studied *in vivo*, on lab animals, the hypocholesterolemic, antioxidant, hepatoprotective and prebiotic properties of muffins with pumpkin flour. The results suggested the effectiveness of the addition on the evaluated parameters. However, no indication about sensory properties of the product were reported. Previous research showed the possibility to obtain bread with good organoleptic properties and texture when pumpkin powder was added to wheat dough (de Escalada Pla et al., 2007; Manjula & Suneetha, 2014; Ptitchkina et al., 1998). Marchetti et al. (2018) evaluated the effect of replacing wheat flour by pecan nut expeller meal in a muffin formulation. Nuts are source of unsaturated fatty acids, proteins, fiber and micronutrients; their consumption had been associated with a lower cardiovascular mortality index due to their healthier lipid profile (Marchetti et al., 2018; Yang et al., 2009). Specifically, pecan nuts are rich in fiber, they are a rich source of γ- and poor source of α- tocopherol and contains complex flavonoid substances that are recognized for their effective inhibition of lipid oxidation in foods and possibly in biological systems (Haddad et al., 2006; Kornsteiner et al., 2006; Pinheiro do Prado et al., 2009). Pecan nut expeller meal incorporated into baked products as muffins in replacement of wheat flour could improve their profile and fiber content, it would also increase mineral content and it could be a good industrial and nutritional alternative to wheat flour in bakery products (Marchetti et al., 2018) (Table 9.1).

9.5 Breakfast Cereals

The global breakfast cereal market is principally segmented by type of products. Actually, breakfast cereals falls in two main categories: hot cereals, which could require further heating or cooking before consumption and cold ready-to-eat cereals (Tribelhorn, 1991). The manufacturing of breakfast cereals involves several methods to produce a wide range of products. A general breakfast cereal process include

Table 9.1 Effects of the use of functional ingredients on nutritional and functional properties of bakery products

	Products	Functional ingredients	Main effects on nutritional and functional properties	References
Cereal, pseudocereal and legume flours	Biscuits (gluten free)	Buckwheat Sorghum Lentil Chickpea	↓ predicted glycaemic index	Di Cairano et al. 2021; Di Cairano, Condelli, Caruso, et al., 2021
	Biscuits	Acha (*Digitaris exilis* (Kippist) Stapf), a typical West African crop, and kidney bean (*Phaseolus vulgaris* L.)	↑ fiber ↑ calcium ↑ magnesium	Inyang et al. (2018)
	Biscuits	Purple wheat	↑ phenolic compounds ↑ antochyanins	Pasqualone et al. (2015)
Other vegetable flours	Bread	Malted finger millet Sesame seeds Moringa leaves Cumin seeds	↑ calcium content	Agrahar-Murugkar and Dixit-Bajpai (2020)
	Bread	Red bell pepper	↑ antioxidant activity ↑ mineral ↑ fiber	Kaur et al. (2020)
	Biscuits (gluten free)	Platain flour	↑ resistant starch ↓ starch digestibility	García-Solís et al. (2018)
	Biscuits (gluten free)	Moringa leaf powder	↑ dietary fiber ↑ protein ↓ starch digestibility	Giuberti et al. (2021)
	Biscuits	Dried moringa leaf	↑ protein content ↑ iron content	Hedhili et al. (2021)
	Biscuits	Unripe banana flour	↑ resistant starch ↑vitamin conten	Mabogo et al. (2021)
	Biscuits	Banana flour and prickly pear (Opuntia ficus-indica (L.) Mill) peel flour	↑ resistant starch ↑ antioxidant activity ↑ total phenolic content ↑ flavonoids	Mahloko et al. (2019)
	Biscuits	Pumpkin powder	↑ fiber ↑ antioxidants ↑ fatty acids ↑ vitamins content	Yadav et al. (2010)
	Muffins	Pumpkin flour	Positive effects on the hypocholesterolemic, antioxidant, hepatoprotective and prebiotic properties of lab animals	Dyshlyuk et al. (2017)

(continued)

Table 9.1 (continued)

	Products	Functional ingredients	Main effects on nutritional and functional properties	References
Agro food by-products	Bread	Seaweeds	↑ phenolic compounds ↑ pigments ↑ antioxidant activity	Amoriello et al. (2021)
	Bread	Pomegranate whole fruit bagasse	↑ antioxidative potential, ↑ mineral content, especially copper	Bhol et al., 2016
	Bread	Artichoke stem powder	↑ in vitro bioaccessibility of polyphenols	Colantuono et al. (2018)
	Bread	Brewer spent grains	↑ dietary fiber ↑ protein content ↑ fat content ↑ minerals	Fărcaş et al. (2014)
	Bread	Pea and broad bean pods	↑ dietary fiber	Fendri et al. (2016)
	Bread	Grape pomace	↑ phenols, ↑ antioxidative activity ↑ dietary fiber	Hayta et al. (2014) Walker et al. (2014)
	Bread	Grape seed flour	↑ total phenolic content	Hoye and Ross (2011)
	Wheat bread	Pumpkin by-products	↑ carotenoids	Kampuse et al. (2015)
	Home made bread	Dry spent yeast	↑ beta-glucans	Martins et al. (2015)
	Whole wheat bread	Ripe mango peels	↑ phenols, ↑ antioxidative activity ↑ dietary fiber	Pathak et al. (2016)
	Whole bread	Cupassu peel (*Theobroma grandiflorum*)	↑ dietary fiber	Salgado et al. (2011)
	Bread	Ground flaxseed hulls	↑ total phenolic content ↑ antioxidant activity ↓ protein digestibility – no effect on in vitro starch digestibility	Sęczyk et al. (2017)
	Bread	Tomato seed meal	↑ protein quality and content	Sogi et al. (2002)
	Loaf bread (fat free)	Orange peel fiber (+ α-amylase)	Successfully obtained fat free bread with fiber	Stoll et al. (2015)
	Bread	Barley rootlets	↑ beta-glucans	Waters et al. (2013)
	Steamed bread	Pineapple peels	↑ dietary fiber	Wu and Shiau (2015)
	Biscuits	Spent coffee ground	↑ melanoidins	Castaldo et al. (2021)

(continued)

Table 9.1 (continued)

	Products	Functional ingredients	Main effects on nutritional and functional properties	References
	Biscuits	Various fruit peels (pomegranate, orange, prickly pear, passion fruit, banana, carrot, apple, watermelon rinds, orange pomace)	↑ fiber ↑ bioactive compounds	Obafaye and Omoba (2018), Bouazizi et al. (2020), Weng et al. (2021), Rahman et al. (2020), Ogo et al. (2021)
	Biscuits	Antioxidant dietary fiber from spent coffee grounds	Possible release of anti-diabetic compounds	Vázquez-Sánchez et al. (2018)
	Crackers	Hibiscus sabdariffa calyxes residue	↑ dietary fiber ↑ phenols ↑ antioxidant activity ↓ protein ↓ fats	Ahmed and Abozed (2015)
	Breadsticks	Red grape pomace	↑ phenolic content ↑ antioxidant activity	Rainero et al. (2022)
Other ingredients	Bread	Rice bran	↑ minerals, ↑ B-group vitamins, ↑ dietary fiber	Ameh et al. (2013)
	Bread	Dried olive pomace	↑ phenols	Cecchi et al. (2022)
	Bread	Spearmint aqueous extract	↑ phenols ↑ antioxidant activity	Shori et al. (2021)
	Romanian wheat flour half white bread	Commercial inulin	↑ phenols ↑ antioxidant activity	Sirbu and Arghire (2017)
	Biscuits	Annealed white sorghum starch	↑ dietary fiber ↓ starch hydrolysis	Cervini et al. (2021)
	Biscuits	*Leuconostoc citreum* TR116 to ferment oat and wheat bran prior to their use in biscuit making	↓ predicted glycemic index	Sahin et al. (2021)
	Crackers	Fermented soybean paste Soy protein isolate	↑ protein content	Nicole et al. (2021)
	Crackers	Germinated lentils extract	↑ protein content ↑ total phenolic content ↑ antioxidant activity	Polat et al. (2020)

the following operations: mixing, cooking, forming, drying, equilibration, texturizing, fortification, finish drying/toasting, packaging. Various production methods exist, but the ingredients utilized are similar and consist in grains or flours mixed with other secondary components such us water, sugar, salts, oil, additives (e.g. flavor agents) and fortificants. As noted above, the manufacture of functional food represents a great opportunity to introduce in the consumers' diet additional beneficial functions over than purely nutritional properties. This can be achieved by applying technological or biotechnological media to increase the concentration, remove, add or modify a specific compound and increase its bioavailability, provided that the actual bio functionality of the compound is demonstrated (Roberfroid, 1999). Breakfast cereals could be promising carriers of these benefits since they are used worldwide as staple foods, consumed daily and representing a considerable daily source of nutrients.

In recent years, many attempts have been made to obtain functional breakfasts cereals, taking into account the operating conditions of the production process. The strategies adopted for the functionalization of this category of products are essentially two: the partial substitution of traditional raw materials and the enrichment with bioactive and functional substances. As regard the first strategy, many efforts from the scientific community were made to improve the nutritional quality of breakfast cereals by replacing basic raw materials with others that can provide nutrients which are known to be deficient in traditional ingredients. For example, Ukeyima et al. (2021) replaced traditional flours with maize grits, partially defatted peanut flour and beetroot flour to produce functional flakes. The results showed an increase in vitamin content and *in vitro*protein digestibility of the functional flakes over control ones, as well as a favorable sensory evaluation. In a similar study, Okafor and Usman (2014) evaluated the physical and functional properties of breakfast cereals made with maize, African yam bean (*Sphenostylis stenocarpa*), defatted coconut cake and sorghum malt extract. It was obtained an increase of *in vitro* protein digestibility given specifically by the addition of defatted coconut fiber in the formulations. Lemmens et al. (2021) characterized breakfast flakes made with sprouted wheat. Sprouted wheat was used with the aim of increasing the mineral bio accessibility and providing intrinsic sweetness The resulting flakes had a darker color and a higher density, but their textural properties had a good overall acceptance. Oliveira et al. (2018), tested the replacement of traditional flours in the production of extruded breakfast cereals with whole grain wheat flour and jabuticaba (*Myrciaria cauliflora*) peel, obtaining an hardness and crispness decrease for all the experimental formulations evaluated after soaking in milk, but also acceptable results in regards of their general crispness. Color parameters, density and hardness were the parameters able to discriminate experimental samples from control sample.

Lots of research was also made based on the assumption that improvements of the nutritional and technological properties of breakfast cereals could occur by adding ingredients with proven functional properties (e.g. dietary fibers, proteins, antioxidants). In this regards, Ferreira et al. (2021) reported the incorporation of inulin, a dietary fiber, in order to improve the technological and nutritional properties of breakfast cereals made with corn grits. Different operating conditions were tested (e.g. the moisture of the 'corn grits and inulin' mixture, the amount of inulin added) and the results demonstrated that the addition of inulin did not damage the physical properties of breakfast cereals, such as density, expansion ratio, instrumental force and color. In addition, inulin did not affect the overall acceptance of products, and had a positive impact on them. Extrudates with this fiber presented also a high fiber level and moderate glycemic load, in contrast to the control products that presented low fiber level and high glycemic load.

However, the enrichment of breakfast cereals represents a challenge of considerable difficulty when taking into account the operations needed for their production. Actually, the basic problem related to the addition of bioactive compounds in processed foods is the sensitivity of the compound to changes in different parameters such as pH, temperature, light exposure, oxygen and mechanical stress during processing. One of the unit operations that affects most breakfast cereals – especially ready to eat cereals – is the extrusion. The high pressure/temperature conditions of extrusion can strongly damage a large variety of bioactive compounds, such as vitamins, carotenoids, polyphenols, and prebiotics. There are several approaches explored in literature to overcome this issue. Specifically, in regards to the incorporation of carotenoids, taking into account their great solubilization in non-polar solvents, they can be included into organic solutions in order to increase their bioavailability. In addition, this type of protection system also results to have a greater stability to lipid oxidation and the degradation of carotenoids by avoiding the formation of radical species that can react with carotenoids. The dissolution in oil and the addition in the final steps of extrusion is the strategy put in place by Emin et al. (2012) to reduce the loss of β-carotene during the operation. They obtained an increase of the retention of approximately 40%. Other studies demonstrated the effectiveness of heated protein-carbohydrate system, in which carotenoids were mixed before the addition to the process, to guarantee their stability in extruded foods (Ying et al., 2015). The presence of antioxidant substances, such as polyphenols, flavonoids, tocopherols and vitamins (C and E) can also improve the stability of carotenoids during extrusion. This method was evaluated with optimal responses in the retention of carotenoids by Kolniak-Ostek et al. (2017), who formulated extruded corn products made with flour with a high phenolic content (e.g., pumpkin tissue, amaranth seeds). Good responses were also obtained by Obradović et al. (2015) who added ascorbic acid to improve the retention of carotenoids present in pumpkin powder incorporated in an extrusion corn-based formulation. Moreover, microencapsulation, a proper protection technique, has been also reported to improve carotenoid retention during the extrusion operation. The covering of the bioactive compound by layers of protective materials, generally biopolymers, represents the most valuable strategy to obtain the preservation of the molecule itself and its bio-functionality (Tachaprutinun et al., 2009). Favaro-Trindade et al. (2020), tested microencapsulation by complex coacervation as a tool to produce extruded cereals functionalized with proanthocyanidin-rich cinnamon extract, obtaining products with greater protection of the bioactive compound during processing and storage.

Some papers reported effective attempts of functionalization that combines the replacement of traditional flours and the enrichment with nutraceutical ingredients for the production of functional breakfast cereals. For example, Emelike et al. (2020) evaluated the nutritional composition, the functional and organoleptic properties of breakfast cereals processed using as raw materials acha, wheat, cashew (*Anacardium occidentale*) kernel and prawn. Acha grains and cashew kernels were processed into flour and prawn was processed into powder. Acha flour was used as substitute of wheat flour whereas a constant percentage of cashew kernel flour and prawn was used as an enrichment in formulations tasted. The produced breakfast cereals resulted to be rich in protein, fat, carbohydrates and ash contents demonstrating an appropriate combination of raw materials. In addition, the observed decrease in moisture content at increasing replacement levels of acha flour suggests another favorable characteristics of this functional product, that is an extended shelf-life, if appropriate embalming is carried out. In another study, carried out by Asmoro and co-workers (Asmoro et al., 2021), the formulation of breakfast cereals was tested using Mocaf cassava flour as a gluten-free substitute of wheat flour and moringa leaves flour as a functional ingredient to add nutritional value to the product. The resulting products showed improved nutritional profiles especially in terms of ash content, but also an increase of fat content – up to 8.2% – a factor that actually does not meet the demand for healthier products.

9.6 Pasta

Improvements of pasta nutritional quality may occur by incorporating ingredients that promote increased protein content and functional properties, such as adding ingredients rich in fiber and/or with high antioxidant activity (Kaur et al., 2021; De Pasquale et al., 2021). Unsurprisingly, numerous efforts to increase the nutritional value of pasta have already been made. Detailed research using different cereals (Montalbano et al., 2016), legumes (Petitot et al., 2010), pseudocereal flours (Lorusso et al., 2017), as well as fish, seafood, and meat products (Liu et al., 2016; Calanche et al., 2019), were published. It is also interesting the possibility to re–use typical agro–industrial by–products such as brewers' spent grain, grape pomace, celery root sugar beet, okara, olive pomace, and tomato skin, as recently reviewed by Bianchi et al. (2021), and the chance to increase the nutritional value of pasta by using edible insects (Çabuk & Yılmaz, 2020). It should also be underlined that the use of flours different from wheat is considered an attractive strategy by the food industry to meet the demand for new products coming from the health–conscious consumers who pay greater attention to the management of available resources (De Pasquale et al., 2021). Detailed, in some pasta products, the replacement of refined durum wheat with other cereals, pseudocereals, or wholegrain cereals has been evaluated. Through this expedient, ingredients rich in phytochemicals and other nutrients are added to pasta (Wang et al., 2021). Carcea et al. (2017) reported that pasta made of pseudocereal species such as emmer, oats, and buckwheat had higher protein, fiber, and polyphenols than cereal products (even whole meal cereal products). Similarly, Schoenlechner et al. (2010) reported a higher concentration of total folate in gluten–free pasta produced from pseudocereals such as amaranth, quinoa, and buckwheat. Different are the health benefits delivered using pseudocereal flour in pasta production. For example, the addition of buckwheat in pasta resulted in a more balanced amino acid content, higher phenolic compounds content, and antioxidant capacity than control pasta. However, the overall quality of cooked spaghetti, evaluated as swelling index and cooking loss, was reduced by the introduction of buckwheat due to a not well–formed gluten network resulting in the leaching out of starch granules during cooking leading to a sticky product (Biney & Beta, 2014). Interesting are the results of a clinical study conducted by Khan et al. (2015) who explored the acute effect of pasta produced with 30% red or white sorghum flour in healthy subjects. The consumption of the experimental pasta samples determined the increase in plasma polyphenols, antioxidant capacity and superoxide dismutase activity with a reduction of the protein carbonyl level was significantly lower in the subject compared to 100% semolina pasta.

Due to the higher amounts of dietary fiber and bioactive micronutrients compared to the corresponding refined cereal, wholegrain cereals are usually used in pasta production (Kristensen et al., 2010). Baiano et al. (2006) used whole durum wheat to replace refined durum flours, found a significant increase in total dietary fiber content. However, as West et al. (2013) reported, the use of wholegrain wheat for pasta production, apart from providing higher fiber, minerals, and antioxidant capacity, could determine changes in the sensory profile. Whole grain pasta tends to be darker in color, rougher, heavier in texture, and prone to developing off–flavors over time. For this purpose, researchers have used extracted soluble and insoluble dietary fiber to create functional pasta (Tudorica et al., 2002). Results show that both the type of dietary fiber influenced the quality of uncooked and cooked pasta in terms of physical, chemical, textural, and nutritional properties. However, it should be noted that the addition of soluble dietary fiber significantly reduced the *in vitro* glucose release from pasta. However, different fiber types may specifically affect the pasta structure and nutritional value. From research conducted by Bustos et al. (2011) emerged that the inclusion of inulin, a soluble fiber, in pasta should be avoided since, because of its cooking loss, it negatively influenced the technological quality of pasta during the boiling process. Instead, the addition of oat bran had a positive effect on the cooking properties of the final products up to 5.0% of substitution and, in the same way, the inclusion of resistant starch resulted in pasta with improved final quality, mild cooking losses, and increased hardness. Thus, the results suggested that using insoluble fiber, the nutritional quality of pasta could be enhanced without negatively affect its sensory properties. Interestingly, prebiotic and probiotic substances may also be introduced in pasta. In this regard, Fares et al. (2015) produced a functional pasta using a durum wheat flour rich in polyphenols, with added barley beta–glucan and *Bacillus coagulans* GBI–30, 6086. The obtained product had a glycemic index, measured in healthy volunteers, of 59.7 and sufficient probiotic strain in cooked pasta (about 9.0 log CFU/100 g). Recently, in a similar product was assessed the effect of a consumption of the synbiotic functional pasta on 41 healthy overweight or obese volunteers during a 12–week single–blind, parallel, randomized, placebo–controlled dietary intervention study. The research demonstrated how the daily consumption of symbiotic pasta could be a valuable strategy to obtain positive effects on the athero–protective function of serum HDL (Favari et al., 2020).

Different chemical compositions and technological properties characterize legumes compared to wheat. They are an excellent source of proteins, with a high amount of essential amino acids, and they contain carbohydrates and appropriate levels of vitamins and minerals (Foschia et al., 2017). Many legumes also contain dietary fiber and resistant starch, which reduces the glycemic response and provides health benefits (Punia, Dhull et al., 2019; Punia, Siroha et al., 2019; Punia, Sandhu, & Kaur, 2020; Punia, Dhull, et al., 2020). Since legumes are a better source of essential amino acids, particularly lysine, the enrichment of cereal flour with pulses can improve the nutritional quality of the resulting product, increasing the protein content and contributing to a better balance in the protein profile (Bouchenak & Lamri-Senhadji, 2013). Food enrichment with pulses, flours, or their proteins has beneficial effects on health, well–being, and potential disease prevention. Petitot et al. (2010) prepared nutritionally enhanced semolina spaghetti enriching flour with high amounts (35%) of legume flour (split pea or fava bean) comparing them with standard semolina spaghetti. Similarly, Laleg et al. (2017) prepared functional pasta partially or completely replacing wheat flour with fava flour to increase the protein content and improve its quality. It was found that the protein content of pasta increased significantly with the increase in fava flour enrichment and, at the same time, a linear increase in the cooking loss was reported. Moreover, enriched pasta is generally high in polyphenol content and antioxidant capacity. Sęczyk et al. (2016) reported that carob flour at 5% significantly increased total phenolic compound and antioxidant activity of the pasta. Unfortunately, the researchers did not report the pasta glycemic index reduction generally observed due to legume flour inclusion. About that, Turco et al. (2019) correlated the low glycemic index observed in pasta produced with different inclusion levels of pea flour, red lentil flour, grass pea and chickpea flour firstly to the higher pasta's protein content and secondarily to the starch composition, fiber and polyphenols content. Goñi and Valentín-Gamazo (2003) confirmed such results also in an in vivo study. It should be underlined that the use of flours different from wheat is considered an attractive strategy by the food industry to meet the demand for new products coming from the health–conscious consumers who pay greater attention to the management of available resources (De Pasquale et al., 2021).

It must be reported that legumes also contain species–specific antinutritional factors such as phytic acid, condensed tannins, and trypsin inhibitors, which can interfere with protein or mineral absorption. Several treatments such as microwave, extrusion, steaming, boiling, and roasting have been proposed to reduce the legumes antinutritional factor. Among the others, selected lactic acid bacteria fermentation was reported to be more efficient. In a recent study, De Pasquale et al. (2021) concluded that fermentation with selected starters might be used to overcome some of the legume flour drawbacks, positively affecting protein digestibility, glycemic index, antinutritional factor degradation, and antioxidant activity. Another major source of protein is fish. Specifically, fish is an excellent source of high–value protein, essential amino acids, and unsaturated fatty acids, especially eicosapentaenoic acid (EPA) and docosahexaenoic acid (DHA). In addition, fish contains micronutrients such as vitamins (A, D, B6, and B12) and many minerals (iron, zinc, iodine, selenium, potassium, and sodium). Because of this, there are different studies dedicated to increasing the nutritional value of pasta products by adding fish and seafood protein concentrate. Among the others, Desai et al. (2018a) evaluated the effect of semolina replacement with protein powder from fish (*Pseudophycis bachus*) on pasta nutraceutical, nutritional, and physicochemical characteristics. The authors reported that supplementation with 5–20% of fish powder was an effective way to increase the essential amino acid and protein content as well as the phenolic content and the antioxidant activity. In addition, a reduction of the glucose release of enriched pasta was reported. The amount of rapidly digestible starch in pasta supplemented with fish powder was lower than the control. Probably the fish powder may have created a protein network around the starch granules reducing the accessibility of α–amylase to starch and affecting, in this way, the ability of the enzyme to hydrolyze the starch into glucose. The same authors also reported a decrease in the n6:n3 ratio from 19:1 (the control pasta) to 3:1 when the 20% of salmon fish powder was incorporated in pasta (Desai et al., 2018b). Moreover, Ramya et al. (2015) evaluated the influence of freeze-dried shrimp (*Penaeus monodon*) meat at different levels (2.5%, 5%, and 10%; w/w) in pasta processing and Goes et al. (2016) produced noodles with tilapia (*Oreochromis niloticus*). Prabhasankar et al. (2009) produced a functional pasta with different levels of wakame (*Undaria pinnatifida*) seaweed. Different levels of the Japanese seaweed (0–30%; w/w) were substituted to wheat flour, and a significant improvement in the protein, fat, ash, and fiber content in fortified pasta samples were observed. From a sensorial point of view, pasta with 10% of seaweed had a higher acceptance rating by the panelists. Increasing the powder concentration up to the 20%, deep modification in the sensory profile was observed in saltiness and aroma.

Due to the high biological value of protein and the concentration of some micronutrients such as iron, selenium, vitamins A, B12, and folic acid that are needed for good health, the feasibility to use a beef emulsion at three different levels of 15, 30 and 45% (w/w) to develop a pasta with enhanced nutritional profile has been evaluated (Liu et al., 2016). Pasta with added meat had a lower glycemic index compared to the control and enhanced amino acid profile. Specifically, in functional pasta, five essential amino acids (leucine, lysine, methionine, threonine, tryptophan) increased significantly with increasing meat addition.

More and more attention is arousing the possibility of increasing the nutritional quality of pasta by using edible insect flour. In fact, because of the significant content of protein, vitamins, and minerals, the Food and Agriculture Organization of the United Nations (FAO) recommends insect flour consumption. Crickets, silkworms, mealy larvae, and ants are some examples. The incorporation of 5% cricket powder flour into pasta was recently assessed. The authors demonstrated how the introduction of insect proteins into pasta significantly influenced cooking and technological properties. Overall, pasta sensory evaluation showed that the product fortified with cricket flour satisfied consumer expectations, exhibiting no significant difference compared to control–wheat pasta (Duda et al., 2019). Instead, Biró et al. (2019) evaluated silkworm–enriched buckwheat pasta technological and sensory properties. Their results revealed that silkworm powder increased buckwheat pasta nutritional value and that the 5% and 10% enriched pasta had a higher protein content, with reduced energy value. According to technological analyses, silkworm powder reduces optimum cooking time while increasing pasta acidity over storage. The consumer sensory analysis of pasta determined that 10% enriched pasta with silkworm received the highest overall acceptance, while the lowest values were assigned to the control product. According to the reported results, it is possible to define insect powder as a suitable material for enriching pasta producing a healthy alternative (Table 9.2).

Table 9.2 Effects of the use of functional ingredients on nutritional and functional properties of pasta

	Products	Functional ingredients	Main effects on nutritional properties	References
Cereal, pseudocereal and legume flous	Pasta	Emmer Oats Buckwheat	↑ protein, ↑ fiber ↑ polyphenols	Carcea et al. (2017)
	Pasta	Wheat flour rich in polyphenols Beta-glucans *Bacillus coagulans* GBI–30, 6086	Glycemic index: 59.7, probiotic pasta	Fares et al. (2015)
	Pasta	Red and white sorghum	↑ plasma polyphenols, ↑ antioxidant capacity, ↑ superoxide dismutase activity	Khan et al. (2015)
	Pasta	Fava bean	↑ in vitro protein digestion	Laleg et al. (2017)
	Pasta	Barley	↓ starch hydrolysis ↑ antioxidant activity ↑ beta-glucans	Montalbano et al. (2016)
	Pasta	Split pea and fava bean	↑ polyphenol content, ↑ antioxidant capacity	Petitot et al. (2010)
	Pasta (gluten free)	Amaranth Quinoa Buckwheat	↑ folate content	Schoenlechner et al. (2010)
	Pasta	Pea Red lentil Grass pea Chickpea	↓ glycemic index	Turco et al. (2019)
Other vegetable flours	Pasta	Carob	↑ phenolic compounds ↑ antioxidant activity	Sęczyk et al. (2016)

(continued)

Table 9.2 (continued)

	Products	Functional ingredients	Main effects on nutritional properties	References
Other ingredients	Bukwheat pasta	Silkworm powder	↑ protein content	Biró et al. (2019)
	Pasta	Lactic acid bacteria fermentation	Overcoming of some drawbacks related to the use of legume flours	De Pasquale et al. (2021)
	Pasta	Protein powder from fish (*Pseudophycis bachus*)	↑ essential amino acids, ↑ protein content, ↓ glucose release	Desai et al. (2018a)
	Pasta	Salmon fish powder	↓ n6: n3 ratio	Desai et al. (2018b)
	Noodles	Tilapia (*Oreochromis niloticus*) protein concentrate	↑ protein ↑ ash ↑ fat ↑minerals	Goes et al. (2016)
	Pasta	Beef emulsion	↓ glycemic index ↑ amino acids profile	Liu et al. (2016)
	Pasta	Freeze dried shrimp meat	↑ protein content ↓ in vitro protein digestibility ↓ glycemic index	Ramya et al. (2015)

9.7 Conclusion

Functionalization of baked goods, breakfast cereals and pasta is a field of promising developments and challenges. The use of new ingredients alongside with the ones generally contemplated in their recipes could contribute to the obtaining of new functional products. These products functionalized with high added value ingredients could be an excellent carrier of nutrients with beneficial effects on human health thanks to their wide consumption.

Due to the variety of ingredients that can be used it is not easy to always maintain or predict the quality characteristic of the products functionalized with new ingredients. The selection of the raw materials and their amount should be done carefully; it should be always kept in mind that the sensory properties could be impaired. Generally, lower replacement level of wheat flour help to reach a good compromise between liking and functional properties. In light of this, it is crucial to understand the effect of the ingredients on texture and sensorial properties. In principle, the addition of high added value ingredients is promising but it should not be neglected that the bio functionality of the incorporated molecules should be better tested prior to stating that a product is functional. This should be a common goal of all future works dealing with formulation of novel functional foods.

References

Abdel-Aal, E. S., & Hucl, P. (2002). Amino acid composition and in vitro protein digestibility of selected ancient wheats and their end products. *Journal of Food Composition and Analysis, 15*(6), 737–747.

Agama-Acevedo, E., Islas-Hernández, J. J., Pacheco-Vargas, G., Osorio-Díaz, P., & Bello-Pérez, L. A. (2012). Starch digestibility and glycemic index of cookies partially substituted with unripe banana flour. *LWT-Food Science and Technology, 46*(1), 177–182.

Agrahar-Murugkar, D., & Dixit-Bajpai, P. (2020). Physicochemical, textural, color, nutritional, scanning electron microscopy and sensorial characterization of calcium-rich breads fortified with sesame, malted finger millet, cumin and moringa leaves. *Nutrition & Food Science, 50*(1), 47–60.

Ahmed, Z. S., & Abozed, S. S. (2015). Functional and antioxidant properties of novel snack crackers incorporated with Hibiscus sabdariffa by-product. *Journal of Advanced Research, 6*(1), 79–87.

Ainsworth, P., İbanoğlu, Ş., Plunkett, A., İbanoğlu, E., & Stojceska, V. (2007). Effect of brewers spent grain addition and screw speed on the selected physical and nutritional properties of an extruded snack. *Journal of Food Engineering, 81*(4), 702–709.

Ajila, C. M., Leelavathi, K. U. J. S., & Rao, U. P. (2008). Improvement of dietary fiber content and antioxidant properties in soft dough biscuits with the incorporation of mango peel powder. *Journal of Cereal Science, 48*(2), 319–326.

Alvarez-Jubete, L., Arendt, E. K., & Gallagher, E. (2009). Nutritive value and chemical composition of pseudocereals as gluten-free ingredients. *International Journal of Food Sciences and Nutrition, 60*(sup4), 240–257.

Amarowicz, R., Estrella, I., Hernández, T., Robredo, S., Troszyńska, A., Kosińska, A., & Pegg, R. B. (2010). Free radical-scavenging capacity, antioxidant activity, and phenolic composition of green lentil (Lens culinaris). *Food Chemistry, 121*(3), 705–711.

Ameh, M. O., Gernah, D. I., & Igbabul, B. D. (2013). Physico-chemical and sensory evaluation of wheat bread supplemented with stabilized undefatted rice bran. *Food and Nutrition Sciences, 4*(09), 43.

Amoriello, T., Mellara, F., Galli, V., Amoriello, M., & Ciccoritti, R. (2020). Technological properties and consumer acceptability of bakery products enriched with brewers' spent grains. *Foods, 9*(10), 1492.

Amoriello, T., Mellara, F., Amoriello, M., Ceccarelli, D., & Ciccoritti, R. (2021). Powdered seaweeds as a valuable ingredient for functional breads. *European Food Research and Technology, 247*(10), 2431–2443.

Angelino, D., Cossu, M., Marti, A., Zanoletti, M., Chiavaroli, L., Brighenti, F., … Martini, D. (2017). Bioaccessibility and bioavailability of phenolic compounds in bread: A review. *Food & Function, 8*(7), 2368–2393.

Asmoro, W. N., Tari, I. N. A., Widyastuti, R., & Kurnia, S. C. (2021). Addition of moringa oleifera lam. Leaves flour for increasing the nutritional value of modified cassava flour–based breakfast cereal. *E3S Web of Conferences, 226*(2), 00033.

Baiano, A., Conte, A., & Del Nobile, M. A. (2006). Influence of drying temperature on the spaghetti cooking quality. *Journal of Food Engineering, 76*(3), 341–347.

Batifoulier, F., Verny, M. A., Chanliaud, E., Rémésy, C., & Demigné, C. (2005). Effect of different breadmaking methods on thiamine, riboflavin and pyridoxine contents of wheat bread. *Journal of Cereal Science, 42*(1), 101–108.

Betoret, E., & Rosell, C. M. (2020). Enrichment of bread with fruits and vegetables: Trends and strategies to increase functionality. *Cereal Chemistry, 97*(1), 9–19.

Bhol, S., Lanka, D., & Bosco, S. J. D. (2016). Quality characteristics and antioxidant properties of breads incorporated with pomegranate whole fruit bagasse. *Journal of Food Science and Technology, 53*(3), 1717–1721.

Bianchi, F., Tolve, R., Rainero, G., Bordiga, M., Brennan, C. S., & Simonato, B. (2021). Technological, nutritional and sensory properties of pasta fortified with agro-industrial by-products: A review. *International Journal of Food Science & Technology, 56*(9), 4356–4366.

Biney, K., & Beta, T. (2014). Phenolic profile and carbohydrate digestibility of durum spaghetti enriched with buckwheat flour and bran. *LWT-Food Science and Technology, 57*(2), 569–579.

Biró, B., Fodor, R., Szedljak, I., Pásztor-Huszár, K., & Gere, A. (2019). Buckwheat-pasta enriched with silkworm powder: Technological analysis and sensory evaluation. *LWT-Food Science and Technology, 116*, 108542.

Bouazizi, S., Montevecchi, G., Antonelli, A., & Hamdi, M. (2020). Effects of prickly pear (Opuntia ficus-indica L.) peel flour as an innovative ingredient in biscuits formulation. *LWT-Food Science and Technology, 124*, 109155.

Bouchenak, M., & Lamri-Senhadji, M. (2013). Nutritional quality of legumes, and their role in cardiometabolic risk prevention: A review. *Journal of Medicinal Food, 16*(3), 185–198.

Bustos, M. C., Pérez, G. T., & León, A. E. (2011). Effect of four types of dietary fiber on the technological quality of pasta. *Food Science and Technology International, 17*(3), 213–221.

Çabuk, B., & Yılmaz, B. (2020). Fortification of traditional egg pasta (erişte) with edible insects: Nutritional quality, cooking properties and sensory characteristics evaluation. *Journal of Food Science and Technology, 57*(7), 2750–2757.

Calanche, J., Beltrán, H., Marquina, P., Roncalés, P., & Beltrán, J. A. (2019). Eating fish in another way: Development of functional pasta with added concentrates of farmed sea bass (Dicentrarchus labrax). *Cereal Chemistry, 96*(5), 856–865.

Caldwell, E. F., Kadan, R. S., Wrigley, C., Corke, H., & Walker, C. E. (2004). Breakfast cereals. *Encyclopedia of Grain Science, 1*, 201–206.

Carcea, M., Narducci, V., Turfani, V., & Giannini, V. (2017). Polyphenols in raw and cooked cereals/pseudocereals/legume pasta and couscous. *Foods, 6*(9), 80.

Castaldo, L., Lombardi, S., Gaspari, A., Rubino, M., Izzo, L., Narváez, A., … Grosso, M. (2021). In vitro bioaccessibility and antioxidant activity of polyphenolic compounds from spent coffee grounds-enriched cookies. *Foods, 10*(8), 1837.

Cecchi, L., Migliorini, M., Giambanelli, E., Canuti, V., Bellumori, M., Mulinacci, N., & Zanoni, B. (2022). Exploitation of virgin olive oil by-products (Olea europaea L.): phenolic and volatile compounds transformations phenomena in fresh two-phase olive pomace ('alperujo') under different storage conditions. *Journal of the Science of Food and Agriculture, 102*(6), 2515–2525.

Cervini, M., Frustace, A., Garrido, G. D., Rocchetti, G., & Giuberti, G. (2021). Nutritional, physical and sensory characteristics of gluten-free biscuits incorporated with a novel resistant starch ingredient. *Heliyon, 7*(3), e06562.

Chen, C. C., Hsu, J. D., Wang, S. F., Chiang, H. C., Yang, M. Y., Kao, E. S., … Wang, C. J. (2003). Hibiscus sabdariffa extract inhibits the development of atherosclerosis in cholesterol-fed rabbits. *Journal of Agricultural and Food Chemistry, 51*(18), 5472–5477.

Ciudad-Mulero, M., Barros, L., Fernandes, Â., Ferreira, I. C. F. R., Callejo, M., Matallana-González, M., Fernández-Ruiz, V., … Carrillo, J. M. (2020). Potential health claims of durum and bread wheat flours as functional ingredients. *Nutrients, 12*(2), 504.

Coda, R., Rizzello, C. G., & Gobbetti, M. (2010). Use of sourdough fermentation and pseudo-cereals and leguminous flours for the making of a functional bread enriched of γ-aminobutyric acid (GABA). *International Journal of Food Microbiology, 137*(2–3), 236–245.

Colantuono, A., Ferracane, R., & Vitaglione, P. (2016). In vitro bioaccessibility and functional properties of polyphenols from pomegranate peels and pomegranate peels-enriched cookies. *Food & Function, 7*(10), 4247–4258.

Colantuono, A., Ferracane, R., & Vitaglione, P. (2018). Potential bioaccessibility and functionality of polyphenols and cynaropicrin from breads enriched with artichoke stem. *Food Chemistry, 245*, 838–844.

de Escalada Pla, M. F., Ponce, N. M., Stortz, C. A., Gerschenson, L. N., & Rojas, A. M. (2007). Composition and functional properties of enriched fiber products obtained from pump-

kin (Cucurbita moschata Duchesne ex Poiret). *LWT-Food Science and Technology, 40*(7), 1176–1185.
de Pinho Ferreira Guine, R., & dos Reis Correia, P. M. (Eds.). (2013). *Engineering aspects of cereal and cereal-based products*. CRC Press.
De Pasquale, I., Verni, M., Verardo, V., Gómez-Caravaca, A. M., & Rizzello, C. G. (2021). Nutritional and functional advantages of the use of fermented black chickpea flour for semolina-pasta fortification. *Foods, 10*(1), 182.
Desai, A. S., Brennan, M. A., & Brennan, C. S. (2018a). Effect of fortification with fish (Pseudophycis bachus) powder on nutritional quality of durum wheat pasta. *Foods, 7*(4), 62.
Desai, A. S., Brennan, M. A., & Brennan, C. S. (2018b). Amino acid and fatty acid profile and digestible indispensable amino acid score of pasta fortified with salmon (Oncorhynchus tshawytscha) powder. *European Food Research and Technology, 244*(10), 1729–1739.
Di Cairano, M., Caruso, M. C., Galgano, F., Favati, F., Ekere, N., & Tchuenbou-Magaia, F. (2021). Effect of sucrose replacement and resistant starch addition on textural properties of gluten-free doughs and biscuits. *European Food Research and Technology, 247*(3), 707–718.
Di Cairano, M., Condelli, N., Caruso, M. C., Cela, N., Tolve, R., & Galgano, F. (2021). Use of under-exploited flours for the reduction of glycaemic index of gluten-free biscuits: Physicochemical and sensory characterization. *Food and Bioprocess Technology, 14*, 1490–1502.
Di Cairano, M., Condelli, N., Cela, N., Sportiello, L., Caruso, M. C., & Galgano, F. (2022). Formulation of gluten-free biscuits with reduced glycaemic index: Focus on in vitro glucose release, physical and sensory properties. *LWT, 154*(112654), 1–9.
do Prado, A. C. P., Aragão, A. M., Fett, R., & Block, J. M. (2009). Antioxidant properties of Pecan nut [Carya illinoinensis (Wangenh.) C. Koch] shell infusion. *Grasas y Aceites, 60*(4), 330–335.
Duda, A., Adamczak, J., Chełmińska, P., Juszkiewicz, J., & Kowalczewski, P. (2019). Quality and nutritional/textural properties of durum wheat pasta enriched with cricket powder. *Foods, 8*(2), 46.
Dueñas, M., Martínez-Villaluenga, C., Limón, R. I., Peñas, E., & Frias, J. (2015). Effect of germination and elicitation on phenolic composition and bioactivity of kidney beans. *Food Research International, 70*, 55–63.
Dyshlyuk, L., Babich, O., Prosekov, A., Ivanova, S., Pavsky, V., & Yang, Y. (2017). In vivo study of medical and biological properties of functional bakery products with the addition of pumpkin flour. *Bioactive Carbohydrates and Dietary Fibre, 12*, 20–24.
Emelike, N. J. T., Achinewhu, S. C., & Ebere, C. O. (2020). Nutritional composition, functional and organoleptic properties of breakfast cereals formulated from acha, wheat, cashew kernel and prawn. *European Journal of Agriculture and Food Sciences, 2*(5), 1-6.
Emin, M. A., Mayer-Miebach, E., & Schuchmann, H. P. (2012). Retention of β-carotene as a model substance for lipophilic phytochemicals during extrusion cooking. *LWT-Food Science and Technology, 48*(2), 302–307.
Erkan, S. B., Gürler, H. N., Bilgin, D. G., Germec, M., & Turhan, I. (2020). Production and characterization of tempehs from different sources of legume by Rhizopus oligosporus. *LWT, 119*, 108880.
Eshak, N. S. (2016). Sensory evaluation and nutritional value of balady flat bread supplemented with banana peels as a natural source of dietary fiber. *Annals of Agricultural Sciences, 61*(2), 229–235.
European Commission. (2012). Regulation (EU) no. 432/2012 establishing a list of permitted health claims. *European Food and Feed Law Review, 7*, 162.
Fărcaş, A. C., Socaci, S. A., Tofană, M., Mureşan, C., Mudura, E., Salanţă, L., & Scrob, S. (2014). Nutritional properties and volatile profile of brewer's spent grain supplemented bread. *Romanian Biotechnological Letters, 19*(5), 9705–9714.
Fares, C., Menga, V., Martina, A., Pellegrini, N., Scazzina, F., & Torriani, S. (2015). Nutritional profile and cooking quality of a new functional pasta naturally enriched in phenolic acids, added with β-glucan and Bacillus coagulans GBI-30, 6086. *Journal of Cereal Science, 65*, 260–266.

Favari, E., Angelino, D., Cipollari, E., Adorni, M. P., Zimetti, F., Bernini, F., ... Pellegrini, N. (2020). Functional pasta consumption in healthy volunteers modulates ABCG1-mediated cholesterol efflux capacity of HDL. *Nutrition, Metabolism and Cardiovascular Diseases, 30*(10), 1768–1776.

Favaro-Trindade, C. S., Patel, B., Silva, M. P., Comunian, T. A., Federici, E., Jones, O. G., & Campanella, O. H. (2020). Microencapsulation as a tool to producing an extruded functional food. *LWT, 128*, 109433.

Fendri, L. B., Chaari, F., Maaloul, M., Kallel, F., Abdelkafi, L., Chaabouni, S. E., & Ghribi-Aydi, D. (2016). Wheat bread enrichment by pea and broad bean pods fibers: Effect on dough rheology and bread quality. *LWT, 73*, 584–591.

Ferreira, S. M., Capriles, V. D., & Conti-Silva, A. C. (2021). Inulin as an ingredient for improvement of glycemic response and sensory acceptance of breakfast cereals. *Food Hydrocolloids, 114*, 106582.

Foschia, M., Horstmann, S. W., Arendt, E. K., & Zannini, E. (2017). Legumes as functional ingredients in gluten-free bakery and pasta products. *Annual Review of Food Science and Technology, 8*, 75–96.

Garcia-Serna, E., Martinez-Saez, N., Mesias, M., Morales, F. J., & del Castillo, M. D. (2014). Use of coffee silverskin and stevia to improve the formulation of biscuits. *Polish Journal of Food and Nutrition Sciences, 64*(4), 243-251.

García-Solís, S. E., Bello-Pérez, L. A., Agama-Acevedo, E., & Flores-Silva, P. C. (2018). Plantain flour: A potential nutraceutical ingredient to increase fiber and reduce starch digestibility of gluten-free cookies. *Starch-Stärke, 70*(1–2), 1700107.

Gibney, M. J., Barr, S. I., Bellisle, F., Drewnowski, A., Fagt, S., Livingstone, B., ... Hopkins, S. (2018). Breakfast in human nutrition: The international breakfast research initiative. *Nutrients, 10*(5), 559.

Giuberti, G., Bresciani, A., Cervini, M., Frustace, A., & Marti, A. (2021). Moringa oleifera L. leaf powder as ingredient in gluten-free biscuits: nutritional and physicochemical characteristics. *European Food Research and Technology, 247*(3), 687–694.

Goes, E. S. D. R., Souza, M. L. R. D., Michka, J. M. G., Kimura, K. S., Lara, J. A. F. D., Delbem, A. C. B., & Gasparino, E. (2016). Fresh pasta enrichment with protein concentrate of tilapia: Nutritional and sensory characteristics. *Food Science and Technology, 36*, 76–82.

Goñi, I., & Valentín-Gamazo, C. (2003). Chickpea flour ingredient slows glycemic response to pasta in healthy volunteers. *Food Chemistry, 81*(4), 511–515.

Grasso, S., & Asioli, D. (2020). Consumer preferences for upcycled ingredients: A case study with biscuits. *Food Quality and Preference, 84*, 103951.

Haddad, E., Jambazian, P., Karunia, M., Tanzman, J., & Sabaté, J. (2006). A pecan-enriched diet increases γ-tocopherol/cholesterol and decreases thiobarbituric acid reactive substances in plasma of adults. *Nutrition Research, 26*(8), 397–402.

Hayta, M., Özuğur, G., Etgü, H., & Şeker, İ. T. (2014). Effect of grape (V itis Vinifera L.) pomace on the quality, total phenolic content and anti-radical activity of bread. *Journal of Food Processing and Preservation, 38*(3), 980–986.

Hedhili, A., Lubbers, S., Bou-Maroun, E., Griffon, F., Akinyemi, B. E., Husson, F., & Valentin, D. (2021). Moringa Oleifera supplemented biscuits: Nutritional values and consumer segmentation. *South African Journal of Botany, 138*, 406–414.

Hoye, Jr, C., & Ross, C. F. (2011). Total phenolic content, consumer acceptance, and instrumental analysis of bread made with grape seed flour. *Journal of Food Science, 76*(7), S428–S436.

Hu, G., Huang, S., Cao, S., & Ma, Z. (2009). Effect of enrichment with hemicellulose from rice bran on chemical and functional properties of bread. *Food Chemistry, 115*(3), 839–842.

Ibrahim, U. K., Salleh, R. M., & Maqsood-ul-Haque, S. N. S. (2015). Bread towards functional food: An overview. *International Journal of Food Engineering, 1*(1), 39–43.

Inyang, U. E., Daniel, E. A., & Bello, F. A. (2018). Production and quality evaluation of functional biscuits from whole wheat flour supplemented with acha (fonio) and kidney bean flours. *Asian Journal of Agriculture and Food Sciences, 6*(6), 193-201.

Kampuse, S., Ozola, L., Straumite, E., & Galoburda, R. (2015). Quality parameters of wheat bread enriched with pumpkin (Cucurbita moschata) by-products. *Acta Universitatis Cinbinesis, Series E: Food Technology, 19*(2), 3-14.

Kaur, R., Kaur, K., Wagh, R. V., Kaur, A., & Aggarwal, P. (2020). Red bell pepper (Capsicum annuum L.): Optimization of drying conditions and preparation of functional bread. *Journal of Food Science, 85*(8), 2340–2349.

Kaur, M., Dhaliwal, M., Kaur, H., Singh, M., Bangar, S. P., Kumar, M., & Pandiselvam, R. (2021). Preparation of antioxidant-rich tricolor pasta using microwave processed orange pomace and cucumber peel powder: A study on nutraceutical, textural, color and sensory attributes. *Journal of Texture Studies,* 1-10.

Khan, I., Yousif, A. M., Johnson, S. K., & Gamlath, S. (2015). Acute effect of sorghum flour-containing pasta on plasma total polyphenols, antioxidant capacity and oxidative stress markers in healthy subjects: A randomised controlled trial. *Clinical Nutrition, 34*(3), 415–421.

Kolanowski, W., & Laufenberg, G. (2006). Enrichment of food products with polyunsaturated fatty acids by fish oil addition. *European Food Research and Technology, 222*(3), 472–477.

Kolniak-Ostek, J., Kita, A., Pęksa, A., Wawrzyniak, A., Hamułka, J., Jeznach, M., … Jariene, E. (2017). Analysis of the content of bioactive compounds in selected flours and enriched extruded corn products. *Journal of Food Composition and Analysis, 64*, 147–155.

Kornsteiner, M., Wagner, K. H., & Elmadfa, I. (2006). Tocopherols and total phenolics in 10 different nut types. *Food Chemistry, 98*(2), 381–387.

Kristensen, M., Jensen, M. G., Riboldi, G., Petronio, M., Bügel, S., Toubro, S., … Astrup, A. (2010). Wholegrain vs. refined wheat bread and pasta. Effect on postprandial glycemia, appetite, and subsequent ad libitum energy intake in young healthy adults. *Appetite, 54*(1), 163–169.

Kurek, M., & Wyrwisz, J. (2015). The application of dietary fiber in bread products. *Journal of Food Processing and Technology, 6*(5), 447–450.

Laleg, K., Barron, C., Cordelle, S., Schlich, P., Walrand, S., & Micard, V. (2017). How the structure, nutritional and sensory attributes of pasta made from legume flour is affected by the proportion of legume protein. *LWT-Food Science and Technology, 79*, 471–478.

Lemmens, E., Deleu, L. J., De Brier, N., Smolders, E., & Delcour, J. A. (2021). Mineral bioaccessibility and intrinsic saccharides in breakfast flakes manufactured from sprouted wheat. *LWT-Food Science and Technology, 143*, 111079.

Lindenmeier, M., & Hofmann, T. (2004). Influence of baking conditions and precursor supplementation on the amounts of the antioxidant pronyl-L-lysine in bakery products. *Journal of Agricultural and Food Chemistry, 52*(2), 350–354.

Liu, L. Y., Wu, K. L., Jen, Y. W., & Yang, M. H. (2007). Effect of sweet potato leaf and stem addition on dough properties and bread quality. *Food Science and Technology International, 13*(3), 239–244.

Liu, T., Hamid, N., Kantono, K., Pereira, L., Farouk, M. M., & Knowles, S. O. (2016). Effects of meat addition on pasta structure, nutrition and in vitro digestibility. *Food Chemistry, 213*, 108–114.

Livesey, G., Taylor, R., Livesey, H. F., Buyken, A. E., Jenkins, D. J., Augustin, L. S., … Brand-Miller, J. C. (2019). Dietary glycemic index and load and the risk of type 2 diabetes: A systematic review and updated meta-analyses of prospective cohort studies. *Nutrients, 11*(6), 1280.

Lockyer, S., & Nugent, A. P. (2017). Health effects of resistant starch. *Nutrition Bulletin, 42*(1), 10–41.

Lorusso, A., Verni, M., Montemurro, M., Coda, R., Gobbetti, M., & Rizzello, C. G. (2017). Use of fermented quinoa flour for pasta making and evaluation of the technological and nutritional features. *LWT-Food Science and Technology, 78*, 215–221.

Mabogo, F. A., Mashau, M. E., & Ramashia, S. E. (2021). Effect of partial replacement of wheat flour with unripe banana flour on the functional, thermal, and physicochemical characteristics of flour and biscuits. *International Food Research Journal, 28*(1), 138–147.

Mahadevan, N., & Kamboj, P. (2009). Hibiscus sabdariffa Linn. – an overview. *Natural Product Radiance, 8*, 77–83.

Mahloko, L. M., Silungwe, H., Mashau, M. E., & Kgatla, T. E. (2019). Bioactive compounds, antioxidant activity and physical characteristics of wheat-prickly pear and banana biscuits. *Heliyon, 5*(10), e02479.

Manjula, K., & Suneetha, C. (2014). Formulation and development of functional confectionery by incorporating pumpkin juice. *International Journal of Food, Agriculture & Veterinary Sciences, 4*, 47–52.

Manley, D. (2000). *Technology of biscuits, crackers and cookies*. Woodhead Publishing Ltd.

Marchetti, L., Califano, A. N., & Andrés, S. C. (2018). Partial replacement of wheat flour by pecan nut expeller meal on bakery products. Effect on muffins quality. *LWT-Food Science and Technology, 95*, 85–91.

Martinez-Saez, N., García, A. T., Pérez, I. D., Rebollo-Hernanz, M., Mesías, M., Morales, F. J., … Del Castillo, M. D. (2017). Use of spent coffee grounds as food ingredient in bakery products. *Food Chemistry, 216*, 114–122.

Martins, V. B., & Netto, F. M. (2006). Physicochemical and functional properties of soy protein isolate as a function of water activity and storage. *Food Research International, 39*(2), 145–153.

Martins, Z. E., Erben, M., Gallardo, A. E., Silva, R., Barbosa, I., Pinho, O., & Ferreira, I. M. (2015). Effect of spent yeast fortification on physical parameters, volatiles and sensorial characteristics of home-made bread. *International Journal of Food Science & Technology, 50*(8), 1855–1863.

Martins, Z. E., Pinho, O., & Ferreira, I. M. P. L. V. O. (2017). Food industry by-products used as functional ingredients of bakery products. *Trends in Food Science & Technology, 67*, 106–128.

Marlett, J. A. (2001). Dietary fiber and cardiovascular disease. In *Handbook of dietary fiber* (pp. 17–30). CRC Press.

Michels, N., De Henauw, S., Beghin, L., Cuenca-García, M., Gonzalez-Gross, M., Hallstrom, L., … Claessens, M. (2016). Ready-to-eat cereals improve nutrient, milk and fruit intake at breakfast in European adolescents. *European Journal of Nutrition, 55*(2), 771–779.

Miskelly, D. (2017). Optimisation of end-product quality for the consumer. In *Cereal grains* (pp. 653–688). Woodhead Publishing, pp 149–150.

Molinari, R., Costantini, L., Timperio, A. M., Lelli, V., Bonafaccia, F., Bonafaccia, G., & Merendino, N. (2018). Tartary buckwheat malt as ingredient of gluten-free cookies. *Journal of Cereal Science, 80*, 37–43.

Mollakhalili-Meybodi, N., Arab, M., Nematollahi, A., & Khaneghah, A. M. (2021). *Prebiotic wheat bread: Technological, sensorial and nutritional perspectives and challenges, 149*, 111823.

Montalbano, A., Tesoriere, L., Diana, P., Barraja, P., Carbone, A., Spanò, V., ... & Cirrincione, G. (2016). Quality characteristics and in vitro digestibility study of barley flour enriched ditalini pasta. *LWT-Food Science and Technology, 72*, 223–228.

Nehra, M., Siroha, A. K., Punia, S., & Kumar, S. (2021). Process standardization for bread preparation using composite blend of wheat and pearl millet: Nutritional, antioxidant and sensory approach. *Current Research in Nutrition and Food Science Journal, 9*(2), 1–10.

Nicole, T. Z. H., Nichelle, T. S., Elizabeth, T. E., & Yuliarti, O. (2021). Formulation of functional crackers enriched with fermented soybean (tempeh) paste: Rheological and microstructural properties. *Future Foods, 4*, 100050, 1–9.

Nout, M. R. (2009). Rich nutrition from the poorest – Cereal fermentations in Africa and Asia. *Food Microbiology, 26*(7), 685–692.

Obafaye, R. O., & Omoba, O. S. (2018). Orange peel flour: A potential source of antioxidant and dietary fiber in pearl-millet biscuit. *Journal of Food Biochemistry, 42*(4), e12523.

Obradović, V., Babić, J., Šubarić, D., Jozinović, A., Ačkar, Đ., & Klarić, I. (2015). Influence of dried Hokkaido pumpkin and ascorbic acid addition on chemical properties and colour of corn extrudates. *Food Chemistry, 183*, 136–143.

Ogo, A. O., Ajekwe, D. J., Enenche, D. E., & Obochi, G. O. (2021). Quality evaluation of novel biscuits made from wheat supplemented with watermelon rinds and orange pomace flour blends. *Food and Nutrition Sciences, 12*(3), 332–341.

Okafor, G. I., & Usman, G. O. (2014). Physical and functional properties of breakfast cereals from maize, African yam bean and coconut cake. *Agro-Science, 13*(2), 7–16.

Oliveira, L. C., Alencar, N. M., & Steel, C. J. (2018). Improvement of sensorial and technological characteristics of extruded breakfast cereals enriched with whole grain wheat flour and jabuticaba (Myrciaria cauliflora) peel. *LWT-Food Science & Technology, 90*, 207–214.

Oyeyinka, A. T., & Oyeyinka, S. A. (2018). Moringa oleifera as a food fortificant: Recent trends and prospects. *Journal of the Saudi Society of Agricultural Sciences, 17*(2), 127–136.

Pasqualone, A., Bianco, A. M., Paradiso, V. M., Summo, C., Gambacorta, G., Caponio, F., & Blanco, A. (2015). Production and characterization of functional biscuits obtained from purple wheat. *Food Chemistry, 180*, 64–70.

Pathak, D., Majumdar, J., Raychaudhuri, U., & Chakraborty, R. (2016). Characterization of physicochemical properties in whole wheat bread after incorporation of ripe mango peel. *Journal of Food Measurement and Characterization, 10*(3), 554–561.

Pellegrini, N., & Agostoni, C. (2015). Nutritional aspects of gluten-free products. *Journal of the Science of Food and Agriculture, 95*(12), 2380–2385.

Petitot, M., Boyer, L., Minier, C., & Micard, V. (2010). Fortification of pasta with split pea and faba bean flours: Pasta processing and quality evaluation. *Food Research International, 43*(2), 634–641.

Polat, H., Capar, T. D., Inanir, C., Ekici, L., & Yalcin, H. (2020). Formulation of functional crackers enriched with germinated lentil extract: A response surface methodology Box-Behnken design. *LWT-Food Science & Technology, 123*, 109065.

Pourmohammadi, K., Abedi, E., Farahmandi, S., Mahmoudi, M. R., Hashemi, S. M. B., & Torri, L. (2019). Modeling the effects of corn and wheat resistant starch on texture properties and quality of resistant starch-enrichment dough and biscuit. *Journal of Food Process Engineering, 42*(2), e12962.

Prabhasankar, P., Ganesan, P., Bhaskar, N., Hirose, A., Stephen, N., Gowda, L. R., … Miyashita, K. J. F. C. (2009). Edible Japanese seaweed, wakame (Undaria pinnatifida) as an ingredient in pasta: Chemical, functional and structural evaluation. *Food Chemistry, 115*(2), 501–508.

Ptitchkina, N. M., Novokreschonova, L. V., Piskunova, G. V., & Morris, E. R. (1998). Large enhancements in loaf volume and organoleptic acceptability of wheat bread by small additions of pumpkin powder: Possible role of acetylated pectin in stabilising gas-cell structure. *Food Hydrocolloids, 12*(3), 333–337.

Punia Bangar, S., Sandhu, K. S., Purewal, S. S., Kaur, M., Kaur, P., Siroha, A. K., … Kumar, M. (2021). Fermented barley bran: An improvement in phenolic compounds and antioxidant properties. *Journal of Food Processing and Preservation, e15543*.

Punia, S., Sandhu, K. S., Siroha, A. K., & Dhull, S. B. (2019). Omega 3-metabolism, absorption, bioavailability and health benefits – A review. *Pharma Nutrition, 10*, 100162.

Punia, S., Dhull, S. B., Sandhu, K. S., & Kaur, M. (2019). Faba bean (Vicia faba) starch: Structure, properties, and in vitro digestibility—A review. *Legume Science, 1*(1), e18.

Punia, S., Siroha, A. K., Sandhu, K. S., & Kaur, M. (2019). Rheological and pasting behavior of OSA modified mungbean starches and its utilization in cake formulation as fat replacer. *International Journal of Biological Macromolecules, 128*, 230–236.

Punia, S., Sandhu, K. S., & Kaur, M. (2020). Quantification of phenolic acids and antioxidant potential of wheat rusks as influenced by partial replacement with barley flour. *Journal of Food Science and Technology, 57*(10), 3782–3791.

Punia, S., Dhull, S. B., Sandhu, K. S., Kaur, M., & Purewal, S. S. (2020). Kidney bean (Phaseolus vulgaris) starch: A review. *Legume Science, 2*(3), e52.

Punia, S., Kumar, M., Siroha, A. K., & Purewal, S. S. (2021). Rice bran oil: Emerging trends in extraction, health benefit, and its industrial application. *Rice Science, 28*(3), 217–232.

Punia, S., Sandhu, K. S., Grasso, S., Purewal, S. S., Kaur, M., Siroha, A. K., … Kumar, M. (2021). Aspergillus oryzae fermented rice bran: A byproduct with enhanced bioactive compounds and antioxidant potential. *Foods, 10*(1), 70.

Pyanikova, E. A., Kovaleva, A. E., Ovchinnikova, E. V., & Afanasieva, L. A. (2021, March). Researching the possibility of using recycled apple raw materials to create functional food products. In IOP Conference Series: Earth and Environmental Science, *677*(3), 032030, IOP Publishing.

Rahman, N., Uddin, M. B., Quader, M. F. B., & Bakar, M. A. (2020). Optimization of mixed peels from banana, carrot and apple to develop high fiber biscuit. *International Journal of Natural and Social Sciences, 7*(1), 21–25.

Rainero, G., Bianchi, F., Rizzi, C., Cervini, M., Giuberti, G., & Simonato, B. (2022). Breadsticks fortification with red grape pomace: Effect on nutritional, technological, and sensory properties. *Journal of the Science of Food and Agriculture, 102*(6), 2545–2552.

Ramya, N. S., Prabhasankar, P., Gowda, L. R., Modi, V. K., & Bhaskar, N. (2015). Influence of freeze-dried shrimp meat in pasta processing qualities of Indian T. durum wheat. *Journal of Aquatic Food Product Technology, 24*(6), 582–596.

Roberfroid, M. B. (1999). What is beneficial for health? The concept of functional food. *Food and Chemical Toxicology, 37*(9–10), 1039–1041.

Sahin, A. W., Coffey, A., & Zannini, E. (2021). Functionalisation of wheat and oat bran using single-strain fermentation and its impact on techno-functional and nutritional properties of biscuits. *European Food Research and Technology*, 247, 1825–1837.

Salehi, F., & Aghajanzadeh, S. (2020). Effect of dried fruits and vegetables powder on cakes quality: A review. *Trends in Food Science & Technology, 95*, 162–172.

Salgado, J. M., Rodrigues, B. S., Donado-Pestana, C. M., dos Santos Dias, C. T., & Morzelle, M. C. (2011). Cupuassu (Theobroma grandiflorum) peel as potential source of dietary fiber and phytochemicals in whole-bread preparations. *Plant Foods for Human Nutrition, 66*(4), 384–390.

Sandhu, K. S., & Punia, S. (2017). Enhancement of bioactive compounds in barley cultivars by solid substrate fermentation. *Journal of Food Measurement and Characterization, 11*(3), 1355–1361.

Sanjukta, S., & Rai, A. K. (2016). Production of bioactive peptides during soybean fermentation and their potential health benefits. *Trends in Food Science & Technology, 50*, 1–10.

Schoenlechner, R., Wendner, M., Siebenhandl-Ehn, S., & Berghofer, E. (2010). Pseudocereals as alternative sources for high folate content in staple foods. *Journal of Cereal Science, 52*(3), 475–479.

Schober, T. J., O'brien, C. M., McCarthy, D., Darnedde, A., & Arendt, E. K. (2003). Influence of gluten-free flour mixes and fat powders on the quality of gluten-free biscuits. *European Food Research and Technology, 216*(5), 369–376.

Sęczyk, Ł., Świeca, M., & Gawlik-Dziki, U. (2016). Effect of carob (Ceratonia siliqua L.) flour on the antioxidant potential, nutritional quality, and sensory characteristics of fortified durum wheat pasta. *Food Chemistry, 194*, 637–642.

Sęczyk, Ł., Świeca, M., Dziki, D., Anders, A., & Gawlik-Dziki, U. (2017). Antioxidant, nutritional and functional characteristics of wheat bread enriched with ground flaxseed hulls. *Food Chemistry, 214*, 32–38.

Shori, A. B., Kee, L. A., & Baba, A. S. (2021). Total phenols, antioxidant activity and sensory evaluation of bread fortified with spearmint. *Arabian Journal for Science and Engineering, 46*(6), 5257–5264.

Sidhu, J. S., Kabir, Y., & Huffman, F. G. (2007). Functional foods from cereal grains. *International Journal of Food Properties, 10*(2), 231–244.

Simonato, B., Trevisan, S., Tolve, R., Favati, F., & Pasini, G. (2019). Pasta fortification with olive pomace: Effects on the technological characteristics and nutritional properties. *LWT, 114*, 108368.

Singh, B., Singh, J. P., Kaur, A., & Singh, N. (2017). Phenolic composition and antioxidant potential of grain legume seeds: A review. *Food Research International, 101*, 1–16.

Sirbu, A., & Arghire, C. (2017). Functional bread: Effect of inulin-type products addition on dough rheology and bread quality. *Journal of Cereal Science, 75*, 220–227.

Sogi, D. S., Garg, S. K., & Bawa, A. S. (2002). Functional properties of seed meals and protein concentrates from tomato-processing waste. *Journal of Food Science, 67*(8), 2997–3001.

Sridhar, K., Sharma, M., Choudhary, A., Dikkala, P. K., & Narsaiah, K. (2021). Fish and garlic oils hybridized microcapsules: Fortification in functional bread. *Journal of Food Processing and Preservation, 45*(4), e15346.

Stamataki, N. S., Yanni, A. E., & Karathanos, V. T. (2016). Non-cereal ingredients for the attenuation of glycaemic response to bread: A review of the clinical evidence. *Food & Function, 7*(7), 2926–2936.

Stoll, L., Flôres, S. H., & Thys, R. C. S. (2015). Fibra de casca de laranja como substituto de gordura em pão de forma. *Ciência Rural, 45*, 567–573.

Sun-Waterhouse, D., Teoh, A., Massarotto, C., Wibisono, R., & Wadhwa, S. (2010). Comparative analysis of fruit-based functional snack bars. *Food Chemistry, 119*(4), 1369–1379.

Tachaprutinun, A., Udomsup, T., Luadthong, C., & Wanichwecharungruang, S. (2009). Preventing the thermal degradation of astaxanthin through nanoencapsulation. *International Journal of Pharmaceutics, 374*(1–2), 119–124.

Thejasri, V., Hymavathi, T. V., Roberts, T. P., Anusha, B., & Devi, S. S. (2017). Sensory, physico-chemical and nutritional properties of gluten free biscuits formulated with Quinoa (Chenopodium quinoa Willd.), Foxtail Millet (Setaria italica) and hydrocolloids. *International Journal of Current Microbiology and Applied Sciences, 6*(8), 1710–1721.

Tribelhorn, R. E. (1991). Breakfast cereals. In *Handbook of cereal science and technology* (pp. 741–751). Marcel Dekker.

Tudorica, C. M., Kuri, V., & Brennan, C. S. (2002). Nutritional and physicochemical characteristics of dietary fiber enriched pasta. *Journal of Agricultural and Food Chemistry, 50*(2), 347–356.

Turco, I., Bacchetti, T., Morresi, C., Padalino, L., & Ferretti, G. (2019). Polyphenols and the glycaemic index of legume pasta. *Food & Function, 10*(9), 5931–5938.

Ukeyima, T. M., Akor, I. A., & Kyenge, B. (2021). In vitro digestibility, nutritional and sensory quality of extruded breakfast cereal from maize grits, partially defatted peanut and beetroot flour. *Asian Food Science Journal, 20*(8), 66–75.

Vázquez-Sánchez, K., Martinez-Saez, N., Rebollo-Hernanz, M., Del Castillo, M. D., Gaytán-Martínez, M., & Campos-Vega, R. (2018). In vitro health promoting properties of antioxidant dietary fiber extracted from spent coffee (Coffee arabica L.) grounds. *Food Chemistry, 261*, 253–259.

Vega-López, S., Venn, B. J., & Slavin, J. L. (2018). Relevance of the glycemic index and glycemic load for body weight, diabetes, and cardiovascular disease. *Nutrients, 10*(10), 1361.

Walker, R., Tseng, A., Cavender, G., Ross, A., & Zhao, Y. (2014). Physicochemical, nutritional, and sensory qualities of wine grape pomace fortified baked goods. *Journal of Food Science, 79*(9), S1811–S1822.

Wang, J., Brennan, M. A., Serventi, L., & Brennan, C. S. (2021). Impact of functional vegetable ingredients on the technical and nutritional quality of pasta. *Critical Reviews in Food Science and Nutrition*, 1–12.

Waters, D. M., Kingston, W., Jacob, F., Titze, J., Arendt, E. K., & Zannini, E. (2013). Wheat bread biofortification with rootlets, a malting by-product. *Journal of the Science of Food and Agriculture, 93*(10), 2372–2383.

Weisstaub, A. R., Salinas, M. V., Correa, M. J., Barchuk, M., Berg, G., & Zuleta, A. (2018). Effects of the intake of white wheat bread added with garlic and resistant starch: Action on calcium bioavailability and metabolic parameters of growing Wistar rats. *Food & Function, 9*(11), 5707–5714.

Weng, M., Li, Y., Wu, L., Zheng, H., Lai, P., Tang, B., & Luo, X. (2021). Effects of passion fruit peel flour as a dietary fibre resource on biscuit quality. *Food Science and Technology, 41*, 65–73.

West, R., Seetharaman, K., & Duizer, L. M. (2013). Effect of drying profile and whole grain content on flavour and texture of pasta. *Journal of Cereal Science, 58*(1), 82–88.

Wirkijowska, A., Zarzycki, P., Sobota, A., Nawrocka, A., Blicharz-Kania, A., & Andrejko, D. (2020). The possibility of using by-products from the flaxseed industry for functional bread production. *LWT, 118*, 108860.

Wu, M. Y., & Shiau, S. Y. (2015). Effect of the amount and particle size of pineapple peel fiber on dough rheology and steamed bread quality. *Journal of Food Processing and Preservation, 39*(6), 549–558.

Wunsch, N. G. (2020). Market value of breakfast cereals worldwide 2016–2022. In *Breakfast cereals: Market value worldwide 2022*. Statista.

Yadav, M., Jain, S., Tomar, R., Prasad, G. B. K. S., & Yadav, H. (2010). Medicinal and biological potential of pumpkin: An updated review. *Nutrition Research Reviews, 23*(2), 184–190.

Ying, D., Cheng, L. J., Chibracq, G., Sanguansri, L., Oiseth, S. K., & Augustin, M. A. (2015). The format of β-carotene delivery affects its stability during extrusion. *LWT-Food Science and Technology, 60*(1), 1–7.

Zhao, H., Chen, J., Hemar, Y., & Cui, B. (2020). Improvement of the rheological and textural properties of calcium sulfate-induced soy protein isolate gels by the incorporation of different polysaccharides. *Food Chemistry, 310*, 125983.

Chapter 10
Cereal Grain-Based Milks and Their Potential Health Properties

Khongsak Srikaeo

10.1 Introduction

Cereals are plants that belong to the monocot Gramineae family. They include common wheat (*Triticum aestivum* L.), durum wheat (*Triticum durum* Desf.), maize (*Zea mays* L.), oats (*Avena sativa* L.), rye (*Secale cereale* L.), rice (*Oryza sativa* L.), barley (*Hordeum vulgare* L.), sorghum (*Sorghum bicolor* L. Moench), pearl millet (*Pennisetum glaucum* L.), finger millet (*Eleusine coracana* L.), teff (*Eragrostis tef*), and triticale (x *Triticosecale* Wittmack). Pseudo-cereals such as quinoa (*Chenopodium quinoa* Willd), amaranth (*Amaranthus* sp.), buckwheat (*Fagopyrum esculentum* Moench), and chia (*Salvia hispanica* L.) do not belong to the Gramineae family. However, due to their nutrient-rich, gluten-free properties (Thakur et al., 2021), they can be used in a similar manner to cereal grains. Cereals are grown over 73% of the world's harvested land and account for nearly 60% of global food production, providing essential nutrients such as carbohydrates, proteins, dietary fiber, minerals, and vitamins (Basinskiene & Cizeikiene, 2020; Sandhu et al., 2017; Charalampopoulos et al., 2002).

Cereal grain-based milk is classified as a group in plant-based milk alternatives which are fluids made from the breakdown of plant materials, extracted in water and further homogenization which imitates cow's milk in several properties (Sethi et al., 2016). Currently, plant sources such as cereals, legumes, seeds, nuts, and pseudo-cereals can be used to commercially produce plant-based milks (Outi Elina Mäkinen et al., 2016). It should be noted that the use of the term "milk" to refer to the cereal/plant-based alternatives to traditional cow's milk is controversial and is protected by legislation in many countries. In these points, plant-based fluid products are usually

K. Srikaeo (✉)
Faculty of Food and Agricultural Technology, Pibulsongkram Rajabhat University, Muang Phitsanulok, Thailand
e-mail: khongsak@live.psru.ac.th

S. Punia Bangar, A. Kumar Siroha (eds.), *Functional Cereals and Cereal Foods*,
https://doi.org/10.1007/978-3-031-05611-6_10

referred to as drinks, beverages, dairy alternatives, or some other name rather than "milk." The U.S. Food and Drug Administration is currently examining whether the producers of plant-based beverages should be allowed to refer to them as "milk" (McClements et al., 2019). According to EU 1308/2013 regulation, it is not possible to use the term "milk" for plant-based drinks. Only what is obtained by milking can be called "milk"; so, except for almond and coconut milk, all the other products can be named as "beverage" or "drink" (Verduci et al., 2019). Current regulations vary depending on countries.

The market for plant-based milk replacements has grown steadily in recent years, reaching an annual volume of around US$1.8 billion in the United States alone. From a global viewpoint, the global market is expected to exceed US$26 billion by 2023, with a compound annual growth rate of more than 10%. Different reasons and consumer demands are driving the growing popularity for plant-based milk replacements. Health-related issues such as lactose intolerance and milk allergies, consumer worries about cow's milk hormones and cholesterol, ethical debates over animal use, environmental issues, and shifts in lifestyle toward vegetarian and vegan foods are among them (Craig, 2010; Crittenden & Bennett, 2005; Epstein, 1990; Hughes, 1995; Rotz et al., 2010). Plant-based milk alternatives are also widely marketed as health-promoting products. Consequently, leading dairy companies are currently adding plant-based milk alternatives to their product line (Tangyu et al., 2019).

Cereals are considered nutraceuticals and functional foods. This is due to the health-promoting components, such as dietary fibers, vitamins, minerals, and antioxidants. The major phytochemicals available in cereal grains are phenolic acids, flavones, phytic acid, flavonoids, coumarins, and terpenes. Cereal germs are good sources of ferulic acid, phytic acid, glutathione, and phytosterols.

Functional foods have piqued the interest of not just food scientists, but also consumers, regulatory authorities, food producers, and processors, due to recent implications of a role in illness prevention and treatment, as well as rising healthcare expenses (Sidhu et al., 2007). Because cereal grains are such an important ingredient in the production of functional foods. This chapter examines the technological aspects of making functional cereal-based beverages and the health benefits they provide, with a focus on cereal-grain based milks.

10.2 Production of Cereal-Based Milks

In general, the most important ingredients used to formulate cereal-based milks as well as other plant-based milk substitutes are plant sources, water, emulsifiers, and additives. There are two types of cereal grain-based milk: (i) milk-like cereal-based juice, which preserves the color and texture of the original cereal, such as corn milk, and (ii) resembles cow's milk in terms of color and texture, e.g. rice milk and oat milk, generally with a particle size distribution of approximately 5–20 μm (Sethi et al., 2016).

The manufacturing processes of cereal/plant-based milk alternatives should ideally produce a final product that has some similarities to cow's milk in terms of their composition and structure (McClements et al., 2019). In particular, the colloidal nature of cow's milk as well as natural sugars and salts found in milks play a crucial role in producing its desirable quality attributes and flavor profile (Singh, 2006). Plant-based products typically consist of colloidal fat and/or protein dispersed within an aqueous solution containing other ingredients. In general, depending on the desired final products, two main approaches have been developed to produce cereal/plant based milks: (i) breaking down the natural structures of particular plant materials to release oil bodies or other colloidal matter (Sethi et al., 2016), and (ii) the fabrication of simulated fat globules using plant-based ingredients (Do et al., 2018). The breaking down process is the most frequently used processing operations while the latter is a modern process designed to facilitate specialty plant-based beverages. The general manufacturing process of cereal grain-based milk is illustrated in Fig. 10.1.

Today, numerous commercial plant-based milk products are readily available, with some of the most popular being almond, coconut, flaxseed, oat, rice, and soy milks. All cereals could be used to produce cereal-based milks. However, the popular cereals being used for commercial cereal-based milks are corn, rice, and oat. Wheat, barley, millet, and sorghum milks are also produced but mostly in small scale for niche markets. Most commercial processes for the production of cereal-based milks employ the breaking down process.

Corn milk, also referred to as corn juice, is a type of cereal milk that has only recently appeared on the market in some Asian countries such as China and Thailand. Corn milk is rich in starch, protein, dietary fiber and other bioactive ingredients and is widely believed to help in preventing cardiovascular diseases, obesity and diabetes due to its rich bioactive components (Charunuch et al., 2003; Haggard et al., 2018). Corn milk uses whole grain corn, often sweet corn, as the main raw material and is typically manufactured by precooking, milling, blending, enzymatic hydrolysis (saccharification), homogenization, and then pasteurization (Omueti et al., 2000; Prabhavat et al., 1999). Alternatively, small-scale food manufactures may produce fresh corn milk by boiling fresh corn kernels in water, milling and then filtration. Unlike other cereal milk such as rice and oat milk, corn milk has a yellow color with a thick consistency. Although corn milk is now a fast-growing beverage on the market, current research on corn milk is still limited and focuses mainly on the formulation and processing optimization of corn milk production (Prabhavat et al., 1999).

Rice milk or sometimes called rice soup is a traditional beverage from East Asian countries. Traditionally, rice milk refers to the rice water remaining as a co-product, after cooking/boiling rice, or as the layer of liquid floating on top of rice porridge, which is commonly consumed during meals to quench thirst and stimulate the appetite (Xiong et al., 2020). Today, a number of new rice milk products have been developed and commercialized worldwide (Cheowtiraku et al., 2001; Xiong et al., 2020). Commercial rice milk is often made from whole grain brown rice and is typically manufactured by grinding/milling, mixing with water, followed by starch hydrolysis, filtration and fortification, and then homogenization (Cheowtiraku et al.,

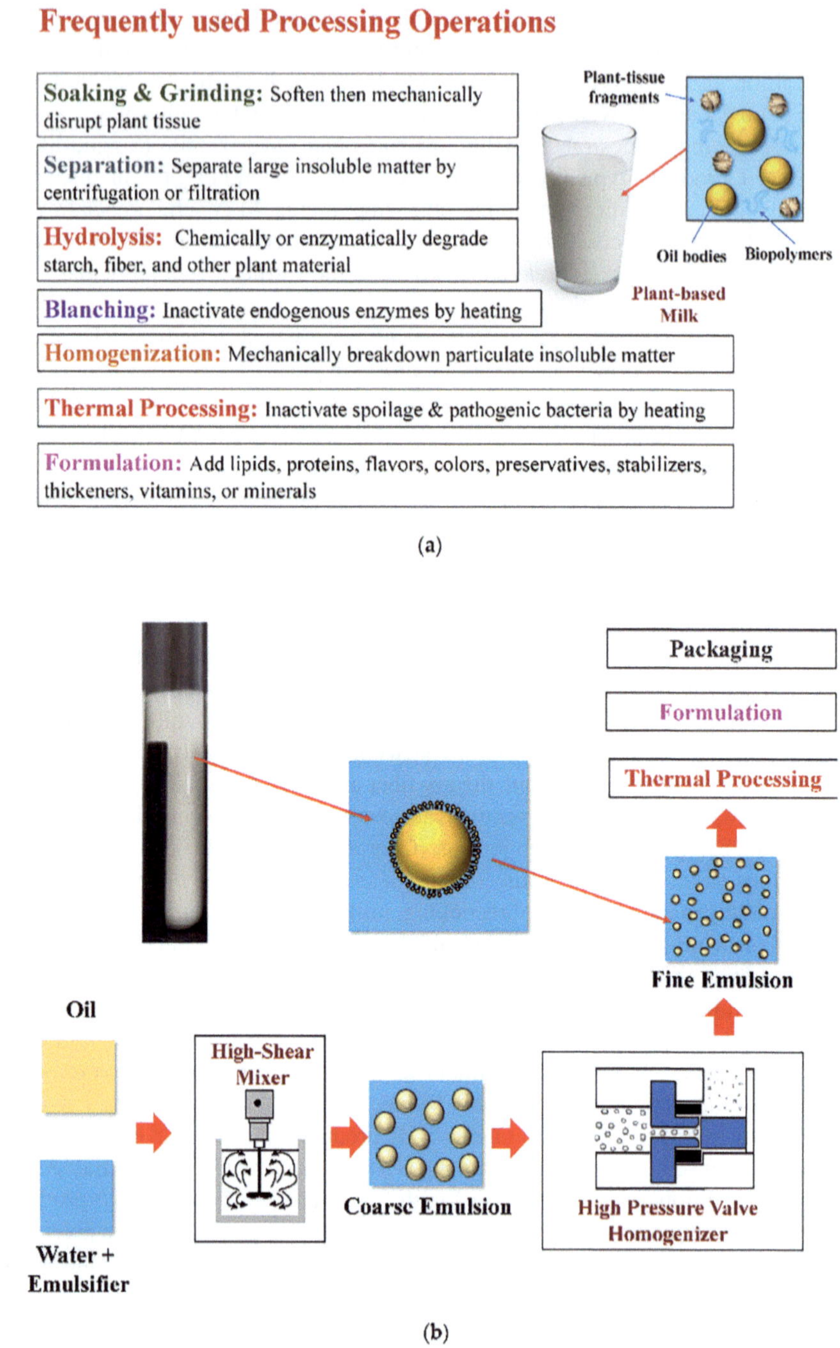

Fig. 10.1 The general steps for the production of cereal/plant-based milks: (**a**) the commonly used process and (**b**) fabrication using plant-based ingredients. (Source: McClements, 2020. Open Access)

2001). Young paddy (immature rice grains) can be used to produce young rice milks. This type of products is commercially available in Thailand and some Asian countries. General production steps involve the water extraction of green paddy, about 8 days after flowering, followed by filtration, formulation, homogenization, and finally pasteurization or UHT (Fig. 10.2). Young rice exhibited higher bioactive compounds such as phenolics, flavonoids and ferulic and soluble fiber than mature rice (Jiamyangyuen et al., 2017; Lin & Lai, 2011). During rice grain development, the highest protein content was found in the milky stage. The rice proteins obtained during flowery-to-milky stage supported the hypoallergic and easily to digest plant protein (Pantoa et al., 2020).

Rice milk is lactose free, making them suitable for patients suffering from lactose intolerance. It can also act as a perfect alternative in the case of patients with allergy issues caused by beans and nuts. Therefore, it is considered to be the most hypoallergenic of all plant-based milks and has been suggested as the best alternative to cow's milk for people with multiple allergies (Bridges, 2018; Vanga & Raghavan, 2018). In addition to the bioactive compounds, rice and its products are also rich sources of B vitamins. A recent study found that the thiamin content of rice milk was far higher than that of cow's milk and other dairy products (Lalić et al., 2014). Moreover, rice milk is also rich in selenium and magnesium, which are considered to assist in enhancing the immune system and in preventing bacterial and viral infections (Park et al., 2017; Sohail et al., 2017). However, research has shown that using rice milk as a substitute for cow's milk without sufficient supervision can lead to malnutrition, particularly in infants, due to the nutrient profile differences (Katz et al., 2005; Massa et al., 2001). Kwashiorkor, a type of protein-energy malnutrition was observed in infants that were on a rice based vegan diet. Rice milks that are not fortified, particularly those manufactured at home, are deficient in minerals and vitamins such as calcium and B_{12} unless fortified, as most commercial milks are (Dinu et al., 2017). Further, there has been a concern on high arsenic content of rice milks. Exposure to high arsenic content in long term can cause cancer and various other health effects (Shannon & Rodriguez, 2014).

Oat milk is one of the most popular cereal milks. It was developed in the 1990s by Swedish scientists seeking alternatives for people with lactose intolerance and nut allergies (Önning et al., 1998). Oat milk has become the most successful and

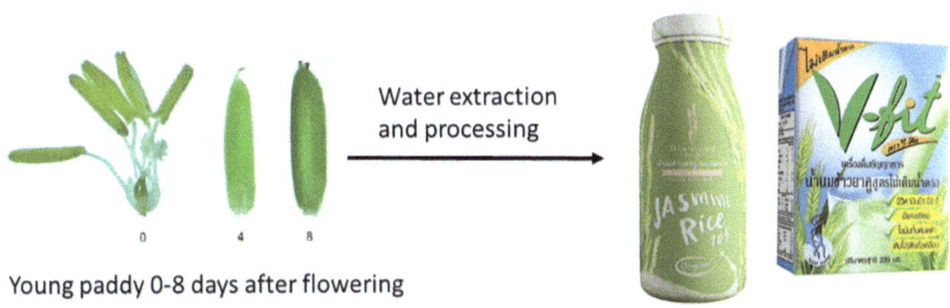

Fig. 10.2 Young rice milks produced from young paddy (immature rice grain)

commercialized cereal milk and is now available under many brands around the world. Oat milk is typically manufactured by milling, mixing with water, enzymatic hydrolysis, filtration and then fortification; homogenization and thermal treatments are also used to improve product stability and shelf life (Deswal et al., 2014). Oats' health advantages are linked to dietary fibers such as β-glucan, functional protein, lipid, carbohydrate components, and phytochemicals found in the oat grain, making oat one of the most promising raw materials for making functional plant-based milk. Oats are a good source of high-quality protein with a well-balanced amino acid profile. The existence of a functionally active component, β-glucan, which has nutraceutical characteristics, has sparked interest in oats. β-glucan, a soluble fiber, can enhance solution viscosity and delay stomach emptying time, as well as increase gastrointestinal transit time, all of which are linked to lower blood glucose levels. Oat fibers have a hypocholesterolemic effect, which could lower total and LDL cholesterol (Truswell, 2002). The U.S. Food and Drug Administration approved the health claim that a diet high in soluble fiber from whole oats (oat bran, oat meal, oat flour) and low in saturated fat and cholesterol may reduce the risk of cardiovascular disease. Antioxidants and polyphenols are also abundant in oats. Oat contains 60% starch, 11–15% total protein, 5–9% lipids, 2.3–8.5% dietary fiber and 0.54% calcium (Rasane et al., 2015).

Oats, like other cereals, have a high concentration of starch, which causes a problem in preparation of a stable emulsion and during the heat processing of oat milk. When starch is heated, it begins to gelatinize, and liquid milk tends to attain a gel-like consistency with a high viscosity, lowering its acceptability. So, to keep the fluidity or beverage-like consistency, hydrolysis of starch is one of the strategies for removing it and preventing gelatinization during thermal treatment (Sethi et al., 2016). Previously, response surface methodology was used to optimize the enzymatic production process of oat milk (Deswal et al., 2014). In addition, oats contain a high amount of phytic acid, an antinutrient. In oats, phytate phosphorus makes up 48.7–70.9% of total phosphorus (Frølich & Nyman, 1988). Oat milks could be treated with phytase to improve the nutritional content by releasing inorganic phosphate from phytic acid (Zhang et al., 2007). Enrichment could also be used to restore nutrients that are lost during processing. Despite its nutritional benefits, oat milk, like rice milk, lacks calcium, a vital vitamin for growth and development. As a result, it must be supplemented before being consumed as a milk substitute (Sethi et al., 2016).

10.3 Processing Techniques for Cereal-Based Milks

Each ingredient used for the production of cereal-based milks must be carefully selected to produce a final product that has the desired functional attributes and properties. Water quality is very important. Commercial sources of water usually have different types and levels of minerals and pH values. They may also contain organic matters that can interact with other ingredients or interfere with emulsifiers used to formulate emulsions. Consequently, the treatment of the water source by various techniques such as thermal processing, filtering, or reverse osmosis before it is used in the production of cereal-based milks is a crucial step (McClements et al., 2019). Also, it is important to acknowledge that the bioavailability of the nutritional contents of some cereals varies significantly from one raw material to another; however, in many cases, it is not precisely known (Bridges, 2018). Hence, blending of two or more cereals or with other plant sources have been suggested to enhance its nutritional or sensory properties. In addition, fortification with vitamins and minerals can also be applied.

10.3.1 Effects of Starch

It should be emphasized that cereals are starchy materials and starch in cereals are gelatinized during sterilization or pasteurization. This causes technical problems in downstream processing (Outi Elina Mäkinen et al., 2016). When heated above the gelatinization temperature (55–65 °C), gelatinized starch creates a thick slurry. As a consequence, gelatinized starch must be liquefied, and saccharified with amylases or a malt enzyme extract to prevent this in subsequent processing processes. To hydrolyze the starch, both α- and β-amylases can be used until the required level of sweetness and viscosity is achieved. The saccharification process may be done before or after the coarse particle removal. To liquefy the starch, α-amylase enzyme is needed. It attacks the starch's amorphous structure, hydrolyzes the α-1,4-glycosidic bonds of amylose and amylopectin at arbitrary distances and releases water-soluble dextrins, which gelatinize at higher, more appropriate temperatures. The mouthfeel and digestibility of the cereal beverages may be adjusted by adding the hemicellulase and the protease enzymes. The percentages of the different sugars in the final product are determined by the enzymes used for saccharification. β-Amylase breaks down dextrins into maltose and maltotriose. While glucoamylase hydrolyzes dextrins into glucose. Through the combination of these enzymes, the cereal-based milks acquire a sweetness that may be regulated optimally in respect of enzymes' perception and intensity (Basinskiene & Cizeikiene, 2020).

10.3.2 Stability of the Products

Similar to other plant-based milk, cereal milk is a colloidal suspension or emulsion composed of large particles, including starch granules, proteins and fibers. The cereal grain size and its composition are not as stable as that of a single emulsion due to the disintegration of the cereal particles (Mäkinen et al., 2015). Therefore, cereal milks require more process attention to overcome the stability, texture, nutrition, and sensory constrains. The emulsion's stability is determined by droplet size and aggregation. Cereal-grain milks are prepared by dissolving plant components;

hence particle composition and size are not as consistent as in bovine milk. The manner by which cereals and grains are milled will have a significant impact on particle size. The presence of larger particles will impact the relative stability of the milks and may cause issues with wetting the dry materials and heat transfer, resulting in ineffective thermal destruction of microbes (Durand et al., 2003). Moreover, if the particles are too big, this can lead to poor sensory experience, such as sandy, rough or chalky mouthfeel. Also, during storage, the larger particles in the milk may sink due to gravity and result in precipitation, sedimentation or delamination, which can compromise the product quality (Sethi et al., 2016). Different manufacturing strategies have been developed to solve this problem, of which homogenization is the most commonly used (Deswal et al., 2014). Homogenization can reduce the particle size and ensure the particles are evenly distributed, thereby reducing sedimentation and improving the stability of the beverage. Ultra high pressure homogenization (UHPH), an emerging and more advanced homogenization technology that employs up to 400 MPa pressure, is another promising technology that can be utilized for plant-based milk. UHPH can produce smaller and more uniform sized particles and more effectively improve the stability and quality of plant-based milk (Briviba et al., 2016). It can also achieve microbial reduction and enzyme inactivation, with an effect similar to that of pasteurization (Cruz et al., 2007). Currently, there is no research conducted on UHPH treated cereal milk, which warrants further investigation. Moreover, another promising technology for homogenization is ultrasound processing. High-intensity ultrasound treatment can be applied to modify the functional properties, such as viscosity and gelation, of biopolymers, providing smaller size droplets of emulsion (Lu et al., 2019).

10.3.3 The Use of Stabilizers and Emulsifiers

Adding stabilizers or emulsifiers is another strategy to improve the cereal milk stability. The stability of cereal-based milks is determined by three factors: particle size, emulsion formation, and protein solubility. The choice of stabilizers used for the improvement of emulsion stability depends on the composition of the beverages. If the ingredients contain significant amount of protein, the selection depends on product's pH. For high acid beverages ($pH < 4.6$), high ester pectin or carboxy methyl cellulose may be used. For neutral pH, carrageenan, colloidal microcrystalline cellulose, and gellan gum may be used. In low protein ingredients such as most cereals, if oil ring is present on top of the beverage's bottle, propylene glycol alginate, gum Arabic, and most emulsifiers may be selected. If pulp is settling on the bottom of beverage's bottle, the mixtures of xanthum, guar, locust bean, gellan gums, and alginate may be required (Sethi et al., 2016). A recent study investigated the effect of incorporating stabilizer, xanthan gum, on the physical stability of rice milk, and found that rice milk was stabilized by incorporating 0.1% xanthan gum (Koyama & Kitamura, 2014).

10.3.4 Shelf-Life of the Products

Commercial cereal-based milk substitutes are pasteurized or UHT treated to extend the shelf life of the product. Heat, on the other hand, can alter protein characteristics, which can affect stability, as well as flavor, aroma, and color (Mäkinen et al., 2016). Excessive heating caused severe effects on nutrients, especially vitamins and amino acids, browning and formation of cooked flavor (Kwok & Niranjan, 1995). Pasteurization is operated at temperatures below 100 °C, which destroys enough microorganisms to enable a shelf life of about 1 week at refrigerated conditions. In the UHT process, the product is heated to 135–150 °C for a few seconds, yielding a sterile product that can be kept unrefrigerated (Basinskiene & Cizeikiene, 2020). Though, thermal treatments are well applied to cereal/plant-based milks but some limitations such as high starch content in cereals require the exploration of modern techniques e.g. non-thermal processing technologies to extend the shelf-life. Currently, little scientific literature on the application of non-thermal technologies on the production of cereal/plant-based milks is available. High-pressure throttling, ultra-high pressure homogenization (UHPH), high pressure processing technologies are one of the promising technologies and they have been investigated in soy milk (Cruz et al., 2007). The effect of UHPH in combination with conventional thermal treatment was investigated in almond milk (Valencia-Flores et al., 2013). The UHPH treatment was found to provide high quality products when compared with those treated by pasteurization and UHT techniques. Pulsed electric field technique has been investigated for its potential in microbial reduction of cow's milk (Valizadeh et al., 2009). There is still a need for more investigation of other appropriate non-thermal technologies as well as their preservative effects on plant-based milks (Sethi et al., 2016).

More in-depth reviews with regards to the production of cereal/plant-based milk alternatives are available: with particular focus on the breaking down process (McClements et al., 2019; Sethi et al., 2016), and fabrication of simulated fat globules (Do et al., 2018; McClements, 2020).

10.4 Processes for Improving Nutritional Qualities of Cereal-Based Milks

Cereal milk is lactose-free, hypoallergenic, and contains bioactive compounds with potential health-promoting properties, which is a promising alternative to cow's milk. However, it is generally not nutritionally balanced compared to cow's milk and faces a number of technological challenges related to processing and product stability. Compared to cow's milk, cereal milk is rich in carbohydrates and dietary fiber but generally lacks calcium, iron, vitamin A and certain amino acids, and contains less protein content, and therefore cereal milk cannot completely replace cow's milk for children under 5 years of age (Sethi et al., 2016). The aforementioned nutrients must be enhanced to ensure that the final product is a nutritional replacement for cow's milk. Besides, processing may result in a loss of nutrients, especially minerals and vitamins, so enrichment/fortification should also be performed to restore and balance these nutrients lost during processing or absent from the raw materials. In addition, blending two or more types of cereals or plant-based milk, to have a final product with high nutritive properties comparable with cow's milk, is an important step for the production of cereal-based milks (Xiong et al., 2020).

10.4.1 Blending for Nutritional Balance and Improvement in Sensory Acceptability

The corresponding plant-based beverages vary greatly in composition and flavor, depending on the specific plant materials. Plant-based milk alternatives should preferably resemble the physical, nutritional and organoleptic properties of cow's milk. Researchers and developers in academia and industry must overcome a number of obstacles to attain this goal (Tangyu et al., 2019). Therefore, blending two or more types of plant-based milks, to obtain a product with enhanced nutritive values comparable with cow's milk, is one of the simple yet efficient ways. Figure 10.3 shows the macronutrient composition, functional components and limiting factors of

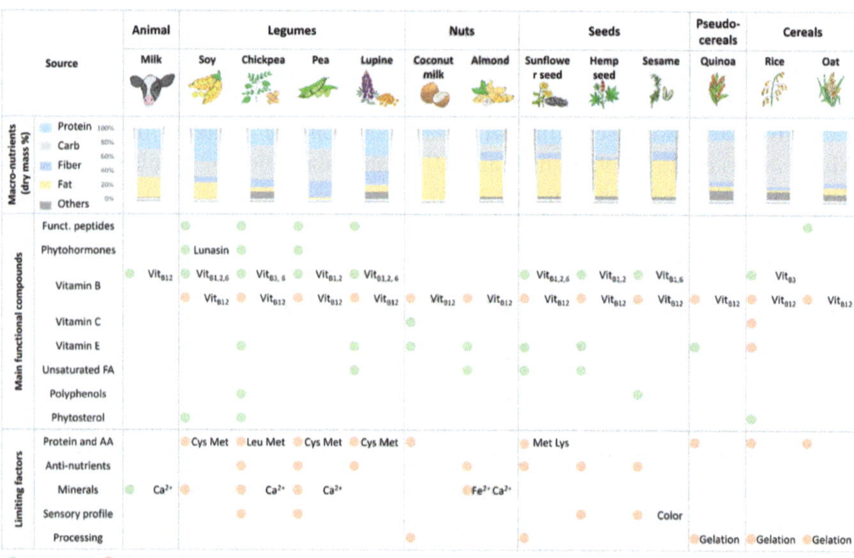

Fig. 10.3 Macronutrient composition, functional components and limiting factors of cereals and other common plants used for the production of plant-based milk alternatives. (Source: Tangyu et al., 2019. Open Access)

cereals in comparison with other common plants used for plant-based milk alternative production.

Recently, various multigrain plant-based milks are available in the market. Figure 10.4 shows the commercial soy milk mixed with colored brown rice, black sesame, malt extract and barley, to enhance its nutritional values. In addition, modern food manufacturers with advanced processing technology can formulate a variety of cereal/plant-based milk alternatives with desired properties based on the market demand.

It has been reported that blending of cereal milks with legume and nut milks helped in improving quality and wider acceptability of plant-based milks. Cereal proteins which are deficient in lysine, when blended with legumes which are deficient in sulfur amino acids, will have complementary amino acid profiles (Duranti, 2006). In theory, combining various plant-based proteins could compensate for the lower anabolic capacity of these protein sources, which means that mixing plant-based protein sources such as cereals and legumes in the same food could improve essential amino acids composition, ensuring that its profile meets the body's needs and even prove more efficient than fortification with free limiting amino acids (Berrazaga et al., 2019). The free essential amino acids used to fortify plant-based

Fig. 10.4 Example of commercial soy milk mixed with purple brown rice, black sesame, malt extract and barley

proteins could be digested and absorbed faster than their constitutive amino acids due to the postprandial desynchronization effect (Dardevet et al., 2012).

10.4.2 Fortification of Plant-Based Milk

Cereal/plant-based diets are deficient in some important macronutrients and micronutrients. These include high-quality proteins, omega-3 fatty acids, vitamin B12, vitamin D, iron, calcium, and iodine (Obeid et al., 2019). For long term, deficiencies in these nutrients could lead to health problems, especially in infants and the elderly (Hunt, 2019; Sebastiani et al., 2019). Therefore, it would be beneficial to supplement plant-based diets with these critical nutrients, to overcome the deficiencies. In addition, additional health benefits associated with fortifying cereal-based milks with nutraceuticals, such as carotenoids, curcuminoids, and polyphenols could also be obtained (Assadpour & Mahdi Jafari, 2019). Cereal/plant-based milk substitutes are an excellent vehicle for supplementing the human diet with important nutrients and nutraceuticals. They are easily integrated into a person's daily routine, for example, as a beverage, creamer, or component of a breakfast. They are colloidal dispersions that can be enhanced with bioactive substances of various molecular polarities: hydrophilic, hydrophobic, or amphiphilic. They can also be engineered to enhance the bioavailability of the bioactive substances enclosed. They must, however, be carefully designed to ensure that the bioactive compounds remain stable throughout the product's shelf life, do not degrade the product's quality, and are highly bioavailable following consumption (McClements, 2020).

There are a variety of alternative strategies for fortifying plant-based milk substitutes, which vary according to the formulation methods employed and the physicochemical properties of the bioactive compounds used (McClements et al., 2019). If a plant-based milk is produced by homogenization of bulk oil and water phases, hydrophobic bioactive compounds can easily be dissolved in the oil phase prior to emulsion formation, whereas hydrophilic bioactive compounds can be administered either before or after emulsification (McClements, 2018). Vitamins that are oil-soluble can be added into emulsions stabilized by other kinds of plant-based emulsifiers, such as lecithin (Mehmood et al., 2019).

If plant-based milk is produced by the breakdown of the typical cellular structure of plant components, then distinct procedures are necessary. Hydrophilic bioactive compounds are frequently distributed or dissolved easily within the aqueous phase of plant-based milk. Nonetheless, certain bioactive substances may involve further precautions. For instance, calcium often has to be incorporated in a colloidal form or small solid particles; this is to avoid aggregation of any anionic constituents, such as proteins, in the product promoted by free calcium ions. Hydrophobic bioactive compounds can be added into premade plant-based milk substitutes using a variety of methods (Maurya et al., 2020). To begin, it may be possible to simply homogenize a bioactive-fortified oil phase with plant-based milk to create bioactive-fortified oil droplets. Any surface-active chemicals naturally present in the aqueous phase of plant-based milk substitutes, such as proteins or phospholipids, will stabilize these oil droplets. Second, it is also possible to create a separate plant-based emulsion containing the hydrophobic bioactive compounds dissolved in the oil phase (McClements, 2018), and then combine this with the plant-based milk at final processing step. In this case, the hydrophobic bioactive compounds may diffuse through the aqueous phase so that they are distributed consistently throughout all the oil droplets. Third, a pH-shift approach can be employed to load some hydrophobic bioactive compounds into plant-based milk substitutes. For example, curcumin which is a hydrophobic nutraceutical could be loaded into commercial plant-based milk substitutes using a pH-shift procedure (Zheng et al., 2019). The curcumin is water-insoluble under neutral or acidic conditions (pH $\leq$ 7) but highly water-soluble under strongly alkaline conditions (pH 12). Hence, a slightly acidic plant-based milk can be added to an alkaline curcumin solution, causing the curcumin to migrate from the aqueous phase into the oil bodies. Nonetheless, this latter technique is suitable to a limited number of bioactive compounds that are highly soluble under certain solution conditions but not others (McClements, 2020).

Emulsion-based delivery techniques are extremely adaptable for loading hydrophobic bioactive compounds into premade plant-based milk substitutes (McClements, 2015). Previous research has established that the composition and structure of emulsion-based delivery systems have a significant effect on the stability and bioavailability of encapsulated hydrophobic bioactive compounds (McClements, 2018). The nature of the oil phase, the size of the oil droplets, and the type of emulsifiers all have a significant effect on the bio accessibility, transformation, and/or absorption of hydrophobic bioactive compounds (McClements et al., 2015). Consequently, their attributes can be tailored to create plant-based milk substitutes in which the encapsulated bioactive compounds maintain their stability throughout storage yet are released in a highly bioavailable form in the gastrointestinal system following ingestion (McClements, 2020).

10.4.3 Fermentation for Improvement of Nutrition and Sensory Acceptability

As mentioned earlier, commercial cereal/plant-based milk alternatives are typically fortified with necessary additives such as vitamins, amino acids, and minerals (Sethi et al., 2016). However, leading food and beverage firms have committed to omitting artificial ingredients from their products – clean label foods and beverages are no longer a trend, but are increasingly becoming an expectation. As a result, natural plant-based milk replacements that are nutritionally equivalent to and taste as good as animal-derived milk without blending are of great interest (Asioli et al., 2017). Fermentation is an appealing method for accomplishing this goal. For centuries, cereals such as maize, wheat, rice, and sorghum have been fermented (Sandhu & Punia, 2017; Dhull et al., 2020, 2021). It is also reported that fermented, wheat, barley and rice bran may improve the bioactive profile of various food products (Sandhu et al., 2016; Punia Bangar et al., 2021a, b). Fermentation enhances the palatability, aroma, flavor, and taste of cereal beverages by generating organic acids and volatile compounds such as lactic acid, acetic acid, propionic acid, and acetaldehyde (Setta et al., 2020). Notably that cereal fermented beverages can be classified as alcoholic and non-alcoholic beverages. Alcoholic fermented cereal beverages are not covered here in this chapter.

Plant materials are required for the growth of microorganisms (Espirito-Santo et al., 2014; Peyer et al., 2016). The most widely used microbes for plant-based fermentation are lactic acid bacteria (LAB), bacilli and yeast (Jeske et al., 2017b). These microorganisms, which are primarily used as monocultures, have been shown to contain traits that increase critical nutritional and/or sensory aspects (Tangyu et al., 2019).

The growth of microbes during fermentation can enhance protein content, improve protein solubility and amino acid composition and availability. Notably, during fermentation, some microbial strains synthesize vitamins, including vitamin K and B group vitamins (LeBlanc et al., 2011, 2013). The ability of yeast to produce vitamin B_2 is well recognized (Steinkraus, 1997). In comparison to synthetic fortification, vitamins produced by microorganisms are largely acknowledged as safer, more natural and more environmentally friendly (LeBlanc et al., 2011).

Fermentation, alone or in combination with other processes such as cooking, sprouting, and soaking, can significantly lower the amount of anti-nutrients in plant-based milks such as tannins, phytates, and cyanides (Soetan & Oyewole, 2009). For instance, LABs can produce phytases and maintain optimal pH conditions for these enzymes, which subsequently catalyze the hydrolysis of phytates to myo-inositol and phosphate, thereby increasing digestibility and mineral bioavailability (Rekha & Vijayalakshmi, 2010). As an example, fermentation of finger millet considerably reduced a number of undesired anti-nutrients (phytates, tannins, and trypsin inhibitor) while increasing mineral extractability and digestibility (Antony & Chandra, 1998).

Fermentation has the potential to enhance the sensory profile of plant-based milk substitutes, resulting in desirable volatile flavors. For instance, during cereal-based fermentation, diacetyl (2,3-butanedione) is released, imparting a pleasant butterscotch aroma (Peyer et al., 2016). Acetaldehyde concentration is increased during cereal fermentation, delivering a pungent, fruity (green apple) flavor with sweet notes (Sethi et al., 2016). Changes in the amounts of amino acids during fermentation can affect the flavor and taste of plant-based milk replacements (Tangyu et al., 2019).

Mixed-culturefermentation exhibits synergistic effects that enhance the qualities of products. Though, fermentation of cereals for the production of probiotic beverages in most cases employ single cereal substrate and culture as delivery vehicle for potentially probiotic lactic acid bacteria (Angelov et al., 2006). It was discovered that mixed culture fermentations on mixed grain substrates had a significant effect on the organoleptic qualities of the products, but not on the cell populations. Malt was found to be the optimal substrate for cell growth (as a single and mixed media), which could be related to its chemical composition. Low pH and bacterium strains may be the primary factors restricting microbial growth (Rathore et al., 2012).

10.4.4 Sprouted Grains as Ingredients for Plant-Based Milks

Recent favorable consumer impressions of sprouted cereals have resulted in the production of new foods and beverages. There is no globally recognized definition or regulation of "sprouting" at the moment. Frequently, the terms "germination," "sprouting," and "malting" are used interchangeably. Indeed, malting is a particular type of sprouting that results in fermentable extracts for the brewing and distilling industries (Hübner & Arendt, 2013). In the scutellum and aleurone cells, sprouting initiates the synthesis of starch-degrading enzymes such as α-amylase and α-glucosidase. The overall starch content of cereals falls by 5–15% as a result of sprouting, depending on the cereal type, sprouting time, and temperature (Tian et al., 2010). While sprouting causes protein hydrolysis, it has no discernible effect on the total protein concentration (Cáceres et al., 2014). Sprouting alters the composition and amount of insoluble and soluble dietary fiber, which can be utilized to increase or decrease dietary fiber content (Teixeira et al., 2016).

At the moment, the health benefits of germinated or sprouted cereals can only be determined by examining the biochemical changes that occur during sprouting and associating them with prospective health benefits. The direct effect of such biochemical changes on in vivo health markers is limited (Lemmens et al., 2019). Nonetheless, it may be summarized that sprouting induced the changes of macronutrients as a consequence of biologically activated grains. A variety of bioactive compounds is affected by germination. While some compounds, such as antioxidants and total phenolics are increased, but others like β-glucans are degraded. Minerals are released from their phytate chelates and become available for intestinal absorption, and vitamins and γ-amino butyric acid are synthesized and accumulate. Sprouted grains also improved their functional properties such as absorption capacity as well as emulsion and foaming capacity which enable their applications in foods and beverages. Sprouting which is a natural way could significantly enhance the nutritional, functional, and bioactive compounds of grains, as well as improve palatability (Hübner & Arendt, 2013; Lemmens et al., 2019; Singh & Sharma, 2017).

Though, all whole grain cereals can be germinated and used for beverage ingredients. For commercial cereal-based milks, sprouted or germinated brown rice has become widely accepted and various product formulae are available (Fig. 10.5). The production process employs similar process as used for other cereal-based milks, the breakdown process in which germinated brown rice is used as the main ingredient (Beaulieu et al., 2020; Patil & Khan, 2011).

The utilizations of sprouted seeds from other cereals such as corn, wheat and millet for beverages rely on organoleptic properties and consumer acceptance of the final products. Current research on germinated cereal products focusses on fermented cereal beverages that could enhance both functional and sensorial attributes.

Fig. 10.5 Commercial sprouted or germinated brown rice milks

For example, a novel synbiotic beverage based on millet sprouts with a mixed cultures was developed (Mohammadi et al., 2021). Others included probiotic sprouted wheat based milk (Sharma et al., 2014), and sprouted sorghum beverages (Laetitia et al., 2005). In addition, sprouted cereal grains can be processed into flour and then used as the ingredient for the production of beverages such as fermented beverages from sprouted wheat flour (Shokoohi et al., 2015), and beverage from sprouted oat flour (Aparicio-García et al., 2021).

10.4.5 Antinutrients in Cereals

Antinutrients are associated with compounds of natural or synthetic origin, which interfere with nutrient absorption, thereby reducing nutrient intake, digestion, and utilization to produce adverse effects. Sensitivity of antinutrients differs in each person and proper food processing is suggested to reduce antinutritional factors (Popova & Mihaylova, 2019). Antinutrients are found in higher concentrations especially in uncooked cereals, legumes, and nuts. The most common antinutrients in cereal/plant-based foods are phytic acids, saponins, tannins, α-amylose inhibitors, trypsin inhibitors, protease inhibitors, lectins, oxalates, exorphins, and goitrogens (Reyes-Jurado et al., 2021).

Lectins which are commonly found in cereals (0.5–7.3 mg/100 g) could bind or modify carbohydrates (glycoproteins, glycolipids, and polysaccharides), as well as evade the immune system and travel through the body, causing diseases in the small intestine (Popova & Mihaylova, 2019). Trypsin inhibitors which could reduce protein absorption are found in the legume family at a significant amount, less in cereals. Tannins can be found in cereals, such as sorghum, at notable amounts (Palacios et al., 2021). They could affect the digestion of nutrients and preventing the body from absorbing beneficial bioavailable substances. They can also bind and shrink proteins, inactivate digestive enzymes, and reduce protein digestibility (Popova & Mihaylova, 2019). Phytic acid or phytate or myo-inositol hexaphosphate (IP6) is mainly found in the outermost layer of cereals, in the range is 50–74 mg/100 g. They are heat stable, since they do not degrade during cooking, and could cause low mineral bioavailability, solubility, and functionality, as well as protein and carbohydrate digestibility (Petroski & Minich, 2020). Saponins which are toxic in high concentrations and can affect the absorption of nutrients are commonly found in legumes, less in cereals. Oxalates content in cereals and nuts are commonly in the range of 35–490 mg/100 g. Oxalates cannot be metabolized outside the urinary tract once they are processed through the digestive system. The signs and symptoms that may occur with small or high amounts of oxalates are enteric and primary hyperoxaluria; burning of the eyes, ears, mouth, and throat; abdominal pain; muscle weakness; nausea; and diarrhea. Calcium oxalate can accumulate as kidney stones, thereby affecting renal health (Popova & Mihaylova, 2019; Reyes-Jurado et al., 2021).

Although, cereals could contain significant amount of antinutrients, fortunately, most of them are removed from cereals through dehulling, soaking, fermenting, or thermal treatment (cooking) (Reyes-Jurado et al., 2021). The combination of these treatments is more effective in reducing or eliminating the antinutrients and this commonly takes place in the production of cereal-based milks as previously described. In addition, when administered in doses ranging from moderate to high in the diet, or when administered in isolation, antinutrients may cause negative health impacts. Severe effects may be observed by some individuals who are more susceptible to these antinutrient compounds. These compounds, on the other hand, are rarely consumed in their isolated or high concentration forms. Cereal/plant-based diets that contain these compounds also contain thousands of additional compounds in the food matrix, many of which act as a counterbalance to the antinutrients' potential effects. Antinutrient compounds may, in some cases, act as therapeutic agents for a variety of conditions. Thus, additional research incorporating these variables is necessary before conclusive conclusions about the antinutrients' adverse effects in their whole food form can be drawn (Petroski & Minich, 2020).

10.5 Health Benefits of Cereal-Based Milks

The health and functional properties of cereals and cereal products are mainly attributed to their bioactive phytochemicals. Cereal grains have a unique phytochemical profile that is complementary to fruits and vegetables (Liu, 2007). The main bioactive phytochemicals in the whole grain are phenolic compounds, dietary fibers, carotenoids, tocols, phytosterol, γ-oryzanol and phytic acid. Although there are several excellent reviews and book chapters have discussed the bioactive phytochemicals in most cereals (Björck et al., 2012; Granato et al., 2010; Liu, 2007; Okarter & Liu, 2010; Slavin, 2004; Srikaeo et al., 2017; Srikaeo, 2020), this chapter provides important updates and highlights the main categories and the most abundant sub-categories of bioactive phytochemicals found in cereals.

10.5.1 Phenolic Compounds

Phenolic compounds, or polyphenols, are a large group of chemical compounds consisting of one or more benzene rings with hydroxyl groups, such as phenolic acids, flavonoids, coumarins, stilbenes, etc. These compounds are ubiquitous in plants and therefore an unavoidable part of human diet (Dykes & Rooney, 2007). Phenolic compounds possess many physiological functions including antioxidant, anti-inflammatory, anti-thrombotic, anti-allergenic, anti-microbial, anti-carcinogenic effects, etc., with the potential to reduce the risk of a wide range of chronic diseases and promote human health (Dai & Mumper, 2010; Fardet et al., 2008). The antioxidant effect has been the most widely studied health benefits and is considered as the major role of phenolic compounds. Many phenolic compounds have been demonstrated to have potent antioxidant activity in cell culture, in vitro and in vivo studies (Fernandez-Panchon et al., 2008). The mechanisms of the antioxidant action of phenolic compounds are highly complex. It has been suggested that phenolics exert the antioxidant effects via the following three main actions: (1) direct scavenging of free radicals; (2) preventing the formation of free radicals by inhibiting oxidant enzymes or chelating some transition metals; (3) upregulation of antioxidants and detoxifying enzymes. There is abundant evidence that the intake of plant phenolic compounds is inversely associated with the risk of many oxidative stress led diseases, including cardiovascular diseases and cancers (Fardet et al., 2008; Scalbert et al., 2005; Weichselbaum & Buttriss, 2010). Phenolic compounds are the major source of antioxidants in cereals, and the most common phenolic compounds are phenolic acids and flavonoids, in which sorghum (87–2960 mg GAE/100 g) and buckwheat (506–1642 mg GAE/100 g) have the most abundant phenolic compounds among the major cereals (Fardet et al., 2008; Goufo & Trindade, 2014; Guo et al., 2011; Liu et al., 2019).

10.5.1.1 Phenolic Acids

Phenolicacids are a significant class of phenolic compounds found in all cereals. They are classified into two subclasses: hydroxybenzoic acids and hydroxycinnamic acids. The primary phenolic acids in cereals are hydroxybenzoic acids, which include salicylic and phydroxybenzoic acids, and hydroxycinnamic acids, which include ferulic, *p*-coumaric, sinapic, and caffeic acids. Among the major cereals, sorghum (39–285 mg/100 g) and millet (61–391 mg/100 g) have the highest concentration of phenolic acids (Dykes & Rooney, 2007; Xiong et al., 2019a, b). Phenolic acids exist in both free and bound forms. Most phenolic acids are found in the bound form in cereals, which are covalently bound to cell walls and are an integral part of the cell wall structures, requiring acid and alkaline conditions to break the covalent bonds for extraction; whereas free phenolic acids are found mainly in the outer layer of pericarp and are extractable by organic solvents (López-Alarcón & Denicola, 2013; Robbins, 2003). Ferulic acid is the most abundant phenolic acid in all common cereals (except buckwheat) and is found mainly in the bound form in the cell wall of the bran fraction (Fardet et al., 2008; Liu et al., 2019). Wheat and corn have been reported to contain the highest concentration of ferulic acid and can account for up to 90% of the total phenolic contents of the whole grains (Boz, 2015; Lempereur et al., 1997). However, in buckwheat, ferulic acid is only detected either in trace amounts in common buckwheat, or not detected in tatary buckwheat (Liu et al., 2019). Tartary buckwheat has about six times more phenolic acid content than common buckwheat, and salicylic acid has been reported to be the most abundant phenolic acid (>90% of the total phenolic acids). The primary health effect of phenolic acids is considered to be the antioxidant activity (Heleno et al., 2015).

10.5.1.2 Flavonoids

The main flavonoids (subclasses) reported in cereals include anthocyanidins, flavones, flavonols, falvan-3-ols and flavanones. Among the major cereals, sorghum has the most diverse and abundant flavonoids (Xiong et al., 2020). Anthocyanidins, a group of water soluble pigments (orange, red to purple, blue color), has been the most studied flavonoids in cereals. The anthocyanidin content varies greatly among cereals. High concentrations are found in cereals with pigmented pericarp, such as orange-red to purple-blue or black colored corn, rice, wheat, barley and sorghum (Goufo & Trindade, 2014; Idehen et al., 2017; Rouf et al., 2016; Xiong et al., 2020). Sorghum contains a rare type of anthocyanidins, called 3-deoxyanthocyanidins, which lacks an OH group at C-3 position. 3-Deoxyanthocyanidins are more stable than anthocyanidins, and have been proposed as potential multifunctional natural colorants (Xiong et al., 2019a, b). Other flavonoids in cereals are also widely present in fruits and vegetables. Antioxidant, anti-inflammatory, anti-cancer and antimutagenic effects, and modulation of certain cellular enzymes have been suggested as the main healthbenefits of flavonoids (Panche et al., 2016).

10.5.1.3 Other Phenolics

Condensed tannins (proanthocyanidins or procyanidins) are a group of polymeric phenolic compounds. In cereals, they are composed primarily of flavan-3-ol monomers and are mainly found in some pigmented cereal grains such as brown sorghum, red finger millets and dark-colored barley (Dykes & Rooney, 2007; Salar et al., 2017). In sorghum, condensed tannins have a high molecular weight and a high degree of polymerization, and are not commonly found in other major cereals; but are very abundant in sorghums with pigmented testa (type III high tannin sorghums) (Xiong et al., 2020). Condensed tannins are known for their bitter and astringent taste and have long been considered as anti-nutritional compounds due to their ability to bind to some proteins, amino acids and minerals and thus reduce their bioavailability (Xiong et al., 2020). Nevertheless, they have been demonstrated to possess potential multiple health effects. These compounds exhibit much higher antioxidant capacity in vitro than monomeric phenolics, with potential gastro protective and cholesterol lowering effects, and have been suggested as an ideal dietary ingredient for people with obesity and diabetes (Smeriglio et al., 2017; Xiong et al., 2020).

Lignans are a group of phytoestrogens present in many plants. They are found in the bran of cereal grain and are relatively abundant in oat and rye (Durazzo et al., 2013; Rodríguez-García et al., 2019). The main plant lignans include secoisolariciresinol, matairesinol, lariciresinol, pinoresinol and syringaresinol. These lignans can be converted into mammalian lignans (in the colon by microflora), which have strong antioxidant and weak estrogenic activities, and are believed to promote colon health and reduce the risk of hormone related cancers (Adlercreutz, 2007).

Alkylresorcinols are a group of phenolic lipids. They are located in the bran fraction of cereal grain and have been reported in some cereals like corn, wheat, barley, millet and rye, where rye has the highest amount of alkylresorcinols among them (Fardet et al., 2008; Mattila et al., 2005; Ross et al., 2003). Alkylresorcinols have shown anti-microbial activity and antioxidant activity in vitro, and they are considered as lipophilic antioxidants, although less efficient than vitamin E (Ross et al., 2004).

Avenanthramides are a group of phenolic alkaloids unique to oat. The main avenanthramides reported in oat are avenanthramide A, B and C, and these compounds have been demonstrated to have antioxidant, anti-inflammatory and anti-atherogenic activities *in vitro* and/or *in vivo* (Meydani, 2009).

10.5.2 Dietary Fibers

Dietary fiber is another large family of bioactive phytochemicals and one of the major and common health-promoting components of cereals. Dietary fibers are located in both bran and endosperm fractions of cereal grain, and the main dietary fiber components in cereals include cellulose, arabinoxylan, lignin, β-glucan, resistant starch and inulin (Dhingra et al., 2012). The health effects of dietary fibers have been extensively studied and well documented. Whole grain dietary fibers have multiple physiological effects, including anti-glycemic, antioxidant, hormone regulation, prebiotic, and modulation of intestinal microbiota. High dietary fiber intake has been linked to a reduced risk of certain diseases such as gastrointestinal diseases, cardiovascular diseases, diabetes and obesity (Aune et al., 2011; Charalampopoulos et al., 2002; Smith & Tucker, 2011; Threapleton et al., 2013). β-glucan, arabinoxylan and resistant starch are the most widely studied dietary fibers in cereals. β-glucans are a group of polysaccharides composed of glucose units. They are important dietary fibers, mainly located in the cell walls of the aleurone and endosperm, and are found in all major cereals (Ermawar et al., 2015). Among the major cereals, oat and barley contain the highest amount of β-glucans where they are concentrated in the aleurone layer and endosperm, respectively (Ahmad et al., 2012). The main physiological effects of β-glucans include lowering blood cholesterol, regulating blood glucose and enhancing the immune system (Wood, 2010). β-Glucans are water-soluble and have high viscosity which can bind cholesterol and bile acids in the gastrointestinal tract, thus facilitating their elimination in the body (Othman et al., 2011). It has been recommended by the U.S. Food and Drug Administration that consumption of 3 g or more β-glucans, from barley and oat, per day, can help to lower blood cholesterol levels and reduce the risk of coronary heart disease. Arabinoxylans are a type of hemicellulose consisting of a linear xylose backbone chain with arabinose substitutions. Arabinoxylans have been reported in all major cereals, and rye has the highest arabinoxylan content (Vinkx & Delcour, 1996). Arabinoxylans are mainly located in the cell walls of both bran and endosperm fractions of grains. They are the second most abundant biopolymers, after cellulose, in many cereals and form a major source of dietary fiber in the diet (Saeed et al., 2011). The unique feature of arabinoxylans is the presence of phenolic acids, such as ferulic acid, which are covalently linked to xyloses and arabinoses. The ferulated arabinoxylans have prebiotic, antioxidant, and anti-cancer properties and have gained much attention in the pharmaceutical industry (Mendez-Encinas et al., 2018).

Resistant starch, also considered a member of the dietary fiber group, is a type of starch commonly found in cereal grains and legumes. It resists the digestion in human digestive tract, passes into the large intestine. Resistant starch can be categorized into five types (RS1–5), some of which occur naturally in foods, as found in cereal grains, and some of which are produced or modified commercially, and incorporated into food products (Lockyer & Nugent, 2017). Resistant starch has various benefits to colonic and metabolic health (Fuentes-Zaragoza et al., 2010). Like soluble fiber, resistant starch is a substrate for the colonic microflora, forming metabolites including the short-chain fatty acids, mainly acetic, propionic and butyric acid. The short-chain fatty acids, particularly butyric acid, appear to be particularly important for the colonic epithelium's health (Liljeberg Elmståhl, 2002). A high amylose found in the starch is associated with an increase in resistant starch content. Processing cereal grains may result in either increase or decrease in the resistant starch content, depending on the conditions (Alsaffar, 2011).

10.5.3 Carotenoids

Carotenoids are a group of natural pigments widely present in plants, responsible for yellow, orange and red color. In cereals, carotenoids are more evenly distributed within the grain compared to other bioactive compounds such as phenolic compounds, and significant amounts of carotenoids are present in the endosperm, which results in the yellow color of the endosperm (Trono, 2019). The main carotenoids found in cereals are lutein, zeaxanthin, α-carotene, β-carotene and β-cryptoxanthin, and the composition and concentration of these carotenoids in cereals vary considerably. In general, cereals have a lower carotenoid content compared to fruits and vegetables, but some cereals with dark yellow or golden color pigmented endosperm, such as high carotenoid corn, golden rice, foxtail millet, contain high levels of carotenoids (Mellado-Ortega & Hornero-Méndez, 2015; Ndolo & Beta, 2013; Zhai et al., 2016). Among the major cereals, corn is the best source of carotenoids, ranging from 0.9 to 15.6 mg/100 g of the whole grain. Lutein, zeaxanthin and β-carotene are the most abundant carotenoids in corn (Trono, 2019). Carotenoids have received much attention due to their pro-vitamin and lipid-soluble antioxidant properties, and have been reported to be responsible for several physiological functions, including supporting eye, skin and cardiovascular health and enhancing the immune system (Fiedor & Burda, 2014).

10.5.4 Tocols

Tocols (vitamin E analogs), including tocopherols and tocotrienols, are naturally occurring lipid-soluble antioxidants in fruits, vegetables and cereal grains. In cereals, tocols are present in all major cereals and are mostly located in the germ fraction of the grain (Ko et al., 2003). Tocols are generally found at moderate levels in cereals, but substantial amounts have been reported in corn, wheat, barley, oat and rye (Chung et al., 2013; Tiwari & Cummins, 2009; Zielinski et al., 2001). The composition and concentration of tocols vary considerably among cereals. There are six main tocols in cereals, which are α-, β- and γ-tocopherol and α-, β- and γ-tocotrienol. Among these, α-tocopherol is the most common one in all major cereals; and high levels of α-tocopherol (in corn), γ-tocopherol (in corn and buckwheat), and α-tocotrienol (in barley and oat) have been reported (Asharani et al., 2010; Ehrenbergerová et al., 2006; Tiwari & Cummins, 2009). The health effect of tocols has been a subject of extensive research. The most important health functions of tocols are the antioxidant activity and the maintenance of membrane integrity, and it has been shown to protect against some degenerative diseases, including cardiovascular diseases and cancers (Fardet et al., 2008; Zielinski et al., 2001).

10.5.5 Phytosterols

Phytosterols are plant sterols and stanols, having a structure similar to cholesterol. Phytosterols are structural components of plant cell membranes, located primarily in the bran fraction of cereal grains and found in all major cereals (Jiang & Wang, 2005; Lampi et al., 2004; Nurmi et al., 2012) Phytosterols in cereals exist in free form, esterified with fatty acids or phenolic acids or conjugated with glycosides, and cereal grains are considered to be the major dietary source of phytosterols (Piironen et al., 2002). The main phytosterols in cereals are sitosterol, campesterol, stigmasterol, sitostanol and campestanol; sitosterol is the predominant phytosterol in all major cereals. Phytosterols are one of the bioactive compounds currently under active research, and the cholesterol lowering effect is the main health function that has been well documented (Ostlund, 2004).

10.5.6 γ-Oryzanol

γ-Oryzanol, originally presumed to be a single chemical compound and unique to rice, but is now known to be a mixture of triterpene and ferulated phytosterol and has also been detected in other cereals such as corn, wheat, barley and rye, but at much lower levels compared to rice (Goufo & Trindade, 2014). γ-Oryzanol is mainly located in the bran of the whole grain. In rice, the γ-oryzanol content is about 20–122 mg/100 g in the whole grain and 185–421 mg/100 g in the bran fraction (Goufo & Trindade, 2014; Kim et al., 2015; Sing et al., 2015). It is a major functional component in rice bran oil. γ-Oryzanol has been demonstrated to have significant antioxidant activity both in vitro and in vivo (Juliano et al., 2005; Minatel et al., 2016), as well as potential anti-inflammatory, anti-cancer and anti-diabetic health functions (Minatel et al., 2016).

10.5.7 Phytic Acid

Phytic acid (inositol hexakisphosphate), or phytate when in salt form, is the principal storage compound for phosphorus in many plants. In cereal grains, phytic acid is mainly located in the bran fraction, in particular the aleurone layer and germ of the grain. Phytic acid contains more than 70% of total phosphorus in the whole grain and accounts for 1–7% of the whole grain dry weight (Fardet et al., 2008; Goufo & Trindade, 2014). Phytic acid has long been considered as an anti-nutritional compound due to its strong ability to chelate minerals such as iron, calcium, magnesium and zinc, thus affecting their intestinal bioavailability (Kumar et al., 2010). Despite the anti-nutritional effect, phytic acid has also been reported to have multiple positive physiological functions, including antioxidant activity (suppressing iron catalyzed oxidation), interruption of cellular signal transduction, and promotion of DNA repair, etc. (Kumar et al., 2010; Oatway et al., 2001). Consumption of phytic acid can promote colon health, and provide protection against cardiovascular diseases, diabetes and a number of oxidation mediated cancers (Nissar et al., 2017).

It should be noted with regard to the limitations of the narrative review on antinutritional compounds. To begin, human clinical trial research on the effects of antinutritional compounds in whole foods is limited and, in some cases, does not always yield conclusive results. Epidemiological and observational studies must be used in lieu of clinical trials, though they are typically limited in their applicability due to uncontrolled variables. Second, the majority of research on antinutritional components is conducted on isolated compounds in animal models that do not reflect a balanced diet. The limitations of research are exacerbated even further by the frequently synergistic nature of food, the effects of cooking and processing, and the bio-individuality of study participants. Additional research incorporating these variables is required before definitive conclusions about the ill-effects of these compounds in their whole food form can be drawn.

10.6 Cereal-Based Milks Versus Cow's Milk

When compared to cow's milk, cereal milk is lactose-free, hypoallergenic and contains bioactive components with health-promoting properties as well as being economical and sustainable. However, cereal milk faces some problems related to nutritional imbalance, poor sensory experience, processing difficulty and emulsion stability (Xiong et al., 2020). Sensory and processing difficulties have been discussed earlier. In terms of nutritional imbalance, regarding the proteins, cereal proteins are normally of lower nutritional quality compared to animal-derived proteins because of limiting amino acids especially lysine, and poor protein digestibility (Outi Elina Mäkinen et al., 2016). The nutritional inferiority of cereal-based milks was reported to represent a risk, especially when given to younger children as cow's milk substitutes, without recognizing the peculiar differences (Verduci et al., 2019). In this regard, these beverages have been inappropriately consumed as substitutes for infant formula, particularly in the case of alleged cow's milk protein allergy (CMPA). Cereal/plant-based beverage composition does not adhere to European guidelines; they are low-energy beverages with insufficient protein, vitamins, and minerals for early childhood. They may result in a severe nutritional deficiency in infants. Between 2008 and 2011, nine cases of severe nutritional deficiency caused by vegetable drink consumption (aged 4–14 months) were reported (Le Louer et al., 2014). Regarding the lipid profile, except for coconut milk, most cereal/plant-based milks generally have low saturated fatty acid levels; however, some products have energy-producing levels comparable to whole cow's milk, owing to sugars and other carbohydrates (Singhal et al., 2017). Certain products also contain added sugars and sweeteners, and there is a difference in the carbohydrate profile of vegetable drinks due to the absence of lactose and galactose. A recent research finding has led to a different glycemic index (GI) in plant-based beverages pointing out high levels in rice and coconut drinks (GI > 96) due to a high glucose content; oat drink (GI = 59) due to β-glucan content; high GI for most rice-based drinks (Jeske et al., 2017a). Due to the low protein, vitamin (B_{12}, B_2, D, and E) and mineral content (particularly calcium) in the majority of cereal/plant-based beverages, they are typically fortified. Nonetheless, fortified plant-based beverages and cow's milk have strikingly different nutritional properties owing to the fact that some nutrients' bioavailability varies significantly (Singhal et al., 2017).

In comparison to cow's milk, rice-based milk contains fewer lipids (particularly polyunsaturated fatty acids) and proteins. It contains more vitamin A and D. Due to the fact that it frequently contains arsenic, it is not recommended for infants and younger children. Oat milk contains less fat, protein, and calcium than cow's milk. It contains an antinutrient that impairs the absorption of certain nutrients. It also has cholesterol-lowering properties due to its unique fiber content (Verduci et al., 2019).

Moreover, the research on the health effects of the final cereal milks/beverages is very limited, but research on the health effects of raw cereal grains, grain extracts, or some phytochemicals isolated from cereal grains have been well documented (Fardet et al., 2008; Liu, 2007; Xiong et al., 2020). Although food processing may alter or damage some phytochemicals and affects the potential health benefits, which is inevitable; good processing methods and practices can not only minimize the losses, but also reinforce some phytochemicals by the formation of new bioactive compounds, the release of bound bioactive phytochemicals, and the increase in their bioavailability (Ragaee et al., 2014).

10.7 Safety Concerns of Cereal-Based Milks

There is no doubt that cereal-based milks are beneficial for human health. As previously stated, the functional properties of cereals and cereal products are primarily attributed to their bioactive phytochemicals. However, there are a few points to emphasize regarding the safety of cereal-based milk products. Allergies and toxic contaminants are examples of these.

10.7.1 Allergies

The eight most common food allergens are eggs, milk, peanuts, tree nuts, soy, wheat, crustacean shellfish, and fish (Divya et al., 2020). Lactose intolerance, milk allergies, ethical and environmental issues of cow's milk drive the changes in lifestyle toward a healthier and more plant-based diet. Rice milk is considered to be the most hypoallergenic of all plant-based milks. It has been suggested as the best alternative to cow's milk for people with multiple allergies such as lactose intolerance, soy or nut allergies (Bridges, 2018). Only a few cases of an immediate, hypersensitive reaction to rice consumption have been reported among Asians, for whom rice is the primary staple food (Fiocchi et al., 2003). Rice proteins reported as potential allergens include those with molecular weight of 9 kDa, 14–16 kDa, 25–26 kDA, 30–33 kDa, 52 kDa, 55 kDa, 60 kDa, 63 kDa, and 90 kDa, corresponding to non-specific lipid transfer protein-1, α-amylase/trypsin inhibitor family proteins, α-globulin, glyoxalase 1, globulin-like protein, protein disulfide isomerase, granule-bound starch synthetase, embryo globulin, and α-glucosidase, respectively (Pantoa et al., 2020). Still, rice and rice products are considered as a low allergenic food.

10.7.2 Toxic Metals and Metalloids

Metals and metalloids that are toxic are abundant in the Earth's crust. Inorganic arsenic and cadmium are classified as the group-1 carcinogens due to their toxicity, prevalence, and potential for human exposure (Davis et al., 2017; Deng et al., 2019). Cereals such as wheat, rice, barley, and maize are the primary foods that contain toxic metalloids. Rice, which is consumed by people in a wide variety of countries, is a significant source of arsenic and cadmium exposure (Deng et al., 2019). Rice is a plant that can absorb approximately tenfold the amount of arsenic absorbed by other cereals. Arsenic is primarily known for its genotoxic and carcinogenic properties, with inorganic arsenic being approximately 100 times more toxic than organic arsenic. Inorganic arsenic is primarily found mostly in water in specific geographic areas and in rice and rice products (Sun et al., 2008). The majority of inorganic arsenic in rice is concentrated in the bran layers, which contain up to ten to twenty times the amount found in whole grain. As a result, the risk associated with the consumption of products made from rice bran or whole grains, such as rice beverages, is significantly greater than the risk associated with polished rice (Meharg et al., 2008a, b). Twenty epidemiologic studies, including 18 observational studies and two human experimental studies, reported that rice consumption was consistently associated with arsenic biomarkers, and the association was clearly demonstrated in experimental studies (Davis et al., 2017).

Currently, there are no EU, US, or WHO limits for either total or inorganic arsenic in any food, including rice. The only country that regulates the level of inorganic arsenic in rice is China where the maximum contaminant level permitted is 0.15 mg/kg (Zhu et al., 2008). However, because inorganic arsenic is a non-threshold carcinogen, any exposure carries a risk, and no safe intake level can be established (Hite, 2013). The Joint Food and Agriculture Organization of the United Nations/WHO Expert Committee on Food Additives determined the lower limit of the benchmark dose confidence limit (BMDL) based on a 0.5% increase in the incidence of lung cancer to be 3.0 μg/kg body weight per day. The Contaminants in the Food Chain (CONTAM) Panel of the European Food Safety Authority (EFSA) determined the BMDL for a 1% extra risk for lung, skin, and bladder cancers and skin lesions to be in a range from 0.3 to 8 μg/kg body weight per day. Using this range instead of a single reference point has been recommended in the risk characterization process for inorganic arsenic intake (Hojsak et al., 2015).

Based on the available evidence at present, several recommendations have been made. Some of these related to rice-based milks such as: (1) inorganic arsenic intake in infancy and childhood should be as low as possible; (2) rice drinks should not be used in infants and young children, and (3) inorganic arsenic exposure from food can be reduced by including a variety of grains such as oat, barley, wheat, maize, and rice (Hojsak et al., 2015).

10.7.3 Mycotoxins

Cereals can be infected in the field and during storage by a variety of toxigenic fungi, which have the ability to accumulate mycotoxins in cereals. Mycotoxins are stable chemical compounds produced by various fungi that can be transferred from raw cereal grains to finished food products, posing a significant health risk to the consumer (Milani & Maleki, 2014). The major occurrences of mycotoxins in cereals are aflatoxins, fumonisins, ochratoxins, deoxynivalenol, and zearalenone. Integrated management system throughout the food supply chain is recommended for the safe production of cereal-based milks. Control parameters for processing commodities prone to mycotoxin contamination would include harvesting time, temperature, moisture content during storage and transportation, selection of agricultural products prior to processing, processing/decontamination conditions, temperature, chemical addition, and final product storage and transportation (Lopez-Garcia et al., 1999). Two cereal-based drinks produced from corn and sorghum were assessed for the fate of mycotoxins during processing. They reported that mycotoxin levels were higher in the maize-based than in the sorghum-based drinks. Processing steps drastically reduced concentrations of all mycotoxins, up to 99.99%. They concluded that both cereal drinks therefore represent a safe food (Ezekiel et al., 2015).

10.8 Summary and Future Perspectives

Cereal grain-based milks are functional beverages that are gaining global traction. While all cereals can be processed into cereal-based beverages, oat, rice, and corn are particularly popular for commercial products. In comparison to cow's milk, cereal milk is lactose-free, hypoallergenic, and contains beneficial bioactive components. Cereal milk, on the other hand, has a number of issues related to nutritional imbalance, sensory experience, processing difficulty, and emulsion stability. Despite this, advanced processing technologies are being developed, providing manufacturers with an excellent opportunity to address these issues.

Cereal-based milks' health and functional properties are largely attributed to the bioactive phytochemicals found in cereal grains. Phenolic compounds, dietary fibers, carotenoids, tocols, phytosterol, γ-oryzanol, and phytic acid are the primary bioactive phytochemicals found in whole grains. It should be emphasized that while research on the health effects of finished cereal milks/beverages is scarce, research on the health effects of raw cereal grains, grain extracts, or some phytochemicals isolated from cereal grains is extensive. There is no doubt that cereal-based milks provide health benefits. However, concerns should be raised about antinutrients and contaminants, such as inorganic arsenic in rice-based beverages. Human clinical trial research on the health benefits of cereal-based milks as whole foods is urgently needed to reach a conclusive conclusion in terms of health.

References

Adlercreutz, H. (2007). Lignans and human health. *Critical Reviews in Clinical Laboratory Sciences, 44*(5–6), 483–525. https://doi.org/10.1080/10408360701612942

Ahmad, A., Munir, B., Abrar, M., Bashir, S., Adnan, M., & Tabassum, T. (2012). Perspective of β-glucan as functional ingredient for food industry. *Journal of Nutrition & Food Sciences, 2*(2), 133–139.

Alsaffar, A. A. (2011). Effect of food processing on the resistant starch content of cereals and cereal products – A review. *International Journal of Food Science & Technology, 46*(3), 455–462.

Angelov, A., Gotcheva, V., Kuncheva, R., & Hristozova, T. (2006). Development of a new oat-based probiotic drink. *International Journal of Food Microbiology, 112*(1), 75–80.

Antony, U., & Chandra, T. S. (1998). Antinutrient reduction and enhancement in protein, starch, and mineral availability in fermented flour of finger millet (Eleusine coracana). *Journal of Agricultural and Food Chemistry, 46*(7), 2578–2582.

Aparicio-García, N., Martínez-Villaluenga, C., Frias, J., & Peñas, E. (2021). Production and characterization of a novel gluten-free fermented beverage based on sprouted oat flour. *Food, 10*(1), 139.

Asharani, V. T., Jayadeep, A., & Malleshi, N. G. (2010). Natural antioxidants in edible flours of selected small millets. *International Journal of Food Properties, 13*(1), 41–50. https://doi.org/10.1080/10942910802163105

Asioli, D., Aschemann-Witzel, J., Caputo, V., Vecchio, R., Annunziata, A., Næs, T., & Varela, P. (2017). Making sense of the "clean label" trends: A review of consumer food choice behavior and discussion of industry implications. *Food Research International, 99*, 58–71. https://doi.org/10.1016/j.foodres.2017.07.022

Assadpour, E., & Mahdi Jafari, S. (2019). A systematic review on nanoencapsulation of food bioactive ingredients and nutraceuticals by various nanocarriers. *Critical Reviews in Food Science and Nutrition, 59*(19), 3129–3151. https://doi.org/10.1080/10408398.2018.1484687

Aune, D., Chan, D. S. M., Lau, R., Vieira, R., Greenwood, D. C., Kampman, E., & Norat, T. (2011). Dietary fibre, whole grains, and risk of colorectal cancer: Systematic review and dose-response meta-analysis of prospective studies. *BMJ, 343*, d6617. https://doi.org/10.1136/bmj.d6617

Basinskiene, L., & Cizeikiene, D. (2020). *Chapter 3 – Cereal-based nonalcoholic beverages* (C. M. B. T.-T. in N. B. Galanakis, Ed.). https://doi.org/10.1016/B978-0-12-816938-4.00003-3

Beaulieu, J. C., Reed, S. S., Obando-Ulloa, J. M., & McClung, A. M. (2020). Green processing protocol for germinating and wet milling brown rice for beverage formulations: Sprouting, milling and gelatinization effects. *Food Science & Nutrition, 8*(5), 2445–2457. https://doi.org/10.1002/fsn3.1534

Berrazaga, I., Micard, V., Gueugneau, M., & Walrand, S. (2019). The role of the anabolic properties of plant- versus animal-based protein sources in supporting muscle mass maintenance: A critical review. *Nutrients, 11*. https://doi.org/10.3390/nu11081825

Björck, I., Östman, E., Kristensen, M., Anson, N. M., Price, R. K., Haenen, G. R. M. M., … Mykkänen, H. (2012). Cereal grains for nutrition and health benefits: Overview of results from in vitro, animal and human studies in the HEALTHGRAIN project. *Trends in Food Science & Technology, 25*(2), 87–100.

Boz, H. (2015). Ferulic acid in cereals-a review. *Czech Journal of Food Sciences, 33*(1), 1–7.

Bridges, M. (2018). Moo-ove over, cow's milk: The rise of plant-based dairy alternatives. *Practical Gastroenterology, 21*.

Briviba, K., Gräf, V., Walz, E., Guamis, B., & Butz, P. (2016). Ultra high pressure homogenization of almond milk: Physico-chemical and physiological effects. *Food Chemistry, 192*, 82–89. https://doi.org/10.1016/j.foodchem.2015.06.063

Cáceres, P. J., Martínez-Villaluenga, C., Amigo, L., & Frias, J. (2014). Assessment on proximate composition, dietary fiber, phytic acid and protein hydrolysis of germinated Ecuatorian brown rice. *Plant Foods for Human Nutrition, 69*(3), 261–267. https://doi.org/10.1007/s11130-014-0433-x

Charalampopoulos, D., Wang, R., Pandiella, S. S., & Webb, C. (2002). Application of cereals and cereal components in functional foods: A review. *International Journal of Food Microbiology, 79*(1), 131–141. https://doi.org/10.1016/S0168-1605(02)00187-3

Charunuch, C., Boonyasirikool, P., & Tiengpook, C. (2003). Using of extrusion process for preparation of instant cereal beverage powders based on corn and soybean. *Agriculture and Natural Resources, 37*(1), 72–83.

Cheowtiraku, C., Chaijany, P., Jirathumkitku, P., Kriangkraiphipha, K., Ampunsang, M., & Ruangcha, D. (2001). A study on rice-milk production. *AU Journal of Technology, 4*(3), 143–156.

Chung, I. M., Yong, S. J., Lee, J., & Kim, S. H. (2013). Effect of genotype and cultivation location on β-sitosterol and α-, β-, γ-, and δ-tocopherols in sorghum. *Food Research International, 51*(2), 971–976. https://doi.org/10.1016/j.foodres.2013.02.027

Craig, W. J. (2010). Nutrition concerns and health effects of vegetarian diets. *Nutrition in Clinical Practice, 25*(6), 613–620. https://doi.org/10.1177/0884533610385707

Crittenden, R. G., & Bennett, L. E. (2005). Cow's milk allergy: A complex disorder. *Journal of the American College of Nutrition, 24*(sup6), 582S–591S. https://doi.org/10.1080/07315724.2005.10719507

Cruz, N., Capellas, M., Hernández, M., Trujillo, A. J., Guamis, B., & Ferragut, V. (2007). Ultra high pressure homogenization of soymilk: Microbiological, physicochemical and microstructural characteristics. *Food Research International, 40*(6), 725–732. https://doi.org/10.1016/j.foodres.2007.01.003

Dai, J., & Mumper, R. J. (2010). Plant phenolics: Extraction, analysis and their antioxidant and anticancer properties. *Molecules, 15*. https://doi.org/10.3390/molecules15107313

Dardevet, D., Rémond, D., Peyron, M.-A., Papet, I., Savary-Auzeloux, I., & Mosoni, L. (2012). Muscle wasting and resistance of muscle anabolism: The "anabolic threshold concept" for adapted nutritional strategies during sarcopenia. *The Scientific World Journal, 2012*, 269531.

Davis, M. A., Signes-Pastor, A. J., Argos, M., Slaughter, F., Pendergrast, C., Punshon, T., … Karagas, M. R. (2017). Assessment of human dietary exposure to arsenic through rice. *Science of the Total Environment, 586*, 1237–1244. https://doi.org/10.1016/j.scitotenv.2017.02.119

Deng, F., Yu, M., Martinoia, E., & Song, W.-Y. (2019). Ideal cereals with lower arsenic and cadmium by accurately enhancing vacuolar sequestration capacity. *Frontiers in Genetics, 10*, 322. Retrieve from https://www.frontiersin.org/article/10.3389/fgene.2019.00322

Deswal, A., Deora, N. S., & Mishra, H. N. (2014). Effect of concentration and temperature on the rheological properties of oat milk. *Food and Bioprocess Technology, 7*(8), 2451–2459. https://doi.org/10.1007/s11947-014-1332-8

Dhingra, D., Michael, M., Rajput, H., & Patil, R. T. (2012). Dietary fibre in foods: A review. *Journal of Food Science and Technology, 49*(3), 255–266. https://doi.org/10.1007/s13197-011-0365-5
Dhull, S. B., Punia, S., Kidwai, M. K., Kaur, M., Chawla, P., Purewal, S. S., … Palthania, S. (2020). Solid-state fermentation of lentil (Lens culinaris L.) with Aspergillus awamori: Effect on phenolic compounds, mineral content, and their bioavailability. *Legume Science, 2*(3), e37.
Dhull, S. B., Punia, S., Kumar, R., Kumar, M., Nain, K. B., Jangra, K., & Chudamani, C. (2021). Solid state fermentation of fenugreek (Trigonella foenum-graecum): Implications on bioactive compounds, mineral content and in vitro bioavailability. *Journal of Food Science and Technology, 58*(5), 1927–1936.
Dinu, M., Abbate, R., Gensini, G. F., Casini, A., & Sofi, F. (2017). Vegetarian, vegan diets and multiple health outcomes: A systematic review with meta-analysis of observational studies. *Critical Reviews in Food Science and Nutrition, 57*(17), 3640–3649. https://doi.org/10.1080/10408398.2016.1138447
Divya, S., Pavadee, P., Milind, P., & Deepak, K. (2020). Food allergy: A review. *Pediatric Annals, 49*(1), e50–e58. https://doi.org/10.3928/19382359-20191206-01
Do, D. T., Singh, J., Oey, I., & Singh, H. (2018). Biomimetic plant foods: Structural design and functionality. *Trends in Food Science & Technology, 82*, 46–59. https://doi.org/10.1016/j.tifs.2018.09.010
Durand, A., Franks, G. V., & Hosken, R. W. (2003). Particle sizes and stability of UHT bovine, cereal and grain milks. *Food Hydrocolloids, 17*(5), 671–678. https://doi.org/10.1016/S0268-005X(03)00012-2
Duranti, M. (2006). Grain legume proteins and nutraceutical properties. *Fitoterapia, 77*(2), 67–82.
Durazzo, A., Zaccaria, M., Polito, A., Maiani, G., & Carcea, M. (2013). Lignan content in cereals, buckwheat and derived foods. *Food, 2*. https://doi.org/10.3390/foods2010053
Dykes, L., & Rooney, L. W. (2007). Phenolic compounds in cereal grains and their health benefits. *Cereal Foods World, 52*(3), 105–111.
Ehrenbergerová, J., Belcrediová, N., Prýma, J., Vaculová, K., & Newman, C. W. (2006). Effect of cultivar, year grown, and cropping system on the content of tocopherols and tocotrienols in grains of hulled and hulless barley. *Plant Foods for Human Nutrition, 61*(3), 145–150. https://doi.org/10.1007/s11130-006-0024-6
Epstein, S. S. (1990). Potential public health hazards of biosynthetic milk hormones. *International Journal of Health Services, 20*(1), 73–84. https://doi.org/10.2190/PRTT-HT8G-4FNQ-ATJJ
Ermawar, R. A., Collins, H. M., Byrt, C. S., Betts, N. S., Henderson, M., Shirley, N. J., … Burton, R. A. (2015). Distribution, structure and biosynthetic gene families of (1,3;1,4)-β-glucan in Sorghum bicolor. *Journal of Integrative Plant Biology, 57*(4), 429–445. https://doi.org/10.1111/jipb.12338
Espirito-Santo, A. P. d., Mouquet-Rivier, C., Humblot, C., Cazevieille, C., Icard-Verniѐre, C., Soccol, C. R., & Guyot, J.-P. (2014). Influence of cofermentation by amylolytic Lactobacillus strains and probiotic bacteria on the fermentation process, viscosity and microstructure of gruels made of rice, soy milk and passion fruit fiber. *Food Research International, 57*, 104–113. https://doi.org/10.1016/j.foodres.2014.01.028
Ezekiel, C. N., Abia, W. A., Ogara, I. M., Sulyok, M., Warth, B., & Krska, R. (2015). Fate of mycotoxins in two popular traditional cereal-based beverages (kunu-zaki and pito) from rural Nigeria. *LWT – Food Science and Technology, 60*(1), 137–141. https://doi.org/10.1016/j.lwt.2014.08.018
Fardet, A., Rock, E., & Rémésy, C. (2008). Is the in vitro antioxidant potential of whole-grain cereals and cereal products well reflected in vivo? *Journal of Cereal Science, 48*(2), 258–276.
Fernandez-Panchon, M. S., Villano, D., Troncoso, A. M., & Garcia-Parrilla, M. C. (2008). Antioxidant activity of phenolic compounds: From in vitro results to in vivo evidence. *Critical Reviews in Food Science and Nutrition, 48*(7), 649–671. https://doi.org/10.1080/10408390701761845
Fiedor, J., & Burda, K. (2014). Potential role of carotenoids as antioxidants in human health and disease. *Nutrients, 6*(2), 466–488.

Fiocchi, A., Travaini, M., D'Auria, E., Banderali, G., Bernardo, L., & Riva, E. (2003). Tolerance to a rice hydrolysate formula in children allergic to cow's milk and soy. *Clinical & Experimental Allergy, 33*(11), 1576–1580. https://doi.org/10.1046/j.1365-2222.2003.01781.x

Frølich, W., & Nyman, M. (1988). Minerals, phytate and dietary fibre in different fractions of oat-grain. *Journal of Cereal Science, 7*(1), 73–82. https://doi.org/10.1016/S0733-5210(88)80061-4

Fuentes-Zaragoza, E., Riquelme-Navarrete, M. J., Sánchez-Zapata, E., & Pérez-Álvarez, J. A. (2010). Resistant starch as functional ingredient: A review. *Food Research International, 43*(4), 931–942. https://doi.org/10.1016/j.foodres.2010.02.004

Goufo, P., & Trindade, H. (2014). Rice antioxidants: Phenolic acids, flavonoids, anthocyanins, proanthocyanidins, tocopherols, tocotrienols, γ-oryzanol, and phytic acid. *Food Science & Nutrition, 2*(2), 75–104. https://doi.org/10.1002/fsn3.86

Granato, D., Branco, G. F., Nazzaro, F., Cruz, A. G., & Faria, J. A. F. J. A. F. (2010). Functional foods and nondairy probiotic food development: Trends, concepts, and products. *Comprehensive Reviews in Food Science and Food Safety, 9*(3), 292–302. https://doi.org/10.1111/j.1541-4337.2010.00110.x

Guo, X.-D., Ma, Y.-J., Parry, J., Gao, J.-M., Yu, L.-L., & Wang, M. (2011). Phenolics content and antioxidant activity of tartary buckwheat from different locations. *Molecules, 16*. https://doi.org/10.3390/molecules16129850

Haggard, S., Luna-Vital, D., West, L., Juvik, J. A., Chatham, L., Paulsmeyer, M., & Gonzalez de Mejia, E. (2018). Comparison of chemical, color stability, and phenolic composition from pericarp of nine colored corn unique varieties in a beverage model. *Food Research International, 105*, 286–297. https://doi.org/10.1016/j.foodres.2017.11.038

Heleno, S. A., Martins, A., Queiroz, M. J. R. P., & Ferreira, I. C. F. R. (2015). Bioactivity of phenolic acids: Metabolites versus parent compounds: A review. *Food Chemistry, 173*, 501–513. https://doi.org/10.1016/j.foodchem.2014.10.057

Hite, A. H. (2013). Arsenic and rice: A call for regulation. *Nutrition, 29*(1), 353–354. https://doi.org/10.1016/j.nut.2012.10.001

Hojsak, I., Braegger, C., Bronsky, J., Campoy, C., Colomb, V., Decsi, T., … Nutrition, for the E. C. on. (2015). Arsenic in rice: A cause for concern. *Journal of Pediatric Gastroenterology and Nutrition, 60*(1). Retrieved from https://journals.lww.com/jpgn/Fulltext/2015/01000/Arsenic_in_Rice__A_Cause_for_Concern.28.aspx

Hübner, F., & Arendt, E. K. (2013). Germination of cereal grains as a way to improve the nutritional value: A review. *Critical Reviews in Food Science and Nutrition, 53*(8), 853–861. https://doi.org/10.1080/10408398.2011.562060

Hughes, D. (1995). Animal welfare. *British Food Journal, 97*(10), 3–7. https://doi.org/10.1108/00070709510104529

Hunt, M. W. (2019). Veganism and children: Physical and social well-being. *Journal of Agricultural and Environmental Ethics, 32*(2), 269–291. https://doi.org/10.1007/s10806-019-09773-4

Idehen, E., Tang, Y., & Sang, S. (2017). Bioactive phytochemicals in barley. *Journal of Food and Drug Analysis, 25*(1), 148–161. https://doi.org/10.1016/j.jfda.2016.08.002

Jeske, S., Zannini, E., & Arendt, E. K. (2017a). Evaluation of physicochemical and glycaemic properties of commercial plant-based milk substitutes. *Plant Foods for Human Nutrition, 72*(1), 26–33.

Jeske, S., Zannini, E., & Arendt, E. K. (2017b). Past, present and future: The strength of plant-based dairy substitutes based on gluten-free raw materials. *Foodservice Research International*. https://doi.org/10.1016/j.foodres.2017.03.045

Jiamyangyuen, S., Nuengchamnong, N., & Ngamdee, P. (2017). Bioactivity and chemical components of Thai rice in five stages of grain development. *Journal of Cereal Science, 74*, 136–144. https://doi.org/10.1016/j.jcs.2017.01.021

Jiang, Y., & Wang, T. (2005). Phytosterols in cereal by-products. *Journal of the American Oil Chemists' Society, 82*(6), 439–444.

Juliano, C., Cossu, M., Alamanni, M. C., & Piu, L. (2005). Antioxidant activity of gamma-oryzanol: Mechanism of action and its effect on oxidative stability of pharmaceutical oils. *International Journal of Pharmaceutics, 299*(1–2), 146–154.

Katz, K. A., Mahlberg, M. H., Honig, P. J., & Yan, A. C. (2005). Rice nightmare: Kwashiorkor in 2 Philadelphia-area infants fed rice dream beverage. *Journal of the American Academy of Dermatology, 52*(5, Suppl), S69–S72. https://doi.org/10.1016/j.jaad.2004.07.056

Kim, H. W., Kim, J. B., Cho, S.-M., Cho, I. K., Li, Q. X., Jang, H.-H., … Hwang, K.-A. (2015). Characterization and quantification of γ-oryzanol in grains of 16 Korean rice varieties. *International Journal of Food Sciences and Nutrition, 66*(2), 166–174.

Ko, S., Kim, C., Kim, H., Kim, C., Chung, S., Tae, B., & Kim, I. (2003). Tocol levels in milling fractions of some cereal grains and soybean. *Journal of the American Oil Chemists' Society, 80*(6), 585–589.

Koyama, M., & Kitamura, Y. (2014). Development of a new rice beverage by improving the physical stability of rice slurry. *Journal of Food Engineering, 131*, 89–95. https://doi.org/10.1016/j.jfoodeng.2014.01.030

Kumar, V., Sinha, A. K., Makkar, H. P. S., & Becker, K. (2010). Dietary roles of phytate and phytase in human nutrition: A review. *Food Chemistry, 120*(4), 945–959.

Kwok, K.-C., & Niranjan, K. (1995). Review: Effect of thermal processing on soymilk. *International Journal of Food Science & Technology, 30*(3), 263–295. https://doi.org/10.1111/j.1365-2621.1995.tb01377.x

Laetitia, M.-M., Joseph, H. D., Joseph, D., & Christian, M. (2005). Physical, chemical and microbiological changes during natural fermentation of\" gowé\", a sprouted or non sprouted sorghum beverage from West-Africa. *African Journal of Biotechnology, 4*(6), 487–496.

Lalić, J., Denić, M., Sunarić, S., Kocić, G., Trutić, N., Mitić, S., & Jovanović, T. (2014). Assessment of thiamine content in some dairy products and rice milk. *CyTA Journal of Food, 12*(3), 203–209. https://doi.org/10.1080/19476337.2013.814713

Lampi, A., Moreau, R. A., Piironen, V., & Hicks, K. B. (2004). Pearling barley and rye to produce phytosterol-rich fractions. *Lipids, 39*(8), 783–787.

Le Louer, B., Lemale, J., Garcette, K., Orzechowski, C., Chalvon, A., Girardet, J. P., & Tounian, P. (2014). Severe nutritional deficiencies in young infants with inappropriate plant milk consumption. *Archives de pediatrie : organe officiel de la Societe francaise de pediatrie, 21*(5), 483–488. https://doi.org/10.1016/j.arcped.2014.02.027

LeBlanc, J. G., Laiño, J. E., del Valle, M. J., Vannini, V., Van Sinderen, D., Taranto, M. P., … Sesma, F. (2011). B-Group vitamin production by lactic acid bacteria–current knowledge and potential applications. *Journal of Applied Microbiology, 111*(6), 1297–1309.

LeBlanc, J. G., Milani, C., de Giori, G. S., Sesma, F., van Sinderen, D., & Ventura, M. (2013). Bacteria as vitamin suppliers to their host: A gut microbiota perspective. *Current Opinion in Biotechnology, 24*(2), 160–168. https://doi.org/10.1016/j.copbio.2012.08.005

Lemmens, E., Moroni, A. V., Pagand, J., Heirbaut, P., Ritala, A., Karlen, Y., … Delcour, J. A. (2019). Impact of cereal seed sprouting on its nutritional and technological properties: A critical review. *Comprehensive Reviews in Food Science and Food Safety, 18*(1), 305–328. https://doi.org/10.1111/1541-4337.12414

Lempereur, I., Rouau, X., & Abecassis, J. (1997). Genetic and agronomic variation in arabinoxylan and ferulic acid contents of durum wheat (Triticum durumL.) grain and its milling fractions. *Journal of Cereal Science, 25*(2), 103–110. https://doi.org/10.1006/jcrs.1996.0090

Liljeberg Elmståhl, H. (2002). Resistant starch content in a selection of starchy foods on the Swedish market. *European Journal of Clinical Nutrition, 56*(6), 500–505. https://doi.org/10.1038/sj.ejcn.1601338

Lin, P.-Y., & Lai, H.-M. (2011). Bioactive compounds in rice during grain development. *Food Chemistry, 127*(1), 86–93. https://doi.org/10.1016/j.foodchem.2010.12.092

Liu, R. H. (2007). Whole grain phytochemicals and health. *Journal of Cereal Science, 46*(3), 207–219. https://doi.org/10.1016/j.jcs.2007.06.010

Liu, Y., Cai, C., Yao, Y., & Xu, B. (2019). Alteration of phenolic profiles and antioxidant capacities of common buckwheat and tartary buckwheat produced in China upon thermal processing. *Journal of the Science of Food and Agriculture, 99*(12), 5565–5576. https://doi.org/10.1002/jsfa.9825

Lockyer, S., & Nugent, A. P. (2017). Health effects of resistant starch. *Nutrition Bulletin, 42*(1), 10–41. https://doi.org/10.1111/nbu.12244

López-Alarcón, C., & Denicola, A. (2013). Evaluating the antioxidant capacity of natural products: A review on chemical and cellular-based assays. *Analytica Chimica Acta, 763*, 1–10. https://doi.org/10.1016/j.aca.2012.11.051

Lopez-Garcia, R., Park, D. L., & Phillips, T. D. (1999). Integrated mycotoxin management systems. *Food Nutrition and Agriculture*, 38–48.

Lu, X., Chen, J., Zheng, M., Guo, J., Qi, J., Chen, Y., … Zheng, B. (2019). Effect of high-intensity ultrasound irradiation on the stability and structural features of coconut-grain milk composite systems utilizing maize kernels and starch with different amylose contents. *Ultrasonics Sonochemistry, 55*, 135–148. https://doi.org/10.1016/j.ultsonch.2019.03.003

Mäkinen, O. E., Uniacke-Lowe, T., O'Mahony, J. A., & Arendt, E. K. (2015). Physicochemical and acid gelation properties of commercial UHT-treated plant-based milk substitutes and lactose free bovine milk. *Food Chemistry, 168*, 630–638.

Mäkinen, O. E., Wanhalinna, V., Zannini, E., & Arendt, E. K. (2016). Foods for special dietary needs: Non-dairy plant-based milk substitutes and fermented dairy-type products. *Critical Reviews in Food Science and Nutrition, 56*(3), 339–349. https://doi.org/10.1080/10408398.2012.761950

Massa, G., Vanoppen, A., Gillis, P., Aerssens, P., Alliet, P., & Raes, M. (2001). Protein malnutrition due to replacement of milk by rice drink. *European Journal of Pediatrics, 160*(6), 382–384. https://doi.org/10.1007/s004310100746

Mattila, P., Pihlava, J., & Hellström, J. (2005). Contents of phenolic acids, alkyl- and alkenylresorcinols, and avenanthramides in commercial grain products. *Journal of Agricultural and Food Chemistry, 53*(21), 8290–8295. https://doi.org/10.1021/jf051437z

Maurya, V. K., Bashir, K., & Aggarwal, M. (2020). Vitamin D microencapsulation and fortification: Trends and technologies. *The Journal of Steroid Biochemistry and Molecular Biology, 196*, 105489. https://doi.org/10.1016/j.jsbmb.2019.105489

McClements, D. J. (2015). Nanoscale nutrient delivery systems for food applications: Improving bioactive dispersibility, stability, and bioavailability. *Journal of Food Science, 80*(7), N1602–N1611. https://doi.org/10.1111/1750-3841.12919

McClements, D. J. (2018). Enhanced delivery of lipophilic bioactives using emulsions: A review of major factors affecting vitamin, nutraceutical, and lipid bioaccessibility. *Food & Function, 9*(1), 22–41. https://doi.org/10.1039/C7FO01515A

McClements, D. J. (2020). Development of next-generation nutritionally fortified plant-based milk substitutes: Structural design principles. *Food, 9*. https://doi.org/10.3390/foods9040421

McClements, D. J., Li, F., & Xiao, H. (2015). The nutraceutical bioavailability classification scheme: Classifying nutraceuticals according to factors limiting their oral bioavailability. *Annual Review of Food Science and Technology, 6*, 299–327.

McClements, D. J., Newman, E., & McClements, I. F. (2019). Plant-based milks: A review of the science underpinning their design, fabrication, and performance. *Comprehensive Reviews in Food Science and Food Safety, 18*(6), 2047–2067. https://doi.org/10.1111/1541-4337.12505

Meharg, A. A., Deacon, C., Campbell, R. C. J., Carey, A.-M., Williams, P. N., Feldmann, J., & Raab, A. (2008a). Inorganic arsenic levels in rice milk exceed EU and US drinking water standards. *Journal of Environmental Monitoring, 10*(4), 428–431. https://doi.org/10.1039/B800981C

Meharg, A. A., Sun, G., Williams, P. N., Adomako, E., Deacon, C., Zhu, Y.-G., … Raab, A. (2008b). Inorganic arsenic levels in baby rice are of concern. *Environmental Pollution, 152*(3), 746–749. https://doi.org/10.1016/j.envpol.2008.01.043

Mehmood, T., Ahmed, A., Ahmed, Z., & Ahmad, M. S. (2019). Optimization of soya lecithin and Tween 80 based novel vitamin D nanoemulsions prepared by ultrasonication using response surface methodology. *Food Chemistry, 289*, 664–670. https://doi.org/10.1016/j.foodchem.2019.03.112

Mellado-Ortega, E., & Hornero-Méndez, D. (2015). Carotenoids in cereals: An ancient resource with present and future applications. *Phytochemistry Reviews, 14*(6), 873–890.

Mendez-Encinas, M. A., Carvajal-Millan, E., Rascon-Chu, A., Astiazaran-Garcia, H. F., & Valencia-Rivera, D. E. (2018). Ferulated arabinoxylans and their gels: Functional properties and potential application as antioxidant and anticancer agent. *Oxidative Medicine and Cellular Longevity, 2018*, 2314759. https://doi.org/10.1155/2018/2314759

Meydani, M. (2009). Potential health benefits of avenanthramides of oats. *Nutrition Reviews, 67*(12), 731–735. https://doi.org/10.1111/j.1753-4887.2009.00256.x

Milani, J., & Maleki, G. (2014). Effects of processing on mycotoxin stability in cereals. *Journal of the Science of Food and Agriculture, 94*(12), 2372–2375. https://doi.org/10.1002/jsfa.6600

Minatel, I. O., Francisqueti, F. V., Corrêa, C. R., & Lima, G. P. (2016). Antioxidant activity of γ-Oryzanol: A complex network of interactions. *International Journal of Molecular Sciences, 17*. https://doi.org/10.3390/ijms17081107

Mohammadi, M., Nouri, L., & Mortazavian, A. M. (2021). Development of a functional synbiotic beverage fortified with different cereal sprouts and prebiotics. *Journal of Food Science and Technology, 58*(11), 4185–4193. https://doi.org/10.1007/s13197-020-04887-4

Ndolo, V. U., & Beta, T. (2013). Distribution of carotenoids in endosperm, germ, and aleurone fractions of cereal grain kernels. *Food Chemistry, 139*(1–4), 663–671.

Nissar, J., Ahad, T., Naik, H. R., & Hussain, S. Z. (2017). A review phytic acid: As antinutrient or nutraceutical. *Journal of Pharmacognosy and Phytochemistry, 6*(6), 1554–1560.

Nurmi, T., Lampi, A.-M., Nyström, L., Hemery, Y., Rouau, X., & Piironen, V. (2012). Distribution and composition of phytosterols and steryl ferulates in wheat grain and bran fractions. *Journal of Cereal Science, 56*(2), 379–388.

Oatway, L., Vasanthan, T., & Helm, J. H. (2001). Phytic acid. *Food Reviews International, 17*(4), 419–431.

Obeid, R., Heil, S. G., Verhoeven, M. M. A., van den Heuvel, E. G. H. M., de Groot, L. C. P. G. M., & Eussen, S. J. P. M. (2019). Vitamin B12 intake from animal foods, biomarkers, and health aspects. *Frontiers in Nutrition, 6*, 93. Retrieved from https://www.frontiersin.org/article/10.3389/fnut.2019.00093

Okarter, N., & Liu, R. H. (2010). Health benefits of whole grain phytochemicals. *Critical Reviews in Food Science and Nutrition, 50*(3), 193–208.

Omueti, O., Oguntona, E. B., Jaiyeola, O., & Ashaye, O. A. (2000). Nutritional evaluation of home-prepared soy-corn milk – A protein beverage. *Nutrition & Food Science, 30*(3), 128–132. https://doi.org/10.1108/00346650010319723

Önning, G., Åkesson, B., Öste, R., & Lundquist, I. (1998). Effects of consumption of oat Milk, soya milk, or cow's milk on plasma lipids and antioxidative capacity in healthy subjects. *Annals of Nutrition and Metabolism, 42*(4), 211–220. Retrieved from https://www.karger.com/DOI/10.1159/000012736

Ostlund, R. E., Jr. (2004). Phytosterols and cholesterol metabolism. *Current Opinion in Lipidology, 15*(1), 37–41.

Othman, R. A., Moghadasian, M. H., & Jones, P. J. H. (2011). Cholesterol-lowering effects of oat β-glucan. *Nutrition Reviews, 69*(6), 299–309.

Palacios, C. E., Nagai, A., Torres, P., Rodrigues, J. A., & Salatino, A. (2021). Contents of tannins of cultivars of sorghum cultivated in Brazil, as determined by four quantification methods. *Food Chemistry, 337*, 127970.

Panche, A. N., Diwan, A. D., & Chandra, S. R. (2016). Flavonoids: An overview. *Journal of Nutritional Science, 5*, e47. https://doi.org/10.1017/jns.2016.41

Pantoa, T., Baricevic-Jones, I., Suwannaporn, P., Kadowaki, M., Kubota, M., Roytrakul, S., & Mills, E. N. C. (2020). Young rice protein as a new source of low allergenic plant-base protein. *Journal of Cereal Science, 93*, 102970. https://doi.org/10.1016/j.jcs.2020.102970
Park, H.-Y., Lee, K. W., & Choi, H. D. (2017). Rice bran constituents: Immunomodulatory and therapeutic activities. *Food & Function, 8*(3), 935–943.
Patil, S. B., & Khan, M. K. (2011). Germinated brown rice as a value added rice product: A review. *Journal of Food Science and Technology*. https://doi.org/10.1007/s13197-011-0232-4
Petroski, W., & Minich, D. M. (2020). Is there such a thing as "anti-nutrients"? A narrative review of perceived problematic plant compounds. *Nutrients, 12*. https://doi.org/10.3390/nu12102929
Peyer, L. C., Zannini, E., & Arendt, E. K. (2016). Lactic acid bacteria as sensory biomodulators for fermented cereal-based beverages. *Trends in Food Science & Technology, 54*(Supplement C), 17–25. https://doi.org/10.1016/j.tifs.2016.05.009
Piironen, V., Toivo, J., & Lampi, A.-M. (2002). Plant sterols in cereals and cereal products. *Cereal Chemistry, 79*(1), 148–154. https://doi.org/10.1094/CCHEM.2002.79.1.148
Popova, A., & Mihaylova, D. (2019). Antinutrients in plant-based foods: A review. *The Open Biotechnology Journal, 13*(1), 68.
Prabhavat, S., Kanchanapakornchai, U., Boonyaratanakornkit, M., Tiengpook, C., Boonvisut, S., & Poosimuang, S. (1999). Canned corn beverage from sweet corn kernel and its cob. *Agriculture and Natural Resources, 33*(3), 430–443.
Punia Bangar, S., Sandhu, K. S., Purewal, S. S., Kaur, M., Kaur, P., Siroha, A. K., ... Kumar, M. (2021a). Fermented barley bran: An improvement in phenolic compounds and antioxidant properties. *Journal of Food Processing and Preservation*, e15543.
Punia Bangar, S., Sharma, N., Kumar, M., Ozogul, F., Purewal, S. S., & Trif, M. (2021b). Recent developments in applications of lactic acid bacteria against mycotoxin production and fungal contamination. *Food Bioscience, 44*, 101444.
Ragaee, S., Seetharaman, K., & Abdel-Aal, E.-S. M. (2014). The impact of milling and thermal processing on phenolic compounds in cereal grains. *Critical Reviews in Food Science and Nutrition, 54*(7), 837–849. https://doi.org/10.1080/10408398.2011.610906
Rasane, P., Jha, A., Sabikhi, L., Kumar, A., & Unnikrishnan, V. S. (2015). Nutritional advantages of oats and opportunities for its processing as value added foods – A review. *Journal of Food Science and Technology, 52*(2), 662–675. https://doi.org/10.1007/s13197-013-1072-1
Rathore, S., Salmerón, I., & Pandiella, S. S. (2012). Production of potentially probiotic beverages using single and mixed cereal substrates fermented with lactic acid bacteria cultures. *Food Microbiology, 30*(1), 239–244. https://doi.org/10.1016/j.fm.2011.09.001
Rekha, C. R., & Vijayalakshmi, G. (2010). Bioconversion of isoflavone glycosides to aglycones, mineral bioavailability and vitamin B complex in fermented soymilk by probiotic bacteria and yeast. *Journal of Applied Microbiology, 109*(4), 1198–1208.
Reyes-Jurado, F., Soto-Reyes, N., Dávila-Rodríguez, M., Lorenzo-Leal, A. C., Jiménez-Munguía, M. T., Mani-López, E., & López-Malo, A. (2021). Plant-based milk alternatives: Types, processes, benefits, and characteristics. *Food Reviews International, 1–32*. https://doi.org/10.1080/87559129.2021.1952421
Robbins, R. J. (2003). Phenolic acids in foods: An overview of analytical methodology. *Journal of Agricultural and Food Chemistry, 51*(10), 2866–2887. https://doi.org/10.1021/jf026182t
Rodríguez-García, C., Sánchez-Quesada, C., Toledo, E., Delgado-Rodríguez, M., & Gaforio, J. J. (2019). Naturally lignan-rich foods: A dietary tool for health promotion? *Molecules, 24*. https://doi.org/10.3390/molecules24050917
Ross, A. B., Shepherd, M. J., Schüpphaus, M., Sinclair, V., Alfaro, B., Kamal-Eldin, A., & Åman, P. (2003). Alkylresorcinols in cereals and cereal products. *Journal of Agricultural and Food Chemistry, 51*(14), 4111–4118. https://doi.org/10.1021/jf0340456
Ross, A. B., Kamal-Eldin, A., & Åman, P. (2004). Dietary alkylresorcinols: Absorption, bioactivities, and possible use as biomarkers of whole-grain wheat–and rye–rich foods. *Nutrition Reviews, 62*(3), 81–95. https://doi.org/10.1111/j.1753-4887.2004.tb00029.x

Rotz, C. A., Montes, F., & Chianese, D. S. (2010). The carbon footprint of dairy production systems through partial life cycle assessment. *Journal of Dairy Science, 93*(3), 1266–1282. https://doi.org/10.3168/jds.2009-2162

Rouf Shah, T., Prasad, K., & Kumar, P. (2016). Maize – A potential source of human nutrition and health: A review. *Cogent Food & Agriculture, 2*(1), 1166995. https://doi.org/10.1080/23311932.2016.1166995

Saeed, F., Pasha, I., Anjum, F. M., & Sultan, M. T. (2011). Arabinoxylans and arabinogalactans: A comprehensive treatise. *Critical Reviews in Food Science and Nutrition, 51*(5), 467–476.

Salar, R. K., Purewal, S. S., & Sandhu, K. S. (2017). Relationships between DNA damage protection activity, total phenolic content, condensed tannin content and antioxidant potential among Indian barley cultivars. *Biocatalysis and Agricultural Biotechnology, 11*, 201–206. https://doi.org/10.1016/j.bcab.2017.07.006

Sandhu, K. S., & Punia, S. (2017). Enhancement of bioactive compounds in barley cultivars by solid substrate fermentation. *Journal of Food Measurement and Characterization, 11*(3), 1355–1361.

Sandhu, K. S., Punia, S., & Kaur, M. (2016). Effect of duration of solid state fermentation by Aspergillus awamorinakazawa on antioxidant properties of wheat cultivars. *LWT Food Science and Technology, 71*, 323–328.

Sandhu, K. S., Punia, S., & Kaur, M. (2017). Fermentation in cereals: A tool to enhance bioactive compounds. In S. K. Gahlawat & J. S. Duhan (Eds.), *Plant biotechnology: Recent advancements and developments* (pp. 157–170). Springer.

Scalbert, A., Manach, C., Morand, C., Rémésy, C., & Jiménez, L. (2005). Dietary polyphenols and the prevention of diseases. *Critical Reviews in Food Science and Nutrition, 45*(4), 287–306. https://doi.org/10.1080/1040869059096

Sebastiani, G., Herranz Barbero, A., Borrás-Novell, C., Alsina Casanova, M., Aldecoa-Bilbao, V., Andreu-Fernández, V., … García-Algar, O. (2019). The effects of vegetarian and vegan diet during pregnancy on the health of mothers and offspring. *Nutrients, 11*. https://doi.org/10.3390/nu11030557

Sethi, S., Tyagi, S. K., & Anurag, R. K. (2016). Plant-based milk alternatives an emerging segment of functional beverages: A review. *Journal of Food Science and Technology, 53*(9), 3408–3423. https://doi.org/10.1007/s13197-016-2328-3

Setta, M. C., Matemu, A., & Mbega, E. R. (2020). Potential of probiotics from fermented cereal-based beverages in improving health of poor people in Africa. *Journal of Food Science and Technology, 57*(11), 3935–3946. https://doi.org/10.1007/s13197-020-04432-3

Shannon, R., & Rodriguez, J. M. (2014). Total arsenic in rice milk. *Food Additives & Contaminants: Part B, 7*(1), 54–56. https://doi.org/10.1080/19393210.2013.842941

Sharma, M., Mridula, D., & Gupta, R. K. (2014). Development of sprouted wheat based probiotic beverage. *Journal of Food Science and Technology, 51*(12), 3926–3933. https://doi.org/10.1007/s13197-013-0959-1

Shokoohi, M., Razavi, S. H., Labbafi, M., Vahidinia, A., & Gharibzahedi, S. M. T. (2015). Wheat sprout flour as an attractive substrate for the producing probiotic fermented beverages: Process development and product characterisation. *Quality Assurance & Safety of Crops and Food, 7*(4), 469–475.

Sidhu, J. S., Kabir, Y., & Huffman, F. G. (2007). Functional foods from cereal grains. *International Journal of Food Properties, 10*(2), 231–244. https://doi.org/10.1080/10942910601045289

Sing, S. X., Lee, H. H., Wong, S. C., Bong, C. F. J., & Yiu, P. H. (2015). Ferulic acid, gamma oryzanol and GABA content in whole grain rice and their variation with bran colour. *Emirates Journal of Food and Agriculture, 27*(9), 706–711.

Singh, H. (2006). The milk fat globule membrane – A biophysical system for food applications. *Current Opinion in Colloid & Interface Science, 11*(2), 154–163. https://doi.org/10.1016/j.cocis.2005.11.002

Singh, A., & Sharma, S. (2017). Bioactive components and functional properties of biologically activated cereal grains: A bibliographic review. *Critical Reviews in Food Science and Nutrition, 57*(14), 3051–3071. https://doi.org/10.1080/10408398.2015.1085828

Singhal, S., Baker, R. D., & Baker, S. S. (2017). A comparison of the nutritional value of cow's milk and nondairy beverages. *Journal of Pediatric Gastroenterology and Nutrition, 64*(5), 799–805.

Slavin, J. (2004). Whole grains and human health. *Nutrition Research Reviews, 17*(1), 99–110.

Smeriglio, A., Barreca, D., Bellocco, E., & Trombetta, D. (2017). Proanthocyanidins and hydrolysable tannins: Occurrence, dietary intake and pharmacological effects. *British Journal of Pharmacology, 174*(11), 1244–1262. https://doi.org/10.1111/bph.13630

Smith, C. E., & Tucker, K. L. (2011). Health benefits of cereal fibre: A review of clinical trials. *Nutrition Research Reviews, 24*(1), 118–131. https://doi.org/10.1017/S0954422411000023

Soetan, K. O., & Oyewole, O. E. (2009). The need for adequate processing to reduce the antinutritional factors in plants used as human foods and animal feeds: A review. *African Journal of Food Science, 3*(9), 223–232.

Sohail, M., Rakha, A., Butt, M. S., Iqbal, M. J., & Rashid, S. (2017). Rice bran nutraceutics: A comprehensive review. *Critical Reviews in Food Science and Nutrition, 57*(17), 3771–3780.

Srikaeo, K. (2020). *5 – Biotechnological tools in the production of functional cereal-based Bbverages* (A. M. Grumezescu, & A. M. B. T.-B. P. & B. C. Holban, Eds.). https://doi.org/10.1016/B978-0-12-816678-9.00005-9

Srikaeo, K., Poungsampao, P., & Phuong, N. T. (2017). Utilization of the fine particles obtained from cold pressed vegetable oils: A case study in organic rice bran, sunflower and sesame oils. *Journal of Oleo Science, 66*(1). https://doi.org/10.5650/jos.ess16131

Steinkraus, K. H. (1997). Classification of fermented foods: Worldwide review of household fermentation techniques. *Food Control, 8*(5), 311–317. https://doi.org/10.1016/S0956-7135(97)00050-9

Sun, G. X., Williams, P. N., Carey, A. M., Zhu, Y.-G., Deacon, C., Raab, A., … Meharg, A. A. (2008). Inorganic arsenic in rice bran and its products are an order of magnitude higher than in bulk grain. *Environmental Science & Technology, 42*(19), 7542–7546. https://doi.org/10.1021/es801238p

Tangyu, M., Muller, J., Bolten, C. J., & Wittmann, C. (2019). Fermentation of plant-based milk alternatives for improved flavour and nutritional value. *Applied Microbiology and Biotechnology, 103*(23), 9263–9275. https://doi.org/10.1007/s00253-019-10175-9

Teixeira, C., Nyman, M., Andersson, R., & Alminger, M. (2016). Effects of variety and steeping conditions on some barley components associated with colonic health. *Journal of the Science of Food and Agriculture, 96*(14), 4821–4827.

Thakur, P., Kumar, K., & Dhaliwal, H. S. (2021). Nutritional facts, bio-active components and processing aspects of pseudocereals: A comprehensive review. *Food Bioscience, 42*, 101170. https://doi.org/10.1016/j.fbio.2021.101170

Threapleton, D. E., Greenwood, D. C., Evans, C. E. L., Cleghorn, C. L., Nykjaer, C., Woodhead, C., … Burley, V. J. (2013). Dietary fibre intake and risk of cardiovascular disease: Systematic review and meta-analysis. *BMJ [British Medical Journal], 347*, f6879. https://doi.org/10.1136/bmj.f6879

Tian, B., Xie, B., Shi, J., Wu, J., Cai, Y., Xu, T., … Deng, Q. (2010). Physicochemical changes of oat seeds during germination. *Food Chemistry, 119*(3), 1195–1200. https://doi.org/10.1016/j.foodchem.2009.08.035

Tiwari, U., & Cummins, E. (2009). Nutritional importance and effect of processing on tocols in cereals. *Trends in Food Science & Technology, 20*(11), 511–520. https://doi.org/10.1016/j.tifs.2009.06.001

Trono, D. (2019). Carotenoids in cereal food crops: Composition and retention throughout grain storage and food processing. *Plants, 8*. https://doi.org/10.3390/plants8120551

Truswell, A. S. (2002). Cereal grains and coronary heart disease. *European Journal of Clinical Nutrition, 56*(1), 1–15.

Valencia-Flores, D. C., Hernández-Herrero, M., Guamis, B., & Ferragut, V. (2013). Comparing the effects of ultra-high-pressure homogenization and conventional thermal treatments on the microbiological, physical, and chemical quality of almond beverages. *Journal of Food Science, 78*(2), E199–E205. https://doi.org/10.1111/1750-3841.12029

Valizadeh, R., Kargarsana, H., Shojaei, M., & Mehbodnia, M. (2009). Effect of high intensity pulsed electric fields on microbial inactivation of cow milk. *Journal of Animal and Veterinary Advances, 8*(12), 2638–2643.

Vanga, S. K., & Raghavan, V. (2018). How well do plant based alternatives fare nutritionally compared to cow's milk? *Journal of Food Science and Technology, 55*(1), 10–20. https://doi.org/10.1007/s13197-017-2915-y

Verduci, E., D'Elios, S., Cerrato, L., Comberiati, P., Calvani, M., Palazzo, S., … Peroni, D. G. (2019). Cow's milk substitutes for children: Nutritional aspects of milk from different mammalian species, special formula and plant-based beverages. *Nutrients, 11*. https://doi.org/10.3390/nu11081739

Vinkx, C. J. A., & Delcour, J. A. (1996). Rye (Secale cerealeL.) arabinoxylans: A critical review. *Journal of Cereal Science, 24*(1), 1–14.

Weichselbaum, E., & Buttriss, J. L. (2010). Polyphenols in the diet. *Nutrition Bulletin, 35*(2), 157–164. https://doi.org/10.1111/j.1467-3010.2010.01821.x

Wood, P. J. (2010). REVIEW: Oat and rye β-glucan: Properties and function. *Cereal Chemistry, 87*(4), 315–330. https://doi.org/10.1094/CCHEM-87-4-0315

Xiong, Y., Zhang, P., Luo, J., Johnson, S., & Fang, Z. (2019a). Effect of processing on the phenolic contents, antioxidant activity and volatile compounds of sorghum grain tea. *Journal of Cereal Science, 85*, 6–14. https://doi.org/10.1016/j.jcs.2018.10.012

Xiong, Y., Zhang, P., Warner, R. D., & Fang, Z. (2019b). Sorghum grain: From genotype, nutrition, and phenolic profile to its health benefits and food applications. *Comprehensive Reviews in Food Science and Food Safety, 18*(6), 2025–2046. https://doi.org/10.1111/1541-4337.12506

Xiong, Y., Zhang, P., Warner, R. D., Shen, S., & Fang, Z. (2020). Cereal grain-based functional beverages: From cereal grain bioactive phytochemicals to beverage processing technologies, health benefits and product features. *Critical Reviews in Food Science and Nutrition, 1–25*. https://doi.org/10.1080/10408398.2020.1853037

Zhai, S., Xia, X., & He, Z. (2016). Carotenoids in staple cereals: Metabolism, regulation, and genetic manipulation. *Frontiers in Plant Science, 7*, 1197.

Zhang, H., Önning, G., Triantafyllou, A. Ö., & Öste, R. (2007). Nutritional properties of oat-based beverages as affected by processing and storage. *Journal of the Science of Food and Agriculture, 87*(12), 2294–2301. https://doi.org/10.1002/jsfa.2987

Zheng, B., Zhang, X., Lin, H., & McClements, D. J. (2019). Loading natural emulsions with nutraceuticals using the pH-driven method: Formation & stability of curcumin-loaded soybean oil bodies. *Food & Function, 10*(9), 5473–5484.

Zhu, Y.-G., Williams, P. N., & Meharg, A. A. (2008). Exposure to inorganic arsenic from rice: A global health issue? *Environmental Pollution, 154*(2), 169–171.

Zielinski, H., Ciska, E., & Kozlowska, H. (2001). The cereal grains: Focus on vitamin E. *Czech Journal of Food Sciences, 19*, 182–188.

Chapter 11
Cereal Grain Tea Beverages and Their Potential Health Properties

Prasanna J. Patil, Muhammad Usman, Arshad Mehmood, Shabbir Ahmad, Junaid Haider, Chengnan Zhang, Chao Teng, and Xiuting Li

11.1 Introduction

As with humans, animals get most of their energy from cereal grains and the products made from them. Consequently, their production has escalated dramatically in recent years to fulfill the mounting global demand (Rashwan et al., 2021). Cereals are one of the world's major crops and a vital part of our daily diets. Cereals have been utilized as a staple food for both human ingestion and animal feedstuff ever since the origin of the civilization (Ragaee et al., 2013). Cereals currently contribute to more than 74% of the total worldwide cultivated region and 60% of the world's

Authors Prasanna J. Patil and Muhammad Usman have equally contributed to this chapter.

P. J. Patil · M. Usman
Beijing Advanced Innovation Center for Food Nutrition and Human Health, Beijing Technology and Business University, Beijing, China

School of Food and health, Beijing Technology and Business University, Beijing, China

A. Mehmood · C. Zhang · C. Teng · X. Li (✉)
Beijing Advanced Innovation Center for Food Nutrition and Human Health, Beijing Technology and Business University, Beijing, China

School of Food and health, Beijing Technology and Business University, Beijing, China

Beijing Engineering and Technology Research Center of Food Additives, School of Food and Chemical Technology, Beijing Technology and Business University, Beijing, China
e-mail: lixt@btbu.edu.cn

S. Ahmad
Department of Food Science and Technology, MNS-University of Agriculture, Multan, Pakistan

J. Haider
Tianjin Institute of Industrial Biotechnology, Chinese Academy of Sciences, Tinajin, China

S. Punia Bangar, A. Kumar Siroha (eds.), *Functional Cereals and Cereal Foods*,
https://doi.org/10.1007/978-3-031-05611-6_11

food production [FAO (Food and Agriculture Organization of the United Nations), 2018] in terms of volume. They have become a fundamental component of the human diet because they provide a substantial quantity of energy, proteins, carbohydrates, vitamins, and minerals; they are also high in bioactive phytochemicals that may have health-mending properties. Carotenoids, dietary fibers, phenolic compounds, and phytosterols are all examples of bioactive phytochemicals, normally identified in minor quantities in plants (Kris-Etherton et al., 2002). It is common for these bioactive phytochemicals to serve specialized roles in human physiological activities, for example, antioxidant, anti-inflammatory, hormone modulation, and immune system boosting (Xiong et al., 2019b, 2020a). Cereals are monocots, whereas pseudocereals are dicotyledonous, making them distinct from each other (Ciudad-Mulero et al., 2019). They are, nonetheless, identical to traditional cereals in terms of flavour and processing characteristics (Ispiryan et al., 2021). Buckwheat (*Fagopyrum* spp.), Quinoa (*Chenopodium quinoa*), and Amaranth (*Amaranthus* spp.) are the most explored and frequently cultivated pseudocereals. Rice, sorghum, wheat, corn, oat, rye, barley, millet, and buckwheat are the world's most widely produced grains, as per the FAO (FAO, 2018). Rising customer awareness of the significance of nutrition and health led to a rapidly growing demand for functional foods possessing health promoting properties, which has opened up markets for a wide array of processed foods, comprising beverages with specific health features, such as cereal grain tea, possessing explicit health benefits (Xiong et al., 2020a). Numerous epidemiological studies show that eating whole-grain cereals lowers one's menace of developing numerous chronic diseases, including cardiovascular disease (CVDs), diabetes, and various types of cancer (Gil et al., 2011, Francavilla & Joye, 2020).

The use of cereal grains to produce grain tea beverages has become popular in recent times. Cereal grain-based teas are fetching mounting attention due to their distinct flavour, potential health-promoting features, and caffeine/alcohol-free properties (Xiong et al., 2020a). Some of the most popular cereal grains tea beverages on the market today include Tartary buckwheat tea, barley tea, sorghum tea, rice tea, wheat bran tea, and buckwheat tea; all of which have a distinctive flavour, taste, aroma, and health-promoting characteristics (Qin et al., 2011; Wang et al., 2019; Sun et al., 2020; Wu et al., 2013a; Xiong et al., 2020b). Antinutritional components, encompassing tannins, phytic acid, trypsin inhibitors, and protein cross-linkers, are found in an extensive array of cereal grains, particularly pigmented ones, which exert detrimental effects (Rashwan et al., 2021). The presence of these antinutritional substances further hampers animal and human feed efficacy. A wide range of processing techniques can reduce the antinutritional compounds, for instance, germination, fermentation heating, roasting, soaking, and steaming, and are commonly employed for the processing of cereal grain-based teas. These processing techniques can also enhance the quality of cereal grain tea.

In various countries throughout the world, cereal grains have long been utilized to manufacture a range of health-mending beverages for ingestion. Cereal grain-based beverages, such as teas, have received significantly less attention than those produced from vegetables, fruits, or medicinal herbs. There's a scarcity of literature on the research and development of cereal grain-based teas, and only a few reviews/ books are available based on cereal grain-based teas. Henceforth, the present book chapter aimed to focus on the nutritional importance of cereal grains tea beverages. It also comprehensively explored the bioactive phytochemicals profile in cereal grain and its physiological activity. This book chapter further discussed diverse types of cereal grains tea beverages and their potential health-promoting effects. Moreover, volatile compounds of different cereal grain tea beverages have been comprehensively catalogued. Furthermore, nanotechnological aspects of cereal grain-based beverages have been highlighted. Finally, the challenges and opportunities for enhancing cereal tea beverages' intake for human health wellbeing have been proposed. We believe that this book chapter could provide additional insights for both food researchers and industrialists working on cereal grain beverages, including teas.

11.2 Cereal Grain Tea Beverages – In a Historical Perspective

Homo sapiens are thought to have first emerged between 2,00,000 and 1,00,000 years ago. Water and breast milk were the only beverages they consumed. Other beverages, including tea, coffee, animal milk, beer, and most recent, fruit drinks and soft drinks, have only been in our food for around 11,000 years. Beer was produced and consumed between 4000 and 3500 BCE, according to significant archaeological evidences. At the same time, distilled alcoholic beverages were produced. The Sumerians and Egyptians are credited with being the earliest brewers (Wolf et al., 2008). Beer was made possible by the domestication of crops. Cereals' introduction into the human diet is seen as a significant stride forward in human evolution, as converting grains into staples necessitated extraordinary technical and cooking abilities (Harmon, 2009). One of the main reasons for man's search for fresh beverages was poor water quality. Fermentation was thus used to produce a variety of beverages, principally for their safety (Altay et al., 2013). During the Middle Ages, fermented alcoholic and non-alcoholic beverages were produced. The African and South American continents have generated the majority of grain-based non-alcoholic beverages. Kunan-Zaki prepared from sorghum, Bushera manufactured from sorghum or millet, Mahewu prepared from maize, Gowé prepared from millets, Obiolor prepared from sorghum and millet, and Borde prepared from sorghum, wheat, maize, finger millet, and barley are some of the conventional non-alcoholic African beverages. The last two beverages are prepared from a variety of grains (Solange et al., 2014). Agua-agria, Pozol, Fubá, Champuz, Nap, and Acupe are all fermented non-alcoholic beverages from South America. They're mostly manufactured out of maize (Marshall & Mejia, 2011). Ambil and Kali are two classic Indian fermented drinks. The former is made with finger millet, whereas the latter is created with cooked rice. Ambil is prepared to utilize fermented sour buttermilk (Fernandes et al., 2021). Beverage mixes comprised of malted grains have just lately become popular in the previous 200 years. They were created to be drunk with milk in order

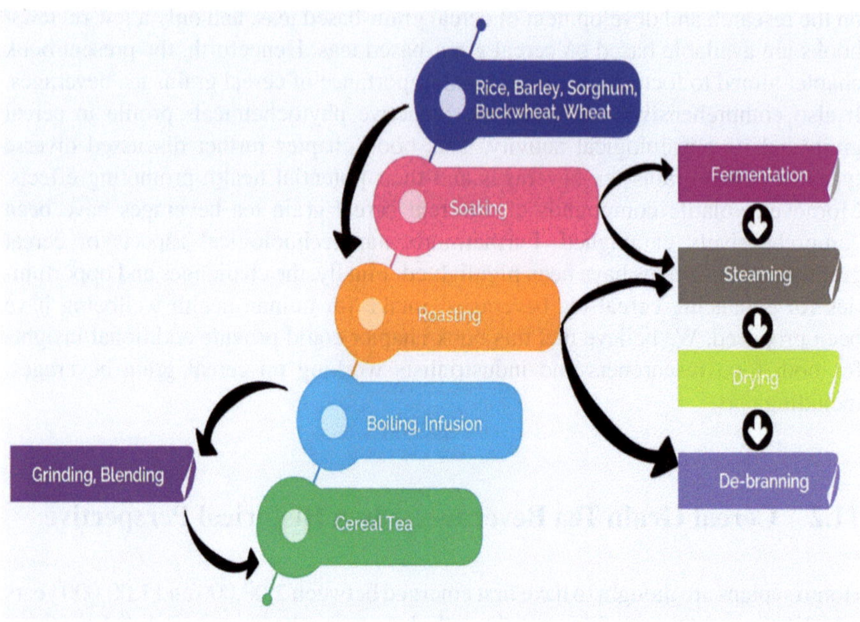

Fig. 11.1 Schematic presentation of cereal tea grain production

to improve the wholesomeness and palatability of the beverage. William Horlick invented the first malt powder in the 1870s. He perfected the drying milk process with malt and wheat in 1882, resulting in an easily soluble product in water. Doctors suggested Horlick's meal to their patients (Fernandes et al., 2021). To speak specifically of cereal grains tea beverages, a variety of cereal grain tea beverages, such as barley, rice, corn, and sorghum teas, have a long history in Asia, particularly China, dating back over 2500 years, and are still consumed in numerous regions of the continent, comprising India, Japan, Korea, and China (particularly Tibet). These teas have now been an integral part of people's daily lives. Figure 11.1 depicts the general techniques for preparing cereal grain teas.

11.3 Bioactive Phytochemicals Profile in the Cereal Grains and Its Physiological Activity

The nutritional, functional, and health properties of cereals and pseudocereals or their products are mostly ascribed to their bioactive compounds. Pseudocereals and cereals have an excellent polyphenols profile balanced to vegetables, fruits, and nuts (Xiong et al., 2020a, b). The major bioactive compounds in the pseudocereal and cereal grains are phytosterols, tocols, phytic acid, γ-oryzanol, carotenoids, dietary fiber, and phenolic compounds (*e.g.*, phenolic acids, flavonoids, polymeric phenolic compounds, phenolic lipids, and phenolic alkaloids). While there are various book chapters and review articles that have presented the polyphenols profile in major cereals (Tokusoglu & Hall, 2011) and pseudocereals (Thakur et al., 2021), this part delivers potential novelties, highlights, and updates of bioactive phytochemicals, and their key sub-categories in the three discussed pseudocereals (amaranth, quinoa, and buckwheat) and eight cereals (rice, sorghum, wheat, oat, corn, millet, rye, and barley), which is essential to reveal the possible functional characteristics of cereal and pseudocereal grain teas. Numerous bioactive as well as anti-nutritional compounds present in different cereal and pseudocereal grains are illustrated in Fig. 11.2a and b. Table 11.1 designates bioactive profile of various cereal and pseudocereal grains.

11.3.1 Phenolic Compounds

Bioactive compounds or polyphenols are a wide range of chemical compounds comprising of multiple benzene rings with hydroxyl groups, including flavonoids, polymeric phenolic compounds (*e.g.*, tannins, lignans), phenolic acids, alkylresorcinols (phenolic lipids) and avenanthramides (phenolic alkaloids) (Campos-Vega & Oomah, 2013). These components are abundant in plants and thus an integral part of human food due to the different biological functions, such as antihypertensive, antioxidant, antimicrobial, anticancer, antiallergic, antithrombotic, anti-inflammatory, antihyperuricemic antidiabetic, antiaging, and cholesterol-lowering activity *in vitro, in vivo,* and in cell culture (Mehmood et al., 2020; Hossen et al., 2020; Masisi et al., 2016; Abdo et al., 2021). Phenolic compounds are the major source of antioxidants in cereals and pseudocereals, and the potential phenolic compounds are flavonoids and phenolic acids, in which buckwheat (506–1642 mg GAE/100 g) and sorghum (87–2960 mg GAE/100 g) have the major phenolic compounds among the various cereals and pseudocereals (Zhu, 2016a, b, 2020; Yang 2010). Total phenolic compounds in millets (2394–3137 μg GAE/g), barley (1567–2322 μg GAE/g), wheat (1650–2095 μg GAE/g) are reported (Siroha et al., 2016; Punia & Sandhu, 2015; Punia et al., 2019).

11.3.1.1 Phenolic Acids

Both pseudocereal and cereal grains are excellent sources of phenolic acids, and there are two basic sub-categories of phenolic acids that embrace hydroxycinnamic (ferulic, sinapic, caffeic, *p*-coumaric, chlorogenic acids) and hydroxybenzoic acids (*p*-hydrobenzoic, vanillic, protocatechuic, salicylic, and gallic acids). It was observed that the millet (61–391 mg/100 g) and sorghum (39–285 mg/100 g) have good amounts of phenolic acids (Xiong et al., 2019b) than other cereals. Phenolic acids are present mainly in the bound form in cereals and are the outer layer of the pericarp, which are easily separable using organic solvents (López-Alarcón &

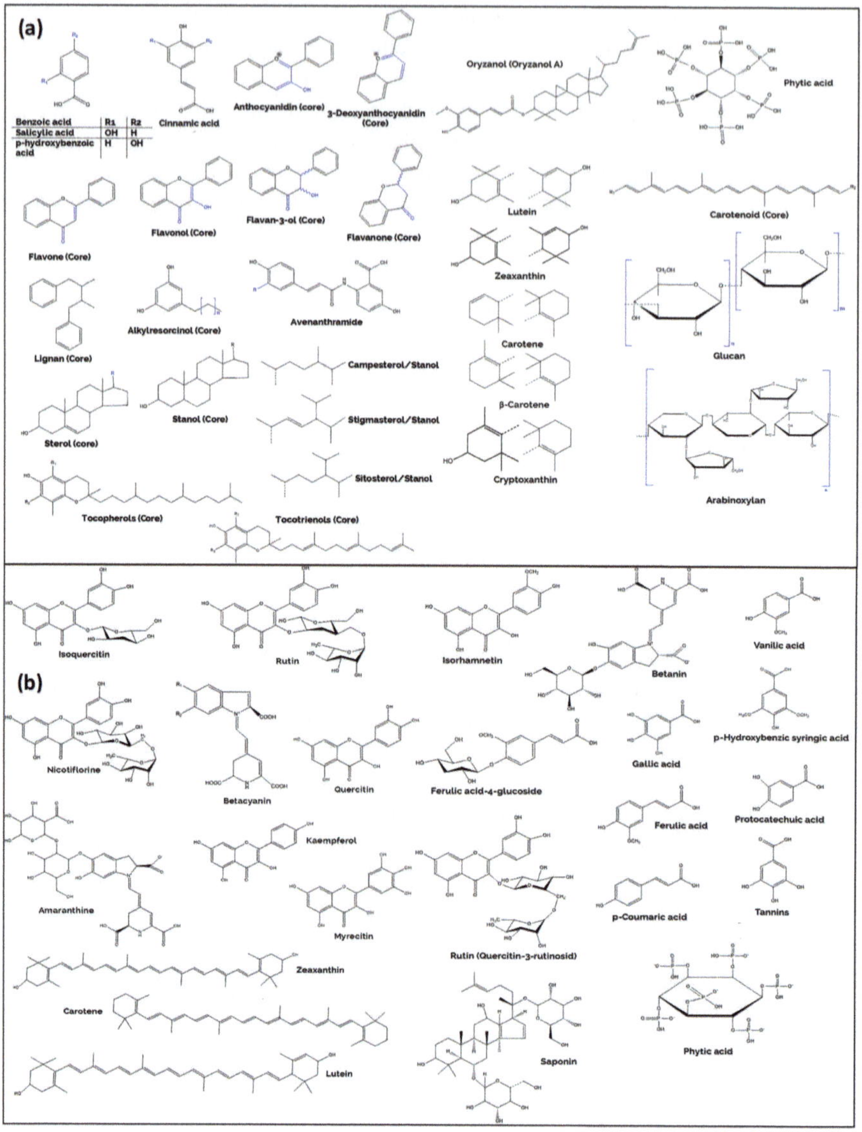

Fig. 11.2 (a, b) Overview of key bioactive phytochemicals as well as anti-nutritional compounds present in cereal and pseudocereal grains along with their chemical structures

Denicola, 2013). Liu et al. (2019) found that ferulic acid is the key phenolic acid in all common cereals and is present in the bound form in the cell wall of the brans' fraction. Moreover, corn and wheat have been observed to hold an excellent amount of ferulic acid (up to 90% of total phenolic concentrations of the entire grains) (Boz, 2015).

Table 11.1 Bioactive phytochemical composition of various cereal and pseudocereal grains

	Corn	Rice	Wheat	Barley	Sorghum	Millet	Oat	Rye	Buckwheat	Reference
Phenolics compounds	38.00–57.04 mg GAE/100 g	6.05 mg/100 g of flour	85.41 mg GAE/g	1.27–1.67 mg GAE/g DW	173.68–1040.73 mg GAE/100 g	300–435 mg catechol eqv./100 g	179–221 mg GAE/g DW	3369.19 mg GAE/kg	10.2 mg/g	Ryan et al. (2007), Siyuan et al. (2018), Yu et al. (2021), Li et al. (2015), Šimić et al. (2017), Punia et al. (2021a, b), Berwal et al. (2016), Kulichová et al. (2019), Kerienė et al. (2015), Zhang et al. (2017) and Kruma et al. (2016)
Phenolic acids	906.13 ± 9.09 Mol/100 g of grain	50.89 ± 1.00 g	516–831 mg/kg DM	991 μg/g DM	445–2850 μg/g	1703.3–1276.8 mg/100 g DW	1320.93–1686.71 mg/kg	216–237 mg/100 g DM	0.391–0.229 mg/g DW	Ryan et al. (2007), Siyuan et al. (2018), Yu et al. (2021), Leváková and Lacko-Bartošová (2017), Horvat et al. (2020), Aura (2016), Vollmannová et al. (2021), Sytar (2015) and Xiong et al. (2019b)
Flavonids	115.4–175.5 mg GAE/100 g DW	177.54 ± 2.89 g	23,124.1 μg GAE/g	48.50 μg/g	15.33–42.84 mg RE/100 g	22.82 ± 0.77 mg CE/100 g	4.37–8.13 mg/g	42.05–203.36 mg Q eqv./kg	22.74 mg /g	Ryan et al. (2007), Siyuan et al. (2018), Yu et al. (2021), Li et al. (2015), Idehen et al. (2017), Sharma et al. (2015), Kulichová et al. (2019), Tariq et al. (2017), Kumar and Kaur (2017), Nawaz et al. (2018), Qin et al. (2010) and Punia et al. (2021a, b)
Carotenoids	15–25 μg/g	100 ng/g	0.78–4.08 mg/kg	2.25–4.54 mg/kg	0.55–0.58 mg/100 g	78–366 mg/100 g	`1.8 mg/kg	2.25–4.54 mg/kg	443.14–509.84 μg/g DW	Zurak et al. (2021), Lamberts and Delcour (2008), Paznocht et al. (2018), Asharani et al. (2010), Abugri et al. (2015), Ndolo and Beta (2013) and Ma et al. (2019)

(continued)

Table 11.1 (continued)

	Corn	Rice	Wheat	Barley	Sorghum	Millet	Oat	Rye	Buckwheat	Reference
Dietary fiber	4–5.2%	0.5 g/100 g	2–7.3 g/100 kg	~17%	6.3%	4.84–5.28%	<7%	15.1%	8.4–10.9%	Katagiri et al. (2021), Obafaye and Omoba (2018), Duchoňová et al. (2021), Huda et al. (2021), Han and Tran (2018) and Rakha (2011)
Tocols	16.5–42.5 μg/g DM	0.10–0.14% (w/w)	1.64%	50.9–61 μg/g	59.8 mg/kg	3.6–4.0 mg/100 g	5.46–61 μg/g	49 μg/g DM	0.50–3.61 mg/100 g	Raguindin et al. (2021), Lerma-García et al. (2009), Asharani et al. (2010), Labuschagne et al. (2017), Suriano et al. (2021) and Becker (2013)
Phytosterols	77–87%	72 mg/100 g	447–830 mg/kg of FW	720–801 mg/kg of FW	46–51 mg/100 g	377.7–495.3 mg/kg	350–491 mg/kg of FW	707–1134 mg/kg of FW	963–1980 mg/kg of FW	Siyuan et al. (2018), Dziedzic et al. (2015), Tolve et al. (2020), Bhandari and Lee (2013) and Awika and Rooney (2004)
c-Oryzanol	28.09–31.0 mg/kg	0.9–2.9% (w/w)	60.9–62.00 mg/kg	2.5–3.81 mg/kg				35.6–75.9 mg/kg		Lerma-Garcia et al. (2009) and Xiong et al. (2020a)
Phytic acid	1.02 g/100 g	40.15–63.31%	0.93 g/100 g	0.97 g/100 g	3–5 mg/g	2.91–3.30%	5.6–7.3 mg/g	1.01 g/100 g	35–38 g/kg	Steadman et al. (2001), Badau et al. (2005), Ratnavathi and Komala (2016), Saastamoinen et al. (1992), Hídvégi and Lásztity (2002) and Perera et al. (2018)

https://scialert.net/fulltext/?doi=ajpnft.2011.1.22
DW Dry Weight, *DM* Dry Mass, *WB* Weight Base, *FW* Fresh Weight

In the case of buckwheat, ferulic acid is only found in common buckwheat (trace amounts) and not found in Tartary buckwheat (Liu et al., 2019). However, various researchers revealed that ferulic acid is detected in Tartary buckwheat with other phenolic acids, for instance, sinapic, caffeic, *p*-coumaric, chlorogenic acids, *p*-hydrobenzoic, vanillic, protocatechuic, salicylic, and gallic acids in all fractions of milled treatments (light flour, bran, hull, and fine bran) (Zhu, 2016b). In contrast to cereals, phenolic acids are mainly identified in the free form in the brans and to a minor amount in the bound form. Comparative analyses exhibited that the major phenolic acids in the fine bran are *p*-hydrobenzoic acid, caffeic acid, chlorogenic acid, and protocatechuic acid up to 360 mg/100 g, 38 mg/100 g, 21 mg/100 g, and 18 mg/100 g, correspondingly. However, protocatechuic acid is abundantly found in the hulls up to 54 mg/100 g (Guo et al., 2012a). Current works on identifying multiple phenolic acids, i.e., ferulic and vanillic acids, and their derivatives in quinoa have been reported (Tang et al., 2015, 2016). These phenolic acids are found in conjugated forms with a concentration of 132–161 mg/kg of quinoa seeds. As a pseudocereal, amaranth is also a rich source of nutrients and is discovered to possess several types of polyphenolic acids in seeds and sprouts, including gallic, vanillic, and *p*-hydrobenzoic acids (de la Rosa et al., 2009; Vollmer et al., 2017). Heleno et al. (2015) revealed that the key health effect of phenolic acids is reported to be antioxidant activity. Punia et al. (2021a, b) and Bangar et al. (2021) reported ascorbic acid, gallic acid, catechin, and vanillin in the fermented rice and barley bran.

11.3.1.2 Flavonoids

In phenolic compounds, flavonoids are considered the abundant family in cereals/pseudocereals and even in other parts of foods. The flavonoids such as flavones, flavanones, anthocyanidins, and falvan-3-ols are documented key sub-categories and are predominantly present in cereal/pseudocereals grains' bran portion (Dykes & Rooney, 2007). It was observed that anthocyanidins, water-soluble pigments, i.e., blue, orange, red, and purple color, are the most reported flavonoids in cereal; besides, their contents vary greatly among cereals. Xiong et al. (2020b) noted that sorghum has the top position among the main cereals in terms of content and superiority of flavonoids. Moreover, they revealed that sorghum found a special kind of anthocyanidins, named 3-deoxyanthocyanidis, that lacks hydroxyl group at the C-3 position, making them more stable than that of anthocyanidins, and have been suggested to be excellent multifunctional organic colorants (Xiong et al., 2019a, c).

Buckwheat is an important pseudocereal, and multiple species of buckwheat found good amounts of flavonoids content. Cai et al. (2016) showed that the flavonoid concentration of Tartary buckwheat is four times greater than that of common buckwheat, and its unpleasant or bitter taste is partially linked to a higher flavonoid content. In contrast, Qin et al. (2010) reported that flavonoid content in Tartary buckwheat flour is ten times higher than that in buckwheat flour altogether. They also evaluated the total flavonoid content of 21 and 18 genotypes for Tartary and common buckwheat from a Chinese collection and observed a huge variation in average total flavonoid content (6.65–22.74 mg/g DW). Mei-Hua-Shan is a variety of Tartary buckwheat that possessed the maximum concentration of total flavonoid content (22.74 mg/g). Rutin (quercetin-3-rutinosid) is an important flavonoid found in buckwheat species. This is also a plant metabolite and is present 10 times greater in the greener portion of the plants than the seeds, which protects from ultraviolet radiation (Zhu, 2016b). Quinoa seeds only had rutin concentrations comparable to buckwheat hulls (329 mg/kg), (Thakur et al., 2021). Amaranth is also a well-known pseudocereal in terms of nutrition, and it has been investigated for its flavonoid profile, including nicotiflorine (4.8–7.2 μg/g flour), rutin (4.0–10.2 μg/g flour), and isoquercitrin (0.3–0.5 μg/g flour) (de la Rosa et al., 2009). Although flavonoids present in pseudocereals and cereals exhibited various health benefits; though, Panche et al. (2016) unveiled that anti-mutagenic, anti-inflammatory, antioxidant, anticancer, and regulation of specific cellular enzymes have been recommended as the main health benefits of flavonoids.

11.3.1.3 Polymeric Phenolic Compounds or Antinutritional Compounds

The term "anti-nutritional compounds" is recognized for their astringent and bitter taste and have widely been documented as compounds that can bind the proteins, amino acids, peptides, minerals, and therefore decrease their absorption, digestion, metabolism, and utilization (Xiong et al., 2020a). Despite the fact mentioned above, they have been exhibited to keep good health claims, possessing much higher antioxidant values and cholesterol-lowering effects *in vitro* than monomeric phenolic compounds. These compounds are also known as excellent dietary materials for people with diabetes and obesity (Smeriglio et al., 2017; Xiong et al., 2020a). Tannins, phytic acid, and saponins are mainly documented anti-nutritional compounds in cereal and pseudocereal grains. The procyanidins (PCs) and proanthocyanidins (PAs) are condensed tannins, which are polymeric phenolic compounds present in all major cereals. They are comprised mainly of flavan-3-ol monomers and are primarily contained in certain pigmented cereal grains such as red finger millet, dark-colored barley, and brown sorghum (Salar et al., 2017). They have been shown a large molecular weight and a high degree of polymerization in sorghum with pigmented testa (type III high tannin sorghums); however, such type is uncommon in other major grains (Xiong et al., 2020a). Also, PAs have been found in pseudocereal seeds and are considered an essential class of oligomeric flavonoids. It was explored that the consumption of PAs from pseudocereals is linked with reducing multiple chronic diseases such as cancer, diabetes, oxidative stress, and inflammation (Zhu, 2019).

Moreover, raw, cooked, and roasted amaranth have been a source of different tannins and inhibited the activity of α-amylase and α-glucosidase up to 50% and 78%, correspondingly (García-Estepa et al., 1999). Besides, buckwheat seeds were observed to be trypsin inhibitors as anti-nutritional compounds, binding with the enzyme chymotrypsin (Krkošková & Mrazova, 2005). Zhang et al. (2012) revealed that buckwheat seed brans have an abundant concentration of phytic acid (35–38 g/kg) and tannins. While, the phytic acid content in quinoa and amaranth is quite low, ranging from 10.5–13.5 g/kg to 2.9–7.9 g/kg, correspondingly (De Ruiz & Bressani, 1990; Koziol, 1992). Phytic acid is also present in cereal grains and is primarily found in the bran fraction, precisely the germ of the grain and aleurone layer (Goufo & Trindade, 2014). Saponins are also considered anti nutritional components (natural detergents) and are glycosylated secondary metabolites present in the pseudocereals. Quinoa seeds are composed of excellent total saponins (6.27–692 mg/kg); however, amaranth comprises only 0.9–4.91 mg/kg of total saponins (Yao et al., 2014).

In many plant kingdoms, lignans are found (in the bran of cereal grain) as a group of phytoestrogens and mainly contained in rye and oat (Rodrıguez-Garcıa et al., 2019; Durazzo et al. 2013). The major plant lignans such as matairesinol, syringaresinol, lariciresinol, secoisolariciresinol, and pinoresinol may be transformed into mammalian lignans in the colon by microflora that has low estrogenic and excellent antioxidant activities. These lignans promote colon health and decrease the chances of hormone-related cancers (Thompson, 1994; Adlercreutz, 2007).

11.3.1.4 Alkylresorcinols (Phenolic Lipids)

Another types of polyphenols present in cereal grains are Alkylresorcinols and are situated in the bran fraction. Among all major cereals, rye has reported the greatest amount of phenolic lipids (Fardet et al., 2008). These phenolic lipids have exhibited good antioxidant and antimicrobial health benefits *in vitro* and are recognized as lipophilic antioxidants; however, their effectiveness is less than vitamin E (Ross et al., 2004).

11.3.1.5 Avenanthramides (Phenolic Alkaloids)

Avenanthramides are another phytochemicals mainly found in the oat, having avenanthramides A, B, and C. They have been shown excellent antiatherogenic, antioxidant, and anti-inflammatory capacities *in vitro* and even *in vivo* (Gani et al., 2012; Meydani, 2009). The discovery of N-trans-feruloyltyramine in Tartary buckwheat was made using high-performance liquid chromatography coupled with photodiode array detector and linear ion trap fourier transform ion cyclotron resonance hybrid mass spectrometry (HPLC-PDA/LTQ-FTICRMS). A study found that this alkaloid had neuroprotective properties (Thangnipon et al., 2012; Ren et al., 2013).

11.3.1.6 Anthraquinones

Anthraquinones are a group of phenolic compounds unique to Tartary buckwheat. Tartary buckwheat seeds have been reported to have six different kinds of anthraquinones, i.e., aloe-emodin, emodin, physcion, aurantio-obtusin, rhein, and chrysophanol. Peng et al. (2013) identified emodin contents (1.72–2.71 mg/kg) in 4 Chinese

Tartary buckwheat genotypes, existing in the seeds as glycosides. These compounds have been demonstrated to have antibacterial, anticancer, and liver protection abilities (Peng et al., 2013; Wu et al., 2015).

11.3.2 Carotenoids

Carotenoids are an important class of natural color pigments (e.g., yellow, red, and orange color) extensively found in cereals and pseudocereals. These compounds are markedly distributed in the endosperm, resulting in the endosperm's yellow color (Trono, 2019). In cereals, α-carotene, lutein, β-carotene, zeaxanthin, and β-cryptoxanthin are major carotenoids and vary in amounts and composition. It was observed that cereals have the least concentration of carotenoids than that of fruits and vegetables; however, some colored cereals with golden and dark yellow-pigmented endosperm, including golden rice, foxtail millet, and high carotenoid corn, possess a significant amounts of carotenoids (Shen et al., 2015; Moreau et al., 2016; Ndolo & Beta, 2013). Trono (2019) noted that corn has the highest content of carotenoids (lutein, β-carotene, and zeaxanthin), ranging from 0.9 to 15.6 mg/100 g of the whole grain.

Quinoa seeds are also found to be rich in carotenoids, particularly lutein, zeaxanthin, and β-carotene ranging from 3.96 to 12.01 mg/kg, 0.31 to 5.37 mg/kg, and 0.26 to mg/kg, correspondingly. On the contrary, amaranth seeds were found to possess low concentrations for lutein and zeaxanthin of 3.55–4.44 mg/kg and 0.14–0.30 mg/kg, respectively (Tang et al., 2016). Germinating Tartary buckwheat under white light for 10 days provides an excellent source of β-carotenee and lutein of 463 μg/g and 666 μg/g correspondingly (Tuan et al., 2013). In the same sprouting condition, the total carotenoid concentration of Tartary buckwheat sprouts was 1.3 mg/g DW. Other carotenoids, such as 13-cis-β-carotene, zeaxanthin, α-carotene, β-cryptoxanthin, and 9-cis-β-carotene, were also found in the small concentrations. The nitrogen pigments, such as betalains, are abundant in pseudocereals. They are synthesized from tyrosine and transformed to L-3, 4-dihydroxyphenylalanine and could be grouped as red-violet colored betacyanins and yellow-orange colored betaxanthins. They've been employed as a natural colorant for a long time and are very effective scavengers to inhibit the oxidation of low-density lipoproteins (LDLs) (Esatbeyoglu et al., 2015). Both the amaranth and quinoa are good in betacyanins such as amaranthine and betanins (1.5–61 mg/kg of grain according to evaluations on the basis of ultra-violet/Vis absorbance result), respectively (Cai et al., 2003; Repo-Carrasco-Valencia et al., 2010). Many researchers revealed that higher dietary consumption of carotenoids is linked with the low rate of chronic diseases and prevents CVDs and ultra-violet induced skin problems, some kinds of cancer, macular degeneration, improve the immune system, and cataracts (Rao & Rao, 2007; Tuan et al., 2013). In addition, they have been exhibited to have lipid-soluble antioxidant and pro-vitamin properties.

11.3.3 Tocols

Tocols are considered as vitamin E analogues. These compounds, including tocotrienols (e.g., α-, β- and γ- tocotrienols) and tocopherols (α-, β- and γ-tocopherols), are organically giving lipid-soluble antioxidants. Tocols are found in all major cereals; in particular, substantial concentrations have been identified in some oat, corn, barley, rye, and even wheat cereals, and their amounts and composition vary significantly among cereals. They are abundantly present in the cereal grains' germ fraction (Zielinska et al., 2013; Ehrenbergerova et al., 2006; Tiwari & Cummins, 2009; Asharani et al., 2010; Chung et al., 2013). Among six major tocols in cereals, the α-tocopherol is the most widely distributed tocol in all studied cereals; and good amounts of α-tocopherol, γ-tocopherol, and α-tocotrienol have been detected in corn, barley, and oat, respectively (Ehrenbergerová et al., 2006; Tiwari & Cummins, 2009; Asharani et al., 2010; Chung et al., 2013). In reference to pseudocereals, four types of tocopherol isoforms, including α-, β-, γ-, and δ-tocopherols found to locate in quinoa and amaranth. In contrast to cereals, pseudocereals have been demonstrated to have high amounts of γ-tocopherol in the range of 57–53 mg/kg DW than that of α- tocopherol ranging from 17 to 26 mg/kg DW. Other two tocopherol isoforms, including β- and δ-tocopherols, are also found in very small amounts, i.e., lower than 5 mg/kg DW (Alvarez-Jubete et al., 2009; Ruales & Nair, 1993). However, Tartary buckwheat seeds are a good source of γ-tocopherol (117 μg/g), δ-tocopherol (7.3 μg/g), and α-tocopherol (2.1 μg/g) (Zhou et al., 2015). Contrarily, other research works revealed that the potential tocopherols in amaranth seeds are δ-tocopherol ranging from 0.01 to 18.79 mg/kg, followed by β-tocotrienol (0.53–43.86 mg/kg), α-tocopherol (1.40–31.6 mg/kg), and γ-tocotrienol (0.53–8.69 mg/kg) (Bruni et al., 2002; Lehmann et al., 1994; Tang et al., 2015). Quinoa seeds also contained small amounts of α- and β-tocotrienol. The biological functions of tocols have been a subject of research interest. Many studies have shown that tocols play a pivotal role in maintaining membrane integrity, antioxidant activity, and some non-communicable diseases, comprising cancers and CVDs (Fardet et al., 2008).

11.3.4 Phytosterols

Both stanols and sterols are considered phytosterols found in plants that have a striking resemblance to cholesterol (in terms of structure) and fundamental compounds of plant cell membranes. They are concentrated in the bran part of cereal grains and are present in all major cereals (Bhandari & Lee, 2013; Gupta et al., 2018). Cereal grains contain phytosterols in their natural state, acetylated (esterified) with phenolic acids or fatty acids, or conjugated with glycosides, and are regarded to be the primary dietary source of phytosterols. In cereals, the most common and abundant phytosterol is sitosterol. Other compounds, such as campesterol, stigmasterol, sitostanol, and campestanol, were also identified in all main cereals. Buckwheat includes a high concentration of phytosterols, which may reach 96 mg/100 g (wet basis) of entire seeds, and β-sitosterol and campesterol were principal and common phytosterols in common buckwheat. Yang et al. (2014) examined the phytosterol content of Tartary buckwheat flour to that of wheat and rice flour and discovered that they contained 74, 34, and 15 mg/100 g of phytosterols, respectively. Moreover, Tartary buckwheat flour had 61, 7, 2, and 5 mg/100 g of β-sitosterol, campesterol, stigmasterol, and avesterol, respectively. Furthermore, daucosterol was extracted from Tartary buckwheat bran (Guo et al., 2012b). Phytosterols are among the bioactive molecules presently under investigation, and their primary health benefit (cholesterol-lowering effect) has been clearly demonstrated (Genser et al., 2012).

11.3.5 γ-Oryzanol

γ-Oryzanol, which was once believed to be a single chemical component specific to rice, has recently been identified as a mixture of phytosterols and triterpene. It has also been observed in other cereals, including wheat, rye, corn, and barley, however, at considerably lower concentrations than that of rice (Goufo & Trindade 2014; Gani et al., 2012; Singh et al., 2015). γ-Oryzanol is primarily found in the bran of the entire grains. Rice contains around 20–122 mg/100 g γ-oryzanol in the entire grains, whereas, in the bran fraction, γ-oryzanol was identified to be in the range of 185–421 mg/100 g. Rice bran oil has a significant amount of this ingredient. Antidiabetic, anti-inflammatory, anticancer, and antioxidant health effects have all been proven for γ-oryzanol *in vitro* and even *in vivo* (Minatel et al., 2016; Suh et al., 2005; Juliano et al., 2005).

11.3.6 Dietary Fiber

Dietary fiber seems to be another big family of phytochemical compounds that promotes human gut health. The nutritional value of grains is greatly influenced by dietary fiber. Resistant starch, non-digestible oligosaccharides, and non-starch polysaccharides are all examples of dietary fiber components (Phillips et al., 2019). Based on water solubility, dietary fiber is traditionally separated into insoluble and soluble parts. Neutral and acid detergent fractions can also be found in the fiber. For the most part, this classification does not apply to the diet of humans. Amaranth, buckwheat, and quinoa had much lower soluble fiber content than insoluble fiber. Dietary fibers are mainly found in both the endosperm and bran layer of cereal grains. The dietary fiber components are β-glucan, cellulose, inulin, lignin, arabinoxylan, and resistant starch (Dhingra et al., 2012; Zhu, 2020). In cereals, the dietary fibers, including resistant starch, β-glucan, and arabinoxylan, have received significant attention and are extensively investigated. β-glucans are polysaccharides that are entirely made up of glucose units. They are significant dietary fibers discovered in all major cereals, primarily in the endosperms and aleurones' cell walls (Ahmad et al., 2012; Ermawar et al., 2015; Sun et al., 2019). Barley and oat, among many of the primary cereals, have the most β-glucan, which is contained in the endosperm and aleurone layer, respectively (Ahmad et al., 2012). The key biological functions of β-glucan are: (1) improving the immune system, (2) regulating blood glucose, and (3) decreasing blood cholesterol because they have larger viscosity and are a water-soluble molecule that can bind bile acids and triglycerides in the gastrointestinal tract (GIT) (Wood, 2010; Sun et al., 2019).

As per the FDA, consuming 3 g of oat and barley-glucan per day lowers blood cholesterol levels and alleviates cardiac failure risk. (FDA, 2008). Arabinoxylans form hemicelluloses composed of a lined xylose backbone chain with arabinose substitutions. All main cereals contain arabinoxylans, although rye has the largest concentration. Arabinoxylans are found mostly in the cell walls of grains' endosperm and bran portions. These are the second most prevalent natural biopolymers in several cereals, behind cellulose, and provide a large amounts of dietary fibers to the diet (Izydorczyk, 2021; Saeed et al., 2011). The distinctive property of arabinoxylans is the availability of phenolic acids, like ferulic acid, that is covalently bonded to arabinose and xyloses. Ferulated arabinoxylans have attained significant consideration in the pharmaceutical industry due to their anticancer, antioxidant, and prebiotic characteristics (Mendez-Encinas et al., 2018). In addition to dietary fiber, resistant starch is a type of starch that restrains digestion in the intestines and moves into the large intestine. It is most typically present in all major cereals and legumes. Many researchers have done a great job of summarizing the health benefits of resistant starch, including intestinal and metabolic functioning (Zhu, 2020; Fuentes-Zaragoza et al., 2010). β-glucans and arabinoxylans are 2 major fiber components found in cereals. Rye and wheat are good in arabinoxylans, while barley and oats encompass good amounts of β-glucans (Cui & Wang, 2009).

Variations in dietary fiber content have been reported among 21 amaranth species in the range of 10–25% (Mustafa et al., 2011; Aguilar et al., 2013; Kurek et al., 2018; Srichuwong et al., 2017). Buckwheat fibers were found to be diverse in composition. For instance, the total fiber content of six commonly used buckwheat genotypes with hulls varied between 20% and 26%. Buckwheat brans included a significant amount of fiber compared to refined flour. The hull of buckwheat had significantly more fiber than that of the brans and groats (Dziadek et al., 2016; Lu et al., 2013; Biel & Maciorowski, 2013; Skrabanja et al., 2004). There have been found fluctuations in the fiber content of quinoa (Miranda et al., 2012; Repo-Carrasco-Valencia & Serna, 2011; Sobota et al., 2020). For instance, the values of the total, insoluble, and soluble fiber contents were 14–20, 10–14, and 3.7–5.9%, correspondingly (Sobota et al., 2020). The fiber contents of quinoa and amaranth entire seeds have been appeared identical generally. Fiber content was higher in quinoa and amaranth seeds than dehulled buckwheat. Amaranth and quinoa showed higher levels of soluble fiber compared with cereals. Amaranth and quinoa fiber compositions were closer to fruits than cereals (Lamothe et al., 2015).

Dietary fiber's health benefits have been thoroughly explored and confirmed. Entire grain dietary fiber offers a variety of biological benefits, such as anti-glycemic, oxidative stress, hormone-regulating, prebiotic, and microbial management. Consumption of a high-fiber diet has been linked to a lower risk of developing certain diseases, including chronic digestive conditions, heart disease, diabetes mellitus, and obesity (Charalampopoulos et al., 2002; Smith & Tucker, 2011; Threapleton et al., 2013; Aune et al., 2011).

11.4 Diverse Types of Cereal/Pseudocereal Grain Tea Beverages and Their Significant Health Claims

11.4.1 Cereal Grain Tea Beverages

As the food industry has become more diversified, cereal grain-based beverages might be an exciting future area of health-promoting functional beverages in our daily lives. Over the last few decades, major cereal crop teas have been the subject of research interest; thereby, wheat bran (Wang et al., 2019) and oat grain (Hao et al., 2018) are being investigated for cereal grain tea production. In the past few years, it has been demonstrated that the health mending effects of cereal tea beverages are strongly correlated with their bioactive phytochemical content. A few of the phytochemicals are involved in various health benefits of cereal tea that have been found, with others still to be discovered. Potential health-promoting effects of various cereal and pseudocereal grains are represented in Fig. 11.3.

11.4.1.1 Barley Tea

Barley tea, commonly referred to as "oriental coffee," is widely prevalent and ingested in numerous countries, comprising Korea, Japan, and China. The roasted entire barley grain has been used to prepare the barley tea historically. It can be consumed cold or hot and is commonly served in tea bags (i.e., in bottles and as ground barley). Moreover, it has a mild, somewhat bitter, and sweet flavour with a nutty and toasted aroma, which is quite prevalent as a thirst-quencher during the whole summertime. Barley tea is generally considered to provide hunger-stimulating, digestion assisting, and weight loss health advantages (Oh et al., 2014; Tatsu et al., 2020). Nowadays, it has been established that barley tea beverages have a variety of biological functions. Alcoholic phenolic extracts and phenolic extracts from the roasted barley tea grain were observed to express magnificent activities *in vitro*, as exhibited by their activity to scavenge radicals [OH^-, 2,2-diphenyl-1-picrylhydrazyl (DPPH), and O_2^-) and Fe^{3+} (chelate transition metal ions) (Omwamba et al., 2013; Omwamba & Hu, 2009; Etoh et al., 2004).

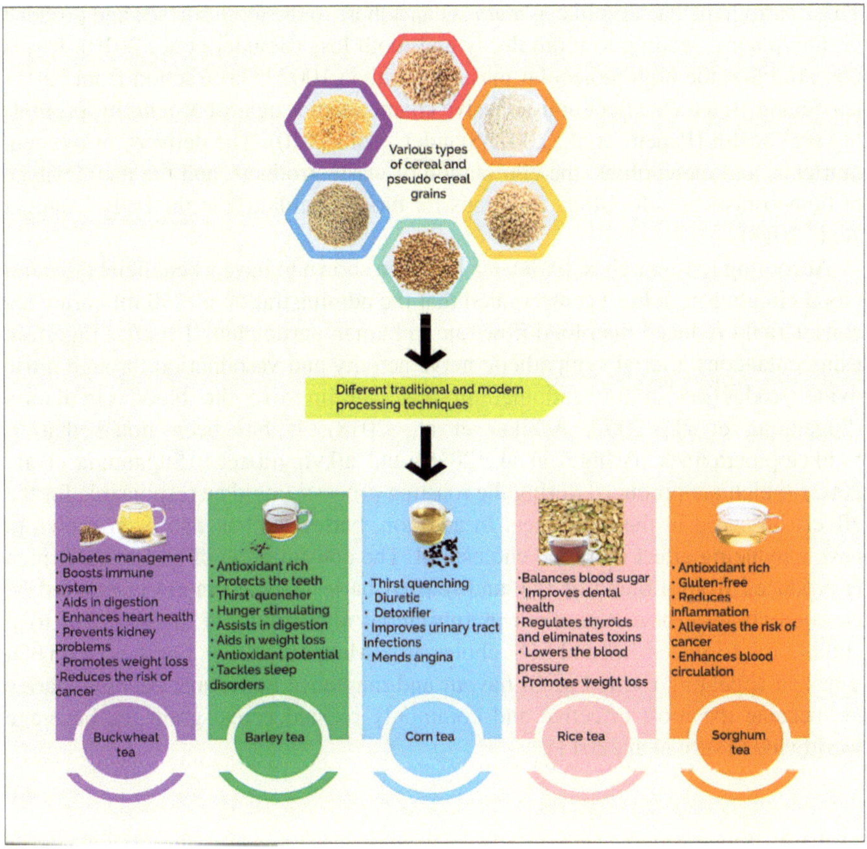

Fig. 11.3 Health promoting effects of various cereal and pseudo cereal grain teas

Moreover, *in vivo* investigations with animal models have demonstrated that barley tea possesses strong antioxidant activity. Omwamba et al. (2013) discovered that elderly mice who were gavage-fed barley tea extract had considerably increased total antioxidant activity in the blood and raised lipid peroxidation inhibition in liver homogenate that was similar to Trolox in terms of antioxidant capacity (a synthetic antioxidant). Furthermore, when compared to untreated mice, the phenolic extract was revealed to enhance the activity of antioxidant enzymes glutathione peroxidase (GSH-Px) and superoxide dismutase (SOD), while reduce the levels of malondialdehyde (MDA) and manoamine oxidase (MAO) in the brain and liver.

Barley tea contains abundant phenolic compounds, encompassing p-coumaric acid, isoamericanol A, 3, 4-dihydroxybenzaldehyde, and quercetin, which exhibited higher antioxidant activity *in vitro* than that of butylated hydroxytoluene (Etoh et al., 2004). Additionally, barley tea has been demonstrated to inhibit the adsorption of a number of *Streptococci mutans* (*S. mutans*) strains to hydroxyapatite, a material that simulates the tooth's surface (Papetti et al., 2007; Stauder et al., 2010).

These cariogenic bacteria like *S. mutans* can adhere to the tooth surface and produce biofilm/plaque, leading to tooth decay and tooth loss (Stauder et al., 2010). It was observed that the high molecular mass (more than 1000 kDa) fraction from barley tea demonstrated excellent anti-adhesive characteristics against *S. mutans*, promoting oral health (Papetti et al., 2007; Stauder et al., 2010). The delivery of oxygen, nutrients, and metabolites, the elimination of waste products, and the maintenance of homeostasis are all critical functions for blood circulation in the body (Ashigai et al., 2018).

According to researches, barley tea has been shown to have a beneficial effect on blood circulation. It has been revealed that the administration of 250 ml barley tea considerably reduced the blood flow rate in human participants 1 h after ingestion using cutaneous arterial sympathetic nerve activity and vasodilation through nitric oxide production in the endothelium that can improve the blood circulation (Suganuma et al., 2002; Ashigai et al., 2018). It has been noted that 2, 5-diketopiperazines (Ashigai et al., 2018) and alkylpyrazines (Suganuma et al., 2002), which are produced during the roasting process, could be responsible for the effects observed in the barley tea. In addition, barley tea has also been shown to have a reducing effect on blood cholesterol. The consumption of barley tea (which is produced from a blend of raw and roasted barley whole grains) in rats fed an elevated cholesterol diet resulted in substantial alleviations in the levels of total cholesterol, triglycerides, and LDL cholesterol (Makpoul, 2016). For the reason that it has an appealing roasted grain flavour and may have nutritional benefits, barley tea became the most effective and commonly roasted cereal grain tea beverage worldwide (Tsuru et al., 2011).

11.4.1.2 Sorghum Tea

Recently, Sorghum tea has gained popularity in China due to its potential health benefits. The high level of phenolic compounds in sorghum has generated considerable interest because of its great variety and abundance compared to other cereals (Xiong et al., 2019b). Sorghum contains a unique combination of condensed tannins and 3-deoxyanthocyanidins. In particular, the water-soluble antioxidant 3-deoxyanthocyanins have promising thermal stability, besides their health benefits, making sorghum a significant raw material for functional foods like tea (Xiong et al. 2019a, b). Wu et al. (2013b) explored the bioactive properties of sorghum tea prepared from sorghum whole grain, produced by traditional methods (soaking, steaming, and roasting), and confirmed that the examined antioxidant activity in sorghum tea was positively associated with the total phenolic, flavonoids, and tannin contents (Wu et al., 2013b). Furthermore, two subsequent studies used similar processing techniques to evaluate the bioactive properties of sorghum tea produced from various genotypes of sorghum whole grains, including white, red, and black (Xiong et al. 2019b, 2020a). Researchers confirmed that black sorghum tea exhibits up to sevenfold higher antioxidant activity and phenolic contents than white and red sorghum tea. Sun et al. (2020) prepared a red sorghum tea by fermenting the whole grain with *Lactobacillus plantarum* or *Saccharomyces cerevisiae* before steaming and roasting it (Sun et al., 2020). They concluded that fermentation has little effect on the antioxidant activity and total phenolic contents of tea, but significantly affects its aroma. Currently, sorghum grain tea research is focused mostly on its processing techniques.

11.4.1.3 Corn Tea

Traditional Korean tea, corn tea, is made from whole corn dry roasting grain and infusing it with boiling water to create a yellow-colored tea. Koreans mostly consume this popular tea throughout the winter as a thirst-quenching, diuretic, and detoxifying beverage (Feng et al., 2009). Today, corn tea is commercially available in Korea and China, and white corn tea product has been patented (Barahona, 2001). Research on corn tea, however, is very limited. In one study, the author investigated the roasting conditions of corn tea and confirmed that 200 °C for 20 min produced the best quality tea (Feng et al., 2009). Similarly, Youn and Chung (2012) also examined the processing conditions of corn tea and revealed that for the best sensory score, 207 °C for 24 min was the optimum temperature (Youn & Chung, 2012).

11.4.1.4 Rice Tea

For over 2500 years in various regions of China, rice tea was used as a traditional beverage. At that time, the general procedure used to prepare rice tea was washing the rice grains, soaking, roasting, and removing bran, browning of grains at low temperature, and at the end, rice grains were boiled in water until they become popped. In rice tea preparation, brown rice is mostly preferred, but white and black rice are also used occasionally. Rice tea has many benefits, like it quenches thirst, helps lower blood pressure, and provides appetite feeling (Adan et al., 2016); today, there are chances to explore these benefits more thoroughly. In recent years, rice has been gaining popularity in Korea and Japan. The researchers (of these countries) are now focusing on optimizing the processing conditions of rice tea with respect to novel rice tea. In Korea, few attempts have been made to improve the taste, nutritional value, and functionality of the rice. Various types of rice, like germinated rough rice, have been used to escalate the phenolic and antioxidant effects of rice tea (Lee et al., 2009), and similarly, black waxy rice (largest embryo) has been used to increase the amino acid contents, antioxidant activity, and phenolic contents of rice tea (Han et al., 2015).

11.4.2 Pseudocereal Grain Tea Beverages

11.4.2.1 Buckwheat Tea

When it comes to psudocereal grain tea, tea made from buckwheat has become famous in Korea, Japan, and China with the major benefit of unique aroma (due to roasting) apart from additional health benefits. Like normal tea processes, hot water is added to roasted buckwheat grains, and sometimes stems and leaves of buckwheat are also used to prepare buckwheat tea. Amongst varieties, one can use both varieties (common and Tartary) for tea; however, tea made from Tartary buckwheat has more phenolic content than the tea prepared from common buckwheat (Liu et al., 2019; Qin et al., 2013). Various types (more than a hundred) of buckwheat teas are available in the market, and these vary from each other based on plant portion used in tea preparation (grains, stems, and leaves, whole grain and whole plant), germination of grains, roasting of grains, branned or de-branned grains, and processing conditions like soaking, steaming, and extrusion (Peng et al., 2015; Guo et al., 2017; Xu et al., 2019). Buckwheat tea has many health effects owing to the high concentration of antioxidants. Functional properties associated with buckwheat tea are the protection of lipids from oxidants (malondialdehyde assay), high 2,2′-azino-bis(3-ethylbenzothiazoline-6-sulfonic acid) ($ABTS^{•+}$) radical scavenging activity but less than ascorbic acid, and very low ferric reducing antioxidant power (Jeong et al., 2011). Extract of buckwheat tea had high resistance against the H_2O_2 that induced PC-12 cells oxidation (Jeong et al., 2011). A high number of flavonoids, especially quercetin and rutin, are present in buckwheat tea that have been highly recommended as a major constituent of antioxidants of buckwheat (Peng et al., 2015; Guo et al., 2017; Jeong et al., 2011).

Part of the buckwheat used as raw material in tea preparation significantly influenced the composition of flavonoids of buckwheat tea. The higher number of flavonoids with the predominance of 7,3′,4′-flavon-3-ol is present in the buckwheat tea prepared from the bran or a whole plant; on the other hand, the tea prepared from the germinated or the whole grain of buckwheat has a comparatively less number of flavonoids, but the concentration of rutin is high in this tea (Guo et al., 2017; Peng et al., 2015). Although rutin and quercetin are both antioxidants, the bioavailability of rutin is lower than quercetin (Manach et al., 1997). Flavonoids of buckwheat have beneficial anti-glycation characteristics. In forming advanced glycation end products (AGEs), buckwheat tea flavonoids (quercetin and rutin) are the most known inhibitors. AGEs are responsible for developing a variety of serious diseases, comprising cardiac disease, cancer, and diabetes (Wu & Yen, 2005). Extract of buckwheat tea formed by the hull of buckwheat has a higher number of flavonoids like orientin, rutin, and vitexin, and this tea has a higher restraining effect in AGEs formation *in vitro*, but as compared to green tea, it has low effects (Zielinska et al., 2013). Although the phenolic content is high in green tea, the tea of buckwheat hull has higher flavonoid contents than green tea (Zielinska et al., 2013), which is highly recommended for the preparation of a blend of fortified tea. Many research attempts have already been done to fortify the grain of buckwheat tea by using different herbal materials like mingri, kudingcha, and zijuan leaves in order to increase health benefits (Xu et al., 2019).

Buckwheat tea fortified with different herbals is healthier and has a high catechin content and total flavonoids content than buckwheat tea only; by lowering the total triglyceride and cholesterol and increasing high-density lipoprotein (HDL) cholesterol level. While buckwheat tea has a higher hypoglycemic effect on alloxan-induced mice (Qin et al., 2014). Buckwheat tea prepared from the blend of different components like jasmine, chrysanthemums, and roses has a greater amount of phenolics and flavonoids and higher antioxidant activity *in vitro*, with strong and a variety of volatile aromatic compounds than the buckwheat tea only (Xu et al., 2019). Buckwheat tea has higher enzyme inhibitory characteristics both in tea soup (aqueous extract) and alcoholic extract of both teas of the whole plant of buckwheat and whole-grain tea that have higher α-glucosidase inhibitory activity *in vitro* (Qin et al., 2013; Guo et al., 2017). The alcoholic extract has higher inhibitory activity as compared to extract of grain aqueous, and the inhibitory activity of extract of whole plant aqueous is lower than extract of whole-grain aqueous. The phenolic concentration of tea and tea composition highly affects the inhibition activity of tea (Qin et al., 2013; Guo et al., 2017). Many studies showed that rutin and quercetin have higher drug-drug interactions by affecting the working of the drug-metabolizing enzymes (Zou et al., 2016). Many studies have noticed that metabolism of the eplerenone rate is highly affected and accelerated by long time intake of buckwheat tea due to the presence of the high amount of rutin and quercetin in tea, but the dose of some drugs like eplerenone can be adjusted while taking different buckwheat based products like tea during the treatment of drug to control the drug-drug interaction (Zou et al., 2016).

While the distinctive malty and roasted aroma of cereal grain tea beverage is one of its most appealing characteristics, however further investigation about odor-active volatile components is required; this information is critical for the enhancement and fortification of the tea's aroma. It will take a great deal of research to learn how to improve the processing parameters for cereal grain tea beverages in order to get the optimal phenolic and bioactive profile.

11.5 Volatile Compounds of Cereal and Pseudocereal Grain Teas

A tea's volatile profile has a significant influence in determining the quality and acceptance of a tea. Cereal grain tea's aroma, which comes from volatile constituents, is a major indicator of the tea's quality (Xiong et al., 2020a; Kim et al., 2021). Cereal grain tea beverages have been found to contain an array of volatile compounds. Table 11.2 highlights the key groups and profiles of volatile components in rice, barley, sorghum, Tartary buckwheat, and wheat bran teas [detected by solid-phase microextraction (SPME) and gas chromatography-mass spectrometry (GC-MS) techniques].

Table 11.2 Summary of volatile profile of various cereal and pseudocereal grain teas

Tea type	Raw material	Processing method	Major groups of volatiles identified (Numbers)												References
			Total	Alcohol	Ester	Carboxylic acid	Aldehyde	Ketone	Alkane	Alkene	Furan	Pyrrole	Pyrazine	Other	
Barley tea	Whole grain	Roasting	32	08	00	01	07	07	00	00	00	00	04	05	Tatsu et al. (2020)
	De-branned grain	Roasting	30	08	00	01	07	07	00	00	00	00	04	05	
	Whole grain	Roasting	71	07	02	00	03	12	00	00	14	06	21	06	Joung et al. (2018)
	Whole grain	Steaming, drying, roasting	70	07	02	00	03	12	00	00	13	06	21	06	
Sorghum tea	Red sorghum whole grain	Soaking, yeast fermentation, steaming, roasting	20	03	05	00	03	01	02	00	00	02	04	00	Sun et al. (2020)
	Red sorghum whole grain	Soaking, lactic acid bacteria Fermentation (*lactobacillus plantarum*), steaming, roasting	17	04	11	00	00	00	02	00	00	00	00	00	

	Red sorghum whole grain	Soaking, steaming, roasting	32	03	08	00	02	01	13	00	01	00	03	01	Xiong et al. (2020a, b)
	Black sorghum whole grain	Soaking, steaming, roasting	30	03	07	00	02	01	13	00	01	00	02	01	
	White sorghum whole grain	Soaking, steaming, roasting	34	03	09	01	02	01	13	00	01	00	03	01	Xiong et al. (2019a, b, c)
Rice tea	Germinated brown rice whole grain	Soaking, germination, roasting	43	02	01	01	06	02	05	00	04	02	14	06	Kim et al. (2021)
	Black rice whole grain	Soaking, steaming, roasting	49	03	02	01	07	03	14	10	02	00	06	01	Wu et al. (2013a)

(continued)

Table 11.2 (continued)

Tea type	Raw material	Processing method	Major groups of volatiles identified (Numbers)												References
			Total	Alcohol	Ester	Carboxylic acid	Aldehyde	Ketone	Alkane	Alkene	Furan	Pyrrole	Pyrazine	Other	
Buckwheat tea	Buckwheat grain	NA	30	03	03	00	05	01	11	03	01	00	01	02	Xu et al. (2019)
	Whole grain	Soaking, steaming, drying, debranning, roasting	34	00	04	00	01	01	03	00	00	02	21	02	Guo et al. (2017)
	Whole plant	Grinding, extruding, drying, roasting	36	00	06	00	04	01	10	00	00	01	08	06	
	NA	NA	39	00	00	00	05	03	01	01	01	02	21	05	Qin et al. (2011)
Wheat bran tea	Wheat bran	Soaking, cellulase treatment, steaming, roasting	40	03	11	03	09	03	06	00	01	00	03	01	Wang et al. (2019)

NA- The information was not given in the study
All data were acquired through solid-phase microextraction gas chromatography/mass spectometry (SPME-GC/MS) analysis

11.5.1 Barley Tea

Earlier attempts were made by Joung et al. (2018), who explored the impact of barley roasting degree on boricha (barley tea) aroma features and the roasting conditions of raw grain and steamed grain in the preparation of barley tea. The volatile profile of tea was discovered to be greatly influenced by roasting. There were 71 and 72 volatiles in the teas generated from raw grain and roasted to a medium level and strong-level, respectively. The prominent volatiles in the medium-level roasted tea were furans, pyrazines, pyrroles, alcohols, and ketones, while the predominant volatiles in the roasted tea were pyrazines, furans, phenols, ketones, and pyrroles. The volatile composition was nearly identical for steamed and medium-level roasted barley tea, but the concentration of these volatile components was significantly higher, indicating that steamed grain barley tea may have a higher quality aroma. It was determined that pyrazine, butyrolactone, ethylpyrazine, and guaiacol were the most important of fifteen volatile chemicals. Boricha's aroma-active chemical intensity was found to be increased with barley roasting degree (Joung et al., 2018). Furthermore, the volatiles extracted from roasted barley tea, generated from either Naked Barley Tea (NBT) or Hulled Barley Tea (HBT), were applied to a comparative aroma extract dilution analysis, yielding 32 odour-active chemicals with flavour dilution factors (FD factors) ranging between 64 and 1024 (Tatsu et al., 2020). The odour-threshold values of 22 and 23 odorants in NBT and HBT, correspondingly, were surpassed by the quantitation of these 32 chemicals in each sample. The HBT had a distinct aroma than the NBT. The odour-active volatiles found in barley tea were carboxylic acids, alcohols, ketones, aldehydes, and pyrazines. Acetic acid was the most prevalent, accounting for 67% and 84% of the volatile content in whole grain tea and debranned tea, respectively. According to the quantitative analysis of the tea's olfactory profile, barley tea had an overall strong malt and roasty aroma, with a weak fatty undertone. Additionally, the odour activity values (OAVs) of individual volatile chemicals were estimated (OAV = volatile concentration/odour-threshold value), and the highest OAVs were found in 2-acetylpyrazine (OAV = 23), 2-methoxyphenol (OAV = 69), 2-acetyl-1-pyrroline (OAV = 19), 3-methylbutanal (OAV = 12), and 2-ethyl-3,5-dimethylpyrazine (OAV = 12). The aroma of barley tea was strongly influenced by these components (Tatsu et al., 2020). Overall, the study found that phenols, specifically 2-methoxyphenol and trans-isoeugenol, were responsible for the higher strength of the smokey note in the total aroma of hulled grain roasted barley tea.

11.5.2 Sorghum Tea

The white sorghum cultivar (Liberty) was utilized in the study led by Xiong et al. (2019b) to prepare a sorghum grain tea using three successive steps of soaking, steaming, and roasting. The phenolic composition of roasted sorghum was identified to be substantially higher than the phenolic content of raw sorghum grain. The sorghum samples yielded 63 volatile chemicals, comprising 2 carboxylic acids, 13 alkanes, 2 aldehydes, 4 ketones, 5 alcohols, 1 phenylenediamine, 15 esters, and 3 pyrazines. The alkane, aldehyde, and ester content escalated considerably, whereas the carboxylic acids, alcohol, and ketone concentration alleviated during the soaking procedure. Steaming, on the other hand, dramatically decreased ester and alcohol levels however significantly boosting alkane and aldehyde concentration. The phenylenediamine and pyrazine compounds were liberated during the roasting procedure. White sorghum tea could have a flowery, sweet, waxy, and nutty aroma based on the volatile components discovered (Xiong et al., 2019b). In another study, the same group of researchers compared MR-Buster (red-colored) and Shawaya Short Black 1 (black-colored) sorghum types to a white sorghum variety (Liberty) for phenolic content, antioxidant activity, and volatile components after grain tea processing stages (Xiong et al., 2020b). Varieties of sorghum had dramatically different volatile profiles. Raw-M (52 volatiles) and Raw-S (51 volatiles) had lower diversity and abundance of volatiles than had been earlier observed for Raw-L (57 volatiles). White and red sorghum teas had more volatile chemicals than black sorghum teas. Alkanes were the most common volatiles in these teas, followed by esters, pyrazines, alcohols, and aldehydes. These teas, on the other hand, differed substantially in volatile intensity and concentration. It is possible that the high quantities of condensed tannins in black sorghum could produce bulk matrixes that could trap volatiles during tea processing (Xiong et al., 2020b). Black sorghum might be preferable choice for manufacturing sorghum tea concerning potential health advantages, whereas white sorghum grain was found to be the best material for sorghum tea production in terms of volatile intensity and diversity. Another study employed *Lactobacillus plantarum* subsp. Argentoratensis (LAB) and *Saccharomyces cerevisiae* to explore changes in antioxidant potential, phenolic content, and volatile profile in sorghum grain throughout the preparation of fermented sorghum grain tea (Sun et al., 2020). Sorghum in its raw and processed forms yielded a total of 53 different volatiles. Esters, phenols, and pyrazines were the most common volatiles. During processing, alkanes, carboxylic acids, ketones, and alcohols were reduced. During soaking, steaming, and fermenting, esters were augmented, but during roasting, they were reduced. Soaking, fermenting, steaming, and roasting were shown to modify the phenolic content, antioxidant capacity, and volatile chemicals considerably (Sun et al., 2020). Each processing step had a considerable impact on the amount and composition of these volatiles, particularly the two forms of fermentations. In terms of volatile profiles, yeast fermentation performed better, with esters, phenols, and pyrazines being the most prominent.

11.5.3 Rice Tea

The volatile composition of whole-grain black rice tea under three different processes: soaking, steaming, and roasting has been investigated by Wu et al. (2013a). The raw and processed black rice contained 94 volatiles. It was found that there were 49 compounds in the raw sample, 54 compounds in the soaking sample, and 37 compounds in the steaming sample, correspondingly. The processing had a considerable impact on the volatile component concentrations. The predominant volatile components were 1-hexanol, nonanal, butylated hydroxytoluene, naphthalene, and 1-octen-3-ol. When compared to processed materials, the volatiles recovered from the raw black rice grain were significantly different. Recently, Kim et al. (2021) scrutinized the impacts of germination and roasting on the physicochemical properties of brown rice. During germination, free amino acids and reducing sugars were found to aid in the non-enzymatic browning reaction and the generation of aroma compounds throughout successive roasting for tea production. Brown rice soaking and germination was found to alter the volatile compositions of tea infusions. Infusions made from brown rice samples that had been soaked, germinated, and roasted yielded 43 volatiles. Alkanes, esters, carboxylic acids, aldehydes, ketones, and amides were the most common volatile chemicals in tea infusions of non-roasted brown rice (Kim et al., 2021). Overall, germination and roasting of brown rice were found to upsurge volatile content. In comparison to a single process, the mixture of germination and roasting resulted in a significant rise in the synthesis of pyrazine and furan compounds in brown rice tea infusion, so tea infusion developed from roasted germinated brown rice samples might have somewhat richer roasted, sweet, floral, and nutty aroma than roasted brown rice tea infusion.

11.5.4 Buckwheat Tea

Tartary buckwheat tea, a valuable and nutritious product, possesses a characteristic malty aroma. Qin et al. (2011) analyzed the volatile components in Tartary buckwheat tea so as to characterize distinctive malty aroma constituents better. There were a total of 77 volatiles detected, 35 among them were evaluated using existing standards. Aroma components (OAV$\geq$10) having a high likelihood of contributing to the Tartary buckwheat tea aroma included: nonanal, trimethylpyrazine, maltol, benzeneacetaldehyde, 2,3-diethyl-5-methylpyrazine, 2,5-dimethyl-4-hydroxy-3(2H)-furanone, 2,5-dimethylpyrazine, and 2-ethyl-5-methylpyrazine. Additionally, nutrients and bioactive substances like niacin, vanillic acid, linoleic acid, butylated hydroxytoluene, and 7-hydroxycoumarin were discovered in this study. Moreover, the volatile profile of two types of Tartary buckwheat tea samples, whole grain tea (WGT) and whole plant tea (WPT) has been studied (Guo et al., 2017). WGT and WPT samples both possessed a distinct malty flavour, although the aroma constituent profiles of the two differed slightly. These samples contained 55 compounds, including alkanes, aldehydes, esters, and pyrazines, but the most prevalent aroma component was 3-ethyl-2,5-dimethyl-pyrazine which was found in 14 samples. In addition, in another study, the aroma constituents of Huantai Tartary buckwheat tea (TBH), three lab-produced scented Tartary buckwheat teas, and tea infusion were evaluated (Xu et al., 2019). The samples included a total of 103 volatile compounds, comprising 6 aldehydes, 8 ketones, 8 esters, 16 alcohols, 56 hydrocarbons, and 9 others. It was discovered that the following 8 constituents were present in all the samples: undecane, dodecane, nonanal, decanal, farnesene, l-caryophyllene, 3, 8-dimethyldecane, and 2,6,10-trimethyldodecane. Tartary buckwheat rose tea (TBR) had 57 aroma constituents, while Tartary buckwheat jasmine tea (TBJ) had 53, both of which were greater than those found in other teas.

11.5.5 Wheat Bran Tea

An important milling byproduct, wheat bran, is rich in nutrients. To better understand the volatile profile of wheat bran tea, Wang et al. (2019) investigated the influence of processing comprising soaking, cellulase treatment, steaming, and roasting. The wheat bran samples contained a total of 40 volatiles, encompassing 1 phenol, 1 furan, 3 alcohols, 3 ketones, 3 pyrazines, 3 carboxylic acids, 6 alkanes, 9 aldehydes, and 11 esters. There was a considerable surge in the total volatile profile following the usage of both soaking and cellulase pre-treatment. After the steaming process, aldehydes and furans became the most abundant aromatic compounds in the bran sample, but their concentrations reduced following the roasting procedure.

Apparently, the volatile profile of different cereals, cereal ingredients (whole grain, bran, etc.), and their processing methods vary vastly. Alkanes, aldehydes, esters, alcohols, and pyrazines are the most common volatile chemicals found in cereal teas. Cereal alcohols are frequently linked to sweet, floral, and fruity aromas, enhancing the tea's aroma. These compounds commonly have high odour threshold values and consequently low intensity, which is linked to unpleasant gasoline or waxy smells (Xiong et al., 2020a). It's also possible for lipid oxidation to produce alkanes, as well as aldehydes, which can cause off-flavors (Bryant & McClung, 2011). Because of their high odour thresholds, esters, which typically have sweet, fruity, and green aromas, may not have much of an impact on the overall aroma of toasted grain teas. Pyrazines are the most important odour compounds in roasted cereal tea owing to the unique roasted, sweet, malty, and nutty aroma they contribute to the flavour and possess low odour threshold values (Xiong et al., 2020a). While the odour-active volatile molecules, particularly those with very pleasant olfactory characteristics, maybe the primary contributors to the tea's aroma, further research is required to examine these chemicals in greater depth.

11.6 Nanotechnological Aspects of Cereal-Based Beverages

A modern paradigm in developing healthy foods is the implementation of nanotechnology to create food-grade nanoparticles (FGNPs). According to the European Commission's definition, engineered nanomaterials are any purposely created materials that contain at least 50% of particles in the number size distribution that possess one or more exterior dimensions that fall within the range of 1–100 nm. To develop foods with augmented nutritional value and qualities that alleviate ailments and decrease prolonged chronic diseases, these materials have piqued tremendous attention in the food industry (Morris, 2011). Lipid-based nanoencapsulation can be utilized as a delivery tool for enzymes, food additives, nutraceuticals, and antimicrobials, among other applications (Trujillo et al., 2016). For ameliorating organoleptic features, escalating absorption and distribution of nutrients, and stabilizing active components, for instance, nutraceuticals, nanoparticles (NPs) can be employed in food sectors, like the sectors of cereal-based fermented drinks (Ranjan et al., 2014; Salmerón et al., 2014). So as to keep up with the ever-changing needs of consumers and the business, the food sector must invest continuously in new research and technology. The functional food market sees new food ingredients with potential health benefits discovered or developed regularly. As a result, it must be added to a food product, and it must be ensured that these constituents approach the intended active site. With nanotechnology's application in the food and bioprocessing industries, the food supply chain stands to gain economically (Neethirajan & Jayas, 2011). Incorporating innovative cereal beverages with FGNPs has great potential because of their design; however, further investigation is necessary to ascertain how they affect the GIT as they travel through it. There aren't many *in vivo* studies assessing FGNPs' impacts on the human digestive system, but the time it takes for FGNP to go through the system after ingestion is an essential consideration. Gastric lipase, hydrochloric acid, and pepsin at pH levels of 1.0–5.0 for periods of 15 min to 3 h expose food to severe conditions in the stomach. In the small intestine, it is exposed to higher pH levels ranging from 6.0 to 7.5 for 2–5 h. It reaches the colon further within 12–24 h with an acidic pH of around 5–7 (Hannon et al., 2015). Assessment of FGNP exposure to the gut pH, bile salts, and time is essential to detect their bioactivity and evaluate whether they may be used as a component in formulating innovative fermented grain beverages. Because of the unique physicochemical and health-promoting characteristics of NPs isolated from cereals, plants, fruits, and vegetables, as well as inorganic FGNP compounds, for example, zinc oxide particles possessing antidiabetic potential (El-Gharbawy et al., 2016; Umrani & Paknikar, 2014), gold NPs displayed promise in developing anticancer drugs (Zhu & Liao, 2015), nanoinorganic metal oxide (Tang & Lv, 2014), and silver NPs (Le Ouay & Stellacci, 2015) that possess good antibacterial features, the discovery of innovative fabrication techniques that will enhance the integration of these NPs into various food matrices is of great significance (Kalakotla et al., 2015). There must be an ability for new FGNPs of organic or inorganic composition, whether organic or inorganic. As a result, the development of nanocapsules and nanogels is a cutting-edge food technology trend. Using these approaches, it will be possible to create new kinds of nutritious foods, such as synbiotics, that supply more bioactive components. The chemicals employed to construct nanomaterials should bear Generally Recognized as Safe (GRAS) status, and the most frequent nanomaterials are biopolymers made from proteins and carbohydrates. A ubiquitous dietary source, polysaccharides, contain gelling qualities and can be used to nanoencapsulate molecules of various physicochemical characteristics and heterogeneous biological entities including phages. With the assistance of enzymes, polysaccharides' physical and chemical properties can be altered so that new shapes can be created by altering their diameter, pore size, and loading. For example, FGNP delivery may be modulated in the gastrointestinal system by controlling how the particle surface interacts with cell surfaces. This is a noteworthy aspect because particle surface control impacts the interaction within the cell surface and may affect FGNP distribution throughout the gastrointestinal system (Santiago & Castro, 2016). As a result, more research in the field of polysaccharides, such as those derived from cereal sources, might be implemented to develop nanocapsules containing bioactive components that could be useful in the production of distinct health-enhancing fermented cereal beverages.

Nonsoluble health-promoting chemicals can now be incorporated into food items thanks to the advances in the delivery technologies, including polymer-based microparticles, micelles, and liposomes in the food industry. Dispersions with at least two phases, one or more dispersed or internal phases, and a continuous or exterior phase, the dispersion medium or vehicle, are referred to as colloidal dispersions (CDs). The particle size of the dispersed phase, rather than the composition, distinguishes these from solutions and coarse dispersions. One or more substances must have at least one dimension between a few nanometers and several micrometres to qualify as colloidal dispersions. Dispersions of self-assembled molecules, solid-in-liquid dispersions, and liquid-in-liquid dispersions are all examples of colloidal delivery systems (Tekiner et al., 2015). Different products can now incorporate lipid molecules thanks to these technologies. This has sparked interest in using seed oils from various plant sources in food compositions, as their components contain specific phytochemicals possessing important antioxidant characteristics (Gumus et al., 2015). Colloidal dispersions in the form of nanoemulsions can be utilised to distribute water-insoluble compounds that were previously unappealing because of their low soluble properties. Cereal-based fermented beverages, such as those produced with nanomolecules with high health qualities, are just one example of the wide variety of healthy foods that can be developed (Salmerón, 2017). The evolution of fermented drinks has taken them from conventional natural fermented products to beverages modified with functional ingredients to mend cardiovascular benefits, and further to fermented beverages that mend gastrointestinal health and may further develop to fermented foods fabricated with precise bioactive NPs. Although the literature review suggests a scarcity of studies on the nanotechnological aspects of cereal-based teas with specific bioactive NPs, this is a huge research gap. Researchers and industrialists working on cereal-based teas have room to develop cereal-based teas employing FGNPs to enhance their health-promoting features and overall applications.

11.7 Challenges and Opportunities for Enhancing Cereal Tea Beverages' Intake for Human Health Wellbeing

While discussing the elevation of cereal-based tea beverages for human health well-being, it must be considered that cereal-based tea beverages are integral components of an individual's overall nutritional status, dietary quality, and lifestyle behaviors

(Gibson et al., 2016). As per social-ecological theories, numerous aspects affect human behavior. Intake of beverages is an example of such behavior. The components in these kinds of theories can be classified into different levels, and they can be used to identify success factors for health promotion (Coughlin & Smith, 2017; Mazarello Paes et al., 2015). Actions to influence cereal-based tea beverages consumption for the sake of good health might therefore take place at a variety of levels. Human personal attitudes, dietary choices and intolerances, physiology, societal beliefs, and health education, among other things, make up a person's psychobiological core. This core is fenced by the community (school, college, universities, faith communities, social gatherings, neighbourhood), interpersonal (friends and family), and national aspects. Availability/accessibility, affordability, and acceptability may perform a role in each of one such aforementioned level. The availability of cereal-based tea beverages is referred to as accessibility. It is primarily concerned with the physical environment and encompasses the food system and the food supply chain, including farm to fork—production to consumption. The economic environment determines the affordability of a product or cereal-based tea beverages and services. The preferences accessible to residents of low-to-middle income nations differ from those of first world/industrialized nations; even within a nation, various income classes are also present. Acceptability is an essential social concept. A community's food culture dictates what, when, and how it eats food and drink. The relative relevance of accessibility/availability, affordability, and acceptability may vary depending on the time of day and the environment in which they are encountered. On the other hand, all these aspects are consistent and are dependent on the context and time in which they occur. Apart from that, modern-day purchasers may be more interested in products that can enhance their overall health, lower their chance of developing diseases, and relieve problems linked to the nutrition including food intolerances and allergies. Presently, consumers are increasingly approaching the idea of wellbeing with a broad perspective, as they have become more health-conscious, owing to the rising prevalence of global pandemics like type 2 diabetes, CVD, cancer, and obesity, as well as other diseases. There is an increased need for nutritional, functional, and appealing food products with clear labels that are also ready to consume. Considering these characteristics, the cereal-based tea beverage industry appears to best meet consumer needs and grow functional product intake. Market figures reveal that functional foods remain a popular choice for consumers regardless of economic conditions. Because of strong sensorial aspects and health-promoting properties, the functional cereal-based tea beverage market has grown significantly and is predicted to continue worldwide. According to investigations, disclosing health-related assertions linked with functional cereal-based tea beverages influences customers' choices for a particular product. Consumers' choices for a particular functional beverage are determined not just by health claims but also by sensory qualities. It was recommended that understanding the significance of certain testing circumstances for functional beverages can assist product developers in reformulating the product through correct trial design, emphasising the consumers' needs since they will decide product acceptance. The critical necessity for educational programs that use accessible and easily understandable language to describe the key nutrients of foodstuffs has already been highlighted.

While the study described before was performed in Brazil, inadequate nutritional information for food goods is a global issue that is significantly associated with education level. Initiatives may place a premium on the demand side. When governments' authorities take the initiative, this may include proper nutrition education for the general public with the goal of enabling consumers to make healthier preferences. In the end, the goal is to alter one's behavior on an individual basis. Food-based dietary guidelines could be the basis of such campaigns. While there are commonalities between countries, a study of North and South American countries' food-based dietary guidelines reveals variances (Montagnese et al., 2017), indicating that low- and middle-income nations place a distinct focus on various issues. Simultaneously, mainstream media (print, radio, and television) have been efficiently employed by industry to boost consumer awareness and demand; however, they are fundamentally profit-driven. Digital technology, including social media platforms and websites, establishes itself as a highly effective channel for communicating behavior change. Despite the widespread promotion of community-based participatory techniques in obesity prevention, a recent literature analysis by Coughlin & Smith (2017) indicated the necessity for multilevel treatments within a hypothetical outline. Isolated efforts are rarely sufficient to have an impact, especially when the goal is to influence long-term behavior change in an individual. Therefore, the data on current consumption (exposures) of various beverages, along with the health status of the population (i.e., the growing of undernutrition and/or the risk of cancers, CVDs), should be used to inform all dietary policies and interventions, whether they are supply- or demand-driven (Singh et al., 2015), in a particular country; hence the FAO/WHO is developing a collaborative online platform called "FAO/WHO GIFT" (http://www.fao.org/nutrition/assessment/food-consumption-database) that will contain information regarding prevailing individual data food and even beverage intake. This understanding of what people consume is expected to serve as a good foundation for developing personalized answers to specific requirements, including those of low- and middle-income nations.

11.8 Conclusions and Future Outlook

Cereals possess a distinctive profile of bioactive phytochemicals, making them an excellent addition to a diet similar to fruits and vegetables. Whole cereal grains contain phenolic compounds, dietary fibres, carotenoids, tocols, c-oryzanol, phytosterols, and phytic acid, as well as other bioactive phytochemicals. The bran fraction contains the majority of these chemicals. However, bran is frequently eliminated throughout the processing of the food since it might cause problems in processing. Certain fermenting bacteria can be suppressed, and the fermentation process is impacted by phenolic substances in cereal bran. Pseudocereals are now being developed as a nutritious alternative to gluten-rich grains to synthesize gluten-free food products. Flavonoids (kaempferol, rutin, quercetin, isoquercitrin, and nicotiflorine), phenolic compounds (vanillic acid and ferulic acid), betalains (betacyanins and amaranthine) and carotenoids (carotene, lutein, zeaxanthin) are significant bioactive constituents in pseudocereals. Pseudocereal grains' bran and hull are high in nutritional fiber. Phenolic acids including ferulic and coumaric acids cross-link a major part of dietary fiber in amaranth and quinoa. The three pseudocereal grains, including amaranth, quinoa, and buckwheat, are more akin to vegetables and fruits in terms of their dietary fibre profile than cereals.

Foods high in bioactive phytochemicals, such as cereal tea, have been demonstrated to positively impact health, and some of these compounds have been recognised, while others have to be discovered. Cereal tea's distinctive malty and roasted aroma is one of its most appealing aspects, but further research is needed to discover the odor-active volatile components, which is critical for improving the tea's aroma. It will need a lot more work to figure out how to process cereal grain tea such that its phenolic and bioactive profiles are as high as possible while still maintaining premium tea quality before it can successfully compete with traditional leaf teas or coffees. Tea beverages enriched with cereal and pseudocereal grains are still in their infancy, but they look to be becoming better and better. It is possible to make better functional cereal and pseudocereal grain tea beverages and supply constant product quality, extend shelf life, and augment nutritional contents to eventually meet consumers' needs.

Dietary diversity has been shown to be linked to adequate micronutrient consumption in many underdeveloped nations. The goal of moderation is to maintain a healthy energy balance, which in turn prevents weight loss and the development of obesity-related diseases. A healthy diet includes beverages since they are essential for hydration and contribute to the intake of nutrients, including gap nutrients. Clean and safe water is the foundation of healthy beverage consumption in both low- and middle-income countries and industrialized countries. Advanced technologies' effects on cereal and pseudocereal grain functional qualities during processing necessitate further research to warrant that these technologies can prevent quality loss and nutritive components from losing their nutritional value. For scientists and businesses alike, the primary issue is how to produce tea beverages from a wide variety of cereal and pseudocereal grains without losing their distinct flavours and other characteristics. In light of the above, it is greatly advised to investigate the sensory qualities of tea beverages made from a combination of cereals, legumes, and fruits. We see a better future for cereal grain teas in the market.

Acknowledgements The work described in this paper was fully supported by a grant from the National Natural Science Foundation of China (No. 31830069, No. 31671793, No. 31201449, No. 31371723), the Beijing Municipal Natural Science Foundation & Beijing Municipal Education Commission (No. KZ202010011018) and the Foundation of Beijing Technology and Business University (No. PXM2017_014213_000036, No. PXM2018_014213_000033).

Conflicts of Interest The authors confirm that they have no known competing financial interests or personal relationships that could have influenced the manuscript.

Authors' Contributions Prasanna J Patil, Muhammad Usman: Conceptualization; Data curation; Visualization; Writing – original draft; Writing – review & editing. Arshad Mehmood, Shabbir Ahmad, Chengnan Zhang, Chao Teng: Writing – review & editing. Xiuting Li: Writing – review & editing; Funding acquisition; and Supervision.

References

Abdo, A. A. A., Zhang, C., Lin, Y., Liang, X., Kaddour, B., Wu, Q., … Li, W. (2021). Xylo-oligosaccharides ameliorate high cholesterol diet induced hypercholesterolemia and modulate sterol excretion and gut microbiota in hamsters. *Journal of Functional Foods, 77*, 104334.

Abugri, D. A., Akudago, J. A., Pritchett, G., Russell, A. E., & McElhenney, W. H. (2015). Comparison of phytochemical compositions of Sorghum bicolor (L) Moench red flour and pale brown leaves. *Journal of Food Science and Nutrition, 1*(003).

Adan, Z., Zhiquan, H., Youming, L., Siming, Z., & Shanbai, X. (2016). The formation and characteristics of volatile odor in roasted rice tea. *Journal of the Chinese Cereals and Oils Association, 31*, 1–6.

Adlercreutz, H. (2007). Lignans and human health. *Critical Reviews in Clinical Laboratory Sciences, 44*(5-6), 483–525.

Aguilar, E. G., Peiretti, E. G., Uñates, M. A., Marchevsky, E. J., Escudero, N. L., & Camiña, J. M. (2013). Amaranth seed varieties. A chemometric approach. *Journal of Food Measurement & Characterization, 7*, 199–206.

Ahmad, A., Munir, B., Abrar, M., Bashir, S., Adnan, M., & Tabassum, T. (2012). Perspective of β-glucan as functional ingredient for food industry. *Journal of Food Science and Nutrition, 2*(2), 133–139.

Altay, F., Karbancıoglu-Güler, F., Daskaya-Dikmen, C., & Heperkan, D. (2013). A review on traditional Turkish fermented non-alcoholic beverages: Microbiota, fermentation process and quality characteristics. *International Journal of Food Microbiology, 167*(1), 44–56.

Alvarez-Jubete, L., Arendt, E., & Gallagher, E. (2009). Nutritive value and chemical composition of pseudocereals as gluten-free ingredients. *International Journal of Food Sciences and Nutrition, 60*(sup4), 240–257.

Asharani, V. T., Jayadeep, A., & Malleshi, N. G. (2010). Natural antioxidants in edible flours of selected small millets. *International Journal of Food Properties, 13*(1), 41–50.

Ashigai, H., Taniguchi, Y., Matsukura, Y., Ikeshima, E., Nakashima, K., Mizutani, M., & Yajima, H. (2018). Roasted barley extract affects blood flow in the rat tail and increases cutaneous blood flow and skin temperature in humans. *Journal of Agricultural and Food Chemistry, 66*(5), 1251–1257.

Aune, D., Chan, D. S., Lau, R., Vieira, R., Greenwood, D. C., Kampman, E., & Norat, T. (2011). Dietary fibre, whole grains, and risk of colorectal cancer: Systematic review and dose-response meta-analysis of prospective studies. *British Medical Journal, 343*, d6617.

Aura, A. M. (2016). Phenolic acids in Rye. In *Rye and health* (pp. 109–119). Elsevier.

Awika, J. M., & Rooney, L. W. (2004). Sorghum phytochemicals and their potential impact on human health. *Phytochemistry, 65*(9), 1199–1221.

Badau, M. H., Nkama, I., & Jideani, I. A. (2005). Phytic acid content and hydrochloric acid extractability of minerals in pearl millet as affected by germination time and cultivar. *Food Chemistry, 92*(3), 425–435.

Bangar, S. P., Sandhu, K. S., Purewal, S. S., Kaur, M., Kaur, P., Siroha, A. K., … Kumar, M. (2021). Fermented barley bran: An improvement in phenolic compounds and antioxidant properties. *Journal of Food Processing and Preservation, 2021, e15543.*

Barahona, N. (2001). *Corn coffee* (U.S. Patent No. US20010026830A1). U.S. Patent and Trademark Office.

Becker, F. (2013). Tocopherols in wheat and rye: A review of research within the subject. *Institutionen för livsmedelsvetenskap*.

Berwal, M. K., Chugh, L. K., Goyal, P., & Kumar, R. (2016). Variability in total phenolic content of pearl millet genotypes. *Journal of Agriculture and Ecology, 1*, 41–49.

Bhandari, S. R., & Lee, Y. S. (2013). The contents of phytosterols, squalene, and vitamin e and the composition of fatty acids of Korean Landrace *Setaria Italica* and sorghum bicolar seeds. *Korean Journal of Plant Resources, 26*(6), 663–672.

Biel, W., & Maciorowski, R. (2013). Evaluation of chemical composition and nutritional quality of buckwheat groat, bran and hull (*Fagopyrum esculentum* Möench L.). *Italian Journal of Food Science, 25*, 384–389.

Boz, H. (2015). Ferulic acid in cereals-a review. *Czech Journal of Food Sciences, 33*(1), 1–7.

Bruni, R., Guerrini, A., Scalia, S., Romagnoli, C., & Sacchetti, G. (2002). Rapid techniques for the extraction of vitamin E isomers from *Amaranthus caudatus* seeds: Ultrasonic and supercritical fluid extraction. *Phytochemical Analysis: An International Journal of Plant Chemical and Biochemical Techniques, 13*(5), 257–261.

Bryant, R. J., & McClung, A. M. (2011). Volatile profiles of aromatic and non-aromatic rice cultivars using SPME/GC–MS. *Food Chemistry, 124*(2), 501–513.

Cai, Y., Sun, M., & Corke, H. (2003). Antioxidant activity of betalains from plants of the Amaranthaceae. *Journal of Agricultural and Food Chemistry, 51*(8), 2288–2294.

Cai, Y., Corke, H., Wang, D., & Li, W. (2016). Buckwheat: Overview. In C. Wringley, H. Corke, K. Seetharaman, & J. Faubion (Eds.), *Encyclopedia of food grains* (Vol. 1, pp. 307–315). Elsevier.

Campos-Vega, R., & Oomah, B. D. (2013). Chemistry and classification of phytochemicals. In *Handbook of plant food phytochemicals: Sources, stability and extraction* (pp. 5–48). Wiley-Blackwell.

Charalampopoulos, D., Wang, R., Pandiella, S. S., & Webb, C. (2002). Application of cereals and cereal components in functional foods: A review. *International Journal of Food Microbiology, 79*(1-2), 131–141.

Chung, I. M., Yong, S. J., Lee, J., & Kim, S. H. (2013). Effect of genotype and cultivation location on β-sitosterol and α-, β-, γ-, and δ-tocopherols in sorghum. *Food Research International, 51*(2), 971–976.

Ciudad Mulero, M., Fernández-Ruiz, V., Matallana-González, M. C., & Morales, P. (2019). Dietary fiber sources and human benefits: The case study of cereal and pseudocereals. In *Advances in food and nutrition research* (Vol. 90, pp. 83–134). Academic.

Coughlin, S. S., & Smith, S. A. (2017). Community-based participatory research to promote healthy diet and nutrition and prevent and control obesity among African-Americans: A literature review. *Journal of Racial and Ethnic Health Disparities, 4*(2), 259–268.

Cui, S. W., & Wang, Q. (2009). Cell wall polysaccharides in cereals: Chemical structures and functional properties. *Structural Chemistry, 20*, 291–297.

de la Rosa, A. B., Fomsgaard, I. S., Laursen, B., Mortensen, A. G., Olvera-Martínez, L., Silva-Sánchez, C., … De León-Rodríguez, A. (2009). Amaranth (*Amaranthus hypochondriacus*) as an alternative crop for sustainable food production: Phenolic acids and flavonoids with potential impact on its nutraceutical quality. *Journal of Cereal Science, 49*(1), 117–121.

De Ruiz, C. A., & Bressani, R. (1990). Effect of germination on the chemical composition and nutritive value of amaranth grain. *Cereal Chemistry, 67*(6), 519–522.

Dhingra, D., Michael, M., Rajput, H., & Patil, R. T. (2012). Dietary fibre in foods: A review. *Journal of Food Science and Technology, 49*(3), 255–266.

Duchoňová, L., Polakovičová, P., Rakická, M., & Šturdík, E. (2021). Characterization and selection of cereals for preparation and utilization of fermented fiber-betaglucan product. *Journal of Microbiology, Biotechnology and Food Sciences, 2021*, 2187–2207.

Durazzo, A., Zaccaria, M., Polito, A., Maiani, G., & Carcea, M. (2013). Lignan content in cereals, buckwheat and derived foods. *Food, 2*(1), 53–63.

Dykes, L., & Rooney, L. W. (2007). Phenolic compounds in cereal grains and their health benefits. *Cereal Foods World, 52*(3), 105–111.

Dziadek, K., Kopeć, A., Pastucha, E., Piątkowska, E., Leszczyńska, T., Pisulewska, E., et al. (2016). Basic chemical composition and bioactive compounds content in selected cultivars of buckwheat whole seeds, dehulled seeds and hulls. *Journal of Cereal Science, 69*, 1–8.

Dziedzic, K., Gorecka, D., Marques, A., Rudzińska, M., & Podolska, G. (2015). Content of phytosterols in raw and roasted buckwheat groats and by-products. *Czech Journal of Food Sciences, 33*(5), 424–430.

Ehrenbergerová, J., Belcrediová, N., Prýma, J., Vaculová, K., & Newman, C. W. (2006). Effect of cultivar, year grown, and cropping system on the content of tocopherols and tocotrienols in grains of hulled and hulless barley. *Plant Foods for Human Nutrition, 61*(3), 145–150.

El-Gharbawy, R. M., Emara, A. M., & Abu-Risha, S. E. S. (2016). Zinc oxide nanoparticles and a standard antidiabetic drug restore the function and structure of beta cells in Type-2 diabetes. *Biomedicine & Pharmacotherapy, 84*, 810–820.

Ermawar, R. A., Collins, H. M., Byrt, C. S., Betts, N. S., Henderson, M., Shirley, N. J., … Burton, R. A. (2015). Distribution, structure and biosynthetic gene families of (1, 3; 1, 4)-β-glucan in Sorghum bicolor. *Journal of Integrative Plant Biology, 57*(4), 429–445.

Esatbeyoglu, T., Wagner, A. E., Schini-Kerth, V. B., & Rimbach, G. (2015). Betanin—A food colorant with biological activity. *Molecular Nutrition & Food Research, 59*(1), 36–47.

Etoh, H., Murakami, K., Yogoh, T., Ishikawa, H., Fukuyama, Y., & Tanaka, H. (2004). Anti-oxidative compounds in barley tea. *Bioscience, Biotechnology, and Biochemistry, 68*(12), 2616–2618.

FAO (Food and Agriculture Organization of the United Nation). (2018). *FAOSTAT, crops*. Retrieved 24 April, 2020, from http://www.fao.org/faostat/en/#data/QC

Fardet, A., Rock, E., & Rémésy, C. (2008). Is the in vitro antioxidant potential of whole-grain cereals and cereal products well reflected in vivo? *Journal of Cereal Science, 48*(2), 258–276.

Feng, R., Zhang, Y. C., Ma, W., & Tian, X. (2009). *Study on the processing technics of transparent type corn tea beverage* (p. 1). Food Science and Technology (China).

Fernandes, C. G., Sonawane, S. K., & Arya, S. S. (2021). Cereal based functional beverages: A review. *Journal of Microbiology, Biotechnology and Food Sciences, 2021*, 914–919.

Food and Drug Administration, HHS. (2008). Food labeling: Health claims; soluble fiber from certain foods and risk of coronary heart disease. Interim final rule. *Federal Register, 73*(37), 9938–9947.

Francavilla, A., & Joye, I. J. (2020). Anthocyanins in whole grain cereals and their potential effect on health. *Nutrients, 12*(10), 2922.

Fuentes-Zaragoza, E., Riquelme-Navarrete, M. J., Sánchez-Zapata, E., & Pérez-Álvarez, J. A. (2010). Resistant starch as functional ingredient: A review. *Food Research International, 43*(4), 931–942.

Gani, A., Wani, S. M., Masoodi, F. A., & Hameed, G. (2012). Whole-grain cereal bioactive compounds and their health benefits: A review. *Journal of Food Processing & Technology, 3*(3), 146–156.

García-Estepa, R. M., Guerra-Hernández, E., & García-Villanova, B. (1999). Phytic acid content in milled cereal products and breads. *Food Research International, 32*(3), 217–221.

Genser, B., Silbernagel, G., De Backer, G., Bruckert, E., Carmena, R., Chapman, M. J., … März, W. (2012). Plant sterols and cardiovascular disease: A systematic review and meta-analysis. *European Heart Journal, 33*(4), 444–451.

Gibson, S. A., Horgan, G. W., Francis, L. E., Gibson, A. A., & Stephen, A. M. (2016). Low calorie beverage consumption is associated with energy and nutrient intakes and diet quality in British adults. *Nutrients, 8*(1), 9.

Gil, A., Ortega, R. M., & Maldonado, J. (2011). Wholegrain cereals and bread: A duet of the Mediterranean diet for the prevention of chronic diseases. *Public Health Nutrition, 14*(12A), 2316–2322.

Goufo, P., & Trindade, H. (2014). Rice antioxidants: Phenolic acids, flavonoids, anthocyanins, proanthocyanidins, tocopherols, tocotrienols, γ-oryzanol, and phytic acid. *Food Science & Nutrition, 2*(2), 75–104.

Gumus, Z. P., Guler, E., Demir, B., Barlas, F. B., Yavuz, M., Colpankan, D., … Timur, S. (2015). Herbal infusions of black seed and wheat germ oil: Their chemical profiles, in vitro bio-investigations and effective formulations as phyto-nanoemulsions. *Colloids and Surfaces B: Biointerfaces, 133*, 73–80.

Guo, X. D., Wu, C. S., Ma, Y. J., Parry, J., Xu, Y. Y., Liu, H., & Wang, M. (2012a). Comparison of milling fractions of tartary buckwheat for their phenolics and antioxidant properties. *Food Research International, 49*, 53–59.

Guo, X. D., Wang, M., Gao, J. M., & Shi, X. W. (2012b). Bioguided fraction of antioxidant activity of ethanol extract from tartary buckwheat bran. *Cereal Chemistry, 89*, 311–315.

Guo, H., Yang, X., Zhou, H., Luo, X., Qin, P., Li, J., & Ren, G. (2017). Comparison of nutritional composition, aroma compounds, and biological activities of two kinds of tartary buckwheat tea. *Journal of Food Science, 82*(7), 1735–1741.

Gupta, A., Sharma, S., & Singh, B. (2018). Processing effect on the bioactive compounds of cereals. In *Trends & prospects in food science & processing technology* (pp. 25–54). Today and Tomorrow's Printers and Publishers.

Han, Y. J., & Tran, T. T. T. (2018). Corn snack with high fiber content: Effects of different fiber types on the product quality. *LWT – Food Science and Technology, 96*, 1–6.

Han, S. I., Seo, W. D., Na, J. E., Park, J. Y., Park, D. S., Cho, J. H., … Nam, M. H. (2015). Characteristics of pop-rice and rice tea using black sticky rice with giant embryo. *Journal of Life Science, 25*(1), 68–74.

Hannon, J. C., Kerry, J., Cruz-Romero, M., Morris, M., & Cummins, E. (2015). Advances and challenges for the use of engineered nanoparticles in food contact materials. *Trends in Food Science & Technology, 43*(1), 43–62.

Hao, J., Wan, Y., Zhang, G., Wu, J., & Zhang, J. (2018). Development of oat tea and evaluation analysis of flavor substances in Qinghai-Tibet plateau. *Cereal and Food Industry, 25*, 40–43.

Harmon, K. (2009). *Humans feasting on grains for at least 100,000 years – Scientific American Blog Network*. Scientific American.

Heleno, S. A., Martins, A., Queiroz, M. J. R., & Ferreira, I. C. (2015). Bioactivity of phenolic acids: Metabolites versus parent compounds: A review. *Food Chemistry, 173*, 501–513.

Hídvégi, M., & Lásztity, R. (2002). Phytic acid content of cereals and legumes and interaction with proteins. *Periodica Polytechnica Chemical Engineering, 46*(1–2), 59–64.

Horvat, D., Šimić, G., Drezner, G., Lalić, A., Ledenčan, T., Tucak, M., … Zdunić, Z. (2020). Phenolic acid profiles and antioxidant activity of major cereal crops. *Antioxidants, 9*(6), 527.

Hossen, I., Hua, W., Ting, L., Mehmood, A., Jingyi, S., Duoxia, X., … Junsong, X. (2020). Phytochemicals and inflammatory bowel disease: A review. *Critical Reviews in Food Science and Nutrition, 60*(8), 1321–1345.

Huda, M. N., Lu, S., Jahan, T., Ding, M., Jha, R., Zhang, K., … Zhou, M. (2021). Treasure from garden: Bioactive compounds of buckwheat. *Food Chemistry, 335*, 127653.

Idehen, E., Tang, Y., & Sang, S. (2017). Bioactive phytochemicals in barley. *Journal of Food and Drug Analysis, 25*(1), 148–161.

Ispiryan, L., Kuktaite, R., Zannini, E., & Arendt, E. K. (2021). Fundamental study on changes in the FODMAP profile of cereals, pseudo-cereals, and pulses during the malting process. *Food Chemistry, 343*, 128549.

Izydorczyk, M. S. (2021). Arabinoxylans. In *Handbook of hydrocolloids* (pp. 399–461). Woodhead Publishing.

Jeong, C. H., Jeong, H. R., Choi, S. G., Shim, K. H., & Heo, H. J. (2011). Neuronal cell protection and antioxidant activities of hot water extract from commercial buckwheat tea. *Korean Journal of Food Preservation, 18*(3), 358–365.

Joung, W. Y., Kim, S. J., Kim, B. G., Hurh, B. S., & Baek, H. H. (2018). Effect of roasting degree of barley on aroma characteristics of boricha. *Korean Journal of Food Science and Technology, 50*(2), 123–131.

Juliano, C., Cossu, M., Alamanni, M. C., & Piu, L. (2005). Antioxidant activity of gamma-oryzanol: Mechanism of action and its effect on oxidative stability of pharmaceutical oils. *International Journal of Pharmaceutics, 299*(1-2), 146–154.

Kalakotla, S., Gottumukkala, K. M., Rani, M., & Pravallika, P. L. (2015). Herbal drugs and herbal mediated silver nano particles as anti diabetics: A new horizon. *International Journal of Pharmaceutical Sciences Review and Research, 31*, 142–148.

Katagiri, R., Goto, A., Shimazu, T., Yamaji, T., Sawada, N., Iwasaki, M., … Japan Public Health Center-based Prospective Study Group. (2021). Dietary fiber intake and risk of gastric cancer: The Japan Public Health Center-based prospective study. *International Journal of Cancer, 148*(11), 2664–2673.

Kerienė, I., Mankevičienė, A., Bliznikas, S., Jablonskytė-Raščė, D., Maikšteniene, S., & Cesnuleviciene, R. (2015). Biologically active phenolic compounds in buckwheat, oats and winter spelt wheat. *Zemdirbyste-Agriculture, 102*(3), 289–296.

Kim, H. J., Han, J. A., Lim, S. T., & Cho, D. H. (2021). Effects of germination and roasting on physicochemical and sensory characteristics of brown rice for tea infusion. *Food Chemistry, 350*, 129240.

Kozioł, M. J. (1992). Chemical composition and nutritional evaluation of quinoa (*Chenopodium quinoa* Willd.). *Journal of Food Composition and Analysis, 5*(1), 35–68.

Kris-Etherton, P. M., Hecker, K. D., Bonanome, A., Coval, S. M., Binkoski, A. E., Hilpert, K. F., … Etherton, T. D. (2002). Bioactive compounds in foods: Their role in the prevention of cardiovascular disease and cancer. *The American Journal of Medicine, 113*(9), 71–88.

Krkošková, B., & Mrazova, Z. (2005). Prophylactic components of buckwheat. *Food Research International, 38*(5), 561–568.

Kruma, Z., Tomsone, L., Galoburda, R., Straumite, E., Kronberga, A., & Åssveen, M. (2016). Total phenols and antioxidant capacity of hull-less barley and hull-less oats. *Agronomy Research, 14*(S2), 1361–1371.

Kulichová, K., Sokol, J., Nemeček, P., Maliarová, M., Maliar, T., Havrlentová, M., & Kraic, J. (2019). Phenolic compounds and biological activities of rye (*Secale cereale* L.) grains. *Open Chemistry, 17*(1), 988–999.

Kumar, H., & Kaur, C. (2017). A comprehensive evaluation of total phenolics, flavonoids content and in-vitro antioxidant capacity of selected 18 cereal crops. *International Journal of Pure & Applied Bioscience, 5*(2), 569–574.

Kurek, M. A., Karp, S., Wyrwisz, J., & Niu, Y. (2018). Physicochemical properties of dietary fibers extracted from gluten-free sources: Quinoa (*Chenopodium quinoa*), amaranth (*Amaranthus caudatus*) and millet (*Panicum miliaceum*). *Food Hydrocolloids, 85*, 321–330.

Labuschagne, M., Mkhatywa, N., Johansson, E., Wentzel, B., & Van Biljon, A. (2017). The content of tocols in South African wheat; impact on nutritional benefits. *Food, 6*(11), 95.

Lamberts, L., & Delcour, J. A. (2008). Carotenoids in raw and parboiled brown and milled rice. *Journal of Agricultural and Food Chemistry, 56*(24), 11914–11919.

Lamothe, L. M., Srichuwong, S., Reuhs, B. L., & Hamaker, B. R. (2015). Quinoa (Chenopodium quinoa W.) and amaranth (*Amaranthus caudatus* L.) provide dietary fibres high in pectic substances and xyloglucans. *Food Chemistry, 167*, 490–496.

Le Ouay, B., & Stellacci, F. (2015). Antibacterial activity of silver nanoparticles: A surface science insight. *Nano Today, 10*(3), 339–354.

Lee, S. H., Lee, Y. R., Hwang, I. G., Woo, K. S., Kim, K. H., Kim, K. J., & Jeong, H. S. (2009). Antioxidant activities and quality characteristics of germinated rough rice tea according to roasting temperature, time and leaching condition. *Korean Journal of Food Science and Technology, 41*(4), 386–391.

Lehmann, J. W., Putnam, D. H., & Qureshi, A. A. (1994). Vitamin E isomers in grain amaranths (*Amaranthus* spp.). *Lipids, 29*(3), 177–181.

Lerma-García, M. J., Herrero-Martínez, J. M., Simó-Alfonso, E. F., Mendonça, C. R., & Ramis-Ramos, G. (2009). Composition, industrial processing and applications of rice bran γ-oryzanol. *Food Chemistry, 115*(2), 389–404.

Leváková, Ľ., & Lacko-Bartošová, M. (2017). Phenolic acids and antioxidant activity of wheat species: A review. *Agriculture/Pol'nohospodárstvo, 63*(3), 92–101.

Li, Y., Ma, D., Sun, D., Wang, C., Zhang, J., Xie, Y., & Guo, T. (2015). Total phenolic, flavonoid content, and antioxidant activity of flour, noodles, and steamed bread made from different colored wheat grains by three milling methods. *The Crop Journal, 3*(4), 328–334.

Liu, Y., Cai, C., Yao, Y., & Xu, B. (2019). Alteration of phenolic profiles and antioxidant capacities of common buckwheat and tartary buckwheat produced in China upon thermal processing. *Journal of the Science of Food and Agriculture, 99*(12), 5565–5576.

López-Alarcón, C., & Denicola, A. (2013). Evaluating the antioxidant capacity of natural products: A review on chemical and cellular-based assays. *Analytica Chimica Acta, 763*, 1–10.

Lu, L., Murphy, K., & Baik, B. K. (2013). Genotypic variation in nutritional composition of buckwheat groats and husks. *Cereal Chemistry, 90*, 132–137.

Ma, W., Kim, J. K., Jia, C., Yin, F., Kim, H. J., Akram, W., … Li, X. (2019). Comparative transcriptome and metabolic profiling analysis of buckwheat (*Fagopyrum tataricum* (L.) Gaertn.) under salinity stress. *Metabolites, 9*(10), 225.

Makpoul, K. R. (2016). Influence of barley coffee on blood cholesterol level. *World Journal of Pharmaceutical Research, 5*, 1–12.

Manach, C., Morand, C., Demigné, C., Texier, O., Régérat, F., & Rémésy, C. (1997). Bioavailability of rutin and quercetin in rats. *FEBS Letters, 409*(1), 12–16.

Marshall, E., & Mejia, D. (2011). *Traditional fermented food and beverages for improved livelihoods* (Diversification Booklet, (21)). FAO.

Masisi, K., Beta, T., & Moghadasian, M. H. (2016). Antioxidant properties of diverse cereal grains: A review on in vitro and in vivo studies. *Food Chemistry, 196*, 90–97.

Mazarello Paes, V., Hesketh, K., O'Malley, C., Moore, H., Summerbell, C., Griffin, S., … Lakshman, R. (2015). Determinants of sugar-sweetened beverage consumption in young children: A systematic review. *Obesity Reviews, 16*(11), 903–913.

Mehmood, A., Zhao, L., Ishaq, M., Xin, W., Zhao, L., Wang, C., … Xu, M. (2020). Anti-hyperuricemic potential of stevia (*Stevia rebaudiana* Bertoni) residue extract in hyperuricemic mice. *Food & Function, 11*(7), 6387–6406.

Mendez-Encinas, M. A., Carvajal-Millan, E., Rascon-Chu, A., Astiazaran-Garcia, H. F., & Valencia-Rivera, D. E. (2018). Ferulated arabinoxylans and their gels: Functional properties and potential application as antioxidant and anticancer agent. *Oxidative Medicine and Cellular Longevity, 2018*, 1–22.

Meydani, M. (2009). Potential health benefits of avenanthramides of oats. *Nutrition Reviews, 67*(12), 731–735.

Minatel, I. O., Francisqueti, F. V., Corrêa, C. R., & Lima, G. P. P. (2016). Antioxidant activity of γ-oryzanol: A complex network of interactions. *International Journal of Molecular Sciences, 17*(8), 1107.

Miranda, M., Vega-Gálvez, A., Martinez, E., López, J., Rodríguez, M. J., Henríquez, K., et al. (2012). Genetic diversity and comparison of physicochemical and nutritional characteristics of six quinoa (*Chenopodium quinoa* willd.) genotypes cultivated in Chile. *Ciência e Tecnologia de Alimentos, 32*, 835–843.

Montagnese, C., Santarpia, L., Iavarone, F., Strangio, F., Caldara, A. R., Silvestri, E., … Pasanisi, F. (2017). North and south American countries food-based dietary guidelines: A comparison. *Nutrition, 42*, 51–63.

Moreau, R. A., Harron, A. F., Powell, M. J., & Hoyt, J. L. (2016). A comparison of the levels of oil, carotenoids, and lipolytic enzyme activities in modern lines and hybrids of grain sorghum. *Journal of the American Oil Chemists' Society, 93*(4), 569–573.

Morris, V. J. (2011). Emerging roles of engineered nanomaterials in the food industry. *Trends in Biotechnology, 29*(10), 509–516.

Mustafa, A. F., Seguin, P., & Gélinas, B. (2011). Chemical composition, dietary fibre, tannins and minerals of grain amaranth genotypes. *International Journal of Food Sciences and Nutrition, 62*, 750–754.

Nawaz, H., Muzaffar, S., Aslam, M., & Ahmad, S. (2018). Phytochemical composition: Antioxidant potential and biological activities of corn. In *Corn-production and human health in changing climate* (pp. 49–68). IntechOpen.

Ndolo, V. U., & Beta, T. (2013). Distribution of carotenoids in endosperm, germ, and aleurone fractions of cereal grain kernels. *Food Chemistry, 139*(1-4), 663–671.

Neethirajan, S., & Jayas, D. S. (2011). Nanotechnology for the food and bioprocessing industries. *Food and Bioprocess Technology, 4*(1), 39–47.

Obafaye, R. O., & Omoba, O. S. (2018). Orange peel flour: A potential source of antioxidant and dietary fiber in pearl-millet biscuit. *Journal of Food Biochemistry, 42*(4), e12523.

Oh, S., Yi, B., Ka, H. J., Song, J., Park, J., Jung, J., … Lee, J. (2014). Evaluation of in vitro antioxidant properties of roasted hulled barley (*Hordeum vulgare* L.). *Food Science and Biotechnology, 23*(4), 1073–1079.

Omwamba, M., & Hu, Q. (2009). Antioxidant capacity and antioxidative compounds in barley (*Hordeum vulgare* L.) grain optimized using response surface methodology in hot air roasting. *European Food Research and Technology, 229*(6), 907–914.

Omwamba, M., Li, F., Sun, G., & Hu, Q. (2013). Antioxidant effect of roasted barley (*Hordeum vulgare* L.) grain extract towards oxidative stress in vitro and in vivo. *Food and Nutrition Sciences, 4*(8), 139–146.

Panche, A. N., Diwan, A. D., & Chandra, S. R. (2016). Flavonoids: An overview. *Journal of Nutritional Science, 5*, e47.

Papetti, A., Pruzzo, C., Daglia, M., Grisoli, P., Bacciaglia, A., Repetto, B., … Gazzani, G. (2007). Effect of barley coffee on the adhesive properties of oral *Streptococci*. *Journal of Agricultural and Food Chemistry, 55*(2), 278–284.

Paznocht, L., Kotíková, Z., Šulc, M., Lachman, J., Orsák, M., Eliášová, M., & Martinek, P. (2018). Free and esterified carotenoids in pigmented wheat, tritordeum and barley grains. *Food Chemistry, 240*, 670–678.

Peng, L. X., Wang, J. B., Hu, L. X., Zhao, J. L., Xiang, D. B., Zou, L., & Zhao, G. (2013). Rapid and simple method for the determination of emodin in tartary buckwheat (*Fagopyrum tataricum*) by high-performance liquid chromatography coupled to a diode array detector. *Journal of Agricultural and Food Chemistry, 61*, 854–857.

Peng, L. X., Zou, L., Wang, J. B., Zhao, J. L., Xiang, D. B., & Zhao, G. (2015). Flavonoids, antioxidant activity and aroma compounds analysis from different kinds of tartary buckwheat tea. *Indian Journal of Pharmaceutical Sciences, 77*(6), 661.

Perera, I., Seneweera, S., & Hirotsu, N. (2018). Manipulating the phytic acid content of rice grain toward improving micronutrient bioavailability. *Rice, 11*(1), 1–13.

Phillips, K. M., Haytowitz, D. B., & Pehrsson, P. R. (2019). Implications of two different methods for analyzing total dietary fiber in foods for food composition databases. *Journal of Food Composition and Analysis, 84*, 103253.

Punia, S., & Sandhu, K. S. (2015). Functional and antioxidant properties of different milling fractions of indian barley cultivars. *Carpathian Journal of Food Science & Technology, 7*(4), 19–27.

Punia, S., Sandhu, K. S., & Siroha, A. K. (2019). Difference in protein content of wheat (Triticum aestivum L.): Effect on functional, pasting, color and antioxidant properties. *Journal of the Saudi society of. Agricultural Sciences, 18*(4), 378–384.

Punia, H., Tokas, J., Malik, A., & Sangwan, S. (2021a). Characterization of phenolic compounds and antioxidant activity in sorghum [Sorghum bicolor (L.) Moench] grains. *Cereal Research Communications, 49*, 1–11.

Punia, S., Sandhu, K. S., Grasso, S., Purewal, S. S., Kaur, M., Siroha, A. K., … Kumar, M. (2021b). Aspergillus oryzae fermented rice bran: A byproduct with enhanced bioactive compounds and antioxidant potential. *Food, 10*(1), 70.

Qin, P., Wang, Q., Shan, F., Hou, Z., & Ren, G. (2010). Nutritional composition and flavonoids content of flour from different buckwheat cultivars. *International Journal of Food Science & Technology, 45*, 951–958.
Qin, P., Ma, T., Wu, L., Shan, F., & Ren, G. (2011). Identification of tartary buckwheat tea aroma compounds with gas chromatography-mass spectrometry. *Journal of Food Science, 76*(6), S401–S407.
Qin, P., Wu, L., Yao, Y., & Ren, G. (2013). Changes in phytochemical compositions, antioxidant and α-glucosidase inhibitory activities during the processing of tartary buckwheat tea. *Food Research International, 50*(2), 562–567.
Qin, L., Sui, X., Zeng, H., & Xu, Z. (2014). Fortification of the health benefit of buckwheat (*Fagopyrum tataricum*) tea. *Journal of Food Processing and Preservation, 38*(4), 1882–1889.
Ragaee, S., Gamel, T., Seethraman, K., & Abdel-Aal, E. S. M. (2013). Food grains. In *Handbook of plant food phytochemicals: Sources, stability and extraction* (pp. 138–162). Wiley.
Raguindin, P. F., Itodo, O. A., Stoyanov, J., Dejanovic, G. M., Gamba, M., Asllanaj, E., … Kern, H. (2021). A systematic review of phytochemicals in oat and buckwheat. *Food Chemistry, 338*, 127982.
Rakha, A. (2011). *Characterisation of dietary fibre in cereal grains and products* (Vol. 2011, p. 82). Department of Food Science, Swedish University of Agricultural Sciences.
Ranjan, S., Dasgupta, N., Chakraborty, A. R., Samuel, S. M., Ramalingam, C., Shanker, R., & Kumar, A. (2014). Nanoscience and nanotechnologies in food industries: Opportunities and research trends. *Journal of Nanoparticle Research, 16*(6), 1–23.
Rao, A. V., & Rao, L. G. (2007). Carotenoids and human health. *Pharmacological Research, 55*(3), 207–216.
Rashwan, A. K., Yones, H. A., Karim, N., Taha, E. M., & Chen, W. (2021). Potential processing technologies for developing sorghum-based food products: An update and comprehensive review. *Trends in Food Science & Technology, 110*, 168–182.
Ratnavathi, C. V., & Komala, V. V. (2016). Sorghum grain quality. In *Sorghum biochemistry* (pp. 1–61). Academic.
Ren, Q., Wu, C., Ren, Y., & Zhang, J. (2013). Characterization and identification of the chemical constituents from tartary buckwheat (*Fagopyrum tataricum* Gaertn) by high performance liquid chromatography/photodiode array detector/linear ion trap FTICR hybrid mass spectrometry. *Food Chemistry, 136*, 1377–1389.
Repo-Carrasco-Valencia, R. A. M., & Serna, L. A. (2011). Quinoa (*Chenopodium quinoa*, Willd.) as a source of dietary fiber and other functional components. *Ciência e Tecnologia de Alimentos, 31*, 225–230.
Repo-Carrasco-Valencia, R. A., Encina, C. R., Binaghi, M. J., Greco, C. B., & Ronayne de Ferrer, P. A. (2010). Effects of roasting and boiling of quinoa, kiwicha and ka˜niwa on composition and availability of minerals in vitro. *Journal of the Science of Food and Agriculture, 90*(12), 2068–2073.
Rodríguez-García, C., Sánchez-Quesada, C., Toledo, E., Delgado-Rodríguez, M., & Gaforio, J. J. (2019). Naturally lignan-rich foods: A dietary tool for health promotion? *Molecules, 24*(5), 917.
Ross, A. B., Kamal-Eldin, A., & Åman, P. (2004). Dietary alkylresorcinols: Absorption, bioactivities, and possible use as biomarkers of whole-grain wheat–and rye–rich foods. *Nutrition Reviews, 62*(3), 81–95.
Ruales, J., & Nair, B. M. (1993). Content of fat, vitamins and minerals in quinoa (*Chenopodium quinoa*, Willd) seeds. *Food Chemistry, 48*(2), 131–136.
Ryan, D., Kendall, M., & Robards, K. (2007). Bioactivity of oats as it relates to cardiovascular disease. *Nutrition Research Reviews, 20*(2), 147–162.
Saastamoinen, M., Plaami, S., & Kumpulainen, J. (1992). β-Glucan and phytic acid content of oats cultivated in Finland. *Acta Agriculturae Scandinavica B-Plant Soil Sciences, 42*(1), 6–11.
Saeed, F., Pasha, I., Anjum, F. M., & Sultan, M. T. (2011). Arabinoxylans and arabinogalactans: A comprehensive treatise. *Critical Reviews in Food Science & Nutrition, 51*(5), 467–476.

Salar, R. K., Purewal, S. S., & Sandhu, K. S. (2017). Relationships between DNA damage protection activity, total phenolic content, condensed tannin content and antioxidant potential among Indian barley cultivars. *Biocatalysis and Agricultural Biotechnology, 11*, 201–206.

Salmerón, I., Thomas, K., & Pandiella, S. S. (2014). Effect of substrate composition and inoculum on the fermentation kinetics and flavour compound profiles of potentially non-dairy probiotic formulations. *LWT-Food Science and Technology, 55*(1), 240–247.

Salmerón, I. (2017). Fermented cereal beverages: from probiotic prebiotic and synbiotic towards Nanoscience designed healthy drinks. *Letters in Applied Microbiology*, 65(2), 114–124.

Santiago, L. G., & Castro, G. R. (2016). Novel technologies for the encapsulation of bioactive food compounds. *Current Opinion in Food Science, 7*, 78–85.

Sharma, S., Saxena, D. C., & Riar, C. S. (2015). Antioxidant activity, total phenolics, flavonoids and antinutritional characteristics of germinated foxtail millet (*Setaria italica*). *Cogent Food & Agriculture, 1*(1), 1081728.

Shen, R., Yang, S., Zhao, G., Shen, Q., & Diao, X. (2015). Identification of carotenoids in foxtail millet (*Setaria italica*) and the effects of cooking methods on carotenoid content. *Journal of Cereal Science, 61*, 86–93.

Šimić, G., Horvat, D., Dvojković, K., Abičić, I., Vuletić, M. V., Tucak, M., & Lalić, A. (2017). Evaluation of total phenolic content and antioxidant activity of malting and hulless barley grain and malt extracts. *Czech Journal of Food Sciences, 35*(1), 73–78.

Singh, G. M., Micha, R., Khatibzadeh, S., Shi, P., Lim, S., Andrews, K. G., … Global Burden of Diseases Nutrition and Chronic Diseases Expert Group (NutriCoDE). (2015). Global, regional, and national consumption of sugar-sweetened beverages, fruit juices, and milk: A systematic assessment of beverage intake in 187 countries. *PLoS One, 10*(8), e0124845

Siroha, A. K., Sandhu, K. S., & Kaur, M. (2016). Physicochemical, functional and antioxidant properties of flour from pearl millet varieties grown in India. *Journal of Food Measurement and Characterization, 10*(2), 311–318.

Siyuan, S., Tong, L., & Liu, R. (2018). Corn phytochemicals and their health benefits. *Food Science and Human Wellness, 7*(3), 185–195.

Skrabanja, V., Kreft, I., Golob, T., Modic, M., Ikeda, S., Ikeda, K., et al. (2004). Nutrient content in buckwheat milling fractions. *Cereal Chemistry, 81*, 172–176.

Smeriglio, A., Barreca, D., Bellocco, E., & Trombetta, D. (2017). Proanthocyanidins and hydrolysable tannins: Occurrence, dietary intake and pharmacological effects. *British Journal of Pharmacology, 174*(11), 1244–1262.

Smith, C. E., & Tucker, K. L. (2011). Health benefits of cereal fibre: A review of clinical trials. *Nutrition Research Reviews, 24*(1), 118–131.

Sobota, A., Świeca, M., Gęsiński, K., Wirkijowska, A., & Bochnak, J. (2020). Yellowcoated quinoa (*Chenopodium quinoa* Willd) – Physicochemical, nutritional, and antioxidant properties. *Journal of the Science of Food and Agriculture, 100*, 2035–2042.

Solange, A., Georgette, K., Gilbert, F., Marcellin, D. K., & Bassirou, B. (2014). Review on African traditional cereal beverages. *American Journal of Research Communication, 2*(5), 103–153.

Srichuwong, S., Curti, D., Austin, S., King, R., Lamothe, L., & Gloria-Hernandez, H. (2017). Physicochemical properties and starch digestibility of whole grain sorghums, millet, quinoa and amaranth flours, as affected by starch and non-starch constituents. *Food Chemistry, 233*, 1–10.

Stauder, M., Papetti, A., Daglia, M., Vezzulli, L., Gazzani, G., Varaldo, P. E., & Pruzzo, C. (2010). Inhibitory activity by barley coffee components towards Streptococcus mutans biofilm. *Current Microbiology, 61*(5), 417–421.

Steadman, K. J., Burgoon, M. S., Lewis, B. A., Edwardson, S. E., & Obendorf, R. L. (2001). Minerals, phytic acid, tannin and rutin in buckwheat seed milling fractions. *Journal of the Science of Food and Agriculture, 81*(11), 1094–1100.

Suganuma, H., Inakuma, T., & Kikuchi, Y. (2002). Amelioratory effect of barley tea drinking on blood fluidity. *Journal of Nutritional Science and Vitaminology, 48*(2), 165–168.

Suh, M. H., Yoo, S. H., Chang, P. S., & Lee, H. G. (2005). Antioxidative activity of microencapsulated γ-oryzanol on high cholesterol-fed rats. *Journal of Agricultural and Food Chemistry, 53*(25), 9747–9750.

Sun, N. X., Tong, L. T., Liang, T. T., Wang, L. L., Liu, L. Y., Zhou, X. R., & Zhou, S. M. (2019). Effect of oat and tartary buckwheat–based food on cholesterol–lowering and gut microbiota in hypercholesterolemic hamsters. *Journal of Oleo Science, 68*, ess18221.

Sun, H., Wang, H., Zhang, P., Ajlouni, S., & Fang, Z. (2020). Changes in phenolic content, antioxidant activity, and volatile compounds during processing of fermented sorghum grain tea. *Cereal Chemistry, 97*(3), 612–625.

Suriano, S., Balconi, C., Valoti, P., & Redaelli, R. (2021). Comparison of total polyphenols, profile anthocyanins, color analysis, carotenoids and tocols in pigmented maize. *LWT – Food Science and Technology, 144*, 111257.

Sytar, O. (2015). Phenolic acids in the inflorescences of different varieties of buckwheat and their antioxidant activity. *Journal of King Saud University-Science, 27*(2), 136–142.

Tang, Z. X., & Lv, B. F. (2014). MgO nanoparticles as antibacterial agent: Preparation and activity. *Brazilian Journal of Chemical Engineering, 31*, 591–601.

Tang, Y., Li, X., Chen, P. X., Zhang, B., Hernandez, M., Zhang, H., Marcone, M. F., Liu, R., & Tsao, R. (2015). Characterisation of fatty acid, carotenoid, tocopherol/tocotrienol compositions and antioxidant activities in seeds of three *Chenopodium quinoa* Willd. genotypes. *Food Chemistry, 174*, 502–508.

Tang, Y., Zhang, B., Li, X., Chen, P. X., Zhang, H., Liu, R., & Tsao, R. (2016). Bound phenolics of quinoa seeds released by acid, alkaline, and enzymatic treatments and their antioxidant and α-glucosidase and pancreatic lipase inhibitory effects. *Journal of Agricultural and Food Chemistry, 64*(8), 1712–1719.

Tariq, H., Awan, S. I., Razzaq, T., Azim, R., Arif, S., Habib, S., & Khan, M. F. (2017). Comparitive analysis of total flavonoid content in various wheat (*Triticum aestivum* L.) landraces in rainfed conditions of Rawalakot. *International Journal of Research in Pharmacy and Science, 7*(3), 23–30.

Tatsu, S., Matsuo, Y., Nakahara, K., Hofmann, T., & Steinhaus, M. (2020). Key odorants in Japanese roasted barley tea (Mugi-Cha)—differences between roasted barley tea prepared from naked barley and roasted barley tea prepared from hulled barley. *Journal of Agricultural and Food Chemistry, 68*(9), 2728–2737.

Tekiner, I. H., Mutlu, H., Algıngil, S., & Dincerler, E. (2015). Considerations for Nanosciences in food science and nutrition: "Enhanced food properties". *Recent Patents on Food, Nutrition & Agriculture, 7*(1), 3–8.

Thakur, P., Kumar, K., & Dhaliwal, H. S. (2021). Nutritional facts, bio-active components and processing aspects of pseudocereals: A comprehensive review. *Food Bioscience, 42*, 101170.

Thangnipon, W., Suwanna, N., Kitiyanant, N., Soi-ampornkul, R., Tuchinda, P., Munyoo, B., & Nobsathian, S. (2012). Protective role of N-trans-feruloyltyramine against b-amyloid peptide-induced neurotoxicity in rat cultured cortical neurons. *Neuroscience Letters, 513*, 229–232.

Thompson, L. U. (1994). Antioxidants and hormone-mediated health benefits of whole grains. *Critical Reviews in Food Science & Nutrition, 34*(5-6), 473–497.

Threapleton, D. E., Greenwood, D. C., Evans, C. E., Cleghorn, C. L., Nykjaer, C., Woodhead, C., … Burley, V. J. (2013). Dietary fibre intake and risk of cardiovascular disease: Systematic review and meta-analysis. *British Medical Journal, 347*, f6879.

Tiwari, U., & Cummins, E. (2009). Nutritional importance and effect of processing on tocols in cereals. *Trends in Food Science & Technology, 20*(11–12), 511–520.

Tokusoglu, Ö., & Hall, C. A., III (Eds.). (2011). *Fruit and cereal bioactives: Sources, chemistry, and applications*. CRC Press.

Tolve, R., Cela, N., Condelli, N., Di Cairano, M., Caruso, M. C., & Galgano, F. (2020). Microencapsulation as a tool for the formulation of functional foods: The phytosterols' case study. *Food, 9*(4), 470.

Trono, D. (2019). Carotenoids in cereal food crops: Composition and retention throughout grain storage and food processing. *Plants, 8*(12), 551.
Trujillo, L. E., Ávalos, R., Granda, S., Guerra, L. S., & País-Chanfrau, J. M. (2016). Nanotechnology applications for food and bioprocessing industries. *Biology and Medicine, 8*(3), 1.
Tsuru, K., Hida, K., Haraguchi, Y., & Taniguchi, M. (2011). U.S. Patent Application No. 12/923,570.
Tuan, P. A., Thwe, A. A., Kim, Y. B., Kim, J. K., Kim, S. J., Lee, S., … Park, S. U. (2013). Effects of white, blue, and red light-emitting diodes on carotenoid biosynthetic gene expression levels and carotenoid accumulation in sprouts of tartary buckwheat (*Fagopyrum tataricum* Gaertn.). *Journal of Agricultural and Food Chemistry, 61*, 12356–12361.
Umrani, R. D., & Paknikar, K. M. (2014). Zinc oxide nanoparticles show antidiabetic activity in streptozotocin-induced type 1 and 2 diabetic rats. *Nanomedicine, 9*(1), 89–104.
Vollmannová, A., Musilová, J., Lidiková, J., Árvay, J., Šnirc, M., Tóth, T., … Germ, M. (2021). Concentrations of phenolic acids are differently genetically determined in leaves, flowers, and grain of common buckwheat (*Fagopyrum esculentum* Moench). *Plants, 10*(6), 1142.
Vollmer, M., Schröter, D., Esders, S., Neugart, S., Farquharson, F. M., Duncan, S. H., Schreiner, M., Louis, P., Maul, R., & Rohn, S. (2017). Chlorogenic acid versus amaranth's caffeoylisocitric acid–Gut microbial degradation of caffeic acid derivatives. *Food Research International, 100*, 375–384.
Wang, H., Sun, H., Zhang, P., & Fang, Z. (2019). Effects of processing on the phenolic contents, antioxidant activity and volatile profile of wheat bran tea. *International Journal of Food Science & Technology, 54*(12), 3156–3165.
Wolf, A., Bray, G. A., & Popkin, B. M. (2008). A short history of beverages and how our body treats them. *Obesity Reviews, 9*(2), 151–164.
Wood, P. J. (2010). Oat and rye β-glucan: Properties and function. *Cereal Chemistry, 87*(4), 315–330.
Wu, C. H., & Yen, G. C. (2005). Inhibitory effect of naturally occurring flavonoids on the formation of advanced glycation endproducts. *Journal of Agricultural and Food Chemistry, 53*(8), 3167–3173.
Wu, L., Zhai, M., Yao, Y., Dong, C., Shuang, S., & Ren, G. (2013a). Changes in nutritional constituents, anthocyanins, and volatile compounds during the processing of black rice tea. *Food Science and Biotechnology, 22*(4), 917–923.
Wu, L., Huang, Z., Qin, P., & Ren, G. (2013b). Effects of processing on phytochemical profiles and biological activities for production of sorghum tea. *Food Research International, 53*(2), 678–685.
Wu, X., Ge, X., Liang, S., Lv, Y., & Sun, H. (2015). A novel selective accelerated solvent extraction for effective separation and rapid simultaneous determination of six anthraquinones in Tartary buckwheat and its products by UPLC–DAD. *Food Analytical Methods, 8*, 1124–1132.
Xiong, Y., Zhang, P., Warner, R. D., & Fang, Z. (2019a). 3-Deoxyanthocyanidin colorant: Nature, health, synthesis, and food applications. *Comprehensive Reviews in Food Science and Food Safety, 18*(5), 1533–1549.
Xiong, Y., Zhang, P., Warner, R. D., & Fang, Z. (2019b). Sorghum grain: From genotype, nutrition, and phenolic profile to its health benefits and food applications. *Comprehensive Reviews in Food Science and Food Safety, 18*(6), 2025–2046.
Xiong, Y., Zhang, P., Luo, J., Johnson, S., & Fang, Z. (2019c). Effect of processing on the phenolic contents, antioxidant activity and volatile compounds of sorghum grain tea. *Journal of Cereal Science, 85*, 6–14.
Xiong, Y., Zhang, P., Warner, R. D., Shen, S., & Fang, Z. (2020a). Cereal grain-based functional beverages: From cereal grain bioactive phytochemicals to beverage processing technologies, health benefits and product features. *Critical Reviews in Food Science and Nutrition, 62*, 1–25.
Xiong, Y., Zhang, P., Johnson, S., Luo, J., & Fang, Z. (2020b). Comparison of the phenolic contents, antioxidant activity and volatile compounds of different sorghum varieties during tea processing. *Journal of the Science of Food and Agriculture, 100*(3), 978–985.

Xu, Q., Wang, L., Li, W., Xing, Y., Zhang, P., Wang, Q., … Ma, Y. (2019). Scented tartary buckwheat tea: Aroma components and antioxidant activity. *Molecules, 24*(23), 4368.

Yang, L. (2010). *Chemopreventive potential of sorghum with different phenolic profiles*. Doctoral dissertation, Texas A & M University.

Yang, N., Li, Y. M., Zhang, K., Jiao, R., Ma, K. Y., Zhang, R., … Chen, Z. Y. (2014). Hypocholesterolemic activity of buckwheat flour is mediated by increasing sterol excretion and down-regulation of intestinal NPC1L1 and ACAT2. *Journal of Functional Foods, 6*, 311–318.

Yao, Y., Yang, X., Shi, Z., & Ren, G. (2014). Anti-inflammatory activity of saponins from quinoa (*Chenopodium quinoa* Willd.) seeds in lipopolysaccharide-stimulated RAW 264.7 macrophages cells. *Journal of Food Science, 79*(5), H1018–H1023.

Youn, K. S., & Chung, H. S. (2012). Optimization of the roasting temperature and time for preparation of coffee-like maize beverage using the response surface methodology. *LWT-Food Science and Technology, 46*(1), 305–310.

Yu, X., Yang, T., Qi, Q., Du, Y., Shi, J., Liu, X., … Yan, N. (2021). Comparison of the contents of phenolic compounds including flavonoids and antioxidant activity of rice (*Oryza sativa*) and Chinese wild rice (*Zizania latifolia*). *Food Chemistry, 344*, 128600.

Zhang, Z.-L., Zhou, M.-L., Tang, Y., Li, F.-L., Tang, Y.-X., Shao, J.-R., Xue, W.-T., & Wu, Y.-M. (2012). Bioactive compounds in functional buckwheat food. *Food Research International, 49*(1), 389–395.

Zhang, R., Huang, L., Deng, Y., Chi, J., Zhang, Y., Wei, Z., & Zhang, M. (2017). Phenolic content and antioxidant activity of eight representative sweet corn varieties grown in South China. *International Journal of Food Properties, 20*(12), 3043–3055.

Zhou, X., Hao, T., Zhou, Y., Tang, W., Xiao, Y., Meng, X., & Fang, X. (2015). Relationships between antioxidant compounds and antioxidant activities of tartary buckwheat during germination. *Journal of Food Science and Technology, 52*, 2458–2463.

Zhu, F. (2016a). Buckwheat starch: Structures, properties, and applications. *Trends in Food Science & Technology, 49*, 121–135.

Zhu, F. (2016b). Chemical composition and health effects of Tartary buckwheat. *Food Chemistry, 203*, 231–245.

Zhu, F. (2019). Proanthocyanidins in cereals and pseudocereals. *Critical Reviews in Food Science and Nutrition, 59*(10), 1521–1153.

Zhu, F. (2020). Dietary fiber polysaccharides of amaranth, buckwheat and quinoa grains: A review of chemical structure, biological functions and food uses. *Carbohydrate Polymers, 248*, 116819.

Zhu, Y., & Liao, L. (2015). Applications of nanoparticles for anticancer drug delivery: A review. *Journal of Nanoscience and Nanotechnology, 15*(7), 4753–4773.

Zielinska, D., Szawara-Nowak, D., & Zielinski, H. (2013). Antioxidative and anti-glycation activity of buckwheat hull tea infusion. *International Journal of Food Properties, 16*(1), 228–239.

Zou, L., Jia, K., Li, R., Wang, P., Lin, J. Z., Chen, H. J., … Peng, L. X. (2016). Pharmacokinetic study of eplerenone in rats after long-term coadministration with buckwheat tea. *The Kaohsiung Journal of Medical Sciences, 32*(4), 177–184.

Zurak, D., Grbeša, D., Duvnjak, M., Kiš, G., Međimurec, T., & Kljak, K. (2021). Carotenoid content and bioaccessibility in commercial maize hybrids. *Agriculture, 11*(7), 586.

Chapter 12
Low Glycaemic Index Cereal Grain Functional Foods

Muhammad Usman, Prasanna J. Patil, Devashree N. Patil, Arshad Mehmood, Haroon Shah, Syeda Mahvish Zahra, Zeshan Ali, and Sehrish Nasreen

12.1 Introduction

Carbohydrate-rich foods (rice, barley, rye, and oats) are essential categories that fulfil the energy requirement of different age groups. Carbohydrates possess an available energy value of 17 kJ/g (4 kcal/g), contributing to 40 75% of overall

---

Authors Muhammad Usman and Prasanna J Patil have equally contributed to this chapter.

M. Usman (✉) · P. J. Patil (✉) · H. Shah
Beijing Advanced Innovation Center for Food Nutrition and Human Health, Beijing Technology and Business University, Beijing, China

School of Food and health, Beijing Technology and Business University, Beijing, China

D. N. Patil
Department of Biotechnology, Shivaji University, Kolhapur, Maharashtra, India

A. Mehmood
Beijing Advanced Innovation Center for Food Nutrition and Human Health, Beijing Technology and Business University, Beijing, China

School of Food and health, Beijing Technology and Business University, Beijing, China

Beijing Engineering and Technology Research Center of Food Additives, School of Food and Chemical Technology, Beijing Technology and Business University, Beijing, China

S. M. Zahra
Department of Environmental Design Health and Nutritional Sciences, Allama Iqbal Open University, Islamabad, Pakistan

Z. Ali
Department of Life Sciences, COMSATS University Islamabad, Sahiwal Campus, Sahiwal, Pakistan

S. Nasreen
Doctor of Human Nutrition and Dietetics, Sargodha Medical College, Sargodha, Punjab, Pakistan

S. Punia Bangar, A. Kumar Siroha (eds.), *Functional Cereals and Cereal Foods*,
https://doi.org/10.1007/978-3-031-05611-6_12

energy consumption based on the population (FAO/WHO, 1998). Starch is by far the most prevalent kind of carbohydrate and is present in grains such as wheat, maize, rice, etc. There are numerous health advantages of starchy vegetables, comprising Resistant Starch (RS) and polysaccharide fibers. Only glucose impacts blood sugar levels, causing insulin to be liberated, and the carbohydrates' digestion discharges a variety of sugars in the blood. This has satiety as well as a prebiotic impact on the large intestine (Lockyer & Nugent, 2017). Amylose is a type of a linear polysaccharide consisting of d-glucose units linked by α (1 → 4) glycosidic bonds, whereas amylopectin is larger and possesses more branches, unlike amylose, it also has α (1 → 6) branch points along with α (1 → 4) glycosidic bonds (Ren et al., 2020). With irregular branch points, Amylose has lengthy glucose molecules' unbranched stretches linked together. With a significant number of short-chain branches, amylopectin possesses a high molecular weight. As a rule of thumb, amylose and amylopectin make up 18–33% and 70–80% of cereals, correspondingly, in an estimated ratio of 30:70 (Zhu et al., 2020). The starch content of unpolished (brown) rice ranges from 72% to 82%, while the starch content of polished (milled) rice is around 90%. Different varieties of barley, maize, and rice that have lower than 1% amylose are referred to as waxy. The texture and digestibility of starchy grains are determined by the ratio of amylose to amylopectin. The more branch points in the amylopectin, the greater the surface area of the molecule, which permits starch hydrolysing enzymes to more simply access enzymes involved in the starch hydrolysis (Van Amelsvoort & Weststrate, 1992). For starch production, the enzyme Adenosine 5′ diphosphate pyrophosphorylase is the rate-limiting enzyme. This enzyme uses ADP-glucose to catalyse the production of starch. The chain length can be augmented using starch synthases. The granule-bound starch synthase (SS) I enzyme and the isoforms of the debranching enzymes are necessary for amylose production. Production of amylopectin is carried out by a variety of enzymes, including SS, starch branching and debranching enzymes. Granule-bound SS I and SS I isoforms control the production of starch primarily (Bao et al., 2017). It was detected that postprandial blood glucose and insulin responses were raised in both healthy and diabetic persons after eating starchy foods. Nevertheless, starch digestion rate in the human body influences the response. The notion of Glycemic index (GI) has been established to support diabetic people in choosing their meal, with the commendation that foods possessing a low GI are beneficial to diabetes patients. This term was introduced around 30 years ago; basically, it's a marker that ranks foods on the basis of glucose released in the bloodstream (Englyst et al., 2003). The GI is a physiological scale that ranges from 0 to 100 that measures how much a carbohydrate-rich food elevates blood sugar (glucose) levels. The International Standard Organization also defined GI and established the estimation techniques of GI for different foods and food products. Then, they have classified the foods on the basis of GI value – low (55), medium (55–69), or high (70) GI foods using glucose being a benchmark.

The glycemic response (GR) is another approach that determines the nature of foods, whether it is high GI or low GI foods. This response of food and its products is based on the degree of processing, starch granules' structure and the particle size,

Table 12.1 Glycemic index, carbohydrate amount, glycemic load, and resistant starch of some cereal grains

Cereal types	Glycemic index	Carbohydrate composition (%)	Glycemic load	Resistant starch (%)	References
Barley	34–70	60–70	5–17	0.5–2.9	Punia (2019)
Wheat	52–75	35–75	8–22	0.5–2.5	Birt et al. (2013)
Rice	55–85	70–82	8–25	0.5–2.5	Kumar et al. (2018a)
Maize	46–80	77.46	9–20	3.3–7.2	Ape et al. (2016)

cooking methods, cooling or retrogradation, and amount of other food ingredients, like lipid/fat, protein, the dietary fiber (DF) present in it. Moreover, certain environmental factors can influence the GI, glycemic load (GL), and GR of foods and their products. Furthermore, the foods' GI value also differs from the composition and quantity of carbohydrates found in them. It is observed to impact insulin metabolism and secretion in the human body (Toh et al., 2020). GL is one of the functions of GI and the quantity of carbohydrates ingested. The GL is illustrated as the average value of GI of individual foodstuff multiplied by the quantity (%) of carbohydrate present in the individual food ingested [GL = (GI × quantity of present carbohydrate)/100]. The greater the GL of cereal grains food, the greater is the GI of food or postprandial blood glucose response (Salari-Moghaddam et al., 2019). As an example, the carrot possesses higher amounts of GI, but low GL, whereas starchy foods, including maize, wheat, and rice, possess higher amounts of GI and high GL. Therefore, GL and GR are critical parameters determining foods' overall GI value, as shown in Table 12.1. Hence, the high GI and GL foods should upsurge the menace of obesity and obesity-related diseases (Greenwood et al., 2013). This occurs with the rapid increase of postprandial response, as described in Fig. 12.1. Therefore, low GI food should be recommended, and this is necessary at a time when cereal grains are the staple diet for the people of developed and developing countries, as this book chapter discussed.

12.2 GI of Foods and Their Relationship with Chronic Diseases – In a Historical Perspective

One of the most significant dietary shifts from the antique to the civilized era has been the growing intake of processed carb meals that are low in fiber, which has been associated with increasing diabetes and obesity that have strong coincided with cardiovascular diseases (CVDs) (Ford & Mokdad, 2008; Mokdad et al., 2003). The usage of pharmacological treatments to optimize blood sugar control in type 2 diabetes mellitus (T2DM) has been found to be effectual; however, studies have revealed that eating meals with a lower GI instead of a higher GI can result in

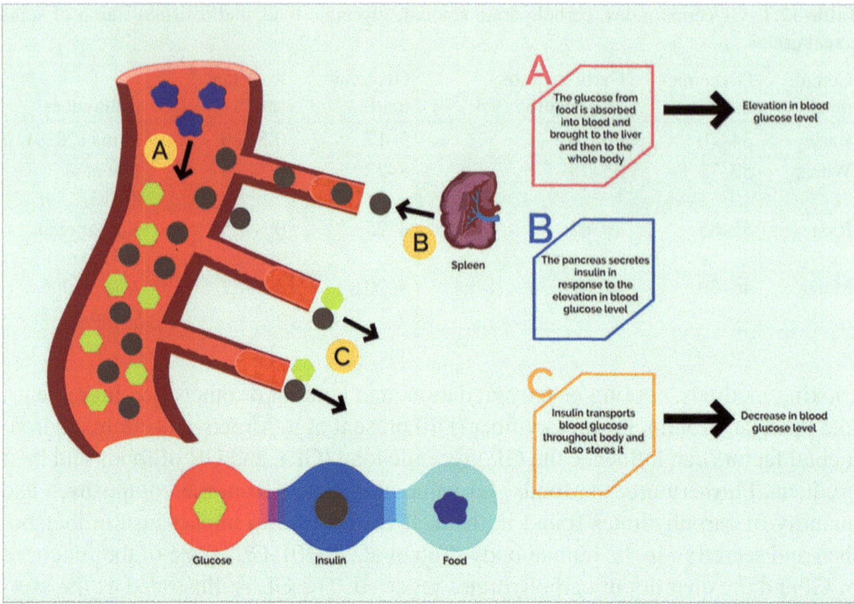

Fig. 12.1 Relationship between blood glucose level and foods

significant benefits in diabetes and obesity control that ultimately lowered the menace of CVDs (Anderson et al., 2004; Barclay et al., 2008; Livesey et al., 2008; Brand-Miller et al., 2003). Besides, it is widely believed that pharmacologic treatments are only a partial solution. In some cases, it may be a problem when used for a long time (Turnbull et al., 2009). Low GI foods' fundamental molecular mechanism may be linked to the delayed glucose absorption found when GI meals are ingested, viscous fiber, and Acarbose, an α-glucoside hydrolase suppressor. Via its suppression of pancreatic amylase and brush border sucrase-isomaltase, Acarbose is able to reduce postprandial glycemia by either uptake of sugar or inhibiting starch digestion in the small intestine (Jenkins et al., 1981a, b Augustin et al., 2015).

Even today, while there is a disagreement on the significance and relevance of GI, in 1981, Jenkins et al. (1981a) published the first article that provided information about lower GI food, which can minimise the risk of diabetes, obesity, CVDs, and even cancer. In 1995, international GI tables were established, which were later updated in 2002 and 2008 (Atkinson et al., 2008) with the GL concept, which was initially implemented by Walter Willett and his colleagues in 1997, that was used in the development of international GI tables (Salmeron et al., 1997a). Extensive epidemiological experiments have demonstrated that combining a low GL and high cereal fiber diet reduces the incidence of T2DM two-fold in both women and men (Salmeron et al., 1997a, b). Presently, similar findings have been validated in women and men; however, women showed a higher attenuation in their risk of heart disease (Livesey et al., 2013). CVDs were also decreased with low GI (Ma et al., 2012) and GL (Mirrahimi et al., 2012) foods with a larger risk reduction in women. In addition, some researchers partially agreed that low GI food could also decrease the risk of certain cancers, comprising colorectal and breast cancers (Barclay et al., 2008; Choi et al., 2012; Turati et al., 2015). These findings continue to be debated due to the absence of *in vivo* studies.

There have also been investigations that have linked changes in causative factors for various ailments to variations in dietary GI. Meta-analyses revealed that lower GI foods markedly controlled low-density lipoprotein cholesterol (Goff et al., 2013) and blood glucose level (Brand-Miller et al., 2003) with a risk of factors of c-reactive protein and plasminogen activator inhibitor-1 (Järvi et al., 1999; Jensen et al., 2008), especially in people having higher body mass index (BMI) (>25 kg/m^2), having both in clinical (Wolever et al., 2008) and epidemiological investigations (Liu et al., 2002). Moreover, acarbose is linked with a normal human diet in the Non-Insulin-Dependent Diabetes Mellitus (STOP NIDDM) trial, demonstrated to decrease new cases of CVDs, hypertension, and diabetes by −49%, −34%, and − 36%, correspondingly (Chiasson et al., 2002; Chiasson et al., 2003). Similar trends were also validated by the investigation of Hanefeld et al. (2004) using acarbose in a meta-analysis of 7 clinical trials. Other studies have been conducted on alpha-glucoside hydrolase inhibitors, and the results have validated their capability to lower post-prandial blood glucose levels and reduce the chance of developing diabetes (Kawamori et al., 2009; Van de Laar et al., 2005). Noticeably, such experiments with inhibitors are not confounded by fiber or other dietary constituents and taken together; they provide compelling validation that low GI foods may exert a protective influence on hard endpoints, like coronary heart disease and T2DM, even in people who do not have diabetes at baseline.

12.3 Estimation of GI: *In Vitro* and *In Vivo* Tests

Food's GI is often tested *in vivo*, in which subjects are administered a specific level of glucose or test food after fasting for a specified amount of time (approximately 10 h) at a laboratory. It is necessary to identify the increase in blood sugar for 2–3 h at periodic intervals. This procedure is time-intensive, expensive, and not accurate at the same time. Therefore, developing an *in vitro* approach to evaluate GI is presently the primary focus of research efforts. The *in vitro* digestion model is a quick screening approach for measuring the foods' GI, and it is a viable substitute for the *in vivo* digestion model in some cases.

12.3.1 In Vitro *Model of GI Estimation*

The *in vitro* approach for estimating GI is rapid, exact, and devoid of variances resulting from differences in metabolic rates across different volunteers (Table 12.2). It is also non-invasive. As a result of contributions from numerous research workers,

Table 12.2 *In vitro* methods for estimation of Glycemic index

Digestion stage	Body organ	Digestion development	Enzymes	Time	Method	References
Oral phase	Mouth	Mechanical disruption, swallowing	α-amylase	10–30 s	The sample of blood from the human body was collected after 10–12 h of fasting. The volunteers were ingested with sample food, and the blood glucose scores were estimated at various time-frames. The instrument, glucometer, was used for estimation/calculation of blood sugar profile from the blood of the tested individual	Araya et al. (2002)
Gastric phase	Stomach	The enzymatic process inside the stomach	Pancreatin	30 min–1 h		Muir and O'Dea (1992)
Intestinal phase	Small intestine	The enzymatic process inside the small intestine	Amyloglucosidase	2 h		Granfeldt et al. (1994)

the *in vitro* approach for assessing GI in cereals and vegetables has steadily improved and evolved (Kumar et al., 2018a, b, Lal et al., 2020a, b, 2021, Singh et al., 2020).

In vitro procedures are intended to imitate the responses that occur within gastrointestinal systems, including the small intestine, stomach, and mouth, in a controlled environment. Aiming to simulate the *in vivo* environment, biological factors should be optimized and improved. To determine starch digestion's rate, foods abundant in carbohydrates are subjected to the digestive enzymes *in vitro* within controlled environments, and the amount of glucose released is determined.

12.3.1.1 Oral Digestion Phase

In some *in vitro* approaches, it is required to chew the test foods by the subjects, or by milling, mincing, grinding or homogenising the test foods by the inclusion of α-amylose for digesting starch molecules at a temperature of 37 °C for 5 min with a pH of 6.9 that almost mimics the human body conditions (Goñi et al., 1997).

12.3.1.2 Gastric Digestion Phase

Proteolysis is a step in the *in vitro* procedures where pepsin is permitted to digest dietary proteins at a pH of 2.5, simulating the gastric phase. This aids in breaking protein-carbohydrate complexes caused by the decomposition of the food's protein constituent, hence facilitating the starch's digestion. Kumar et al. (2018a) treated the dietary sample using pepsin enzyme with specific conditions, including 60 min at a pH of 2.5 and 37 °C temperature on a water bath shaker (120 rpm). Goñi et al. (1997) standardised the *in vitro* gastric digestion protocol for diverse food stuffs, and so this procedure is a slightly modified version of their methodology.

12.3.1.3 Intestinal Digestion Phase

During the intestinal digestion phase, starch is broken down into glucose. In the same way, additional carbohydrates are digested into their integral monosaccharides, which are taken into the bloodstream and cause a jump in the GI. It is widely known that the food samples are treated with α-amylase, and the aliquots are taken at intervals of half an hour up to 3 h from the time of treatment. Jenkins et al. (1984) treated different food products, such as cornflakes, legumes, bread, and potato, with 10 mL of human saliva. After that, the slurry was inserted in a dialysis tube that was floated in an agitated water bath adjusted at 37 °C for 3 h. After 3 h, the sugar concentration levels liberated were determined. The digestibility index was used to determine the ranking of the foods. Some other investigators chopped the food sample and incubated it at 37 °C for 2 h with an enzyme mixture entailing amyloglucosidase, pancreatin, and invertase at pH 5.2 for 2 h. Aliquots of 0.5 mL were obtained at intervals of 30, 60, 90, 120, 150, and 180 min (Kumar et al., 2018a), and the sugar concentration was determined. Upon sample's digestion, the amount of liberated glucose (sugar) was determined using either the dinitrosalicylic acid (DNS) reagent as well as the glucose oxidase/peroxidase (GOPOD) technique. The GOPOD approach, on the other hand, for estimating decreasing sugar (glucose specifically) produces repeatable outcomes. It was possible to calculate the starch hydrolysis index (SHI) by calculating the incremental area under the curve (IAUC).

12.3.2 In Vivo *Model of GI Evaluation*

It has been well documented that the *in vivo*technique to evaluate GI is conducted by measuring the blood sugar levels of a minimum of ten young, healthy volunteers (who have been fasting for approximately 10 h) at 30-min intervals for 4 h after giving them food containing 50 g of d-glucose (in 200 mL of water) or a quantity of experiment food comprising an equivalent number of usable carbohydrates (Table 12.3).

Table 12.3 *In vitro*methods for estimation of Glycemic index

Phase	Mimicking organ	Digestion process	Enzymes	Time	Methodology	References
Oran phase	Mouth	Grinding, macerating, crushing, chopping (Under neutral condition, i.e. pH 6.9)	α-amylase	10–30 s	The free glucose concentration released is measured by the spectrophotometric method at different conditions	Araya et al. (2002)
Gastric phase	Stomach	Enzymatic (Under acidic condition, i.e. pH 2.5)	Pancreatin	30 min–1 h		Granfeldt et al. (1994)
Intestinal phase	Small intestine	Enzymatic, Use of cellophane tubings (it mimics small intestine) under slight alkaline condition, i.e. pH 6.9	α-amylase, Amyloglucosidase	2–5 h		Kumar et al. (2018a)

In order to ensure that they receive the food and glucose at a reasonable rate, the study with the reference food and test food, i.e., cornflakes, breads, and d-glucose, is carried out on different days and then performed twice more to obtain a mean value for each food. Furthermore, the test foods are served with a drink that can be any of the following: tea, coffee, milk, and water, but it should be ensured that it cannot contain any sugar. It is possible that there may be some slight discrepancies in the approach used by different laboratories. When the blood glucose levels are plotted against time, a curve is formed, where the IAUC owing to the ingestion of reference food and test food, can be calculated (Kim et al., 2019). The IAUC is determined using the trapezoidal rule for determining the area.

According to the various methodologies employed, the test foods' GI values containing white bread are roughly 1.4 times higher than those of test foods containing glucose, which are computed employing glucose as the reference food. GI variations between humans and animals may be attributed to variations in how carbohydrates are digested and absorbed in the human and animal systems. When carbohydrate absorption is slower, the rise in blood sugar is less, which results in a lower GI number (Nayak et al., 2014). It is possible to extract GL values that are a more realistic estimate of the capability of food to boost blood glucose levels.

12.3.3 **In Vitro** ***vs.*** **In Vivo** ***Test: Benefits and Drawbacks***

The *in vivo* GI of food samples or reference samples is also based on a person's physiological formation, which includes the absorption, digestion, and metabolism of glucose or other carbohydrate molecules in the gastrointestinal tract (GIT) (Hur et al., 2011). This approach to ranking foods is rather laborious due to the numerous limitations associated with running these studies. This type of research necessitates the use of human individuals of roughly the same age. The quantity of carbohydrates consumed is kept fixed in such tests, and the influence of food consumption on insulin sensitivity is investigated. The primary disadvantage of *in vivo* procedures is their tedious and time-consuming nature. Besides the management of human beings, the biochemistry of individual volunteers complicates such approaches, resulting in a lack of accuracy and consistency (Hirsch et al., 2013). It is critical, thus, to standardise the GI technique and testing settings for that the *in vitro* and *in vivo* approaches. The ethical committee must approve the *in vivo* GI measurement. Additionally, *in vivo* procedures necessitate a robust system and a rather big supply of experimental foods or reference foods. As a result, several companies and institutions have designed *in vitro* enzymatic techniques to estimate GI (Englyst et al., 2003). This requires less time, and the GL value of starchy food products is easily predictable. As a result, it is highly effective as a quick screening approach for an extensive range of starchy carbohydrate food products and assists plant breeders in developing low GI variants of starchy crops. They include maize, sorghum, wheat, buckwheat, rice, chia, quinoa, amaranth, barley, millet, oats, and rye (Kumar et al., 2018a; Lal et al., 2020a, b, 2021).

12.4 Low GI Cereal Grains

With the inclined market demand for processed foods in the form of packaged or readily available burgers, chocolate drinks, and ice-cream shakes that are loaded with simple carbohydrates, there has been an immense upsurge in the rate of obesity and obesity-related chronic disorders, including hyperlipidemia and hyperglycemia (Barclay et al., 2008) (Figs. 12.1 and 12.2). Global estimate of diabetes for 2011–2030 predicted that around 366 million people would be suffering from this disease (Whiting et al., 2011). To limit this exacerbating situation, there is a dire need to introduce low GI foods, especially snacks incorporating low glycemic cereals, eating less but variety of foods and enhanced physical activity induction in the lifestyle of pre-diabetics, diabetics, and people who are at risk of diabetes or obesity (Chiu et al., 2011). Low GI meals, according to Howlett & Ashwell (2008), ameliorate glycemic management by reducing glycated end products and enhancing insulin susceptibility. The GI, on the other hand, does not assess the amount of carbohydrates consumed but rather their quality. Apart from regulating and reducing the overall intake of carbohydrates, the chemical properties and physiological effects of carbohydrates must be considered (Nambiar & Patwardhan, 2015). Cereals are an essential part of daily meals, and cereals provide almost 46–48% of total carbohydrates in the diet (Bondia-Pons et al., 2011). It was found that the parboiled and brown rice possesses a better GI value than that white rice. Moreover, the application of other processing methods, such as chilling and incorporation of legumes, fats/oils, protein, and DF, also reduced the GI value of rice. As discussed

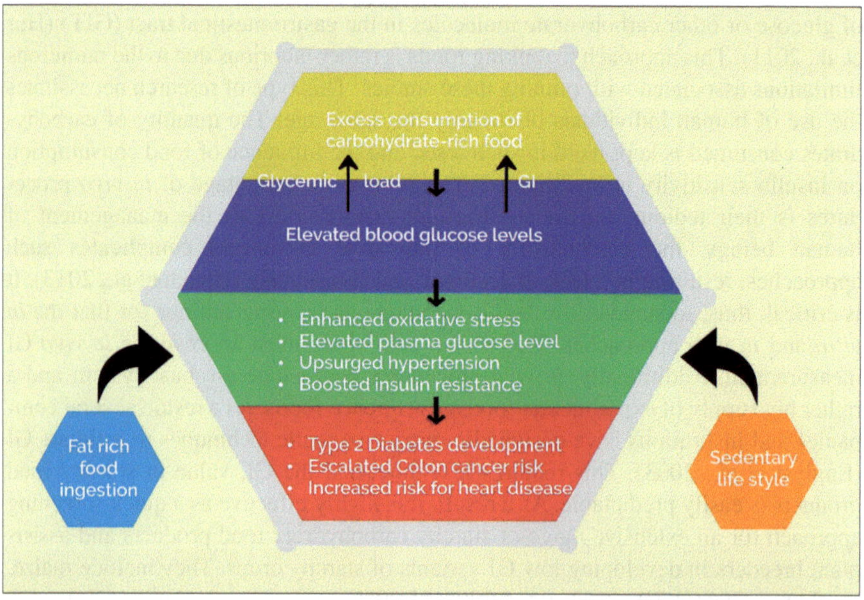

Fig. 12.2 Relationship between carbohydrate-rich foods and risk of chronic diseases

below, wild rice and pearl barley are suitable grain substitutes that fall into the low-GI dietary group; these may complement our main plate quite well in place of white rice.

12.4.1 Rice and GI

Rice variants' low GI can be related to their polyphenol and fiber content. Fibers turn intestinal constituents into a gel-like material that inhibits starch enzymatic activity, resulting in a low GI (Oh et al., 2005). Varying rice cultivars have different types and ratios of starches (amylose and amylopectin) that influence the digestion rate—the larger the proportion of amylose starch, the sluggish the digestion and the lesser the GI. Longer-grain rice with higher amylose content (19–23%) had lesser GI values compared to shorter-grain rice varieties with lower amylose content (12–19%). As a result, the rice's GI can range anywhere from 43 and 96. Zhang et al. (2015) employed 50 g glucose as control and examined blood glucose levels in healthy people (n = 8) after consuming 67 g wild rice within 2 h employing Jenkins and Wolever algorithm to estimate the GI. In animal trials, 50 male Sprague Dawley (SD) rats were arbitrarily allocated to one of five groups with 8 weeks of incessant feeding. They determined that Wild rice has a low GI, though wild rice has the ability to ameliorate the resistance in rats persuaded by a high-fat diet, and replacing 50% refined rice and flour with wild rice being able to mend insulin resistance in rats (Zhang et al., 2015). Similarly, another author stated that replacing brown rice with white rice could assist in alleviating the 24-h glucose and fasting insulin responses within overweight Asian Indians. In contrast to the white rice group, the five-day percentage variation in fasting insulin was 57% lesser for the brown rice group and 54% lesser for the brown rice legume group. Brown rice and brown rice legume diets did not have significantly different glycemic and insulinemic responses (Mohan et al., 2014). In adults with T2DM, Thompson et al. (2012) compared the GR of typical bean and rice meals with rice alone. The randomized crossover experiment included 17 males and females aged 35–70 with type T2DM treated with metformin (n = 14) or diet/exercise (n = 3). After a 12-h fast, 5 different types of beans and rice combinations were given, including control. Compared to rice alone, black and dark red beans with rice reduced the GR (Thompson et al., 2012).

12.4.2 Millet and GI

Millet is a cereal with small-seeded species and rich in bioactive compounds. Various types of millets, viz. foxtail, little, finger, and pearl, showed low GI (54–68) (Mani et al., 1993). Hadimani & Malleshi (1993) documented that different millet varieties have a high DF than wheat and rice. Siroha et al. (2016) reported 2.9–3.8%

fiber content in different millet cultivars. Little millet is beneficial in the management of metabolic syndromes (MetS) (Patil et al., 2015; Geetha et al., 2020). Certain unsaturated fatty acids in millet also exert a hypoglycemic effect. However, caution must be taken during the processing of millet grains to maintain the configuration of fatty acids (Annor et al., 2015). Finger millet is a rich source of bioactive compounds, DF, calcium, and iron; thereby, it possesses a hypoglycemic effect (Shukla & Srivastava, 2011). Phenolics present in finger millet inhibit α-glucosidase and pancreatic amylase that aids in maintaining postprandial blood glucose levels (Chandra et al., 2016). Long term feeding investigation was performed to assess the beneficial features of barnyard millet in T2DM individuals. 15 volunteers were part of an experiment. Out of them, 9 individuals had diabetes, and 6 were non-diabetic. Volunteers were asked to consume dehulled and heat-treated barnyard grains cooked in water with 2 g of green chillies during breakfast, lunch and dinner for 28 days. Blood samples were analyzed before and after the intervention. Results designated that both dehulled and heat-treated barnyard millet had low GI and employed positive impacts on glucose and lipid profile in healthy and diabetic individuals. Low GI was attributed to the synthesis of RS during the processing of millet grains (Ugare et al., 2014). Foxtail millet has moderate GI and low starch digestibility. A self-controlled trial with a final analysis of 64 subjects was conducted for 12 weeks to determine the effectiveness of foxtail millet in managing type T2DM. Out of 64 volunteers, 27 were male, and 34 were female subjects. Volunteers were asked to consume 90 g foxtail millet steamed bread daily for 12 weeks, and 24 h dietary recall was used as an assessment tool, and a blood sample was taken at 0, 6th, and 12th weeks. Intake of foxtail millet resulted in decreased fasting blood glucose levels and 2 h-blood sugar levels in individuals having reduced glucose tolerance throughout intervention time. Foxtail millet slowed gastric emptying and helped diabetics control spikes in blood glucose levels (Ren et al., 2018). 10 healthy people were selected by convenience sampling in another investigation. Individuals were asked to consume 50 g of glucose dissolved in 150 mL of water, and a glucometer analyzed blood glucose levels. In the second step, millet mixed with lean non-vegetarian food like millet bread-chicken sandwiches having 50 g of glucose and GI of 54.6 was consumed, and blood glucose analysis was done. After ingestion of glucose, mean blood glucose levels were 135 mg/dL, 116 mg/dL and 95 mg/dL, while with mixed millets bread chicken sandwich were 110 mg/dL, 96 mg/dL and 88 mg/dL at 60 min, 90 min, and 120 min correspondingly that encouraged combination of millet with non-vegetarian lean foods in vegetarian diabetics to manage hyperglycemia (Unnikrishnan et al., 2018). A cross-sectional study including 150 diabetics in tertiary care hospitals was conducted to determine the effectiveness of millet-based diets. 80 patients consumed a millet diet, and 70 patients consumed a non-millet diet. Fasting blood glucose levels, weight, and BMI were remarkably improved in the group consuming the millet diet (Vedamanickam et al., 2020).

12.4.3 Oats and GI

Oats are considered a functional diet and are used to prepare a variety of functional foods. As human food, oats are mostly taken as oatmeal and their products, including oat cookies and cakes. The major components of oats are β-glucan, protein, starch, and fats (Ahmad et al., 2014). Postprandial glucose level is not solely linked to carbohydrates in a food, but food microstructures, as well as certain nutrients such as protein and fat, are also its determinants (Tosh, 2014). Oats maintain healthy glucose levels in healthy individuals, and consumption of oats is beneficial for diabetic patients because of its positive effects on glucose and lipid profile (Hou et al., 2015). Oats are also helpful for improving glucose intolerance and insulin susceptibility in patients having MetS, overweight, and obese people (Weickert et al., 2006). Oat bran is rich in β-glucan (Jenkins et al., 2002). It's a soluble fiber with rheological properties and is responsible for slow gastric emptying and the formation of short-chain fatty acids (SCFAs) by bacterial fermentation in the lower GIT that positively affects glycemic control (Ames et al., 2021). The β-glucan slows down initial glucose absorption, therefore; giving a flat glucose peak (Tosh, 2014). The β-glucan also down-regulates sodium-potassium co-transporter type 1 and glucose transporter type 2 in epithelial cells of rats (Abbasi et al., 2016). Naked oats possess greater amounts of β-glucan in comparison with common oats. The seed structure of whole-grain oats is also an important determinant of GR besides β-glucan (Tosh, 2014). According to the European Food Safety Authority, Panel on Dietetic Products, Nutrition and Allergies, 4 g of β-glucan from oats must be consumed per meal.

Various processing practices could assist enhance β-glucan's viscosity, but viscosity alone is not adequate for glycemic control. It's also vital to consider particle size and the way fiber is consumed. To check out the impacts of different oat bran products on postprandial glucose levels, a randomized controlled trial with 12 T2DM patients was carried out. In test series 1, volunteers were given glucose load, oat bran crisp, and oat bran flour, exerting 12.5 g glycemic carbohydrate. In contrast, in series 2, 25 g glucose load alone and 25 g glucose load with 30 g oat bran flour and blood investigation were performed prior to fasting and then 15, 30, 45, 60, 90 and 120 min after the meal began. The area under the curve after eating oat bran flour for 120 min was minor compared to ingestion of glucose load. At the same time, no such difference was observed with ingestion of oat bran crisp and glucose load during test series 1. In contrast, in series 2, glucose concentration was lesser at 30, 45, and 60 min and greater at 120 min for glucose load with oat bran flour ingestion than glucose load alone (Tapola et al., 2005). Moreover, processing techniques can change food microstructure affecting GR (Regand et al., 2009). Steel-cut oats have a GI of 55, a large flake of 53, muesli of 56, quick-cooking oats of 71, and instant oats of 75.

GI values of commercial oat breakfast cereal, cereal bar and β-glucan enriched breakfast cereal were found low compared to GI of white bread in research conducted on T2DM (Jenkins et al., 2002). Oats are also rich in phytonutrients such as tocols and avenanthramides with anti-inflammatory and antioxidant properties

(Tosh & Bordenave, 2020). Oat-grain meals must be a part of a glycemic management diet regardless of oat grain type or cooking methods (Zhu et al., 2019).

12.4.4 Rye and GI

Rye is consumed as bread, breakfast cereal and porridge, pastas, hamburgers, and a concoction of rice and steel-cut rye. Rye is rich in DF, vitamins, minerals and other bioactive substances. The amount of DF in the rye is more than that of wheat. Arabinoxylans constitute the soluble fiber of rye. Rye also has β-glucan but in a small proportion when compared to oats (Nordlund et al., 2016). Intakes of soluble fiber positively affect blood sugar profile and improve insulin sensitivity as arabinoxylans have rheological properties. Arabinoxylans in the endosperm are more soluble than those present in the outer layers. Rye also has insoluble fibers such as cellulose and lignin. Insoluble fiber also helps control blood sugar levels by decelerating carbohydrate digestion and absorption. Colonic rye's fermentation is responsible for the SCFAs' synthesis, such as propionate, which also helps to improve glucose tolerance by inhibiting the production of glucose by the liver.

Rye has different varieties. To find out the difference in glucose response of different varieties, a randomized cross-over investigation was piloted. 20 healthy individuals volunteered for the study. White bread was used as a reference, and five varieties of rye viz. Amilo, Evolo, Kaskelott, Picasso, and Vicello were also baked into whole-grain bread. Volunteers were instructed to consume test bread in the morning with approximately a gap of 1 week between each test, take a few slices of white bread in the evening, and consume nothing on the morning test day. Blood samples were taken to analyze blood sugar and serum insulin levels beforehand of the test and 15, 30, 45, 60, 90, 120, 150, and 180 min after the meal began. GI of rye bread made with Vicello and Picasso was lower than white bread. Peak blood glucose levels were also low for bread made with Vicello, Picasso, Amilo, and Evolo than for white bread. Vicello, Picasso, Amilo, and Kaskelott rye bread had reduced insulin indices, and Vicello, Picasso, Amilo, and Evolo rye bread had a decreased insulin response compared to white bread (Rosén et al., 2011).

A randomized cross-over investigation was carried out, including 72 individuals with MetS, divided into 2 groups. Out of 72 individuals, 37 were women, and 35 were men. The duration of the study was 12 weeks. One group was instructed to eat rye bread and pasta, whereas the other was instructed to eat oats, wheat bread, and potatoes. Blood samples were collected and analyzed before and 15, 30, 45, 60, 90, and 120 min after the start of a meal. Blood glucose and insulin concentrations didn't alter much during the 2 h glucose tolerance test, but rye bread and pasta resulted in early insulin response, which decreased the risk of developing T2DM in patients with MetS (Laaksone et al., 2005). A case-control investigation was conducted to check out the risk of T2DM associated with plasma alkylresorcinols, a biomarker associated with whole-grain wheat and rye feeding. Results indicated that plasma alkylresorcinols were not linked with the menace of emerging diabetes, and relative consumption of whole-grain rye to whole-grain wheat was contrariwise correlated to the progress of T2DM (Biskup et al., 2016).

A randomized investigation was accompanied to evaluate the impacts of low GI cereals on glucose tolerance. 12 individuals volunteered for the test. The test was divided into 2 series; in series 1, subjects were asked to consume rye or barley kernel breakfast, and in test series 2, subjects were asked to consume test meals in the evening and night. Blood samples were analyzed before and after the test. Results indicated that rye or barley kernel intake reduced blood glucose levels compared to whole bread (Nilsson et al., 2008). Another randomized cross-over investigation was accomplished to identify the influence of fiber rich rye bread on glucose and insulin metabolism in healthy postmenopausal females. Twenty postmenopausal females were asked to consume high fiber rye bread and white bread as reference for 8 weeks to make up more than or equal to 20% of total energy intake. Blood samples were analyzed before and after the test. Results indicated that although insulin sensitivity was not altered, insulin secretion was increased with the intake of high fiber rye bread, contrary to white bread (Juntunen et al., 2003).

12.4.5 Barley and GI

Barley is gaining renewed interest as an ingredient for production of functional foods due to its high contents of bioactive compounds (Punia & Sandhu, 2015; Sandhu & Punia, 2017). Barley is famous for its low GI attributed to the high concentration of slowly digested starch and β-glucan. Harvesting conditions, type of soil, cultivar, pearling status, cooking or processing conditions, and end food product, especially amylose: amylopectin, causes fluctuation in the GI of barley (Aldughpassi et al., 2012). Chillo et al. (2011) scrutinized the GI of semolina, glucagel barley (GB), and barley balance (BB) variety flours fortified spaghetti. The GI was measured by the expression of the IAUC for the test and reference foods (50 g each) following the trapezoid rule. Results revealed that BB concentrate substantially alleviated the IAUC and GI of spaghetti. Sagnelli et al. (2018) explored the potentials to develop barley-based low GI foods from barley; elevating amylose composition in starch-rich commodities is an alternative approach to attenuate the GI. Amylose-only barley (AO) encompassing 99% amylose starch was engineered. Two *in vitro* gastro-intestinal models were employed for grains and bread from barley and AO; these displayed lesser predicted GIs (pGIs). Barley has comprehensive nutritional potential and has been found of keen interest in developing convenience foods to provide energy as emergency foods (Zahra et al., 2014) and to use as nutraceuticals, especially fruit bars (Zahra et al., 2020). Being low GI food, barley is still an underutilized crop except for utilization in breakfast cereal manufacturing industries. Hussain et al. (2020) explored the potential of barley and water chestnut flours to develop crackers yielding low glycemic products for diabetics or others, using in a combination (100:0, 70:30, 50:50, 30:70 and 0:100). Water chestnut flour (WCF) and barley flour (BF) blends in the ratio of 70:30 yielded good quality low GI crackers.

12.4.6 *Sorghum and GI*

Sorghum contains slowly digesting starch, lipids and proteinous matters, and functional phytonutrients including phenolic components, vitamins, minerals, especially polycosanols, which have been linked to limiting obesity and inflammation to manage diabetes and hypersensitive condition of the body (de Morais Cardoso et al., 2017). Sorghum is mainly utilized in functional foods because of its rich nutritional profile and availability at a lesser price (Lemlioglu-Austin et al., 2012). Foods with lower GI and high resistant starch (RS) aid carbohydrates' absorption and inhibit extreme blood sugar fluctuations (Jenkins et al., 2008; Järvi et al., 1999). Sorghum has high tannins, which are known to develop complexes with starch and proteinous matters making digestibility slowed down, resulting in good food for people with hyperglycemic trends (Le Bourvellec & Renard, 2012). Equivalent outcomes were stated by Magaletta et al. (2010) after intervening with sorghum bran porridge for breakfast. Sorghum contains phenolic antioxidants richly (Vila-Real et al., 2017). These phenolic antioxidants have been known as potential inhibitors of the α-amylase enzyme, that's why they are beneficial in hyperglycemia management (Duodu & Awika, 2019). A comprehensive review comprising 20 studies over 30 years was published by Simnadis and team, which suggests that sorghum is a low glycemic food that has the potential to lower blood sugar, alleviate oxidative stress, and even facilitate health support for human immunodeficiency virus (HIV) patients (Simnadis et al., 2016). Prasad et al. (2015) made a comparison between the interventional diet of sorghum biscuits, upma, poha, and roti with control wheat roti. Results revealed a lower GL for sorghum biscuits, upma, and poha and a higher GL for sorghum roti than wheat roti.

12.5 Impacts on GI by Different Factors

GI is a unique quality of foodstuff that can be impacted by intrinsic, extrinsic, and environmental influences and food mixture. It was observed that the intrinsic factors (i.e., composition of amylopectin and amylose, DF, protein and fats/oils), extrinsic factors (i.e., thermal processing techniques viz. cooking, fermentation, cooling, parboiling and soaking), environmental factors and addition of other food ingredients exhibited strong impacts on GI value of cereal grain functional foods *in vivo* and *in vitro*. The specifics are provided in the sections that follow.

12.5.1 *Intrinsic Factors' Impacts on GI*

12.5.1.1 Amylopectin and Amylose

Wheat, corn, maize, and rice starches are comprised of amylose (20–30%) and amylopectin (70–80%). Amylose is the most important form of RS that is tough to gelatinize throughout cooking. In rice and other cereal grains, the amylopectin: amylose content is a cognitive variable that impacts the rate of carbohydrate digestion and, thus, predicts blood sugar levels and insulin response in rice and other cereal grain functional foods. It is widely believed that the high amylose content markedly helps to reduce the GI, GL, and GR of rice grain and its products. This happens because amylose produces a complex with other food components, particularly lipid contents, that make the food less susceptible to digestion by different enzymes when heated (Kumar et al., 2018b; Kaur et al., 2016).

The ratio of amylose present in cereal grains can provide a good insight into the GI, GL, and GR of cereal grain functional foods. Research on maize crops suggests that the good amount of amylose in both raw and processed forms is linked with a less GI value due to the controlled release of molecules within the human gut and resisting the enzymatic hydrolysis (Rocha et al., 2010). Several studies on cereal grains starch found that the amylopectin: amylose composition and structural configuration possess a strong relationship with the occurrence of RS, having a low hydrolysis index (Lal et al., 2020a, b, 2021). Corn starch with high amylose: amylopectin yields a very low insulinemic and postprandial GR (Granfeldt et al., 1994).

According to the aforementioned reports, it was concluded that rice and other cereal grains have high amylose content, showing low GI than that of high amylopectin cereal grain foods. Therefore, agronomists should produce low GI rice and other cereal grain traits with a high GI. Besides, amylose amount alone does not accurately predict GI, GL, and GR, and other variables also influence carbohydrate metabolism rates and GI, as discussed below.

12.5.1.2 DF

DF is primarily a non-starchy plant fraction that is incapable of being digested by normal gastrointestinal enzymes, including hydrolytic enzymes, due to its composition. DF is an important part of our diet that influences the GI value of functional cereal grains (Lal et al., 2020a, b, 2021). Additionally, DF can pass via the small intestine and be digested in the large intestinal tract and stomach, leading to the formation of short-chain fatty acids (SCFAs) in the human body (acetate, butyrate, propionate, valerate, and caproate) that aid in the modulation of carbohydrates and fatty acid metabolism in the liver. These SCFAs are prone to contribute to various biological functions for treating chronic diseases, including diabetes, colon cancer, hypertension, and so on (Cummings et al., 2004). Trowell et al. (1985) defined DF as "the remnants of plant cells composed of cellulose, hemicellulose, lignin, oligosaccharides, pectins, gums, and waxes that resist human digestive enzymes". The American Association of Cereal Chemists defined DF as "the edible form of a plant part or analogue carbohydrate fermented in whole or in part in the large intestine and is digestive resistant in the small intestine". Moreover, Zhu (2020) characterized the DF into the following categories: non-starch oligosaccharides, non-starch polysaccharides, non-digestible carbohydrates, and lignin complex. In addition, DF can also be classified depending upon the fermentation ability in the human gut and water-solubility in the different solutions. The water-soluble DF is lignin, cellulose, and hemicellulose that have been shown less fermentability in the human gut.

Contrarily, the DFs including mucus, gums, pectin, etc. have been exhibited good fermentability in the human gut. After extensive research, it's clear that DF has a significant impact on intestinal digestion. The physical properties of DF made a small intestine more sensitive to changes in blood sugar and fat absorption, and it made people a little less hungry. Physical and chemical properties including moisture retention, oil holding capacity, and consistency and emulsification activity occur when mixture of water and oil is made (Xie et al., 2017). For young men and women, a total DF intake of 38 g and 25 g/day, respectively, is recommended (Slavin, 2005).

Many drugs and dietary supplements are present in supermarkets produced by chemical modifications. These drugs and dietary supplements are considered DF, including cross-linked starches and digestive resistant dextrin (Oki et al., 2019). DF, found in legumes, vegetables and fruits, and whole grains, affects intermediate metabolism by lowering fat and glucose molecules' absorption from the small intestine of the human body. It was discovered that the higher viscosity of a DF and soluble fiber significantly contributed to lowering the GI value of cereal grain functional foods due to slower digestibility, good RS content, and bran portion.

12.5.1.3 RS

All starches that resist d-glucose digestion once absorbed in the human gut are commonly known as RS (Fuentes-Zaragoza et al., 2010). A method for estimating non-starch polysaccharides *in vitro* was developed in 1982 by Englyst and his colleagues, and they found that some starches are resistant to enzymatic hydrolysis. This phenomenon is termed resistance to enzymatic hydrolysis. Therefore, this type of starch is undigested and passes via the small intestine and then goes to the large intestine, where it is utilised as a raw material for microbial fermentation, resulting to the formation of methane, CO_2, SCFAs, and hydrogen, among other byproducts. RS is currently divided into five categories, which are as follows: (1) type-1 RS is tangibly encapsulated starch within the fibrous cell wall of starch granules, which makes it unavailable to digestive enzymes in the small intestine, (2) type-2 RS is thermostable and may be used in multiple functional foods, including baked goods (Sullivan et al., 2017). Moreover, it is an extremely structured starch that resists digestion by the enzymes. It is found mainly in corn starch than that of other major cereal grains (Zaman & Sarbini, 2016), (3) type-3 RS starch is another type of starch found in cereal grains and occurs when cooked starchy foods are allowed to cool (retrogradation). As a result of the cooling, the gelatinized digestible starch fractions (amylose and amylopectin) are converted into RS through a retrogradation process that lowers their digestibility, (4) type-4 RS is a form of starch with a minimum digestibility due to the procedure of esterification and crosslinking (Zaman & Sarbini, 2016), (5) and at long last, in recent years, the type 5 RS in rice has been identified, in which complexes are formed by the presence of amylose content.

There are two important parameters of starch digestibility that are measured. These are the GI and the RS levels. The RS needs a long time for digestion and negatively affects the GI. Increased RS content in a food or food mixture could be utilized to help attenuate glucose intake (Kumar et al., 2018a). In the small intestine, the postprandial glucose and insulin response are decreased by RS. This has a significant effect on using RS in prediabetes diet plans. Foods with a high RS digest more slowly than foods with a higher concentration of slow- and fast-digesting starch. As a result, RS can be incorporated into preventive and therapeutic initiatives for T2DM and calorie restriction (Zaman & Sarbini, 2016). Factors influencing RS formation include water starch ratio, amylose: amylopectin ratio, stirring, and cooking and retrogradation process (Deepa et al., 2010). After chemical modifications, an increase in RS is reported by many researchers (Siroha & Sandhu, 2018; Siroha et al., 2019; Punia, 2020; Punia et al., 2019).

12.5.1.4 Anti-Nutritional Compounds

Phytic acid (PA) is one of the major anti-nutritional compounds that is almost found in all major cereal grains. It is the most important phosphorus type, accounting for 15% of grain, pulses, and oilseeds (Fig. 12.3). PA biosynthesis involves several enzymes catalysed reactions, catalysing the initial rate-determining phase with Myo-inositol-3-phosphate synthase and the final metabolic pathway with Inositol 1,3,5,6-pentakisphosphate-2-kinase (Ali et al., 2013). In general, genes implicated in the first and last phases of PA production did not escalate total phosphorus but did enhance phosphorus flow into seeds via vegetative organs. Among ruminants and non-ruminants alike, PA has been commonly considered a nutrient inhibitor since it reduces the bioavailability of minerals such as iron, calcium, magnesium, and zinc (Kumar et al., 2017). It also has the additional effect of reducing the efficiency of digestive enzymes by binding to specific amino acid residues prerequisite for catalysis. Monogastric animals, such as humans, will be unable to digest PA because they lack the phytase enzymes necessary in their digestive tracts to use it (Raboy, 2009). PA also serves as an antioxidant, protecting the seeds from oxidative stress, stress tolerance (i.e., biotic and abiotic), and anticancer agent. Furthermore, variations in PA levels might have an impact on the digestion of cereal grains, and starchy functional foods and eventually reduce the blood sugar levels (Kumar et al., 2020b; Saha & Reddy, 2016).

Insufficient research about starch-PA interaction is currently available. However, the two can be combined through phosphate linkages. Foods that are high in PA are essential because a significant effect of PA was found on blood glucose response. Moreover, it can bind the starch molecules in the GIT, thereby lowering the GI value of cereal grain functional foods (Yoon et al., 1983). Therefore, it was stated that the high PA concentration cereal grains are good for the management of diabetes by the action of lowering starch hydrolysis. It also improves intestinal health by decreasing stomach emptying. PA may potentially alter starch digestibility by binding to digestive enzymes or proteins related to starch. Furthermore, in the intestine, PA can interact with Ca^{2+} (a cofactor for amylase activity) and alleviate starch digestibility (Thompson et al., 1987). A high concentration of PA reduces starch hydrolysis

Fig. 12.3 Relationship between glycemic index and phytic acid

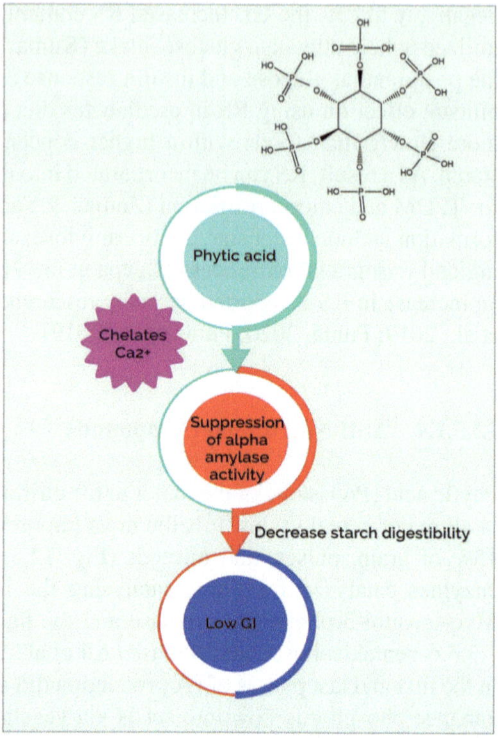

rates, lowering the GI, GL, and GR of cereal grain functional foods. Recently, Kumar et al. (2020b) revealed that the presence of PA in rice markedly reduced the rate of starch digestibility in the GIT and ultimately alleviated the GI value of rice as well.

12.5.1.5 Protein and Lipid Content

The interaction of saccharide-lipid can either assist or inhibit digestion by influencing the characteristics of both starch and lipids. Lipid deposition in plants significantly influences carbon metabolism activities, including starch synthesis, sucrose synthesis, and others. Several efforts were made to raise the lipid content of the diet, which has hampered starch synthesis. Lipid biosynthesis improvements result in a surge in the concentration of soluble carbohydrates (Lal et al., 2020a, b, 2021). Protein is the main component after starch and lipid in the granules. It is primarily found in the type of nucleic acids, different enzymes, and amino acids, among other things. Protein is found in lower concentrations in starchy plants, although it does influence starch production and digestibility. Protein could be a vital constituent of starch granules. In wheat, various proteins are extracted from different parts of the plant, viz. seed, endosperm, whole grain, and flour that are closely connected to starch molecules. Friabilin is a starch granule protein related to the endosperm texture of wheat. Because of the presence of this protein, the texture of wheat grain transforms from soft to hard, leading to slow-down of digestibility of wheat starch (Chichti et al., 2015). Common wheat has higher protein content than durum wheat. Compared to common wheat, durum wheat has a greater amylose concentration and is resistant to starch digestion with a lower GI (Liu et al., 2007). Lipids readily combine with amylose to synthesize additional compounds, with the hydrocarbon fraction of the lipid present inside the helical cavity. This lipid-amylose interaction causes a delay in enzymatic hydrolysis. In recent years, it has been discovered that delaying the digestion of these complexes lowers postprandial serum glucose and insulin response. It has been shown that the quantity of complex formation between wheat flour and oleic acid is inversely proportional to the amount of α-amylase digestion that occurs (Murray et al., 1999). Additionally, another report revealed that the presence of emulsifiers alleviated the *in vitro* digestibility of waxy and non-waxy rice starches (Guraya et al., 1997).

Protein in starchy plants can lower glycemic and insulinemic responses and hinder the hydrolysis of starch. Reports on starch-protein interactions using confocal microscopy and differential scanning calorimetry indicate that protein denaturation and hydrolysis may suppress starch hydration and enzymatic degradation. Additionally, it has the potential to inhibit metabolic key enzymes involved in starch hydrolysis (López-Barn et al., 2017). According to the research findings, the indica rice variety's intrinsic lipids and proteins lower starch hydrolysis rate, thereby lowering GI response and decreasing starch swellability, most likely by creating a protective layer around the starch grain that restricts swelling and reduces the impact on digestive enzymes. When compared to proteins, lipids were more efficient in lowering starch digestibility (Ye et al., 2018). Protein-rich diets stimulate insulin secretion, resulting in lower postprandial blood sugar levels. As a result, the natural protein content of certain foods may explain why their starches are not easily hydrolyzed, resulting in a decreased GI. Pasta prepared with cereals and gluten delays the activity of pancreatic amylases and results in a lower GI.

12.5.2 *Extrinsic Factors' Impacts on GI*

12.5.2.1 Thermal Processing (Cooking, Microwave Cooking, Frying, and Pasteurization)

Starchy cereals, including rice, maize, barley, sorghum, oats, and some others, are normally consumed after processing (Fig. 12.4). The availability and bioaccessibility of the different bioactive phytochemicals and other compounds in foods (i.e. lipid, starch, protein, phenolic acids, flavonoids, DF, and carotenoids) can be modified by processing, viz., frying and cooking (Thakur et al., 2020). When cereal grain functional foods are cooked, the starch granules enlarge owing to the heating

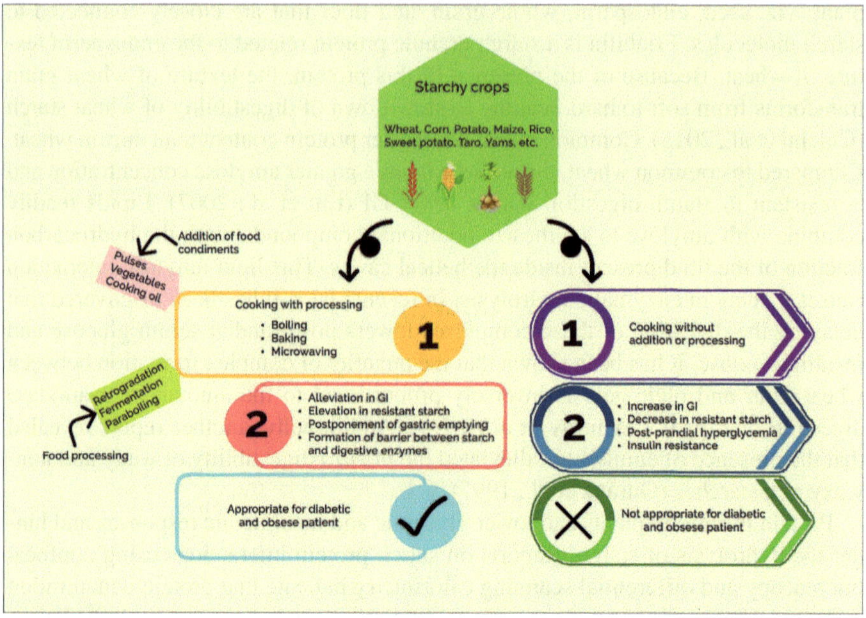

Fig. 12.4 Relationship between food processing and glycemic index

process and hydration. Investigating the functional qualities of starch viz. swelling, solubility, gel formation (commonly known gelatinisation), and retrogradation or refrigeration is essential for the purpose of comprehending the pathway of the GR and GI of starchy cereal grains and crops (Boers et al., 2015; Singh et al., 2020). The cooking of starchy cereal grains also ruptures the starch granules, releasing amylopectin and amylose molecules subjected to enzymatic hydrolysis. It was also noted that in the starchy foods, when cooked at high temperature, the inter-chain and intra-chain hydrogen bonds in amylopectin and amylose were collapsed, and the OH groups of cereal grain starch molecules were uncovered to water molecules. Therefore, the glucan chain should be increased upon gel formation. The GI value of cereal grains and starch hydrolysis rate is dependent on the cooking time of starchy foods. It was found that the longer cooking time or extra-cooking of carbohydrates-rich foods at high temperatures, particularly thermal heating, can increase the GI value of foods. This happens may be due to the greater breakdown of amylose and amylopectin chains.

Kirpitch & Maryniuk (2011) found that pasta processed or cooked for a longer time leads to a higher GI value than lower time (5–10 min). The drying, such as sun-drying and microwave drying, also impacts the GI value of starchy foods. The dried cereal grain flours markedly varied the GI value than that of the unprocessed flour. Likewise, Akyereko et al. (2020) reported that the sun-dried/solar-dried cassava flour was found to possess GI values of 40.20 and 61.11, respectively. It is widely believed that it was partially due to the loss of water content from the foods, leading to reduce the digestibility of starch molecules and apparent starch gelatinization. The type of rice, the degree of polishing, and the conditions of polishing all have an effect on the GI value and starch structure of rice. Thus, it is generally known that polished rice (white basmati rice) has a significantly higher GI value than that of unpolished rice (brown basmati rice). The higher GI value of white basmati rice is attributable to the loss of bioactive phytochemicals upon polishing, mainly concentrated in the bran portion of rice and other cereal grains (Somaratne et al., 2017). Parboiling of rice, for example, increases RS content in rice (which modifies rice's physical and chemical properties) and causes the starch to be digested more slowly, leading to a decline in GR (Kale et al., 2015). Presently, Caballero-Rothar et al. (2022) investigated the effect of cooking on GI with different varieties of maize, and they revealed that cooking with sodium sulfite markedly improved starch hydrolysis in all varieties (13%), showing the effect of disulfide bonds on this property. However, they also revealed that the amylose: starch ratio had no impact on starch digestibility. Baking foods is another cooking method whereby the dry heating process gradually heats the foodstuff at a high temperature. Recently, microwave cooking has been applied for the cooking of various foodstuffs. This new way of cooking is widespread due to its convenience and speed (Rizkalla et al., 2007). So far, the impact of baking on cereal grain starch granules, amylopectin and amylose molecular structure, and composition and digestibility is unexplored; thereby, this is a huge research gap.

12.5.2.2 Chilling/Retrogradation

As we know, cereal grains' starch has a particular chemical arrangement and properties in terms of retrogradation. Retrogradation is the mechanism of re-arrangement of branched-chain amylopectin and amylose during/after cooling, and thereby gelatinisation develops a well-arranged configuration. This mechanism has great importance because the starchy foods have been shown to have significant nutritional value, and subsequently, the retrograded and gelatinised cereal grain starch is less susceptible to enzymatic hydrolysis and targeted delivery of glucose residues in the digestive tract and respiratory system (Wang & Copeland, 2013). Retrogradation occurs within the starch granules of cereal grains. A range of alterations is observed, such as gel formation of paste, increase in viscosity, and turbidity of cereal grain starch molecules. Moreover, in the process of retrogradation, isolated amylose develops the double-helical formation of 70 to 40 residues of glucose molecule by hydrogen bonding (Wang et al., 2015). So far, rice starch has been explored in terms of molecular mechanisms of retrogradation. Li et al. (2014) reported that multiple rice varieties with the differential appearance of SSIIa and Waxy genes (considered for amylopectin and amylose formation, correspondingly) also varied owing to the degree of retrograded starch. The transformation of quickly metabolised starch to gradually metabolised starch is the mechanism underlying the reduction in GI.

12.5.2.3 Parbioling, Washing, and Soaking

The methods of processing viz., washing, soaking, and parboiling can affect the starch properties (i.e., gelatinisation, gel formation, retrogradation, pasting characteristics), nutritive values, and physical attributes of cereal grains that eventually change the GI value of cereal grains (Boers et al., 2015; Paiva et al., 2016). Soaking, a cutting-edge phase during the whole processing of parboiling, can change the arrangement of nutrients and composition of cereal grains by hydration. Another method for modifying starch granules' configuration, molecular alteration, and proximate composition of cereal grains (particularly rice) is steaming. Kale et al. (2017) found that the parboiled rice decreased gelatinisation and augmented hydration at a lower temperature. Recently, the effect of parboiling, polishing hydrothermal treatment, and germination on GI value and bioactive compound of selected landrace rice variety was checked (Kongkachuichai et al., 2020). The findings of this investigation unveiled that polishing reduced the bioactive compounds of rice. In contrast, germination and hydrothermal treatment improved the bioactive compounds (ferulic acid) and decreased the GI value of the selected landrace rice variety. Thus, parboiled rice has medium GI and could be the best candidate for health, followed by germinated parboiled rice and brown rice. Fermentation is another processing and preservation technique responsible for reducing the GI of foods (Rizkalla et al., 2007). Because of the higher oxidative depolymerisation of amylose, the amylose content of fermented cassava dough was higher (Singh et al., 2015). However, the influence of fermentation on cereal grains and cereal grain-based functional foods is still unexplored.

12.5.3 Impact of Diverse Ecological Factors on GI and Starch Digestibility

Starch accumulation is believed to be strongly influenced by the genotype and environment it occurs (Fig. 12.5). Gene, environment, and genotype-environment interaction can play a key role in understanding starch production, just like other essential agronomic variables (Bach et al., 2013). There are a variety of pressures that plants face, and they respond in various ways. The severity of the stress and the genotype's intrinsic sensitivity to stress determine any specific response. The endosperm cell count is frequently lowered during grain development, which reduces the starch content and increases starch digestibility in mature rice grains due to these factors. Stress directly impacts enzyme activity associated with starch production, which modifies the granule structure and lowers the amount of starch at harvest. There's now a general consensus that climate change will negatively impact agriculture in both quantitative and qualitative terms. Rice starch quality characteristics affected by environmental changes, such as milling and cooking quality, GI value, appearance, and taste, are all critical considerations when using this rice.

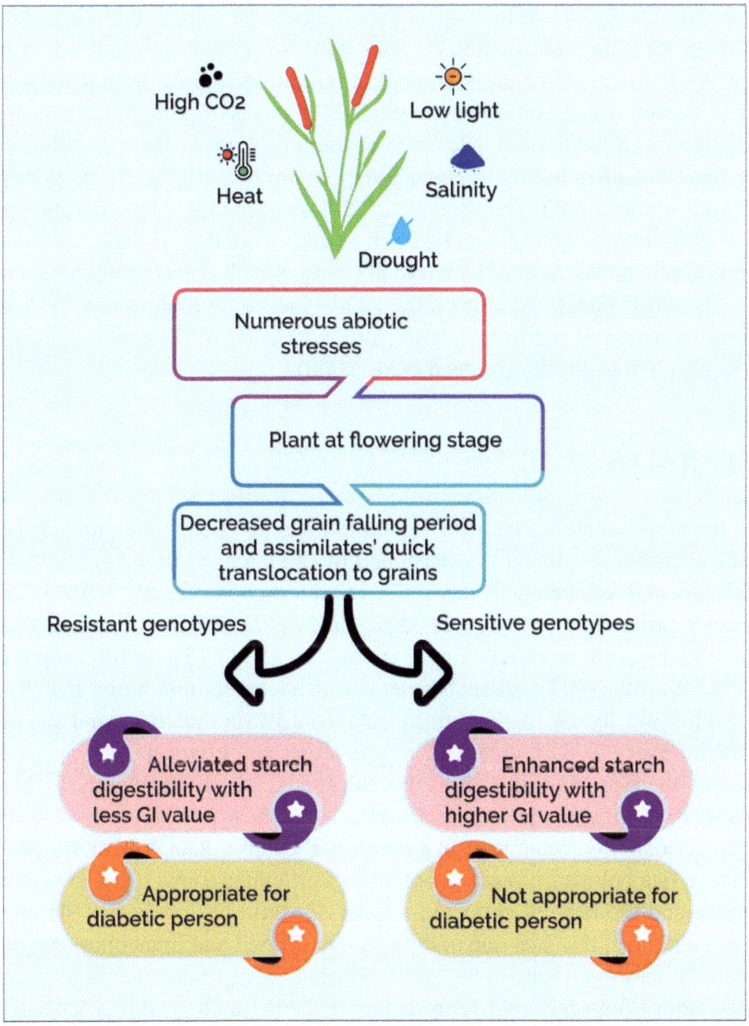

Fig. 12.5 Relationship between biotic and abiotic stresses and glycemic index

12.5.3.1 Cold Stress

Compared to high temperatures, low temperatures have little impact on starch production in cereal crops. Cold stress has been shown to affect starch production in several subtropical species. Cold stress slowed grain filling and prolonged the time of starch production in rice, subsequently changing the starch content. Thus, most of the starch-producing enzyme activity was altered only slightly. Low temperatures activate GBSSI (Granule bound starch synthase I) that possesses good amylose content and may modify the composition of amylose: amylopectin ratio in wheat and rice (Singh et al., 2010), and the activity of GBSSI is controlled at a post-transcriptional level. When wheat experiences cold shock, there is a synthesis of late maturity α-amylase, which elevates its iso-electric point with a less falling number. While the gelatinization temperature of all wholemeal flour samples made from cold-treated wheat grains decreased slightly, this is improbably to have a negative impact on the final result. The current findings suggest that the hypothesis is correct, and it's time to rethink if late maturity α-amylase is a significant contributor to poor end-product quality (Neoh et al., 2020). As a result, cold temperatures affect wheat's vegetative and reproductive growth, resulting in a drop in yield. Additionally, temperature affects the ratio of starch to amylose, which could be due to decreased activity of other enzymes, especially ADP-glucose pyrophosphorylase, starch branching enzyme and sucrose synthase, and increased GBSSI activity, resulting in a shift in starch digestibility (Ahmed et al., 2008).

12.5.3.2 Heat Stress

Temperatures of 20–30 °C are ideal for subtropical and temperate grain crops. The influence on amylose content is determined by the intensity of the genotypes' heat susceptibility and resistance. Compared to other temperate cereals, higher temperatures have a variety of consequences on rice and maize. Wheat is the most vulnerable cereal to temperatures exceeding this range, with 10–15% yearly output losses worldwide (Burrell, 2003). Wheat is susceptible to temperatures above the 20–30 °C range, with yield losses ranging from 10% to 15% in Australia and the United States. With every 5 °C increase in temperature, some wheat cultivars can lose 10–15% of their production (Thitisaksakul et al., 2012). Temperatures between 30 °C and 40 °C attenuated endosperm starch in wheat and barley by 2–33% and 13–33%, respectively, when kept in a controlled environment (Liu et al., 2011). In the japonica rice cultivator, seeds grown at higher temperatures had a lower amount of amylose and the ratio of short to long amylopectin chains than seeds grown at lower temperatures; this was due to the fact that GBSSI and branching enzyme levels were relatively low (Kato et al., 2019). During the reproductive stage of cereals, the unfavourable impact of high temperature (HT) on starch synthesis is particularly severe and prevalent. At the moment, On a genomic level, HT suppresses the expression of numerous genes. SSRGs cause chalky grains and poor starch buildup by activating the gene expression of α-amylase (Huang et al., 2021). Thus, the yield was reported to be lowered at high temperatures, which was linked to decreased activity of starch-producing enzymes. High temperature also showed a negative impact on wheat growth by increasing the level of the isoamylase 2 enzyme. Here, the major reason for starch content alleviation was declined activity and expression of key enzymes involved in converting sucrose to starch (Lu et al., 2019). At high temperatures over 35 °C, dry maize weight was reduced, whereas the content of starch was reduced by 3.0–3.3%, and starch accumulation was reduced by 22.2–21.8% with an increase in glutamate synthase activity (Yang et al., 2018). Raised temperature also impacts endosperm ADP-glucose pyrophosphorylase activity, with thermal inactivation potentially dislodging the part of starch synthase in maize, consequently leading to decreased starch production.

12.5.3.3 Carbon Dioxide (CO_2) Levels

CO_2 levels in the atmosphere have augmented by 35% as a result of the industrial revolution and are expected to increase quadruple by the end of the century. Initially, it was thought that this occurrence would boost crop yields. Higher carbon would be moved to the grain by CO_2, resulting in more starch accumulation. Wheat grown in high CO_2 concentrations of more than 550 ppm had a higher number of grains but no alteration in starch accumulation (Thitisaksakul et al., 2012). When CO_2 levels were augmented from 350 to 700 ppm under ideal conditions, several wheat crop studies predicted an average yield rise of up to 31%. (Lal et al., 2019). Starch is a significant storage carbohydrate enhanced in plants developing in elevated CO_2. The rise in starch likely contributes to the elevated levels of sucrose observed with increased CO_2 due to the overnight conversion of starch to sucrose (Thompson et al., 2017). Plants developed in high CO_2 environments had a higher initial viscosity of starch, which resulted in a harder cooked grain; however, the effect on amylose content varied between experiments, suggesting that it could impact starch digestion and, hence, the GI value of cereal grains. The smaller grain size observed under these conditions would have a negative effect on the quality of rice grain milled under these conditions (Jongkaewwattana & Geng, 2001).

12.5.3.4 Stress in Low Light

Low levels of illumination throughout the winter season are a major abiotic stressor, causing starch biosynthesis to suffer and rice grain yields to plummet. Grain growth and starch synthesis were both reduced by roughly 20% when light intensity (shading) was reduced throughout the grain filling stage. When enzyme activity was assessed on unit grain weight rather than activity per grain, the reduced light intensity did not affect starch phosphorylase. However, shade reduced starch synthase I activity at the subsequent stages of grain development (Mengel & Judel, 1981). Overall, when the amylose concentration in the grain decreases, grain yield decreases, leading to higher GI values in plants exposed to the lower intensity of light (Kumar et al., 2019). Low levels of illumination harmed photosynthate resources in rising cereal grains, resulting in a decrease in starch synthase activity and, as a result, grains that are not completely filled with low starch composition (Li et al., 2006). In the starch biosynthetic pathway, GBSSI is the main enzyme playing a key role in amylose production, a protein found in growing grains (Kumar et al., 2020c), and is one of the most well-studied enzymes in the starch synthesis pathway, with a pivotal role in amylose production that affects GI by changing the amount of amylose and RS in rice grains. However, only a few research investigations have been conducted to investigate the control of the GBSSI gene. The regulation of genes involved in starch production is particularly interesting, which is susceptible to a cyclical alternating (day/night) for production and breakdown in photosynthetic tissues (Tenorio et al., 2003). In *Arabidopsis*, however, a light-mediated regulation of the stable amounts of mRNA of the starch binding enzyme IIa and starch binding enzyme IIb genes, which encode starch branching enzyme isoforms, has been discovered (Tenorio et al., 2003). The waxy gene has been linked to rice GI via a single nucleotide polymorphism (SNP) at the splicing site of intron 1 of GBSSI (G/T), which discovered amylose is significantly associated with RS (Bao et al., 2017). GI values were higher in low-light susceptible genotypes with lower GBSSI expression during the early and late stages of grain growth. This shows a relationship between amylose and starch digestibility.

12.5.3.5 Drought Stress

The largest barrier to agricultural production is water stress during grain development. Drought stress during the grain growth stage is the leading cause of reduced crop yields. More than half (61%) of all crop-producing areas received less than 500 mm of precipitation in 2009, which is significantly below the threshold required for agricultural production farming on arid land (Thitisaksaku et al., 2012). In wheat grains under drought stress, six genes, including AGPL2, AGPS2, PHOL, GBSSI, BEIII, and ISAI, were down-regulated as a result of the decrease in the reformation of sucrose to starch, affecting the starch content (Lu et al., 2019). As water-stressed cereals display fluctuation in starch biosynthetic enzyme activity, lower grain starch composition is one of the key causes that diminish grain production during drought. With the blooming stage of sorghum grains, starch, amylase, and amylopectin accumulation alleviated during drought stress at the mid-late stage of grain filling. It was discovered that the actions of soluble starch biosynthesis pathway, granule-bound starch synthase, starch branching enzyme, and the activities of starch de-branching enzymes were all affected during grain filling, although in various ways (Bing et al., 2014). Hence, drought stress slowed the accumulation of starch to a greater extent. Drought is a common occurrence due to climate change, which impacts the flowering stage, resulting in a maximum yield decrease. Whilst terminal drought stress continues to accelerate grain filling, it also causes a decrease in amylose content, which results in loose packing of starch granules in the endosperm that ultimately results in chalkiness and thus poor milling quality in the grain (Yang et al., 2018). Drought stress significantly reduced waxy maize kernel development, starch accumulation, and starch biosynthetic enzyme activity. The starch granules' shape changed, and the rate of grain filling rate was increased (Guo et al., 2021).

12.5.3.6 Salinity Stress

High salt concentrations in cereal grains induce osmotic stress after a limited exposure and ion toxicity after prolonged exposure. Because of its increased solubility (40 mM is deemed troublesome for agriculture), sodium chloride (NaCl) is the key source of concern. During the day, the starch content of NaCl-stressed rice (*Oryza sativa* L.) seedling dropped. Because the high salt did not influence photosynthetic efficiency or starch breakdown enzyme activity, this impact is still most probably due to starch inhibitory activity production with repression of GBSSI activity. In contrast, other enzymes were less affected (Chen et al., 2008). Cereal grains' (like rice) amylose concentration has been decreased by high salt stress, affecting starch digestion. The amount of change in amylose content was determined by salt quantities and a variety of genotypes, but it was not measured by genotype salt sensitivity. Under field circumstances, 40 mM NaCl reduced grain amylose concentration by 7–11%(Siscar-Lee et al., 1990). Tomato (*Solanum lycopersicum*) fruit sugar buildup was also aided by salinity stress. Here, in the early development of tomato fruits, growth with 160 mM NaCl quadrupled starch accumulation in comparison to control plants, and soluble sugars were enhanced as the fruit grew by increasing activity of ADP-glucose pyrophosphorylase and sucrose transporter expression (Yin et al., 2010).

12.5.4 Impacts on GI of Starchy Grains by the Integration of Diverse Food Constituents

It is widely known that cereals, including rice and wheat, are ingested by numerous people globally and are a significant source of both carbohydrates and energy. Steamed or cooked rice is the most common method of eating rice by combining other food constituents (vegetables, legumes, pulses, nuts, seafoods, and meats). Hence, the GI of rice is likely to differ from the GI of the combined meal that contains more protein, fiber, and fat. According to Virkamäki et al. (1999), the mechanism of T2DM and carbohydrate intake is not simple since other dietary ingredients found in the diet could also alter its ability to boost blood glucose levels. There is some data on the influence of various food components encompassing traditionally eaten fruits, vegetables, cooking fats/oils, pulses, and other ingredients on starch digestion. In the study led by Kumar et al. (2018b), it was identified that the GI and GR values of rice combined with cooking fats/oils and pulses were lesser than that of rice alone in terms of GI variability. Food mixtures with a low GI are better for diabetes patients because they are linked with lower blood sugar levels (Fig. 12.6).

Oils and fats lower the GI and GR through the postponement of gastric emptying and elevating the incretins' release. Human gut-derived hormones and members of the glucagon superfamily (glucagon receptors) are released as a result of nutrient ingestion, notably glucose and fat. In contrast, by boosting insulin release and decreasing stomach emptying, the protein lowers GR. Protein-rich legumes and pulses have long been recognised for their low glycemic GI. Dried legumes have been shown to alleviate the GR in starchy diets (Hätönen et al., 2011). GI values were lower in the meal having the high amylose starch (mung beans) than the meal containing the high amylopectin starch (waxy corn), according to Kabir et al. (1998), when cereal grains starches with varied amylopectin: amylose

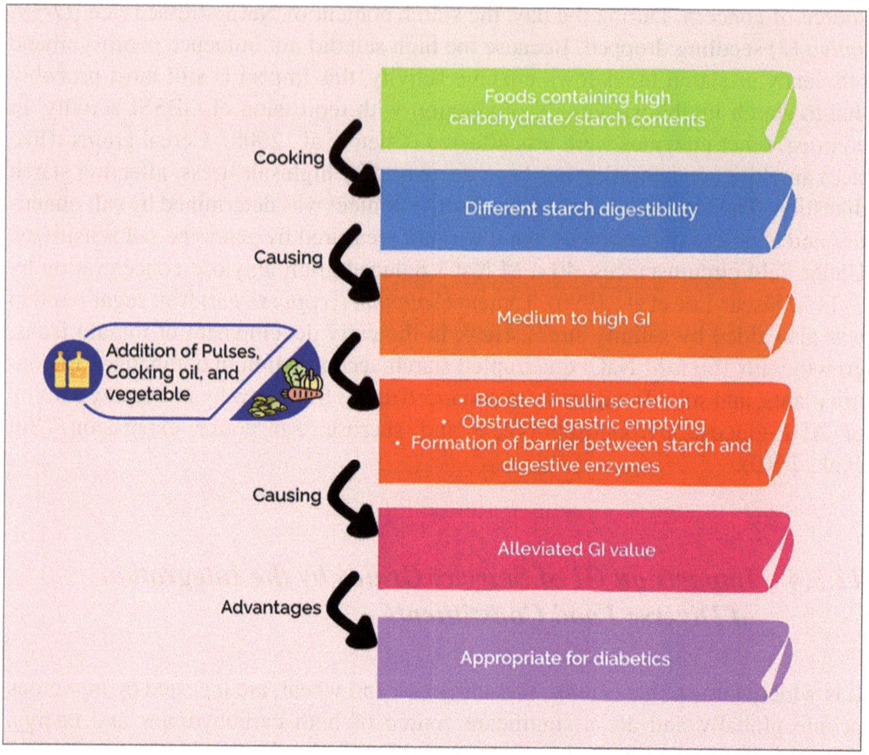

Fig. 12.6 Relationship between different food ingredients and glycemic index

concentrations of 32:68% for mung bean and 0:99% for waxy corn were combined. People with T2DM were advised to consume 60% white rice, with legumes, nuts, and seeds accounting for the remaining 40% of the meal to keep blood glucose levels under control. Non-insulin-dependent diabetics can benefit from the stimulatory impact of chickpea on insulin secretion (Chandalia et al., 1992). Furthermore, pigeon pea was found to have the greatest impact on decreasing rice's GI value when combined with one of six different pulses in a study assessing the GI value of varying food combinations (Kumar et al., 2020a). Fat/oil added with a grain like rice attenuates the GR because the gastric emptying time and incretins' release is postponed. The fiber in carbohydrate-rich diets, on the other hand, inhibits the creation of these complexes *in vitro*, allowing the enzyme and substrate to have a better interaction. Moreover, as rice was cooked, the GI value of the rice grain decreased significantly when clarified butter (ghee) was added (Kumar et al., 2018b). Foods rich in carbohydrates, such as rice, have a lower GR when paired with fats and oils, whilst mustard oil didn't have a considerable impact, possibly because of the glucosinolate's presence. By creating an obstruction between starch and digestive enzymes, the insoluble or less soluble fiber in the combined meals could alleviate the GI value by postponing gastric emptying (Jacobs Jr & Steffen, 2003).

In comparison with other vegetables, cauliflower, fenugreek leaves, and seeds had the most significant GI-lowering influence on rice meals (Kumar et al., 2020a). Several earlier findings revealed that vegetables such as okra and drumstick leaves reduce the GR of foods. When okra (10%) was added to rice noodles, blood glucose levels were significantly reduced (Kang et al., 2018). Even while the GI of drumstick leaves greatly alleviated postprandial hyperglycemia, GI is unrelated to vegetables with low starch content (Hätönen et al., 2011). When mashed potatoes were served with broccoli, the glycemic and insulinemic responses were significantly reduced compared to when potatoes were served alone. The DF and RS that broccoli adds to a potato diet may assist in decreasing the GI value (Ballance et al., 2018).

12.6 Future Perspective, Challenges, and Opportunities

Obesity and obesity-related diseases (i.e., hyperlipidemia and hyperglycemia) and their complications are on the escalation globally. Functional low-GI and GL cereal grain foods and beverages may be implemented into several of these illnesses' preventative and/or management regimens, contributing to preventing worldwide diseases. However, there are still doubts about the GI's universal applicability. A huge question is which features of the person will make it more susceptible to changes in the GI, GL, and GR of the foodstuffs. Over the last two decades, a pattern has emerged in the research of GI that those with augmented insulin resistance evaluated as higher postprandial insulin and higher BMI in the prevalence of obesity-related diseases, including hyperlipidemia and hyperglycemia. It was noted that many cereal grains are considered high GI foods. However, there is a huge opportunity to modify the breeding methods and produce high amylose/resistant starch cereal grains using genetic engineering that will be useful to develop low GI cereal grain functional foods (Caballero-Rothar et al., 2022). Purposely, many efforts were made to develop low GI cereal grains, viz. rice, having GI less than 55. But, the key obstacle in this research is that, in cereal grains like rice and maize, reducing GI to less than 55 is still challenging. Therefore, more work should be needed in the future. A meta-analysis about hyperglycemia diets should be conducted worldwide that international health-related organizations have funded.

The paucity of nutritious low-GI foods and beverages available for purchase is generally perceived as a hurdle to adopting the GI diet (Augustin et al., 2015). Although if they are available, identifying healthy low-GI alternatives among several numbers of food options offered at a typical store is a possible challenge. Certain dietary strategies that enhance blood glucose control while also lowering CHD risk and risk indicators should be prioritized. One of these strategies could be to employ low GI foods to slow down the pace of carbohydrate digestion, absorption, and metabolism. The GI's significance is still being critiqued. To effectively employ GI, techniques for offering customers and healthcare professionals with precise and trustworthy information about the GI of foods must be designed; at the absolute least, this will entail the standardisation and regulation of GI laboratories.

There are substantial disparities in the diets consumed in diverse European countries. This diversity could be advantageous to nutritional studies because this could help to define the role of GI in health. Likewise, the amount of energy generated from carbohydrates and the properties of carbohydrate diets varied significantly among regions. As a result, regional GI databases might well be designed to examine GI exposure levels and disease hazards with greater accuracy. Additional measures are necessary to adequately update the information on diets, perceive current information, and conceptual framework analysis to evaluate the relationships between health and diet and their factors (Augustin et al., 2015). The current book chapter could help cereal-based functional food producers create boosted low-GI functional foods with high nutritional value. In fact, the industrial objective ought to be functional foods with a high nutritional value and a low GI while also addressing certain consumer groups (e.g., CVD or T2DM patients). Further human intervention investigations are vital to better understand the efficacy of low GI functional foods in managing chronic diseases such as CVDs or T2DM. There is potential for the development of functional food products containing these modified low GI ingredients to ensure the health of consumers with CVDs and T2DM, as well as to assist healthy people in maintaining their normal blood glucose levels.

12.7 Conclusion

Diet is a vital component in the management of diabetes (Kastorini et al., 2011). GI can be a good indicator to make healthy food choices (Geetha et al., 2020). Cereal grains are considered the staple diet globally. The intake of low GI cereal grains has significantly reduced the pandemic of obesity and obesity-related diseases (hyperlipidemia and hyperglycemia). The evaluation of GI value has mainly been tested through *in vivo* and *in vitro* methods. It was found that *in vitro* test is easy, quick, and inexpensive than that *in vivo* test. It needs fewer instruments, labour, and infrastructure to estimate a significant number of samples in minimum time. However, sometimes *in vivo* tests are required to validate the health claims of food bioactive compounds. It was observed that the inclusion of other food ingredients could change the GI value of cereal grain functional foods by altering digestibility. Moreover, the intrinsic factors, viz. protein, resistant starch, amylose, amylopectin, anti-nutrient compounds (PA), DF, and lipid are insignificantly linked with the GI value of cereal grain foods. Extrinsic factors such as thermal processing (baking, frying, boiling, and pressure cooking), cooling, parboiling, and polishing (rice) also impact the GI of cereal grain foods. The changes in the value of GI may be due to the chemical, physical and enzymatic modifications of cereal grains. It was found that cooking with high pressure and temperature could increase the GI value of starchy cereal grain functional foods. However, the research works related to this claim are seldom. On the contrary, when white rice is boiled and then chilled, for example, to be used in a rice salad, the digestibility of the starch changes and the rice's GI is reduced, and it comes into the category of a low-GI food. The carbohydrate-rich cereal grain foods are routinely consumed with other food ingredients that eventually affect the RS and GI food mixtures. Moreover, the starch synthesis is extremely altered by abiotic and biotic stress, affecting cereal grain functional foods' GL and GI values. After extensive reading of the research literature, it was concluded that the cereal grain functional foods must be consumed with the combination of different other ingredients of foods, viz. fiber, protein, and unsaturated fat/oil. This aids decrease the postprandial blood glucose level, ultimately reducing the risk of various chronic diseases, including diabetes, obesity, hypertension, CVDs, and even some types of cancer (colorectal cancer). Therefore, it is the need of the hour to choose the ancient practice of eating multigrain cereals with combinations of fiber-rich fruits and vegetables.

Conflicts of Interest The authors confirm that they have no known competing financial interests or personal relationships that could have influenced the manuscript.

Authors' Contributions Muhammad Usman & Prasanna J Patil: Conceptualization; Data curation; Visualization; Writing – original draft; Writing – review & editing. Devashree N Patil, Arshad Mehmood, Haroon Shah, Syeda Mahvish Zahra, Zeshan Ali & Sehrish Nasreen: Writing – review & editing.

References

Abbasi, N. N., Purslow, P. P., Tosh, S. M., & Bakovic, M. (2016). Oat β-glucan depresses SGLT1- and GLUT2-mediated glucose transport in intestinal epithelial cells (IEC-6). *Nutrition Research, 36*(6), 541–552.

Ahmad, M., Dar, Z. A., & Habib, M. (2014). A review on oat (Avena sativa L.) as a dual-purpose crop. *Scientific Research and Essays, 9*(4), 52–59.

Ahmed, N., Maekawa, M., & Tetlow, I. J. (2008). Effects of low temperature on grain filling, amylose content, and activity of starch biosynthesis enzymes in endosperm of basmati rice. *Australian Journal of Agricultural Research, 59*(7), 599–604.

Akyereko, Y. G., Wireko-Manu, F. D., & Oduro, I. (2020). Influence of processing methods on food components and glycaemic index of cassava-based traditional foods. *Journal of Food and Nutrition Sciences, 8*, 6.

Aldughpassi, A., Abdel-Aal, E. S. M., & Wolever, T. M. (2012). Barley cultivar, kernel composition, and processing affect the glycemic index. *The Journal of Nutrition, 142*(9), 1666–1671.

Ali, N., Paul, S., Gayen, D., Sarkar, S. N., Datta, S. K., & Datta, K. (2013). RNAi mediated down regulation of myo-inositol-3-phosphate synthase to generate low phytate rice. *Rice, 6*(1), 1–12.

Ames, N., Malunga, L. N., Mollard, R., Johnson, J., Chu, Y., & Thandapilly, S. J. (2021). Effect of processing on oat β-glucan viscosity, postprandial glycemic response and subjective measures of appetite. *Food & Function, 12*(8), 3672–3679.

Anderson, J. W., Randles, K. M., Kendall, C. W., & Jenkins, D. J. (2004). Carbohydrate and fiber recommendations for individuals with diabetes: A quantitative assessment and meta-analysis of the evidence. *Journal of the American College of Nutrition, 23*(1), 5–17.

Annor, G. A., Marcone, M., Corredig, M., Bertoft, E., & Seetharaman, K. (2015). Effects of the amount and type of fatty acids present in millets on their in vitro starch digestibility and expected glycemic index (eGI). *Journal of Cereal Science, 64*, 76–81.

Ape David, I., Nwongu Nwanyinnaya, A., Uwakwe Ezekiel, I., & Ikedinobi Chinelo, S. (2016). Comparative proximate analysis of maize and sorghum bought from Ogbete main market of Enugu State, Nigeria. *Greener Journal of Agricultural Sciences, 6*(9), 272–375.

Araya, H., Contreras, P., Alvina, M., Vera, G., & Pak, N. (2002). A comparison between an in vitro method to determine carbohydrate digestion rate and the glycemic response in young men. *European Journal of Clinical Nutrition, 56*(8), 735–739.

Atkinson, F. S., Foster-Powell, K., & Brand-Miller, J. C. (2008). International tables of glycemic index and glycemic load values: 2008. *Diabetes Care, 31*(12), 2281–2283.

Augustin, L. S., Kendall, C. W., Jenkins, D. J., Willett, W. C., Astrup, A., Barclay, A. W., … Poli, A. (2015). Glycemic index, glycemic load and glycemic response: An international scientific consensus summit from the international carbohydrate quality consortium (ICQC). *Nutrition, Metabolism and Cardiovascular Diseases, 25*(9), 795–815.

Bach, S., Yada, R. Y., Bizimungu, B., Fan, M., & Sullivan, J. A. (2013). Genotype by environment interaction effects on starch content and digestibility in potato (Solanum tuberosum L.). *Journal of Agricultural and Food Chemistry, 61*(16), 3941–3948.

Ballance, S., Knutsen, S. H., Fosvold, Ø. W., Wickham, M., Trenado, C. D. T., & Monro, J. (2018). Glyceamic and insulinaemic response to mashed potato alone, or with broccoli, broccoli fibre or cellulose in healthy adults. *European Journal of Nutrition, 57*(1), 199–207.

Bao, J., Zhou, X., Xu, F., He, Q., & Park, Y. J. (2017). Genome-wide association study of the resistant starch content in rice grains. *Starch-Stärke, 69*(7–8), 1600343.

Barclay, A. W., Petocz, P., McMillan-Price, J., Flood, V. M., Prvan, T., Mitchell, P., & Brand-Miller, J. C. (2008). Glycemic index, glycemic load, and chronic disease risk—A meta-analysis of observational studies. *The American Journal of Clinical Nutrition, 87*(3), 627–637.

Bing, Y. I., Zhou, Y. F., Gao, M. Y., Zhang, Z., Yi, H. A. N., Yang, G. D., … Huang, R. D. (2014). Effect of drought stress during flowering stage on starch accumulation and starch synthesis enzymes in sorghum grains. *Journal of Integrative Agriculture, 13*(11), 2399–2406.

Birt, D. F., Boylston, T., Hendrich, S., Jane, J. L., Hollis, J., Li, L., … Whitley, E. M. (2013). Resistant starch: Promise for improving human health. *Advances in Nutrition, 4*(6), 587–601.

Biskup, I., Kyrø, C., Marklund, M., Olsen, A., van Dam, R. M., Tjønneland, A., … Landberg, R. (2016). Plasma alkylresorcinols, biomarkers of whole-grain wheat and rye intake, and risk of type 2 diabetes in Scandinavian men and women. *The American Journal of Clinical Nutrition, 104*(1), 88–96.

Boers, H. M., Ten Hoorn, J. S., & Mela, D. J. (2015). A systematic review of the influence of rice characteristics and processing methods on postprandial glycaemic and insulinaemic responses. *British Journal of Nutrition, 114*(7), 1035–1045.

Bondia-Pons, I., Nordlund, E., Mattila, I., Katina, K., Aura, A. M., Kolehmainen, M., … Poutanen, K. (2011). Postprandial differences in the plasma metabolome of healthy Finnish subjects after intake of a sourdough fermented endosperm rye bread versus white wheat bread. *Nutrition Journal, 10*(1), 1–10.

Brand-Miller, J., Hayne, S., Petocz, P., & Colagiuri, S. (2003). Low–glycemic index diets in the management of diabetes: A meta-analysis of randomized controlled trials. *Diabetes Care, 26*(8), 2261–2267.

Burrell, M. M. (2003). Starch: the need for improved quality or quantity-an overview. *Journal of Experimental Botany, 54*(382), 451–456.

Caballero-Rothar, N. N., Borrás, L., & Gerde, J. A. (2022). Physical and chemical kernel traits affect starch digestibility and glycemic index of cooked maize flours. *Food Chemistry, 369*, 130953.

Chandalia, H., Neogi, R., & Mehta, S. (1992). Mechanism of low glycemic index of pulses and pulse-incorporated cereal foods. *Insulin, 66*, 118–116.

Chandra, D., Chandra, S., & Sharma, A. K. (2016). Review of finger millet (Eleusine coracana (L.) Gaertn): A power house of health benefiting nutrients. *Food Science and Human Wellness, 5*(3), 149–155.

Chen, H. J., Chen, J. Y., & Wang, S. J. (2008). Molecular regulation of starch accumulation in rice seedling leaves in response to salt stress. *Acta Physiologiae Plantarum, 30*(2), 135–142.

Chiasson, J. L., Josse, R. G., Gomis, R., Hanefeld, M., Karasik, A., Laakso, M., & STOP-NIDDM Trial Research Group. (2002). Acarbose for prevention of type 2 diabetes mellitus: The STOP-NIDDM randomised trial. *The Lancet, 359*(9323), 2072–2077.

Chiasson, J. L., Josse, R. G., Gomis, R., Hanefeld, M., Karasik, A., Laakso, M., … STOP-NIDDM Trial Research Group. (2003). Acarbose treatment and the risk of cardiovascular disease and hypertension in patients with impaired glucose tolerance: The STOP-NIDDM trial. *JAMA, 290*(4), 486–494.

Chichti, E., George, M., Delenne, J. Y., & Lullien-Pellerin, V. (2015). Changes in the starch-protein interface depending on common wheat grain hardness revealed using atomic force microscopy. *Plant Science, 239*, 1–8.

Chillo, S., Ranawana, D. V., Pratt, M., & Henry, C. J. K. (2011). Glycemic response and glycemic index of semolina spaghetti enriched with barley β-glucan. *Nutrition, 27*(6), 653–658.

Chiu, C. J., Liu, S., Willett, W. C., Wolever, T. M., Brand-Miller, J. C., Barclay, A. W., & Taylor, A. (2011). Informing food choices and health outcomes by use of the dietary glycemic index. *Nutrition Reviews, 69*(4), 231–242.

Choi, Y., Giovannucci, E., & Lee, J. E. (2012). Glycaemic index and glycaemic load in relation to risk of diabetes-related cancers: A meta-analysis. *British Journal of Nutrition, 108*(11), 1934–1947.

Cummings, J. H., Edmond, L. M., & Magee, E. A. (2004). Dietary carbohydrates and health: Do we still need the fibre concept? *Clinical Nutrition Supplements, 1*(2), 5–17.

de Morais Cardoso, L., Pinheiro, S. S., Martino, H. S. D., & Pinheiro Sant'Ana, H. M. (2017). Sorghum (Sorghum bicolor L.): Nutrients, bioactive compounds, and potential impact on human health. *Critical Reviews in Food Science and Nutrition, 57*(2), 372–390.

Deepa, G., Singh, V., & Naidu, K. A. (2010). A comparative study on starch digestibility, glycemic index and resistant starch of pigmented ('Njavara'and 'Jyothi') and a non-pigmented ('IR 64′) rice varieties. *Journal of Food Science and Technology, 47*(6), 644–649.

Duodu, K. G., & Awika, J. M. (2019). Phytochemical-related health-promoting attributes of sorghum and millets. In *Sorghum and Millets* (pp. 225–258). AACC International Press.

Englyst, K. N., Vinoy, S., Englyst, H. N., & Lang, V. (2003). Glycaemic index of cereal products explained by their content of rapidly and slowly available glucose. *British Journal of Nutrition, 89*(3), 329–339.

FAO/WHO. (1998). Carbohydrates in human nutrition: Report of a joint FAO/WHO expert consultation, Rome, 14–18 April 1997. In *Carbohydrates in human nutrition: Report of a joint FAO/WHO expert consultation* (pp. 14–18). World Health Organization: Food and Agriculture Organization of the United Nations.

Ford, E. S., & Mokdad, A. H. (2008). Epidemiology of obesity in the Western hemisphere. *The Journal of Clinical Endocrinology & Metabolism, 93*(11_supplement_1), s1–s8.

Fuentes-Zaragoza, E., Riquelme-Navarrete, M. J., Sánchez-Zapata, E., & Pérez-Álvarez, J. A. (2010). Resistant starch as functional ingredient: A review. *Food Research International, 43*(4), 931–942.

Geetha, K., Yankanchi, G. M., Hulamani, S., & Hiremath, N. (2020). Glycemic index of millet based food mix and its effect on pre diabetic subjects. *Journal of Food Science and Technology, 57*(7), 2732–2738.

Goff, L. M., Cowland, D. E., Hooper, L., & Frost, G. S. (2013). Low glycaemic index diets and blood lipids: A systematic review and meta-analysis of randomised controlled trials. *Nutrition, Metabolism and Cardiovascular Diseases, 23*(1), 1–10.

Goñi, I., Garcia-Alonso, A., & Saura-Calixto, F. (1997). A starch hydrolysis procedure to estimate glycemic index. *Nutrition Research, 17*(3), 427–437.

Granfeldt, Y., Björck, I., Drews, A., & Tovar, J. (1994). An in vitro procedure based on chewing to predict metabolic response to starch in cereal and legume products. *The American Journal of Clinical Nutrition, 59*(3), 777S–777S.

Greenwood, D. C., Threapleton, D. E., Evans, C. E., Cleghorn, C. L., Nykjaer, C., Woodhead, C., & Burley, V. J. (2013). Glycemic index, glycemic load, carbohydrates, and type 2 diabetes: Systematic review and dose–response meta-analysis of prospective studies. *Diabetes Care, 36*(12), 4166–4171.

Guo, J., Qu, L., Hu, Y., Lu, W., & Lu, D. (2021). Proteomics reveals the effects of drought stress on the kernel development and starch formation of waxy maize. *BMC Plant Biology, 21*(1), 1–14.

Guraya, H. S., Kadan, R. S., & Champagne, E. T. (1997). Effect of rice starch-lipid complexes on in vitro digestibility, complexing index, and viscosity. *Cereal Chemistry, 74*(5), 561–565.

Hadimani, N. A., & Malleshi, N. G. (1993). Studies on milling, physico-chemical properties, nutrient composition and dietary fibre content of millets. *Journal of Food Science and Technology (India), 30*(1), 17–20.

Hanefeld, M., Cagatay, M., Petrowitsch, T., Neuser, D., Petzinna, D., & Rupp, M. (2004). Acarbose reduces the risk for myocardial infarction in type 2 diabetic patients: Meta-analysis of seven long-term studies. *European Heart Journal, 25*(1), 10–16.

Hätönen, K. A., Virtamo, J., Eriksson, J. G., Sinkko, H. K., Sundvall, J. E., & Valsta, L. M. (2011). Protein and fat modify the glycaemic and insulinaemic responses to a mashed potato-based meal. *British Journal of Nutrition, 106*(2), 248–253.

Hirsch, S., Barrera, G., Leiva, L., de la Maza, M. P., & Bunout, D. (2013). Variability of glycemic and insulin response to a standard meal, within and between healthy subjects. *Nutricion Hospitalaria, 28*(2), 541–544.

Hou, Q., Li, Y., Li, L., Cheng, G., Sun, X., Li, S., & Tian, H. (2015). The metabolic effects of oats intake in patients with type 2 diabetes: A systematic review and meta-analysis. *Nutrients, 7*(12), 10369–10387.

Howlett, J., & Ashwell, M. (2008). Glycemic response and health: Summary of a workshop. *The American Journal of Clinical Nutrition, 87*(1), 212S–216S.

Huang, L., Tan, H., Zhang, C., Li, Q., & Liu, Q. (2021). Starch biosynthesis in cereal endosperms: An updated review over the last decade. *Plant Communications, 2*(5), 100237.

Hur, S. J., Lim, B. O., Decker, E. A., & McClements, D. J. (2011). In vitro human digestion models for food applications. *Food Chemistry, 125*(1), 1–12.

Hussain, S. Z., Beigh, M., Qadri, T., Ahmad, I., & Naseer, B. (2020). Development of low glycemic index crackers from water chestnut and barley flour. *British Food Journal., 122*, 3–1169.

Jacobs, D. R., Jr., & Steffen, L. M. (2003). Nutrients, foods, and dietary patterns as exposures in research: A framework for food synergy. *The American Journal of Clinical Nutrition, 78*(3), 508S–513S.

Järvi, A. E., Karlström, B. E., Granfeldt, Y. E., Björck, I. E., Asp, N. G., & Vessby, B. O. (1999). Improved glycemic control and lipid profile and normalized fibrinolytic activity on a low-glycemic index diet in type 2 diabetic patients. *Diabetes Care, 22*(1), 10–18.

Jenkins, D. J. A., Taylor, R. H., Goff, D. V., Fielden, H., Misiewicz, J. J., Sarson, D. L., … Alberti, K. G. M. M. (1981a). Scope and specificity of acarbose in slowing carbohydrate absorption in man. *Diabetes, 30*(11), 951–954.

Jenkins, D. J., Wolever, T. M., Taylor, R. H., Barker, H., Fielden, H., Baldwin, J. M., … Goff, D. V. (1981b). Glycemic index of foods: A physiological basis for carbohydrate exchange. *The American Journal of Clinical Nutrition, 34*(3), 362–366.

Jenkins, D. J., Wolever, T. M., Thorne, M. J., Jenkins, A. L., Wong, G. S., Josse, R. G., & Csima, A. (1984). The relationship between glycemic response, digestibility, and factors influencing the dietary habits of diabetics. *The American Journal of Clinical Nutrition, 40*(6), 1175–1191.

Jenkins, A. L., Jenkins, D. J. A., Zdravkovic, U., Würsch, P., & Vuksan, V. (2002). Depression of the glycemic index by high levels of β-glucan fiber in two functional foods tested in type 2 diabetes. *European Journal of Clinical Nutrition, 56*(7), 622–628.

Jenkins, D. J., Kendall, C. W., McKeown-Eyssen, G., Josse, R. G., Silverberg, J., Booth, G. L., … Leiter, L. A. (2008). Effect of a low–glycemic index or a high–cereal fiber diet on type 2 diabetes: A randomized trial. *JAMA, 300*(23), 2742–2753.

Jensen, L., Sloth, B., Krog-Mikkelsen, I., Flint, A., Raben, A., Tholstrup, T., … Astrup, A. (2008). A low-glycemic-index diet reduces plasma plasminogen activator inhibitor-1 activity, but not tissue inhibitor of proteinases-1 or plasminogen activator inhibitor-1 protein, in overweight women. *The American Journal of Clinical Nutrition, 87*(1), 97–105.
Jongkaewwattana, S., & Geng, S. (2001). Inter-relationships amongst grain characteristics, grain-filling parameters and rice (*Oryza sativa* L.) milling quality. *Journal of Agronomy and Crop Science, 187*(4), 223–229.
Juntunen, K. S., Laaksonen, D. E., Poutanen, K. S., Niskanen, L. K., & Mykkänen, H. M. (2003). High-fiber rye bread and insulin secretion and sensitivity in healthy postmenopausal women. *The American Journal of Clinical Nutrition, 77*(2), 385–391.
Kabir, M., Rizkalla, S. W., Champ, M., Luo, J., Boillot, J., Bruzzo, F., & Slama, G. (1998). Dietary amylose-amylopectin starch content affects glucose and lipid metabolism in adipocytes of normal and diabetic rats. *The Journal of Nutrition, 128*(1), 35–42.
Kale, S. J., Jha, S. K., Jha, G. K., Sinha, J. P., & Lal, S. B. (2015). Soaking induced changes in chemical composition, glycemic index and starch characteristics of basmati rice. *Rice Science, 22*(5), 227–236.
Kale, S. J., Kale, P. N., & Jha, S. K. (2017). Effects of parboiling steps on starch characteristics and glycemic index of basmati (PB1121) rice. *International Journal of Agricutural Sciences, 9*(49), 4826–4831.
Kang, M. J., Bae, I. Y., & Lee, H. G. (2018). Rice noodle enriched with okara: Cooking property, texture, and in vitro starch digestibility. *Food Bioscience, 22*, 178–183.
Kastorini, C. M., Milionis, H. J., Esposito, K., Giugliano, D., Goudevenos, J. A., & Panagiotakos, D. B. (2011). The effect of Mediterranean diet on metabolic syndrome and its components: A meta-analysis of 50 studies and 534,906 individuals. *Journal of the American College of Cardiology, 57*(11), 1299–1313.
Kato, K., Suzuki, Y., Hosaka, Y., Takahashi, R., Kodama, I., Sato, K., … Fujita, N. (2019). Effect of high temperature on starch biosynthetic enzymes and starch structure in japonica rice cultivar 'Akitakomachi'(Oryza sativa L.) endosperm and palatability of cooked rice. *Journal of Cereal Science, 87*, 209–214.
Kaur, B., Ranawana, V., & Henry, J. (2016). The glycemic index of rice and rice products: A review, and table of GI values. *Critical Reviews in Food Science and Nutrition, 56*(2), 215–236.
Kawamori, R., Tajima, N., Iwamoto, Y., Kashiwagi, A., Shimamoto, K., Kaku, K., & Voglibose Ph-3 Study Group. (2009). Voglibose for prevention of type 2 diabetes mellitus: A randomised, double-blind trial in Japanese individuals with impaired glucose tolerance. *The Lancet, 373*(9675), 1607–1614.
Kim, J. S., Nam, K., & Chung, S. J. (2019). Effect of nutrient composition in a mixed meal on the postprandial glycemic response in healthy people: A preliminary study. *Nutrition Research and Practice, 13*(2), 126–133.
Kirpitch, A. R., & Maryniuk, M. D. (2011). The 3 R's of glycemic index: Recommendations, research, and the real world. *Clinical Diabetes, 29*(4), 155–159.
Kongkachuichai, R., Charoensiri, R., Meekhruerod, A., & Kettawan, A. (2020). Effect of processing conditions on bioactive compounds and glycemic index of the selected landrace rice variety in pre-diabetes. *Journal of Cereal Science, 94*, 102994.
Kumar, A., Lal, M. K., Kar, S. S., Nayak, L., Ngangkham, U., Samantaray, S., & Sharma, S. G. (2017). Bioavailability of iron and zinc as affected by phytic acid content in rice grain. *Journal of Food Biochemistry, 41*(6), e12413.
Kumar, A., Sahoo, U., Baisakha, B., Okpani, O. A., Ngangkham, U., Parameswaran, C., … Sharma, S. G. (2018a). Resistant starch could be decisive in determining the glycemic index of rice cultivars. *Journal of Cereal Science, 79*, 348–353.
Kumar, A., Sahoo, S., Sahu, S., Nayak, L., Ngangkham, U., Parameswaran, C., … Sharma, S. G. (2018b). Rice with pulses or cooking oils can be used to elicit lower glycemic response. *Journal of Food Composition and Analysis, 71*, 1–7.

Kumar, A., Panda, D., Biswal, M., Dey, P., Behera, L., Baig, M. J., … Sharma, S. (2019). Low light stress influences resistant starch content and glycemic index of rice (O. sativa L). *Starch-Stärke, 71*(5–6), 1800216.

Kumar, A., Panda, P. A., Lal, M. K., Ngangkham, U., Sahu, C., Soren, K. R., … Sharma, S. (2020a). Addition of pulses, cooking oils, and vegetables enhances resistant starch and lowers the glycemic index of rice (Oryza sativa L.). *Starch-Stärke, 72*(9–10), 1900081.

Kumar, A., Sahu, C., Panda, P. A., Biswal, M., Sah, R. P., Lal, M. K., … Sharma, S. (2020b). Phytic acid content may affect starch digestibility and glycemic index value of rice (Oryza sativa L.). *Journal of the Science of Food and Agriculture, 100*(4), 1598–1607.

Kumar, A., Dash, G. K., Barik, M., Panda, P. A., Lal, M. K., Baig, M. J., & Swain, P. (2020c). Effect of drought stress on resistant starch content and glycemic index of rice (Oryza sativa L.). *Starch-Stärke, 72*(11–12), 1900229.

Laaksonen, D. E., Toppinen, L. K., Juntunen, K. S., Autio, K., Liukkonen, K. H., Poutanen, K. S., … Mykkänen, H. M. (2005). Dietary carbohydrate modification enhances insulin secretion in persons with the metabolic syndrome. *The American Journal of Clinical Nutrition, 82*(6), 1218–1227.

Lal, M. K., Vengavasi, K., & Pandey, R. (2019). Interactive effects of low phosphorus and elevated CO 2 on root exudation and nutrient uptake in wheat is modified under Sulphur nutrition. *Plant Physiology Reports, 24*(1), 63–73.

Lal, M. K., Kumar, A., Jena, R., Dutt, S., Thakur, N., Parmar, V., et al. (2020a). Lipids in potato. In *Potato* (pp. 73–85). Springer.

Lal, M. K., Kumar, A., Kumar, A., Raigond, P., Oko, A. O., Thakur, N., et al. (2020b). Dietary fibres in potato. In *Potato* (pp. 37–50). Springer Singapore.

Lal, M. K., Kumar, A., Raigond, P., Dutt, S., Changan, S. S., Chourasia, K. N., … Singh, B. (2021). Impact of starch storage condition on glycemic index and resistant starch of cooked potato (Solanum tuberosum) tubers. *Starch-Stärke, 73*(1–2), 1900281.

Le Bourvellec, C., & Renard, C. M. G. C. (2012). Interactions between polyphenols and macromolecules: Quantification methods and mechanisms. *Critical Reviews in Food Science and Nutrition, 52*(3), 213–248.

Lemlioglu-Austin, D., Turner, N. D., McDonough, C. M., & Rooney, L. W. (2012). Effects of sorghum [Sorghum bicolor (L.) Moench] crude extracts on starch digestibility, estimated glycemic index (EGI), and resistant starch (RS) contents of porridges. *Molecules, 17*(9), 11124–11138.

Li, T., Ohsugi, R., Yamagishi, T., & Sasaki, H. (2006). Effects of weak light on starch accumulation and starch synthesis enzyme activities in rice at the grain filling stage. *Rice Science, 13*(1), 51.

Li, J., Han, W., Xu, J., Xiong, S., & Zhao, S. (2014). Comparison of morphological changes and in vitro starch digestibility of rice cooked by microwave and conductive heating. *Starch-Stärke, 66*(5-6), 549–557.

Liu, S., Manson, J. E., Buring, J. E., Stampfer, M. J., Willett, W. C., & Ridker, P. M. (2002). Relation between a diet with a high glycemic load and plasma concentrations of high-sensitivity C-reactive protein in middle-aged women. *The American Journal of Clinical Nutrition, 75*(3), 492–498.

Liu, Q., Gu, Z., Donner, E., Tetlow, I., & Emes, M. (2007). Investigation of digestibility in vitro and physicochemical properties of A-and B-type starch from soft and hard wheat flour. *Cereal Chemistry, 84*(1), 15–21.

Liu, P., Guo, W., Jiang, Z., Pu, H., Feng, C., Zhu, X., ... & Little, C. R. (2011). Effects of high temperature after anthesis on starch granules in grains of wheat (*Triticum aestivum* L.). *The Journal of Agricultural Science, 149*(2), 159–169.

Livesey, G., Taylor, R., Hulshof, T., & Howlett, J. (2008). Glycemic response and health—A systematic review and meta-analysis: Relations between dietary glycemic properties and health outcomes. *The American Journal of Clinical Nutrition, 87*(1), 258S–268S.

Livesey, G., Taylor, R., Livesey, H., & Liu, S. (2013). Is there a dose-response relation of dietary glycemic load to risk of type 2 diabetes? Meta-analysis of prospective cohort studies. *The American Journal of Clinical Nutrition, 97*(3), 584–596.

Lockyer, S., & Nugent, A. P. (2017). Health effects of resistant starch. *Nutrition Bulletin, 42*(1), 10–41.
López-Barón, N., Gu, Y., Vasanthan, T., & Hoover, R. (2017). Plant proteins mitigate in vitro wheat starch digestibility. *Food Hydrocolloids, 69*, 19–27.
Lu, H., Hu, Y., Wang, C., Liu, W., Ma, G., Han, Q., & Ma, D. (2019). Effects of high temperature and drought stress on the expression of gene encoding enzymes and the activity of key enzymes involved in starch biosynthesis in wheat grains. *Frontiers in plant science*, 1414.
Ma, X. Y., Liu, J. P., & Song, Z. Y. (2012). Glycemic load, glycemic index and risk of cardiovascular diseases: Meta-analyses of prospective studies. *Atherosclerosis, 223*(2), 491–496.
Magaletta, R. L., DiCataldo, S. N., Liu, D., Li, H. L., Borwankar, R. P., & Martini, M. C. (2010). In vitro method for predicting glycemic index of foods using simulated digestion and an artificial neural network. *Cereal Chemistry, 87*(4), 363–369.
Mani, U. V., Prabhu, B. M., Damle, S. S., & Mani, I. (1993). Glycemic index of some commonly consumed foods in Western India. *Asia Pacific Journal of Clinical Nutrition, 2*, 111–114.
Mengel, K., & Judel, G. K. (1981). Effect of light intensity on the activity of starch synthesizing enzymes and starch synthesis in developing wheat grains. *Physiologia Plantarum, 51*(1), 13–18.
Mirrahimi, A., de Souza, R. J., Chiavaroli, L., Sievenpiper, J. L., Beyene, J., Hanley, A. J., ... Jenkins, D. J. (2012). Associations of glycemic index and load with coronary heart disease events: A systematic review and meta-analysis of prospective cohorts. *Journal of the American Heart Association, 1*(5), e000752.
Mohan, V., Spiegelman, D., Sudha, V., Gayathri, R., Hong, B., Praseena, K., ... Krishnaswamy, K. (2014). Effect of brown rice, white rice, and brown rice with legumes on blood glucose and insulin responses in overweight Asian Indians: A randomized controlled trial. *Diabetes Technology & Therapeutics, 16*(5), 317–325.
Mokdad, A. H., Ford, E. S., Bowman, B. A., Dietz, W. H., Vinicor, F., Bales, V. S., & Marks, J. S. (2003). Prevalence of obesity, diabetes, and obesity-related health risk factors, 2001. *JAMA, 289*(1), 76–79.
Muir, J. G., & O'Dea, K. (1992). Measurement of resistant starch: Factors affecting the amount of starch escaping digestion in vitro. *The American Journal of Clinical Nutrition, 56*(1), 123–127.
Murray, I., Sniderman, A. D., Havel, P. J., & Cianflone, K. (1999). Acylation stimulating protein (ASP) deficiency alters postprandial and adipose tissue metabolism in male mice. *Journal of Biological Chemistry, 274*(51), 36219–36225.
Nambiar, V. S., & Patwardhan, T. (2015). Assessment of glycemic index (GI) and glycemic load (GL) for selected traditional millet-based Indian recipes. *International Journal of Diabetes in Developing Countries, 35*(2), 157–162.
Nayak, B., Berrios, J. D. J., & Tang, J. (2014). Impact of food processing on the glycemic index (GI) of potato products. *Food Research International, 56*, 35–46.
Neoh, G. S., Tan, X., Dieters, M., Fox, G., & Gilbert, R. (2020). Effects of cold temperature on starch molecular structure and gelatinization of late-maturity alpha-amylase affected wheat. *Journal of Cereal Science, 92*, 102925.
Nilsson, A. C., Oöstman, E. M., Granfeldt, Y., & Björck, I. M. (2008). Effect of cereal test breakfasts differing in glycemic index and content of indigestible carbohydrates on daylong glucose tolerance in healthy subjects. *The American Journal of Clinical Nutrition, 87*(3), 645–654.
Nordlund, E., Katina, K., Mykkänen, H., & Poutanen, K. (2016). Distinct characteristics of rye and wheat breads impact on their in vitro gastric disintegration and *in vivo* glucose and insulin responses. *Food, 5*(2), 24.
Oh, K., Hu, F. B., Cho, E., Rexrode, K. M., Stampfer, M. J., Manson, J. E., ... Willett, W. C. (2005). Carbohydrate intake, glycemic index, glycemic load, and dietary fiber in relation to risk of stroke in women. *American Journal of Epidemiology, 161*(2), 161–169.
Oki, S., Sato, C., Kubota, N., Nakashima, K., Ishikawa, K., & Akiyama, Y. (2019). Changes in the molecular structure of potato starch conferred digestion resistance by fatty acid addition and heat treatment. *Japan Journal of Food Engineering, 20*, 18534.

Patil, K. B., Chimmad, B. V., & Itagi, S. (2015). Glycemic index and quality evaluation of little millet (Panicum miliare) flakes with enhanced shelf life. *Journal of Food Science and Technology, 52*(9), 6078–6082.

Paiva, F. F., Vanier, N. L., Berrios, J. D. J., Pinto, V. Z., Wood, D., Williams, T., ... & Elias, M. C. (2016). Polishing and parboiling effect on the nutritional and technological properties of pigmented rice.*Food Chemistry, 191*, 105–112.

Prasad, M. P. R., Rao, B. D., Kalpana, K., Rao, M. V., & Patil, J. V. (2015). Glycaemic index and glycaemic load of sorghum products. *Journal of the Science of Food and Agriculture, 95*(8), 1626–1630.

Punia, S. (2020). Barley starch: Structure, properties and in vitro digestibility-a review. *International Journal of Biological Macromolecules, 155*, 868–875.

Punia, S., & Sandhu, K. S. (2015). Functional and antioxidant properties of different milling fractions of Indian barley cultivars. *Carpathian Journal of Food Science & Technology, 7*(4), 19–27.

Punia, S., Siroha, A. K., Sandhu, K. S., & Kaur, M. (2019). Rheological behavior of wheat starch and barley resistant starch (type IV) blends and their starch noodles making potential. *International Journal of Biological Macromolecules, 130*, 595–604.

Raboy, V. (2009). Approaches and challenges to engineering seed phytate and total phosphorus. *Plant Science, 177*, 281–296.

Regand, A., Tosh, S. M., Wolever, T. M., & Wood, P. J. (2009). Physicochemical properties of β-glucan in differently processed oat foods influence glycemic response. *Journal of Agricultural and Food Chemistry, 57*(19), 8831–8838.

Ren, X., Yin, R., Hou, D., Xue, Y., Zhang, M., Diao, X., … Shen, Q. (2018). The glucose-lowering effect of foxtail millet in subjects with impaired glucose tolerance: A self-controlled clinical trial. *Nutrients, 10*(10), 1509.

Ren, J., Chen, S., Li, C., Gu, Z., Cheng, L., Hong, Y., & Li, Z. (2020). A two-stage modification method using 1, 4-α-glucan branching enzyme lowers the in vitro digestibility of corn starch. *Food Chemistry, 305*, 125441.

Rizkalla, S. W., Laromiguiere, M., Champ, M., Bruzzo, F., Boillot, J., & Slama, G. (2007). Effect of baking process on postprandial metabolic consequences: Randomized trials in normal and type 2 diabetic subjects. *European Journal of Clinical Nutrition, 61*(2), 175–183.

Rocha, T. D. S., Carneiro, A. P. D. A., & Franco, C. M. L. (2010). Effect of enzymatic hydrolysis on some physicochemical properties of root and tuber granular starches. *Food Science and Technology, 30*, 544–551.

Rosén, L. A., Östman, E. M., & BjÖrck, I. M. (2011). Postprandial glycemia, insulinemia, and satiety responses in healthy subjects after whole grain rye bread made from different rye varieties. 2. *Journal of Agricultural and Food Chemistry, 59*(22), 12149–12154.

Sagnelli, D., Chessa, S., Mandalari, G., Di Martino, M., Sorndech, W., Mamone, G., … Hebelstrup, K. H. (2018). Low glycaemic index foods from wild barley and amylose-only barley lines. *Journal of Functional Foods, 40*, 408–416.

Saha, A. J., & Reddy, K. S. (2016). Phytic acid: Boon or bane?: Conflict between animal/human nutrition and plant health. *Journal of Food Legumes, 29*(3 & 4), 163–173.

Salari-Moghaddam, A., Keshteli, A. H., Haghighatdoost, F., Esmaillzadeh, A., & Adibi, P. (2019). Dietary glycemic index and glycemic load in relation to general obesity and central adiposity among adults. *Clinical Nutrition, 38*(6), 2936–2942.

Salmerón, J., Ascherio, A., Rimm, E. B., Colditz, G. A., Spiegelman, D., Jenkins, D. J., … Willett, W. C. (1997a). Dietary fiber, glycemic load, and risk of NIDDM in men. *Diabetes Care, 20*(4), 545–550.

Salmerón, J., Manson, J. E., Stampfer, M. J., Colditz, G. A., Wing, A. L., & Willett, W. C. (1997b). Dietary fiber, glycemic load, and risk of non—Insulin-dependent diabetes mellitus in women. *JAMA, 277*(6), 472–477.

Sandhu, K. S., & Punia, S. (2017). Enhancement of bioactive compounds in barley cultivars by solid substrate fermentation. *Journal of Food Measurement and Characterization, 11*(3), 1355–1361.

Shukla, K., & Srivastava, S. (2011). Quality characteristics of finger millet based baby food preparation as affected by its varieties and processing Tecniques. *Journal of Functional and Environmental Botany, 1*(1), 77–84.
Simnadis, T. G., Tapsell, L. C., & Beck, E. J. (2016). Effect of sorghum consumption on health outcomes: A systematic review. *Nutrition Reviews, 74*(11), 690–707.
Singh, J., Dartois, A., & Kaur, L. (2010). Starch digestibility in food matrix: A review. *Trends in Food Science & Technology, 21*(4), 168–180.
Singh, A. K., Rehal, J., Kaur, A., & Jyot, G. (2015). Enhancement of attributes of cereals by germination and fermentation: A review. *Critical Reviews in Food Science and Nutrition, 55*(11), 1575–1589.
Singh, A., Raigond, P., Lal, M. K., Singh, B., Thakur, N., Changan, S. S., … Dutt, S. (2020). Effect of cooking methods on glycemic index and in vitro bioaccessibility of potato (Solanum tuberosum L.) carbohydrates. *LWT, 127*, 109363.
Siroha, A. K., & Sandhu, K. S. (2018). Physicochemical, rheological, morphological, and in vitro digestibility properties of cross-linked starch from pearl millet cultivars. *International Journal of Food Properties, 21*(1), 1371–1385.
Siroha, A. K., Sandhu, K. S., & Kaur, M. (2016). Physicochemical, functional and antioxidant properties of flour from pearl millet varieties grown in India. *Journal of Food Measurement and Characterization, 10*(2), 311–318.
Siroha, A. K., Sandhu, K. S., Kaur, M., & Kaur, V. (2019). Physicochemical, rheological, morphological and in vitro digestibility properties of pearl millet starch modified at varying levels of acetylation. *International Journal of Biological Macromolecules, 131*, 1077–1083.
Siscar-Lee, J. J. H., Juliano, B. O., Qureshi, R. H., & Akbar, M. (1990). Effect of saline soil on grain quality of rices differing in salinity tolerance. *Plant Foods for Human Nutrition, 40*(1), 31–36.
Slavin, J. L. (2005). Dietary fiber and body weight. *Nutrition, 21*(3), 411–418.
Somaratne, G. M., Prasantha, B. D. R., Dunuwila, G. R., Chandrasekara, A., Wijesinghe, D. G. N. G., & Gunasekara, D. C. S. (2017). Effect of polishing on glycemic index and antioxidant properties of red and white basmati rice. *Food Chemistry, 237*, 716–723.
Sullivan, W. R., Hughes, J. G., Cockman, R. W., & Small, D. M. (2017). The effects of temperature on the crystalline properties and resistant starch during storage of white bread. *Food Chemistry, 228*, 57–61.
Tapola, N., Karvonen, H., Niskanen, L., Mikola, M., & Sarkkinen, E. (2005). Glycemic responses of oat bran products in type 2 diabetic patients. *Nutrition, Metabolism and Cardiovascular Diseases, 15*(4), 255–261.
Tenorio, G., Orea, A., Romero, J. M., & Mérida, Á. (2003). Oscillation of mRNA level and activity of granule-bound starch synthase I in Arabidopsis leaves during the day/night cycle. *Plant Molecular Biology, 51*(6), 949–958.
Thakur, N., Raigond, P., Singh, Y., Mishra, T., Singh, B., Lal, M. K., & Dutt, S. (2020). Recent updates on bioaccessibility of phytonutrients. *Trends in Food Science & Technology, 97*, 366–380.
Thitisaksakul, M., Jiménez, R. C., Arias, M. C., & Beckles, D. M. (2012). Effects of environmental factors on cereal starch biosynthesis and composition. *Journal of Cereal Science, 56*(1), 67–80.
Thompson, L. U., Button, C. L., & Jenkins, D. J. (1987). Phytic acid and calcium affect the in vitro rate of navy bean starch digestion and blood glucose response in humans. *The American Journal of Clinical Nutrition, 46*(3), 467–473.
Thompson, S. V., Winham, D. M., & Hutchins, A. M. (2012). Bean and rice meals reduce postprandial glycemic response in adults with type 2 diabetes: A cross-over study. *Nutrition Journal, 11*(1), 1–7.
Thompson, M., Gamage, D., Hirotsu, N., Martin, A., & Seneweera, S. (2017). Effects of elevated carbon dioxide on photosynthesis and carbon partitioning: A perspective on root sugar sensing and hormonal crosstalk. *Frontiers in Physiology, 8*, 578.
Toh, D. W. K., Koh, E. S., & Kim, J. E. (2020). Lowering breakfast glycemic index and glycemic load attenuates postprandial glycemic response: A systematically searched meta-analysis of randomized controlled trials. *Nutrition, 71*, 110634.

Tosh, S. M. (2014). Effects of oats on carbohydrate metabolism.In *Oats Nutrition and Technology*, pp. 281–297. Oxford, UK:Wiley Blackwell.

Tosh, S. M., & Bordenave, N. (2020). Emerging science on benefits of whole grain oat and barley and their soluble dietary fibers for heart health, glycemic response, and gut microbiota. *Nutrition Reviews, 78*(Supplement_1), 13–20.

Trowell, H., Burkitt, D., & Heaton, K. (1985). *Definitions of dietaryfibre and fibre-depleted foods dietary fibre-depleted foods anddisease*. Academic.

Turati, F., Galeone, C., Gandini, S., Augustin, L. S., Jenkins, D. J., Pelucchi, C., & La Vecchia, C. (2015). High glycemic index and glycemic load are associated with moderately increased cancer risk. *Molecular Nutrition & Food Research, 59*(7), 1384–1394.

Turnbull, F. M., Abraira, C., Anderson, R. J., Byington, R. P., Chalmers, J. P., Duckworth, W. C., … Woodward, M. (2009). Intensive glucose control and macrovascular outcomes in type 2 diabetes. *Diabetologia, 52*, 2470.

Ugare, R., Chimmad, B., Naik, R., Bharati, P., & Itagi, S. (2014). Glycemic index and significance of barnyard millet (Echinochloa frumentacae) in type II diabetics. *Journal of Food Science and Technology, 51*(2), 392–395.

Unnikrishnan, V., Chithra, R., & Sujitha, S. (2018). Formulation of millets incorporated non-vegetarian foods and study of glycemic index and glycemic load. *International Journal Pure Applied Bioscience, 6*(4), 291–295.

Van Amelsvoort, J. M., & Weststrate, J. A. (1992). Amylose-amylopectin ratio in a meal affects postprandial variables in male volunteers. *The American Journal of Clinical Nutrition, 55*(3), 712–718.

Van De Laar, F. A., Lucassen, P. L., Akkermans, R. P., Van De Lisdonk, E. H., Rutten, G. E., & Van Weel, C. (2005). α-Glucosidase inhibitors for patients with type 2 diabetes: Results from a Cochrane systematic review and meta-analysis. *Diabetes Care, 28*(1), 154–163.

Vedamanickam, R., Anandan, P., Bupesh, G., & Vasanth, S. (2020). Study of millet and non-millet diet on diabetics and associated metabolic syndrome. *Biomedicine, 40*(1), 55–58.

Vila-Real, C., Pimenta-Martins, A., Maina, N., Gomes, A., & Pinto, E. (2017). Nutritional value of indigenous whole grain cereals millet and sorghum. *Nutrition and Food Science International Journal, 4*(1), 555628.

Virkamäki, A., Ueki, K., & Kahn, C. R. (1999). Protein–protein interaction in insulin signaling and the molecular mechanisms of insulin resistance. *The Journal of Clinical Investigation, 103*(7), 931–943.

Wang, S., & Copeland, L. (2013). Molecular disassembly of starch granules during gelatinization and its effect on starch digestibility: a review. *Food & function*, *4*(11), 1564–1580.

Wang, S., Li, C., Copeland, L., Niu, Q., & Wang, S. (2015). Starch retrogradation: A comprehensive review. *Comprehensive Reviews in Food Science and Food Safety, 14*(5), 568–585.

Weickert, M. O., Möhlig, M., Schöfl, C., Arafat, A. M., Otto, B., Viehoff, H., … Pfeiffer, A. F. (2006). Cereal fiber improves whole-body insulin sensitivity in overweight and obese women. *Diabetes Care, 29*(4), 775–780.

Whiting, D. R., Guariguata, L., Weil, C., & Shaw, J. (2011). IDF diabetes atlas: Global estimates of the prevalence of diabetes for 2011 and 2030. *Diabetes Research and Clinical Practice, 94*(3), 311–321.

Wolever, T. M., Gibbs, A. L., Mehling, C., Chiasson, J. L., Connelly, P. W., Josse, R. G., … Ryan, E. A. (2008). The Canadian trial of carbohydrates in Diabetes (CCD), a 1-y controlled trial of low-glycemic-index dietary carbohydrate in type 2 diabetes: No effect on glycated hemoglobin but reduction in C-reactive protein. *The American Journal of Clinical Nutrition, 87*(1), 114–125.

Xie, F., Li, M., Lan, X., Zhang, W., Gong, S., Wu, J., & Wang, Z. (2017). Modification of dietary fibers from purple-fleshed potatoes (Heimeiren) with high hydrostatic pressure and high pressure homogenization processing: A comparative study. *Innovative Food Science & Emerging Technologies, 42*, 157–164.

Yang, H., Gu, X., Ding, M., Lu, W., & Lu, D. (2018). Heat stress during grain filling affects activities of enzymes involved in grain protein and starch synthesis in waxy maize. *Scientific Reports, 8*(1), 1–9.

Ye, J., Hu, X., Luo, S., McClements, D. J., Liang, L., & Liu, C. (2018). Effect of endogenous proteins and lipids on starch digestibility in rice flour. *Food Research International, 106*, 404–409.

Yin, Y. G., Kobayashi, Y., Sanuki, A., Kondo, S., Fukuda, N., Ezura, H., … Matsukura, C. (2010). Salinity induces carbohydrate accumulation and sugar-regulated starch biosynthetic genes in tomato (Solanum lycopersicum L. cv.'micro-tom') fruits in an ABA-and osmotic stress-independent manner. *Journal of Experimental Botany, 61*(2), 563–574.

Yoon, J. H., Thompson, L. U., & Jenkins, D. J. (1983). The effect of phytic acid on in vitro rate of starch digestibility and blood glucose response. *The American Journal of Clinical Nutrition, 38*(6), 835–842.

Zahra, S. M., Nadeem, M., Hussain, S., Qureshi, T. M., Din, A., & Rashid, F. (2014). Development and evaluation of nutri-bars for internally displaced people in humanitarian emergencies. *Journal of Agriculture Research, 52*(2), 217–226.

Zahra, S. M., Hussain, S., Mahmood, S., Kausar, T., Din, G. M. U., Tufail, T., … Shahid, M. Z. (2020). Shelf stable iron fortified fruit bar's development, proximate estimation and organoleptic characterization. *International Journal of Bioscience, 16*(4), 111 135.

Zaman, S. A., & Sarbini, S. R. (2016). The potential of resistant starch as a prebiotic. In *Critical reviews in biotechnology* (Vol. 36, pp. 578–584). Taylor and Francis Ltd.

Zhang, H., Liu, Y., Zhao, J., & Zhai, U. (2015). Determination of the glycemic index of the wild rice and the effects of wild rice on insulin resistance in rats. *Journal of Hygiene Research, 44*(2), 173–178.

Zhu, F. (2020). Dietary fiber polysaccharides of amaranth, buckwheat and quinoa grains: A review of chemical structure, biological functions and food uses. *Carbohydrate Polymers, 248*, 116819.

Zhu, R., Fan, Z., Han, Y., Li, S., Li, G., Wang, L., … Zhao, W. (2019). Acute effects of three cooked non-cereal starchy foods on postprandial glycemic responses and in vitro carbohydrate digestion in comparison with whole grains: A randomized trial. *Nutrients, 11*(3), 634.

Zhu, J., Yu, W., Zhang, C., Zhu, Y., Xu, J., Li, E., … Liu, Q. (2020). New insights into amylose and amylopectin biosynthesis in rice endosperm. *Carbohydrate Polymers, 230*, 115656.

Chapter 13
High Fibres Functional Products

Aderonke Ibidunni Olagunju and Olufunmilayo Sade Omoba

13.1 Introduction

In the past few decades, eating of fibre has been linked to diverse healthcare benefits (Burkitt et al., 1972; Cleave, 1968; Trowell, 1978). These diverse reports awakened the interest of consumers to the health importance of high fibre functional food products; with resultant increase in epidemiological studies linking the eating of dietary fibre to diseases; and the possible physiological techniques for reducing the threat to diseases. This ultimately influenced the acceptability of dietary fibres as valuable nutrients in healthy eating.

Dietary fibre is referred to as a functional food constituent, due to its ability to impart distinct function to the food, apart from the usual anticipated function, especially when utilized as additive. Many nutritional and health benefits have been ascribed to diets high in fibre; such as decreased incidence of several types of cancer, coronary heart disease, hypertension, stroke, obesity, diabetes, and in some cases gastrointestinal disorder (Lairon et al., 2005; Petruzziello et al., 2006; Weickert & Pfeiffer, 2008). The effects of dietary fibre on these maladies are revealed in diverse ways, which includes: ameliorating the serum lipoprotein, reducing the blood pressure, ameliorating the blood glucose levels, and assisting weight loss (Anderson et al., 2009). Santos et al. (2019), disclosed that the viscosity of the dietary fibre is liable to positively influence cardiovascular (the heart and blood) health, and also regulate the blood glucose.

Dietary fibre is the part of the plant-based food which resist enzymatic digestion by the human digestive enzymes or the undigestible portion of plants that constitute the plant cell wall. These include non-starch polysaccharides (NSP) which are the

A. I. Olagunju (✉) · O. S. Omoba
Department of Food Science and Technology, Federal University of Technology, Akure, Ondo State, Nigeria
e-mail: aiolagunju@futa.edu.ng

S. Punia Bangar, A. Kumar Siroha (eds.), *Functional Cereals and Cereal Foods*,
https://doi.org/10.1007/978-3-031-05611-6_13

hemicellulose and cellulose, others are lignin, oligosaccharides, pectins, gums and waxes. Similarly, American Association of Cereal Chemists (AACC, 2001) referred to dietary fibre as consumable parts of plant or resistant starches (analogous carbohydrates) which resist chemical breakdown (digestion) of the ingested food and assimilation or absorption into the small bowel of human with thorough or incomplete fermentation in the large intestine or bowel. Resistant starches are classified into four (4) categories (based on the features that makes them indigestible), namely: undigestible starch (RS_1), native resistant starch granules (RS_2), retrograded resistant starch (RS_3), and starches that have been chemically modified (RS_4), as reported by Martínez et al. (2010).

Dietary fibre therefore denotes massive variety of biophysically (rheologically) and biochemically different complexes, with assorted outcome on physiological criterion. It is not an entity, but it is an aggregate of complex substances with varied chemical and physical characteristics, which bring into play varied physiological outcomes.

According to the multiple scientific studies carried out with the help of World Health Organization and reported by Reynolds et al. (2019), the positive effects of fibre in the deterrence of chronic diseases was established. According to the authors, the investigation which involved 185 potential studies and 58 clinical trials, recommended 15–30% reduction in the rate of mortality (as a result of these diseases), also, between 25 and 29 grams of fibre was observed to give the essential benefits. However, a recommended daily intake of total dietary fibre of 25–30 grams from food was suggested by the American Heart Association (Van Hort, 1997). The Institute of Medicine and the American Dietetic Association (ADA) similarly, suggested that the dietary fibre daily intake for adult women is 25 g and for male is 38 g; with high fibre food being the source and not dietary supplements (Slavin, 2008), for healthy living. The dietary reference intake values for total fibre by life stages are shown in Table 13.1.

The factors responsible for the array of physiological occurrences displayed by dietary fibres, are: (i) Viscosity of the gastrointestinal contents (Dikeman & Fahey Jr, 2006a, Jenkins et al., 2000); (ii) The biochemical features of the diverse dietary fibres present, and their consequences; (iii) The role of fibre matrix in the food (Englyst & Englyst, 2005); and (iv) The consequence of the dietary fibres on the different microbiota (bifidobacteria) in the large intestine (colon and rectum) and the associated by-products of fermentation.

Dietary fibre is composed of two types: soluble and insoluble fibre, they are integral parts of plant foods (legumes, whole grains and cereals, vegetables, fruits, and nuts or seeds). This classification is based on water solubility as shown on Table 13.2. Pectin, β-glycan and gum are examples of water soluble fibre and are found in barley bran, oat bran, guar, inulin, gum, and psyllium. Soluble fibre ferment completely in the colon (large intestine) and subsequently release short chain fatty acids (SCFAs), mainly butyrate, causing systemic health benefits, such as the protection of the cardiovascular system (Reis et al., 2018; Venegas-Borsellino & Kwon, 2019), and lower blood lipid levels. Fibres from fruits and vegetables possess substantial quantity of soluble dietary fibre, but fibres from cereals comprises of high insoluble

Table 13.1 Dietary reference intake values for total fibre by life stage

	Adequate intake (g/day)	
Life stage group	Male	Female
1–3 years	19	19
4–8 years	25	25
8–13 years	31	26
14–18 years	38	26
19–30 years	38	25
31–50 years	38	25
51–70 years	30	21
Over 70 years	30	21
Pregnancy		28
Lactation		29

Adapted from Slavin (2003)

Table 13.2 Categories of dietary fibre components based on water solubility/fermentability

Features	Fibre component	Description	Major food sources
Water insoluble/ less fermented	Cellulose	Main structural component of plant cell wall. Insoluble in concentrated alkali, soluble in concentrated acid.	Plants (vegetables, sugar beet, various brans)
	Hemicellulose	Cell wall polysaccharides, which contain backbone of β-1, 4 glucosidic linkages. Soluble in dilute alkali.	Cereal grains
	Lignin	Non-carbohydrate cell wall component. Complex cross-linked phenyl propane polymer. Resists bacterial degradation.	Woody plants
Water soluble/well fermented	Pectin	Components of primary cell wall with D-galacturonic acid as principal components. Generally water soluble and gel forming	Fruits, vegetables, legumes, sugar beet, potato
	Gums	Secreted at site of plant injury by specialized secretary cells. Food and pharmaceutical use.	Leguminous seed plants (guar, locust bean), seaweed extracts (carrageenan, alginates), microbial gums (xanthan, gellan)
	Mucilages	Synthesized by plant, prevent desiccation of seed endosperm. Food industry use, hydrophilic, stabilizer.	Plant extracts (gum acacia, gum karaya, gum tragacanth)

Source: Dhingra et al. (2012)

dietary fibre, such as hemicellulose and cellulose (Figuerola et al., 2005). The insoluble fibre (non-fermentable), such as cellulose and lignin are present in wheat bran, brewers spent grain etc.; they increase the bulk of stool, and reduce the fecal time, the mechanical stimulation of the gut (intestine) mucosa is responsible for this. (Reis et al., 2018). Insoluble fibres are also associated to laxative properties (Slavin, 2008).

Soluble and insoluble dietary fibre ratio (SDF/IDF) is of great importance in healthiness and industrial or technological purposes; fibre sources of ratio SDF to IDF of approximately 1: 2 is adjudged suitable in food production (Borchani et al., 2012). The consumption of high fibre functional foods is usually connected to offering assistance to health and reducing the possibility of disease occurrence. This chapter, therefore, intends to give detailed understanding on the importance and quality characteristics of high fibre functional foods.

13.2 High Fibre Functional Products

Advanced research in the utilization of functional food constituents, especially dietary fibre, is employed by the food processing industries in manufacturing high fibre food products capable of offering health benefits apart from their nutritional value. This is particularly essential as consumers choose natural complements in food due to the preconception on the toxicity of synthetic ingredients.

The enormous significance of fibres in food has resulted in the emergence of functional food that are high in dietary fibre from food manufacturing industries; from novel or innovative dietary fibre sources such as fruit processing industry derivatives or Agri-food based dietary fibre and unprocessed cereals otherwise known as whole grains (Abboud et al., 2020; Ayua et al., 2020; Hussain et al., 2020). High fibre functional products have low energy density and low fat content but richer micronutrients; thus consumption of high fibre functional food enhances satiety (Slavin & Green, 2007; Rebello et al., 2016). The concept of functionality in the modern food invention marketing policies is the key motivating force behind the formulation of new food involving the emergence of high fibre functional foods. These include, development of high fibre functional beverages, high fibre functional cereals, wholegrain and fiber-enriched noodles.

13.2.1 High Fibre Functional Beverages, Dairy and Non-dairy Products

The enormous health benefits accrued to dietary fibres and the consumers' increased health consciousness has led to the development of beverages high in fibres. Recent innovation has led to the use of many appropriate fibre ingredients in beverages than from conventional sources (cereals or fruits). Beverages, quench thirst, gives satisfaction and fullness when consumed. Similarly, the included bioactive ingredients (dietary fibres) get to their anticipated active sites quicker and are effortlessly digested. Gums (<80% w/w soluble fibers) and starches are healthy sources of soluble dietary fibres. Gums are hydrocolloids; non- caloric while starches are sources of nutritional energy. Suitability of gums in production of high fibre functional drinks is un-compromisable because they do not impact any taste and smell. Apart from the aforementioned qualities, gums perform other important roles, namely; stabilization of milk proteins at low pH (owing to the ability of caseins to conglomerate at reduced pH), stabilization of emulsions in drinks, offers pleasing mouthfeel and acceptable cloudiness in beverages (Molet-Rodríguez et al., 2018; Teimouri et al., 2018)

Viscous soluble fibres including guar gum (partially hydrolysed and/or natural), psyllium, pectin and oat β-glucan (Rodríguez et al., 2006; Chow et al., 2007; Lyly et al., 2009; Juvonen et al., 2009 Mudgil & Barak, 2016) are mostly used in beverages due their high dispersibility in water. They are highly soluble and clear in solution; this is made possible due to the recent advances in the extraction methods employed (chemical, enzymatic method, membrane filtration, and fermentation) and production of fibres from unconventional sources. Some authors have reported significant increase in the viscosity of beverages with the incorporation of soluble fibre (Juvonen et al., 2009; Lyly et al., 2010; Mudgil & Barak, 2016), and this is associated with the satiety-related qualities of beverages (Lyly et al., 2009, 2010). The viscosity of the chyme (semi-fluid mass of partly digested food) is increased in the stomach and intestine during digestion of soluble fibres, such as oat b-glucan. Thus causing a slowdown in the time it takes the gastric (stomach) to empty i.e. gastric emptying time, this extends the small bowel/intestine passage flow and the rate of nutrient absorption, thereby increasing satiety.

Recent innovation in the formulation of high fibre functional drinks is the use of inulin and oligofructans (fructose polymers) as well as polydextrose. Inulin and oligofructose [also known as fructo-oligosaccharides (FOS)] are food constituents occurring naturally in different percentages in dietary foods, such as onion, garlic, wheat, bananas, and chicory (which comprises ~15–20% of inulin and about 5–10% of oligofructose). Undigested inulin and oligofructose in the digestive tract (upper gastrointestinal tract); exhibit reduced caloric value (Niness, 1999); making it suitable for consumption by diabetics. The connection of the β (2-1) bond and fructose molecules might be responsible for the non – digestion of inulin and oligofructose. Lightowler et al. (2018); Gobinath et al. (2010), revealed that most of the ingested inulin and oligofructose go into the colon to be fermented by the bacteria present in the colon with consequent increase in the generation of colonic short chain fatty acids (SCFAs) and lactate (Tarini & Wolever, 2010), decreased level of plasma triacylglycerides (Brighenti, 2007; Tarini & Wolever, 2010), an enhanced uptake of minerals (Scholz-Ahrens et al., 2007), an inducement of *Bifido Bacterium Longum* and *Acidophilus Lactobacillus* (De Preter et al., 2007) and a complete inducement of the immune system (Lomax & Calder, 2008). The production of SCFAs therefore could moderate the possible occurrence of type 2 diabetes by reducing the blood free fatty acids (FFAs) after meal and positively influencing the hormones of the gut

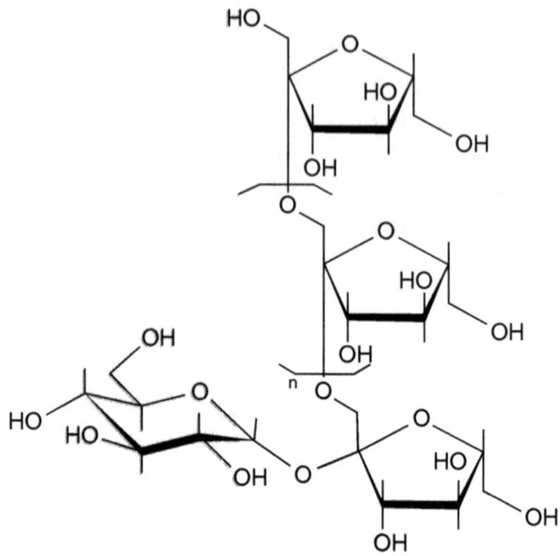

Fig. 13.1 Structure of inulin

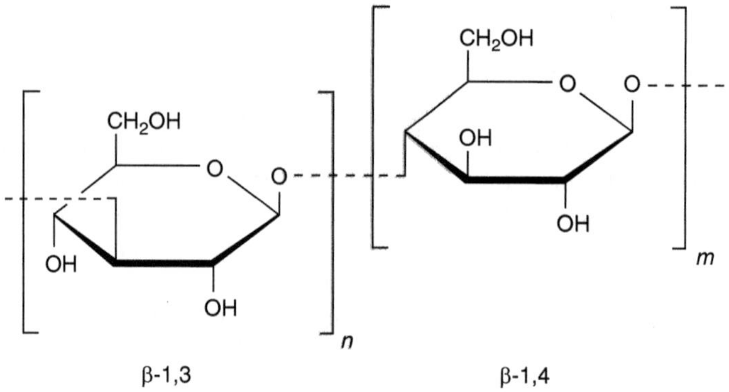

Fig. 13.2 Structure of oat β-glucan

thereby regulating food intake. Figs. 13.1 and13.2 shows the structure of inulin and oat β-glucan.

Yoghurt, is called 'food good for health', addition of varied quantity of dietary fibres into yoghurts, by different authors, significantly affect the sensory attributes, over acceptability and the rheological qualities or parameters of the yoghurts. Hashim et al. (2009) stated that the fortification of yoghurts with date fibre (DF) showed no substantial alterations in the yoghurts acidity, with the increase in pH; yoghurts exhibited firmer texture and darker colour in relation to the control and yoghurts fortified with wheat bran (WB). The sensory scores and overall acceptability decreased with increase in DF (≥4.5%) and WB (≥1.5%). Yildiz and Ozcan (2019) reported that *Streptococcus thermophilus* and *Lactobacillus delbrueckii subsp.* counts increased in fortified yoghurts with purees of carrots, pumpkin, green pea and zucchini. The enriched progression of lactic acid bacteria was attributed to the fibre components, polyphenols and acids existing in the vegetables, resulting in healthier yoghurts. Likewise, fortitification of yoghurts with the vegetable puree influenced the pH, titratable acidity, syneresis, colour, texture and sensory attributes of the fortified yoghurts. The texture indices improved in the yoghurt fortified with puree from carrot in comparison to other vegetable fortified yoghurts. Miocinovic et al. (2018) investigated the possibility of fortifying yoghurts with *triticale* insoluble dietary fibre. Fortified yoghurt (consisting of 1.5% *triticale* insoluble fibres) was adjudged as 'very good' with superior radical scavenging abilities.

De Toledo et al. (2018) reported that passion fruit peel and seeds (PFF) drinkable fortified yoghurts exhibited improved dietary fibre contents, increased viscosity and promoted modifications in the color attributes of yoghurts. Fortified yoghurt (consisting of 2% PFF) was observed as the best practicable food item suitable for commercialization.

13.2.2 High Fibre Functional Baked Products

In spite of the wide acceptance of baked products like bread, cakes and muffins; improving the nutritional properties through incorporation of dietary fibres is challenging due to poor acceptability of such products. However, quite a lot of authors have described the adverse outcome of wheat bran fraction on the technological quality of bread (Almeida et al., 2013; Noort et al., 2010). They concluded that the type of dietary fibre, type of food, particle size and pre -treatment of fibre influences the quality and acceptability of baked products (Almeida et al., 2013; Lebesi & Tzia, 2011; Onipe et al., 2015; Sobota et al., 2015; Sozer et al., 2014). Regardless of the technological challenges facing incorporation of dietary fibres into baked products, the potentials of high fibre baked products as functional foods cannot be overemphasized. Also, the use of whole grain cereal and pseudo – cereal flours as raw materials in baked foods may be an excellent source of fibre, but for the consequential reduced technological and sensorial attributes when compared to the refined flours.

High fibre functional bread produced by incorporating by-products from the flaxseed industry, namely: flax flour (FF) and flax marc (FM) into the bread was investigated by Wirkijowska et al. (2020). Flaxseed, an essential functional food ingredient (rich in α-linolenic acid (ALA), lignans, and dietary fiber). It is an indispensable source of protein of high quality, rich soluble fibre and possess substantial phenolic compounds (Singh et al., 2011). The incorporated by-products of flax seed industry had positive consequence on the crumb moisture content but negatively affected the specific volume and porosity; little or no significant changes were observed in bread fortified with FM. Fortification of bread with flax components especially FM at 10% resulted in bread of improved nutritional value and reduced calorific value and high sensory acceptability.

Current studies have revealed that chia seeds and by-products can be utilized in the production of several baked products including biscuits, bread, cakes (Coelho &

de Mercedes Salas-Mellado, 2015; Felisberto et al., 2015; Hruskova & Svec, 2015; Mesías et al., 2016; Sayed-Ahmad et al., 2018). Chia seeds (*Salvia hispanica L.*) are good sources of protein, fibres (36–40%) and lipids, with polyunsaturated fatty acids (PUFAs, omega-3) constituting more than 60% of total fatty acids (Ayerza & Coates, 2005). It is made up of insoluble dietary fibre (23–46%), soluble dietary fibre (2.5–7.1%) and mucilage (5%) which could function as soluble fibre (Alfredo et al., 2009). Sayed-Ahmad et al. (2018) assessed the nutritional and technological attributes of whole wheat based bread fortified with chia flour. Fortification with chia powders (full fat) at high levels improved the antioxidant properties when compared with chia cakes (defatted). Bread fortified with chia flour enhances its nutrition, particularly when chia cakes was used. Chia - whole wheat bread was darker and highly acceptable.

Similarly, Mesías et al. (2016) reported significant increase in the antioxidant capacity, phenolic compounds, protein, dietary fibre and PUFAs contents, of wheat-based biscuit, with increased inclusion of chia flour resulting in a nutritionally improved biscuit. Authors also reported increased level of impurities (acrylamide, hydroxymethylfurfural and furfural) as a result of processing with higher percentage of chia flour.

By-product from the brewing industry referred to as brewer's spent grain (BSG) is an extracted barley malt remains (residue) solely or in combination with different grains such as corn and rice grits, and obtained after wort production. The chemical and nutritional composition of BSG (dietary fiber- 60–71% and protein content –20%) showed its suitability in pastry/bakery industry and confectionery products (Dabija & Sion, 2005; Mussatto et al., 2006; Santos et al., 2003). Omoba et al. (2013) developed high fibre functional biscuits from blends of plantain flour and BSG. TDF ranged from 52.59% to 68.55%; SDF was between 6.04% and 18.89% whereas IDF ranged from 46.55% to 56.35%.

Similarly, de Oliveira Silva et al. (2018) developed low-carb, high-protein, high-fibre and high isoflavones biscuits from soybean meal and fermented soybean meal as functional raw materials. Soybean meal biscuits exhibited acceptable technological properties and enriched nutritional and functional potentials when compared to wheat biscuits. Soybean meal biscuits showed high sensory scores, whereas fermented soybean meal biscuits showed lower sensory scores.

Obafaye and Omoba (2018) reported the development of high fibre functional biscuits from blends of pearl millet and orange peels (by – products from orange juice industry). The study examined the potential use of orange peel flour (OP) at 5%, 10%, 15%, and 20% substitution in the development of high fibre functional biscuits. As percentage inclusion of OP increases, the dietary fibre contents and antioxidant properties of biscuits also increased. The ratio of IDF to SDF was between 1:2 and 1:3; the ratio is essential because of its physiological consequent linked to consumption of dietary fibre. IDF: SDF ratio is excellent between the ratio of 1–3 according to Hasnaoui et al. (2014).

Omoba et al. (2021) also developed high fibre functional cakes. The raw materials used include orange fleshed sweet potato (OFSP) flour and residue from tiger nut (TN) known as TN fibre (solid waste obtained after the production of "horchata"- vegetable milk from TN). Cakes were observed to be higher in IDF than in SDF (1:4), diets higher in IDF might deliver better chemo-protection in comparison to those with SDF (Bidoli et al., 2013). Also the TDF content of the developed cakes surpassed 6 g/100 g required for products referred to as high fibre products (Goñi & Hervert-Hernández, 2011).

Diez-Sánchez et al. (2020) developed high fibre functional cakes from the blends of wheat flour and black currant pomace. These authors sought to solve the techno-functional challenges negatively influencing the texture and sensory properties of high fibre functional cakes as a result of the pomace using different chemical leavening agents. Cakes baked using pyrophosphate (irrespective of the type of bicarbonate used) were observed to be firm, soft and sweet, and were highly scored in terms of consumer acceptability. Pyrophosphate was therefore concluded as an excellent alternative in baked products to ease the utilization of by-products of fruit and vegetables industries which are high in functional -fibre ingredients.

13.2.3 High Fibre Functional Extruded Products

Extruded products (snacks, breakfast cereals) are typically high in calories and fat; they are not rich in protein and fibre, and are referred to as unhealthy diets. Fortification of extruded products with high dietary fibre ingredients (by-products from food processing plants) might increase the functionality of such products; making it healthier products.

According to Korkerd et al. (2016), by-products from food processing with high contents protein and dietary fibre; soybean meal (defatted), brown rice meal (germinated), and mango peel fibre, were added to 20% (w/w) corn grit, to obtain fortified ready to eat extruded snacks. Formulated snacks has improved nutritional properties; the protein content increased by 2.65 folds, dietary fibre had above three fold increase, total phenolic compound and antioxidant activities increased significantly. The expansion ratio reduced significantly as a result of the increase in protein and dietary fibre, the snacks scored high in the sensory attributes under consideration. Extrusion cooking caused the conversion of IDF to SDF, these may be attributed to the disorganization of bonds (covalent and non-covalent) in carbohydrate and protein moieties, resulting in compact and new soluble remains. Dang and Vasanthan (2019) reported that the increase in SDF after extrusion might be attributed to the solubilization of enzyme resistant starch.

Xie et al., 2021 studied the influence of pectin (dietary fibre) and different methyl esters level on the quality characteristics of extrudate. Authors observed that, the pectin intermingled with the starch particles thereby limiting the water uptake into the amorphous areas of the corn starch and affecting gelatinization. The addition of low methoxyl pectin caused an increase in the viscosity of extrudate while the viscosity decreased in the presence of high methoxyl pectin. Extrudate expansion ratio improved in blends with high methoxyl pectin than extrudate with low methoxyl pectin. These observations might be due to the variation in features or quality of pectins and their influences on the stability of the membranes during formation of bubbles and expansion (Foschia et al., 2013).

Influence of the properties of pomace (chemical and physical); the inclusion levels and processing factors of certain fruits, namely; blueberry, cranberry, apple and grape, pomace – corn starch mixtures were assessed by Wang et al. (2019). The expansion values of pomace revealed apple pomace as the highest at 200 g/kg moisture content. TDF and IDF contents of pomace negatively influenced the expansion ratio while the SDF content and expansion ratio were positively correlated. The composition of the fruit pomace influences level of expansion and hardness of the extrudate (Potter et al., 2013). High level of insoluble fibre result in low expansion values and the high bulk density, while the soluble fibre might not cause any significant alteration or little in the expansion of the extrudate (Robin et al., 2012).

Oliveira et al. (2018) investigated the impact of peel powder inclusion from jabuticaba (*Myrciaria cauliflora*) (JPP) on breakfast cereals (extruded), comprising of corn flour (CF) and/or whole grain wheat flour (WGWF), on the technological/industrial characteristics and sensory acceptability of the breakfast cereals. Breakfast cereal comprising of CF and JPP (CJ) exhibited the maximum expansion and the least bulk density (BD). Breakfast cereal comprising of WGWF and JPP (CWJ), exhibited reduced expansion while JPP enhanced the colour of the extrudate. The sample enhanced with WGWF exhibited the same hardness and crispness in spite of the presence of JPP. The hardness and crispness decreased for all formulations when evaluated after soaking in milk; nevertheless, certain crispness was retained. Authors concluded that JPP enriched the acceptability of WGWF breakfast cereal.

Kamran et al. (2008) formulated breakfast cereal which is ready-to-eat (RTE) comprising of wheat bran-based breakfast with fruits (banana), the addition of soluble fibre gum arabic (Acacia) enhanced the nutritive compositions and physiological qualities of the breakfast cereal. Similarly, Brennan et al. (2008) evaluated the influence of fibres (soluble and insoluble) inclusion into breakfast cereal products that is extruded. Expansion ratio of the breakfast cereal products was not influenced by the addition of dietary fibre (at 5%, 10% and 15%) into the flour bases. The addition of inulin increased the bulk density.

An extruded breakfast cereal comprising of corn and whole peach palm flour was formulated by Santos et al. (2019). Products obtained exhibited high expansion ratio and lower hardness.

Crizel et al. (2015) assessed the influence of adding dietary fibre obtained from orange juice industry byproducts to pasta (fettuccini) at varied concentrations. The pasta comprising of 75 g/kg, had 99% increase in total dietary fibre content in comparison to the control. Pasta increased significantly in carotenoid and phenolic contents with the addition of orange fibre (75 g/kg). The pasta containing 25 g/kg showed no significant different when compared to the control in all the sensory parameters and had the highest acceptance.

13.2.4 High Fibre Functional Meat Products

Meat and food products from meat are poor sources of dietary fibre; production of high fibre functional meat products is a step towards producing healthy products. Kartikawati and Pimomo (2019) revealed the use of untreated rice bran to replace cassava starch in meatballs production to improve the functional qualities. The study concluded that 50% Serang rice bran is the best substitute, it gave an organoleptically superior meatball products, with increased antioxidant activity by 35.78% and increased total phenol contents.

Gupta (2018) replaced meat in chicken products with composite blends of oat and barley flour, potatoes, textured soy protein, and whey protein concentrate. The chicken products exhibited high dietary fibre content, improved mineral contents especially iron, zinc and copper contents, and reduced cholesterol, in relation to control samples. Chicken products made from the addition of two types of flour blends had the highest acceptability.

In another study, Nizhelskaya and Chizhikova (2019) developed semi-finished meat products (cutlets) formed on the basis of meat and vegetable mincemeats, and the vegetable additives such as sprouted wheat grain and dairy wheat flour for a hero – diet. The cutlets with 30% sprouted wheat grain and 26% wallpaper wheat flour were chosen. The developed cutlets had excellent sensory qualities and were made with animal and vegetable proteins suitable for the elderly. Cutlets that have been developed contain calcium, which is necessary to prevent osteoporosis in the elderly.

13.2.5 High Fibre Functional Dough Meals

In vivo studies were carried on high fibre dough meal from plantain enriched with soybean cake and cassava fibre (Famakin et al., 2016), the dough meals were fed to rats and results revealed that rats showed lower postprandial blood glucose response in comparison with rat fed with the control sample (cerolina – a commercial product) and the group treated with a commercially acceptable synthetic antidiabetic drug (metformin). The formulated diets nutritionally compared favourably with cerolina and also in controlling the blood glucose.

Akinjayeju et al. (2020) reported that composite blends of flour from quality protein maize (QPM), soya cake (DSF), whole millet flours (WMF) and cassava starch (CS) for functional dough meal preparation showed high and improved nutritional values with respect to their protein, fibre, amino acids and minerals contents. The *in vivo* studies of the blends reported by Akinjayeju et al. (2020) established that dough meal from the composite flour (76.03% QPM, 16.76% DSF, 7.21% WMF, 5.00% cassava starch, had low glycaemic index (GI) and load (GL), exhibit good anti-diabetic potentials, a good diet for the management and control of diabetic condition.

13.3 Techno-functional and Physiological Properties of High Fibre Functional Products

The dietary importance of fibres are well documented and exceptionally appreciated by consumers, because of their ability to exhibit some techno-functional properties and demonstrate physiological effects of significant health benefits upon consumption. The techno-functional properties of importance include: Hydration properties, which include water and oil holding capacities in addition are, swelling, solubility, viscosity, and antioxidant properties.

13.3.1 Hydration Properties of High Fibre Functional Products and Their Physiological Effects

Hydration properties include the thermal behaviour and kinetics of water absorption/desorption. In high fibre functional products, these include the swelling properties, the water retention capacity, water absorption and porosity. Fibres in high fibre products act differently since fibres are complex and heterogeneous in nature. In high fibre products, they exist as fragments of different sizes, without contaminants, and dietary fibre structure are attached or detached to cell walls. However, specific fibres (especially water soluble) form matrix in the gastrointestinal tracts (GIT) due to their potential to absorb water and swell in aqueous channel (which is influenced by its hydration properties); they retain or trap water and foods/nutrients. The fibre source changes the physical feature of the contents of the gastric and small intestine. The indigestion of the fibre causes bulkiness of the material in the GIT and it remains during the path or flow of the digesta along the small intestine. The water holding capacity of specific fibres are responsible for the increase in volume while the increased viscosity of the intestinal materials are linked to the existence of some fibre sources with viscous polysaccharides such as gums, pectin, *etc*. The alteration in the physical features of the materials in the intestine might delay the gastric emptying, dilute the intestinal enzymes and delay the rate of nutrient absorption and retardation of starch hydrolysis. The aforementioned might bring about reduction in postprandial hyperglycemia and blood cholesterol.

13.3.2 Solubility of High Fibre Functional Products and Their Physiological Effects

Dietary fibres (based on solubility), are categorized as soluble dietary fibre (SDF) and insoluble dietary fibre (IDF), the total dietary fibre (TDF) is the sum of the soluble and insoluble fractions. The soluble dietary and insoluble dietary fibres are linked with certain metabolic and physiological functions in the body of humans.

However, *in vivo* and *in vitro* studies revealed that types of fibres (soluble and insoluble) significantly affect the physiological functions (Harvey & Ferrieres, 2011; Lobley et al., 2013; Zheng et al., 2019). Insoluble dietary fibre passes through the digestive tract undigested; while the soluble is resistant to digestion but can be incompletely or totally fermented into short chain fatty acids (SFCAs) by colonic bacteria in the large intestine/bowel (Harvey & Ferrieres, 2011).

The soluble fibre functions as a prebiotic (by carefully provoking the development and/or interest of beneficial microorganisms in the large bowel (colon) and consequently enhances the host health. It functions as food for the healthy bacteria in the gut (Chawla & Patil, 2010). The consumption of soluble fibre rich diet is linked to the protection or guarding against C-reactive protein (CRP), which is an inflammatory marker, predicting future occurrence of heart disease and diabetes. Ma et al. (2006) and Arranz et al. (2012) established opposite relationships between dietary fibre consumption and CRP concentration values. Similarly, soluble fibre are readily susceptible to fermentation by intestinal bacteria, due to their high water-holding capacity thus, increases the formation of short chain fatty acids (SCFAs) and increased energy availability. The main SCFAs formed are propionate, butyrate, acetate, they are immunomodulators in the inflamed intestine thus exhibit potential therapeutic activities in the diversion of colitis or proctitis, pouchitis, ulcerative colitis (Inflammmatory Bowel Disease – IBD), and antibiotic-associated diarrhoea (Cavaglieri et al., 2003; Pylkas et al., 2005).

Soluble fibres exhibit hypocholesterolemic effects attributed to the repression of the digestion/assimilation of fat and retardation of cholesterologenesis (in the liver) by means of propionate (a SCFA) or microbial products. Similarly, the viscous nature of non – starch polysaccharides (NSPs) reduce postprandial blood glucose and insulin levels in healthy individuals (Kumar et al., 2012; Lobley et al., 2013).

The key consequence of soluble dietary fibres is related to the viscous properties of pectins and gums which are NSPs, as they effect the reduction in the absorption of food/nutrients. The intestinal microbiota formed from the fermentable substrate such as pectin, deposits of the partially fermentable fibre complex such as cellulose and hemicelluloses and the water held, are all liable for the upsurge in the fecal bulk (Habte-Tsion & Kumar, 2018).

Researches have revealed that consuming unprocessed (whole grain) cereals and insoluble fibre are of significant health benefits; for instance wheat bran which embraces cellulose, lignin and some hemicellulose, lessen the danger of cardiovascular diseases and diabetes (Salas-Salvadó et al., 2006; Threapleton et al., 2013). Bran (especially of wheat) in the exterior layer of grain, are excellent springs of nutrients such as minerals, vitamins, antioxidants and other bioactive compounds of essential health benefits, unlike in refined cereal flour. In accordance to the report of Threapleton et al. (2013) there is a negative correlation between the possibility of having coronary heart disease (CHD), cardiovascular disease (CVD) and consumption of insoluble dietary fibre from cereals and vegetables.

Insoluble fibre (long-chain polymers) are indigestible in the gastrointestinal tract and are able to hold water, thereby decreases the time of the intestinal transit and increases the faecal bulk hydration; thereby functioning as a physiological laxative.

Supplementation with 6–40 gram of soluble and insoluble dietary fibre per day has been revealed to cause a reduction in energy intake (Sto et al., 2000Wanders et al., 2011). Lower energy due to the consumption of fibre rich functional products leads to reduction in body weight or mass.

13.3.3 Viscosity of High Fibre Functional Products and Their Physiological Effects

Soluble fibres have capacity to form gels (viscosity); they are well-hydrated, thus forming a viscous mass in the gastrointestinal tract and so reduce absorption rate of the nutrients. The increased viscosity and obstruction of the unstirred water deposit that lines the surface of the intestinal absorptive cells, makes bulk dispersal/diffusion difficult. A decrease in water in the intestine's luminal contents will impede or prevent bile acid absorption in the distal ileum, where the absorption site is located. As a result of this, the bile acids [cholic acid (CA) and chenodeoxycholic acid (CDCA)], that are bound or attached to the fibre, are expelled from the body. Free cholesterol, endogenous metabolites, and xenobiotics, are all secreted from the bile ducts through the production of bile acid. Soluble fibre lowers or reduces serum cholesterol by binding the bile acid to the fibre. It is possible to attribute the metabolic benefits of fibre (lowering serum cholesterol and regulating postpandrial glucose) solely to the viscous fibre in high fibre functional products. (Kumar et al., 2012).

13.3.4 Antioxidant Properties of High Fibre Functional Products and Their Physiological Effects

Dietary fibres obtained from food processing by-products are high in antioxidants, they contain essential antioxidants of significant health importance like polyphenols (rutin, quercetin, chlorogenic acids etc.) and carotenoids. They confer valuable health benefits of dietary fibres and antioxidants and are referred to as antioxidant dietary fibres (ADFs). High fibres functional products with ADFs therefore exhibit both the *physiological consequences* of *dietary fibre* and *antioxidants*. Several authors have exploited the beneficial effects of ADFs in developing different products of valuable health importance. Verma et al. (2013) used guava powder as a source of antioxidant dietary fibre in the production of nuggets from sheep meat; also fortification of bread with whole cereals and seeds to increase the antioxidant dietary fibre (Benítez et al., 2018); the effects of juice by-products as a source of dietary fibre and antioxidants on hepatic steatosis was reported by Amaya-Cruz et al. (2015).

According Saura-Calixto (2011), ADFs discharge the *fiber* matrix in the large intestine/bowel through the *activities* of the bacterial microbiota in the gut, generating metabolites and creating an *environment* that can prevent or slow down damage to cells caused by free radicals. The phenolic antioxidant bioavailability of high fibre functional food is of great importance after consumption, because most high fibre functional food contains substantial amount of polyphenols connected with the dietary fibre. Researches have appraised the evidences showing that the microstructure of food influences the bio-accessibility and bio-availability of numerous nutrients, especially antioxidants (Parada & Aguilera, 2007). The physiological effects of polyphenols would not solely be influenced by their quantity in food, but more importantly their bioavailability after consumption (Manach et al., 2005).

Pérez-Jiménez et al. (2009), observed that polyphenol antioxidants in high fibre matrix (grape antioxidant dietary fibre) are partly bioavailable *in – vivo* as established by an increase in the antioxidant capacity of the blood. According to the authors, dietary fibre constituents and high molecular weight polyphenols (proanthocyanidins) in high fibre functional products may retard the heights of blood antioxidant capacity compared to polyphenol antioxidants in products with low fibre functional products.

In high fibre functional products, the physical confinement of antioxidants within the organized structure and the increased viscosity of chyme, inhibits the peristaltic mixing process (involuntary movement in the digestive tract), that encourages the transport of enzymes, bile salts to unmicellized fat, and soluble antioxidants to the wall of the intestines. The aforementioned are the main factors influencing the absorption/digestion of dietary fibre and antioxidant (Montagne et al., 2003). In the unstirred layer, bile salt binding to fibres and diffusion inhibition/hindrance are minor factors to consider (Palafox-Carlos et al., 2011).

In the small bowel/intestine, the low bioaccssebility of antioxidants due to interactions with indigestible fibres has a significant impact on their bioavailability in high fiber functional food matrices. This is primarily due to the interactions between the antioxidant and the indigestible fibres. Processing and digestion of food may result in the formation of chemical complexes and gel structures in the small intestine that reduces or enhances the bioavailability of antioxidants. This significantly affect the assessment of the therapeutic and nutritional benefits or impacts of many high fibre functional products.

13.4 Conclusion

Dietary fibres are composed of undigested carbohydrates and lignin, naturally found in plants and may exist in their entirety. Fibre is a vital constituent of a plant-based diet, due to the diverse epidemiological studies suggesting an opposite relationship between the habitual consumers of high fibre or fibre rich functional food products and numerous diseases such as coronary heart disease (CHD) and cardiovascular disease (CVD) and diabetes. There are dependable, and potent evidences as fibre in high fibre functional products, are indicators of a healthy dietary pattern. High fibre functional food products primarily contain the indigestible insoluble fibre, with little or no associated metabolic effects; the digestible soluble fibre (secondary component) which are well-hydrated, form gels; thus forming a viscous mass in the gastrointestinal tract, thereby deferring the absorption/adsortion of glucose and bile acids, has positive effects on serum fatty profile (triglycerides and total HDL, and LDL cholesterol) and carbohydrate (glucose) metabolism. Similarly, consumption of fibre (independent of the type of fibre) from high fibre functional products, protects against cardiovascular diseases (coronary heart diseases and cardiovascular disease) and diabetes. Advances in high fibre functional products with antioxidant dietary fibres (ADFs) revealed that they exhibit both the *physiological effects* of *dietary fibres* and *antioxidants*. But, bioaccessibility and bioavailability of the antioxidants as influenced by the food microstructure varied significantly, implying that the most prominent ingested antioxidant through high fibre products might not have the maximum concentration in the serum or as dynamic metabolites in the target site. This is important because of the connections between dietary fibre and antioxidants; which controls their discharge or release from the chyme through gastrointestinal tract. Therefore, to meet with the terms of the current recommendations for daily fibre intake and for maintenance of health, regular habitual intake of high fibre functional products is worthwhile.

References

AACC (American Association of Cereal Chemists). (2001). AACC dietary fiber technical committee. The definition of dietary fiber. *Cereal Foods World, 46*, 112–126.

Abboud, K. Y., Iacomini, M., Simas, F. F., & Cordeiro, L. M. (2020). High methoxyl pectin from the soluble dietary fiber of passion fruit peel forms weak gel without the requirement of sugar addition. *Carbohydrate Polymers, 246*, 116616.

Akinjayeju, O., Ijarotimi, O. S., Awolu, O. O., & Fagbemi, T. N. (2020). Nutritional composition, Glycaemic properties and anti-diabetic potentials of cereal-based soy-fortified flours for functional dough meal in diabetic induced rats. *Journal of Food Science and Nutrition Research, 3*(2), 102–120.

Alfredo, V. O., Gabriel, R. R., Luis, C. G., & David, B. A. (2009). Physicochemical properties of a fibrous fraction from chia (*Salvia hispanica* L.). *LWT-Food Science and Technology, 42*(1), 168–173.

Almeida, E. L., Chang, Y. K., & Steel, C. J. (2013). Dietary fibre sources in bread: Influence on technological quality. *LWT-Food Science and Technology, 50*(2), 545–553.

Amaya-Cruz, D. M., Rodríguez-González, S., Pérez-Ramírez, I. F., Loarca-Piña, G., Amaya-Llano, S., Gallegos-Corona, M. A., & Reynoso-Camacho, R. (2015). Juice by-products as a source of dietary fibre and antioxidants and their effect on hepatic steatosis. *Journal of Functional Foods, 17*, 93–102.

Anderson, J. W., Baird, P., Davis, R. H., Ferreri, S., Knudtson, M., Koraym, A., Waters, V., & Williams, C. L. (2009). Health benefits of dietary fiber. *Nutrition Reviews, 67*(4), 188–205.

Arranz, S., Remon, A. M., Raventos, R. M. L., & Estruch, R. (2012). Effects of dietary fiber intake on cardiovascular risk factors. *Recent Advances in Cardiovascular Risk Factors. In-tech open science/open minds, 978*, 459–488.

Ayerza, R., & Coates, W. (2005). *Chia: Rediscovering a forgotten crop of the Aztecs*. University of Arizona Press.

Ayua, E. O., Kazem, A. E., & Hamaker, B. R. (2020). Whole grain cereal fibers and their support of the gut commensal clostridia for health. *Bioactive Carbohydrates and Dietary Fibre, 24*, 100245.

Benítez, V., Esteban, R. M., Moniz, E., Casado, N., Aguilera, Y., & Mollá, E. (2018). Breads fortified with wholegrain cereals and seeds as source of antioxidant dietary fibre and other bioactive compounds. *Journal of Cereal Science, 82*, 113–120.

Bidoli, E., Pelucchi, C., Polesel, J., Negri, E., Barzan, L., Franchin, G., Franceschi, S., Serraino, D., Paoli, P. D., Vecchia, C. L., & Talamini, R. (2013). Fiber intake and risk of nasopharyngeal carcinoma: A case-control study. *Nutrition and Cancer, 65*(8), 1157–1163.

Borchani, C., Besbes, S., Masmoudi, M., Bouaziz, M. A., Blecker, C., & Attia, H. (2012). Influence of oven-drying temperature on physicochemical and functional properties of date fibre concentrates. *Food and Bioprocess Technology, 5*(5), 1541–1551.

Brennan, M. A., Monro, J. A., & Brennan, C. S. (2008). Effect of inclusion of soluble and insoluble fibres into extruded breakfast cereal products made with reverse screw configuration. *International Journal of Food Science & Technology, 43*(12), 2278–2288.

Brighenti, F. (2007). Dietary fructans and serum triacylglycerols: A meta-analysis of randomized controlled trials. *The Journal of Nutrition, 137*(11), 2552S–2556S.

Burkitt, D. P., Walker, A. R. P., & Painter, N. S. (1972). Effect of dietary fibre on stools and transit-times, and its role in the causation of disease. *The Lancet, 300*(7792), 1408–1411.

Cavaglieri, C. R., Nishiyama, A., Fernandes, L. C., Curi, R., Miles, E. A., & Calder, P. C. (2003). Differential effects of short-chain fatty acids on proliferation and production of pro-and anti-inflammatory cytokines by cultured lymphocytes. *Life Sciences, 73*(13), 1683–1690.

Chawla, R. P., & Patil, G. R. (2010). Soluble dietary fiber. *Comprehensive Reviews in Food Science and Food Safety, 9*(2), 178–196.

Chow, J., Choe, Y. S., Noss, M. J., Robinson, K. J., Dugle, J. E., Acosta, S. H., & Garleb, K. A. (2007). Effect of a viscous fiber-containing nutrition bar on satiety of patients with type 2 diabetes. *Diabetes Research and Clinical Practice, 76*(3), 335–340.

Cleave, T. L. (1968). Sucrose intake and coronary heart-disease. *The Lancet, 292*(7579), 1187.

Coelho, M. S., & de las Mercedes Salas-Mellado, M. (2015). Effects of substituting chia (Salvia hispanica L.) flour or seeds for wheat flour on the quality of the bread. *LWT-Food Science and Technology, 60*(2), 729–736.

Crizel, T. D. M., Rios, A. D. O., Thys, R. C. S., & Flôres, S. H. (2015). Effects of orange by-product fiber incorporation on the functional and technological properties of pasta. *Food Science and Technology, 35*, 546–551.

Dabija, A., & Sion, I. (2005). Spent grains qualitative estimation–by products of beer industry. *A. Dabija, et al. Scientifical Researches Agroalimentary Processes and Technologies, 11*(2), 267–270.

Dang, T. T., & Vasanthan, T. (2019). Modification of rice bran dietary fiber concentrates using enzyme and extrusion cooking. *Food Hydrocolloids, 89*, 773–782.

de Oliveira Silva, F., Miranda, T. G., Justo, T., da Silva Frasão, B., Conte-Junior, C. A., Monteiro, M., & Perrone, D. (2018). Soybean meal and fermented soybean meal as functional ingredients for the production of low-carb, high-protein, high-fiber and high isoflavones biscuits. *LWT, 90*, 224–231.

De Preter, V., Vanhoutte, T., Huys, G., Swings, J., De Vuyst, L., Rutgeerts, P., & Verbeke, K. (2007). Effects of lactobacillus casei Shirota, Bifidobacterium breve, and oligofructose-enriched inulin on colonic nitrogen-protein metabolism in healthy humans. *American Journal of Physiology-Gastrointestinal and Liver Physiology, 292*(1), G358–G368.

De Toledo, N. M. V., De Camargo, A. C., Ramos, P. B. M., Button, D. C., Granato, D., & Canniatti-Brazaca, S. G. (2018). Potentials and pitfalls on the use of passion fruit by-products in drinkable yogurt: Physicochemical, technological, microbiological, and sensory aspects. *Beverages, 4*(3), 47.

Dhingra, D., Michael, M., Rajput, H., & Patil, R. T. (2012). Dietary fibre in foods: A review. *Journal of Food Science and Technology, 49*(3), 255–266.

Diez-Sánchez, E., Llorca, E., Tárrega, A., Fiszman, S., & Hernando, I. (2020). Changing chemical leavening to improve the structural, textural and sensory properties of functional cakes with blackcurrant pomace. *LWT, 127*, 109378.

Dikeman, C. L., & Fahey, G. C., Jr. (2006a). Viscosity as related to dietary fiber: A review. *Critical Reviews in Food Science and Nutrition, 46*(8), 649–663.

Englyst, K. N., & Englyst, H. N. (2005). Carbohydrate bioavailability. *British Journal of Nutrition, 94*(1), 1–11.

Famakin, O., Fatoyinbo, A., Ijarotimi, O. S., Badejo, A. A., & Fagbemi, T. N. (2016). Assessment of nutritional quality, glycaemic index, antidiabetic and sensory properties of plantain (*Musa paradisiaca*)-based functional dough meals. *Journal of Food Science and Technology, 53*(11), 3865–3875.

Felisberto, M. H. F., Wahanik, A. L., Gomes-Ruffi, C. R., Clerici, M. T. P. S., Chang, Y. K., & Steel, C. J. (2015). Use of chia (Salvia hispanica L.) mucilage gel to reduce fat in pound cakes. *LWT-Food Science and Technology, 63*(2), 1049–1055.

Figuerola, F., Hurtado, M. L., Estévez, A. M., Chiffelle, I., & Asenjo, F. (2005). Fibre concentrates from apple pomace and citrus peel as potential fibre sources for food enrichment. *Food Chemistry, 91*(3), 395–401.

Foschia, M., Peressini, D., Sensidoni, A., & Brennan, C. S. (2013). The effects of dietary fibre addition on the quality of common cereal products. *Journal of Cereal Science, 58*(2), 216–227.

Gobinath, D., Madhu, A. N., Prashant, G., Srinivasan, K., & Prapulla, S. G. (2010). Beneficial effect of xylo-oligosaccharides and fructo-oligosaccharides in streptozotocin-induced diabetic rats. *British Journal of Nutrition, 104*(1), 40–47.

Goñi, I., & Hervert-Hernández, D. (2011). By-products from plant foods are sources of dietary fibre and antioxidants. In *Phytochemicals—Bioactivities and impact on health* (pp. 95–116). In-tech.

Gupta, S. (2018). Quality characteristics of functional restructured spent hen meat slices developed by utilizing different binders and extenders. *Food Science and Technology Research, 24*(2), 241–247.

Habte-Tsion, H. M., & Kumar, V. (2018). Nonstarch polysaccharide enzymes—General aspects. In *Enzymes in human and animal nutrition* (pp. 183–209). Academic Press.

Harvey, R., & Ferrieres, D. (2011). In R. A. Harvey (Ed.), *Biochemistry, Lippincott's illustrated reviews* (5th ed.). Lippincott Williams, & Wilkins.

Hashim, I. B., Khalil, A. H., & Afifi, H. S. (2009). Quality characteristics and consumer acceptance of yogurt fortified with date fiber. *Journal of Dairy Science, 92*(11), 5403–5407.

Hasnaoui, N., Wathelet, B., & Jiménez-Araujo, A. (2014). Valorization of pomegranate peel from 12 cultivars: Dietary fibre composition, antioxidant capacity and functional properties. *Food Chemistry, 160*, 196–203.

Hruskova, M., & Svec, I. (2015). Chemical, rheological and bread characteristics of wheat flour influenced by different forms of chia (Salvia hispanica L.). *Emirates Journal of Food and Agriculture, 27*(12), 872–877.

Hussain, S., Jõudu, I., & Bhat, R. (2020). Dietary fiber from underutilized plant resources—A positive approach for valorization of fruit and vegetable wastes. *Sustainability, 12*(13), 5401.

Jenkins, D. J., Kendall, C. W., Axelsen, M., Augustin, L. S., & Vuksan, V. (2000). Viscous and nonviscous fibres, nonabsorbable and low glycaemic index carbohydrates, blood lipids and coronary heart disease. *Current Opinion in Lipidology, 11*(1), 49–56.

Juvonen, K. R., Purhonen, A. K., Salmenkallio-Marttila, M., Lahteenmaki, L., Laaksonen, D. E., Herzig, K. H., Uusitupa, M. I., Poutanen, K. S., & Karhunen, L. J. (2009). Viscosity of oat bran-enriched beverages influences gastrointestinal hormonal responses in healthy humans. *The Journal of Nutrition, 139*(3), 461–466.

Kamran, M., Saleem, N., & Umer, Z. N. (2008). Ready-to-eat (RTE) wheat bran breakfast cereal as a high-fibre diet. *Journal of Food Processing and Preservation, 32*(5), 853–867.

Kartikawati, M., Purnomo, H. (2019). Improving meatball quality using different varieties of rice bran as natural antioxidant. *Food Resource, 3*, 79–85.

Korkerd, S., Wanlapa, S., Puttanlek, C., Uttapap, D., & Rungsardthong, V. (2016). Expansion and functional properties of extruded snacks enriched with nutrition sources from food processing by-products. *Journal of Food Science and Technology, 53*(1), 561–570.

Kumar, V., Sinha, A. K., Makkar, H. P., de Boeck, G., & Becker, K. (2012). Dietary roles of non-starch polysachharides in human nutrition: A review. *Critical Reviews in Food Science and Nutrition, 52*(10), 899–935.

Lairon, D., Arnault, N., Bertrais, S., Planells, R., Clero, E., Hercberg, S., & Boutron-Ruault, M. C. (2005). Dietary fiber intake and risk factors for cardiovascular disease in French adults. *The American Journal of Clinical Nutrition, 82*(6), 1185–1194.

Lebesi, D. M., & Tzia, C. (2011). Effect of the addition of different dietary fiber and edible cereal bran sources on the baking and sensory characteristics of cupcakes. *Food and Bioprocess Technology, 4*(5), 710–722.

Lightowler, H., Thondre, S., Holz, A., & Theis, S. (2018). Replacement of glycaemic carbohydrates by inulin-type fructans from chicory (oligofructose, inulin) reduces the postprandial blood glucose and insulin response to foods: Report of two double-blind, randomized, controlled trials. *European Journal of Nutrition, 57*(3), 1259–1268.

Lobley, G. E., Holtrop, G., Bremner, D. M., Calder, A. G., Milne, E., & Johnstone, A. M. (2013). Impact of short term consumption of diets high in either non-starch polysaccharides or resistant starch in comparison with moderate weight loss on indices of insulin sensitivity in subjects with metabolic syndrome. *Nutrients, 5*(6), 2144–2172.

Lomax, A. R., & Calder, P. C. (2008). Prebiotics, immune function, infection and inflammation: A review of the evidence. *British Journal of Nutrition, 101*(5), 633–658.

Lyly, M., Liukkonen, K. H., Salmenkallio-Marttila, M., Karhunen, L., Poutanen, K., & Lähteenmäki, L. (2009). Fibre in beverages can enhance perceived satiety. *European Journal of Nutrition, 48*(4), 251–258.

Lyly, M., Ohls, N., Lähteenmäki, L., Salmenkallio-Marttila, M., Liukkonen, K. H., Karhunen, L., & Poutanen, K. (2010). The effect of fibre amount, energy level and viscosity of beverages containing oat fibre supplement on perceived satiety. *Food & Nutrition Research, 54*(1), 2149.

Ma, Y., Griffith, J. A., Chasan-Taber, L., Olendzki, B. C., Jackson, E., Stanek III, E. J., ... & Ockene, I. S. (2006). Association between dietary fiber and serum C-reactive protein. *The American journal of clinical nutrition, 83*(4), 760–766.

Manach, C., Williamson, G., Morand, C., Scalbert, A., & Rémésy, C. (2005). Bioavailability and bioefficacy of polyphenols in humans. I. Review of 97 bioavailability studies. *The American Journal of Clinical Nutrition, 81*(1), 230S–242S.

Martínez, I., Kim, J., Duffy, P. R., Schlegel, V. L., & Walter, J. (2010). Resistant starches types 2 and 4 have differential effects on the composition of the fecal microbiota in human subjects. *PloS One, 5*(11), e15046.

Mesías, M., Holgado, F., Márquez-Ruiz, G., & Morales, F. J. (2016). Risk/benefit considerations of a new formulation of wheat-based biscuit supplemented with different amounts of chia flour. *LWT, 73*, 528–535.

Miocinovic, J., Tomic, N., Dojnov, B., Tomasevic, I., Stojanovic, S., Djekic, I., & Vujcic, Z. (2018). Application of new insoluble dietary fibres from triticale as supplement in yoghurt–effects on physico-chemical, rheological and quality properties. *Journal of the Science of Food and Agriculture, 98*(4), 1291–1299.

Molet-Rodríguez, A., Salvia-Trujillo, L., & Martín-Belloso, O. (2018). Beverage emulsions: Key aspects of their formulation and physicochemical stability. *Beverages, 4*(3), 70.

Montagne, L., Pluske, J. R., & Hampson, D. J. (2003). A review of interactions between dietary fibre and the intestinal mucosa, and their consequences on digestive health in young non-ruminant animals. *Animal Feed Science and Technology, 108*(1–4), 95–117.

Mudgil, D., & Barak, S. (2016). Development of functional buttermilk by soluble fibre fortification. *Agro Food Industry Hi Tech, 27*(2), 44–47.

Mussatto, S. I., Dragone, G., & Roberto, I. C. (2006). Brewers' spent grain: Generation, characteristics and potential applications. *Journal of Cereal Science, 43*(1), 1–14.

Niness, K. R. (1999). Inulin and oligofructose: What are they? *The Journal of Nutrition, 129*(7), 1402S–1406S.

Nizhelskaya, K. V., & Chizhikova, O. G. (2019). Development of new types of precooked meat products–cutlets for older people. *Vestnik of Murmansk State Technical University, 21*(3), 488–496.

Noort, M. W., van Haaster, D., Hemery, Y., Schols, H. A., & Hamer, R. J. (2010). The effect of particle size of wheat bran fractions on bread quality–evidence for fibre–protein interactions. *Journal of Cereal Science, 52*(1), 59–64.

Obafaye, R. O., & Omoba, O. S. (2018). Orange peel flour: A potential source of antioxidant and dietary fiber in pearl-millet biscuit. *Journal of food biochemistry, 42*(4), e12523.

Oliveira, L. C., Alencar, N. M., & Steel, C. J. (2018). Improvement of sensorial and technological characteristics of extruded breakfast cereals enriched with whole grain wheat flour and jabuticaba (Myrciaria cauliflora) peel. *LWT, 90*, 207–214.

Omoba, O. S., Awolu, O. O., Olagunju, A. I., & Akomolafe, A. O. (2013). Optimisation of plantain-brewers' spent grain biscuit using response surface methodology. *Journal of Scientific Research and Reports, 2*(2), 665–681.

Omoba, O. S., Olagunju, A. I., Iwaeni, O. O., & Obafaye, O. R. (2021). Effects of tiger nut fiber on the quality characteristics and consumer acceptability of cakes made from orange-fleshed sweet potato flour. *Journal of Culinary Science & Technology, 19*(3), 228–246.

Onipe, O. O., Jideani, A. I., & Beswa, D. (2015). Composition and functionality of wheat bran and its application in some cereal food products. *International Journal of Food Science & Technology, 50*(12), 2509–2518.

Palafox-Carlos, H., Ayala-Zavala, J. F., & González-Aguilar, G. A. (2011). The role of dietary fiber in the bioaccessibility and bioavailability of fruit and vegetable antioxidants. *Journal of Food Science, 76*(1), R6–R15.

Parada, J., & Aguilera, J. M. (2007). Food microstructure affects the bioavailability of several nutrients. *Journal of food science, 72*(2), R21–R32.

Pérez-Jiménez, J., Serrano, J., Tabernero, M., Arranz, S., Díaz-Rubio, M. E., García-Diz, L., Goñi, I., & Saura-Calixto, F. (2009). Bioavailability of phenolic antioxidants associated with dietary fiber: Plasma antioxidant capacity after acute and long-term intake in humans. *Plant Foods for Human Nutrition, 6*(2), 102–107.

Petruzziello, L., Iacopini, F., Bulajic, M., Shah, S., & Costamagna, G. (2006). Uncomplicated diverticular disease of the colon. *Alimentary Pharmacology & Therapeutics, 23*(10), 1379–1391.

Potter, R., Stojceska, V., & Plunkett, A. (2013). The use of fruit powders in extruded snacks suitable for children's diets. *LWT-Food Science and Technology, 51*(2), 537–544.

Pylkas, A. M., Juneja, L. R., & Slavin, J. L. (2005). Comparison of different fibers for in vitro production of short chain fatty acids by intestinal microflora. *Journal of Medicinal Food, 8*(1), 113–116.

Rebello, C. J., O'Neil, C. E., & Greenway, F. L. (2016). Dietary fiber and satiety: The effects of oats on satiety. *Nutrition Reviews, 74*(2), 131–147.

Reis, A. M. D., Fruchtenicht, A. V., Loss, S. H., & Moreira, L. F. (2018). Use of dietary fiber in enteral nutrition in critically ill patients: A systematic review. *Brazilian Journal of Intensive Care, 30*, 358–365.

Reynolds, A., Mann, J., Cummings, J., Winter, N., Mete, E., & Te Morenga, L. (2019). Carbohydrate quality and human health: A series of systematic reviews and meta-analyses. *The Lancet, 393*(10170), 434–445.

Robin, F., Dattinger, S., Boire, A., Forny, L., Horvat, M., Schuchmann, H. P., & Palzer, S. (2012). Elastic properties of extruded starchy melts containing wheat bran using on-line rheology and dynamic mechanical thermal analysis. *Journal of Food Engineering, 109*(3), 414–423.

Rodríguez, R., Jimenez, A., Fernández-Bolanos, J., Guillen, R., & Heredia, A. (2006). Dietary fibre from vegetable products as source of functional ingredients. *Trends in Food Science & Technology, 17*(1), 3–15.

Salas-Salvadó, J., Bulló, M., Pérez-Heras, A., & Ros, E. (2006). Dietary fibre, nuts and cardiovascular diseases. *British Journal of Nutrition, 96*(S2), S45–S51.
Santos, M., Jiménez, J. J., Bartolomé, B., Gómez-Cordovés, C., & Del Nozal, M. J. (2003). Variability of brewer's spent grain within a brewery. *Food Chemistry, 80*(1), 17–21.
Santos, I. L., Schmiele, M., Aguiar, J. P. L., Steel, C. J., Silva, E. P., & Souza, F. D. C. (2019). Evaluation of extruded corn breakfast cereal enriched with whole peach palm (Bactris gasipaes, Kunth) flour. *Food Science and Technology, 40*, 458–464.
Saura-Calixto, F. (2011). Dietary fiber as a carrier of dietary antioxidants: An essential physiological function. *Journal of Agricultural and Food Chemistry, 59*(1), 43–49.
Sayed-Ahmad, B., Talou, T., Straumite, E., Sabovics, M., Kruma, Z., Saad, Z., Hijazi, A., & Merah, O. (2018). Evaluation of nutritional and technological attributes of whole wheat based bread fortified with chia flour. *Foods, 7*(9), 135.
Scholz-Ahrens, K. E., Ade, P., Marten, B., Weber, P., Timm, W., Açil, Y., Glüer, C. C., & Schrezenmeir, J. (2007). Prebiotics, probiotics, and synbiotics affect mineral absorption, bone mineral content, and bone structure. *The Journal of Nutrition, 137*(3), 838S–846S.
Singh, K. K., Mridula, D., Rehal, J., & Barnwal, P. (2011). Flaxseed: A potential source of food, feed and fiber. *Critical Reviews in Food Science and Nutrition, 51*(3), 210–222.
Slavin, J. (2003). Impact of the proposed definition of dietary fiber on nutrient databases. *Journal of Food Composition and Analysis, 16*(3), 287–291.
Slavin, J. L. (2008). Position of the American dietetic association: Health implications of dietary fiber. *Journal of the American Dietetic Association, 108*(10), 1716–1731.
Slavin, J., & Green, H. (2007). Dietary fibre and satiety. *Nutrition Bulletin, 32*, 32–42.
Sobota, A., Rzedzicki, Z., Zarzycki, P., & Kuzawińska, E. (2015). Application of common wheat bran for the industrial production of high-fibre pasta. *International Journal of Food Science & Technology, 50*(1), 111–119.
Sozer, N., Cicerelli, L., Heiniö, R. L., & Poutanen, K. (2014). Effect of wheat bran addition on in vitro starch digestibility, physico-mechanical and sensory properties of biscuits. *Journal of Cereal Science, 60*(1), 105–113.
Sto, G., Birketvedt, J. A., Florholmen, J. R., & Ryttig, K. (2000). Long term effect of fibre supplement and reduced energy intake on body weight and blood lipids in overweight subjects. *Acta Medica (Hradec Králové), 43*(3), 129–132.
Tarini, J., & Wolever, T. M. (2010). The fermentable fibre inulin increases postprandial serum short-chain fatty acids and reduces free-fatty acids and ghrelin in healthy subjects. *Applied Physiology, Nutrition, and Metabolism, 35*(1), 9–16.
Teimouri, S., Abbasi, S., & Scanlon, M. G. (2018). Stabilisation mechanism of various inulins and hydrocolloids: Milk–sour cherry juice mixture. *International Journal of Dairy Technology, 71*(1), 208–215.
Threapleton, D. E., Greenwood, D. C., Evans, C. E., Cleghorn, C. L., Nykjaer, C., Woodhead, C., Cade, J. E., Gale, C. P., & Burley, V. J. (2013). Dietary fibre intake and risk of cardiovascular disease: Systematic review and meta-analysis. *BMJ, 347*, f6879.
Trowell, H. C. (1978). Western diseases, Western diets and fibre. *East African Medical Journal, 55*(6), 283–289.
Venegas-Borsellino, C., & Kwon, M. (2019). Impact of soluble fiber in the microbiome and outcomes in critically ill patients. *Current Nutrition Reports, 8*(4), 347–355.
Verma, A. K., Rajkumar, V., Banerjee, R., Biswas, S., & Das, A. K. (2013). Guava (Psidium guajava L.) powder as an antioxidant dietary fibre in sheep meat nuggets. *Asian-Australasian Journal of Animal Sciences, 26*(6), 886.
Wanders, A. J., van den Borne, J. J., de Graaf, C., Hulshof, T., Jonathan, M. C., Kristensen, M., Mars, M., Schols, H. A., & Feskens, E. J. (2011). Effects of dietary fibre on subjective appetite, energy intake and body weight: A systematic review of randomized controlled trials. *Obesity Reviews, 12*(9), 724–739.
Wang, S., Gu, B. J., & Ganjyal, G. M. (2019). Impacts of the inclusion of various fruit pomace types on the expansion of corn starch extrudates. *LWT, 110*, 223–230.

Weickert, M. O., & Pfeiffer, A. F. (2008). Metabolic effects of dietary fiber consumption and prevention of diabetes. *The Journal of Nutrition, 138*(3), 439–442.
Wirkijowska, A., Zarzycki, P., Sobota, A., Nawrocka, A., Blicharz-Kania, A., & Andrejko, D. (2020). The possibility of using by-products from the flaxseed industry for functional bread production. *LWT, 118*, 108860.
Xie, F., Gu, B. J., Saunders, S. R., & Ganjyal, G. M. (2021). High methoxyl pectin enhances the expansion characteristics of the cornstarch relative to the low methoxyl pectin. *Food Hydrocolloids, 110*, 106131.
Yildiz, E., & Ozcan, T. (2019). Functional and textural properties of vegetable-fibre enriched yoghurt. *International Journal of Dairy Technology, 72*(2), 199–207.
Zheng, Y., Wang, Q., Huang, J., Fang, D., Zhuang, W., Luo, X., Zou, X., Zheng, B., & Cao, H. (2019). Hypoglycemic effect of dietary fibers from bamboo shoot shell: An in vitro and in vivo study. *Food and Chemical Toxicology, 127*, 120–126.

Chapter 14
miRNA-Based Genetic Engineering for Crop Improvement and Production of Functional Foods

Deepansh Sharma, Vanshika, Ajinder Kaur, and Pooja Manchanda

14.1 Introduction

Feeding an ever-increasing population is challenging, especially when crops are vulnerable to a variety of environmental and agronomic stresses that affect their growth and development. In this scenario, incorporating new technologies and ideas to improve agricultural production and yield becomes crucial (Gao et al., 2020). Genetic engineering has shown great promise in crop improvement and ensuring worldwide food security (Kahl & Winter, 1995). The use of gene transfer systems in genetic engineering has resulted in exceptional advances in the production of plants with improved agronomic and quality traits. Besides, the capacity to insert desirable foreign genes or traits into plant cells has also increased our understanding of plant biology (Gasser & Fraley, 1989). In the realm of agricultural crop enhancement, genetic engineering has been stated as one of the most influential technologies (John & Stewart, 2010).

A number of genes in plants work together to provide stress tolerance (Basso et al., 2019; Tang & Chu, 2017). Evidences from multiple microarray profiling efforts and genome-wide high-throughput sequencing suggest that microRNAs (miRNAs) have become increasingly intriguing because of the crucial roles that these play in several areas of a plant's life, including responses to environmental stresses, plant design and growth (Jones-Rhoades et al., 2006; Sunkar, 2010). There has been a large number of growing studies on miRNAs demonstrating their importance in a variety of plant molecular processes. The exploration of google patent

D. Sharma
College of Horticulture and Forestry, Punjab Agricultural University, Ludhiana, Punjab, India

Vanshika · A. Kaur (✉) · P. Manchanda
School of Agricultural Biotechnology, College of Agriculture, Punjab Agricultural University, Ludhiana, Punjab, India
e-mail: ajinder.biotech@pau.edu

S. Punia Bangar, A. Kumar Siroha (eds.), *Functional Cereals and Cereal Foods*,
https://doi.org/10.1007/978-3-031-05611-6_14

search engine (https://patents.google.com/) on August 4th, 2021 revealed that a total of 13,651 patent applications and grants have been filed connected to genetic engineering using plant miRNAs, with China having the most, followed by the United States, the World Intellectual Property Organization, the European Patent Office and some other countries (Patil et al., 2021).

microRNAs (miRNAs) are a type of endogenous single-stranded cDNA molecules that bind to target mRNAs at regions that are partially complementary to the miRNAs involved in the process (Tang & Chu, 2017). *MIR* genes are the genes that encode miRNAs (Gupta et al., 2014). Under the influence of many proteins, these genes are transcribed into primary transcripts (pri-miRNAs), which then undergo post-transcriptional modifications and stabilization to generate precursor microRNAs (pre-miRNAs). The RNA-induced silencing complex (RISC) is formed when pre-miRNA is integrated into the ARGONAUTE (AGO1) protein. The protein expression is inhibited when these RISC complexes are properly paired with an appropriate mRNA (Gao et al., 2020).

It has been shown that many miRNA families are conserved in the plant kingdom. Along with these conserved miRNAs, there are species-specific miRNAs that have evolved recently; hence their functions are yet to be discovered (Willmann & Poethig, 2007). miRNAs are responsible for the regulation of developmental processes in plants, including meristem development, growth of reproductive and vegetative organs, and establishment of organ polarity (Liu et al., 2018), thus any change in the expression of miRNA or the target mRNA leads to physiological and morphological abnormalities. Arabidopsis has been found to be a biologically amenable model plant, allowing researchers to investigate genetic mechanisms underlying miRNA-mediated gene regulation as well as phenotypic effects of disrupting miRNA-mediated gene regulation (Jones-Rhoades et al., 2006). miRNAs and their targets serve as a novel source of transgene candidates. To cite an example, artificial miRNA (amiRNA) technology, a highly targeted and efficient post-transcriptional gene silencing (PTGS) methodology, is being used in a growing number of transgenic investigations. The study of miRNA-target interactions has provided researchers a better understanding of post-transcriptional gene regulation mechanism and numerous molecular pathways that drive plant response to stress (Zhou & Luo, 2013). Important aspects of miRNA regulation as well as potential of miRNAs and their targets for crop improvement by either enhancing plant's response to extrinsic environmental stresses or improving intrinsic yield potential of plants by influencing plant development and changing plant architecture, have been introduced in the following sections.

14.2 microRNA: Genomics

microRNA is one of the classes of small non-coding RNAs (ncRNA) i.e. these do not code for proteins. The miRNAs found in plants are made from definite stem regions of single-stranded hairpin precursors, which have different attributes from that of other ncRNAs (Wang et al., 2019). The length of non-coding miRNAs is generally 22 nucleotides (Berezikov et al., 2006). About half of the miRNAs found are present in clusters and transcribed by the genes as polycistronic primary transcripts (Ghosh, 2011). The first miRNAs to be discovered were encoded by *lin-4* and *let-7* genes of *Caenorhabditis elegans* (roundworm). These were discovered due to mutations in above mentioned genes resulting in defects during transition from larva to adult phase of *C. elegans* (Phelps-Durr, 2010; Wightman et al., 1993). The discovery of miRNAs in animals triggered scientists to look for miRNAs in plants. In 2002, many research groups autonomously discovered more than 100 miRNAs in *Arabidopsis* using direct cloning technique. Presently, direct cloning technique has been replaced by deep sequencing technique for miRNA identification. A number of miRNAs are extremely conserved in plants and are found in tobacco (*Nicotiana*), rice (*Oryza sativa*), and maize (*Zea mays*) (Sun, 2012).

Various miRNAs have been found to play distinct roles in plant species using deep sequencing technique. Easy availability, cost efficiency along with speed and ease of sequencing has led to vast research on miRNA discovery (Kozomara & Griffiths-Jones, 2014). For example, the deep sequencing technique has been used to identify miRNA that plays an important role in fruit ripening in tomato (Moxon et al., 2008).

The direct cloning is also known as shot-gun molecular cloning, and is used for identification of small RNAs by cDNA cloning of RNA transcripts. The identification of miRNAs can be carried out by combining direct cloning with parallel sequencing technology (Liu et al., 2009). The small RNAs are isolated from the tissue/cell of interest and separated by gel electrophoresis, sequenced and miRNAs are identified from these small RNAs (Wang & Li, 2007).

14.2.1 Origin

Various studies prove that many microRNA families are evolutionarily conserved in main classes of kingdom Plantae, including mosses, gymnosperms, monocots and dicots. Hence, gene expression regulated by microRNAs seems to exist from the beginning of plant evolution and conserved in a functional manner for over 425 million years (concerning the evolutionary origins, various mechanisms have been proposed) (Sun, 2012). The very ancient origin of microRNA is evident from some microRNA families conserved in mosses, for example miR156, miR160, miR319 and miR390 regulating ancestral transcriptional factors, which control cell division, basic meristem functions, hormonal control or organ polarity and separation (Voinnet, 2009). In plants, some old microRNA encoding genes were lost during evolution and new microRNA genes emerged. Usually, the young microRNA genes are species-specific and have low expression levels. It has been reported that about half of microRNA gene families present in angiosperm ancestry have been lost over time in some of the plant species studied.

Furthermore, there are various mechanisms proposed for the origin of miRNAs, which include:

1. Duplication of pre-existing miRNA gene/ protein coding gene
2. Generation from transposable elements
3. Formation of hairpin structures during genome evolution

The first two are more likely to contribute to miRNA origin in plants, whereas the third one is more common in animals (Zhang & Wang, 2015).

14.2.2 Biogenesis

There are more than 100 *MIR* genes in most plants, which are usually found in intergenic regions of genome (Rogers & Chen, 2013). Biogenesis process of microRNA involves three major steps, namely

- Transcription
- Post-transcriptional processing
- Stabilization of microRNA

The process starts in the nucleus where *MIR* gene is transcribed by the action of DNA-dependent RNA polymerase II (Fig. 14.1) The transcript formed is called primary microRNA i.e. pri-miRNA that is capped and polyadenylated at 5′ and 3′ ends, respectively. This pri-miRNA has hairpin-like structures, which is recognized and cleaved by various proteins (Gao et al., 2020).

The post-transcriptional processing takes place at the dicing body (d-body) by the action of dicing complex that comprises of various proteins. DAWDLE (DDL) is aRNA binding protein that stabilizes pri-miRNA by interacting with Dicer-like 1 protein (DCL1) and cleaves off the guanine cap at 5′ end and polyadenine tail at 3′ end as shown in Fig. 14.1 (Ha & Kim, 2014; Rogers & Chen, 2013). The stem-loop hairpin-like structure is processed by DCL1 (a RNase III family enzyme) by sequential cleavage, hence forming the precursor microRNA (pre-miRNA) (Yu et al., 2017). The multiple protein complexes including the double-stranded RNA binding HYPONASTIC LEAVES 1 (HYL1), G-patch structural protein (TGH) and zinc-finger protein SERRATE (SE) work together to form miRNA/miRNA* duplex, where miRNA is guide strand and miRNA* is passenger strand (Zhang & Wang, 2015). The mature miRNA that will regulate the gene expressions is the guide strand and the other strand that will be degraded is the passenger strand (Meijer et al., 2014). The processed miRNA/miRNA* duplex is 2′-O-methylated at its 3′ end by the action of HUA Enhancer 1 (HEN1) to prevent uridylation and breakdown of miRNA (Fig. 14.1). The methylated miRNA/miRNA* duplex is then transported in the cytoplasm from the nucleus with the help of HASTY (HST), and helicase separates the miRNA/miRNA* duplex to form the mature miRNA, and the miRNA* is degraded (Gao et al., 2020; Ha & Kim, 2014; Yu et al., 2017; Zhang & Wang, 2015). The resulting mature miRNA incorporates into ARGONAUTE protein (AGO1),

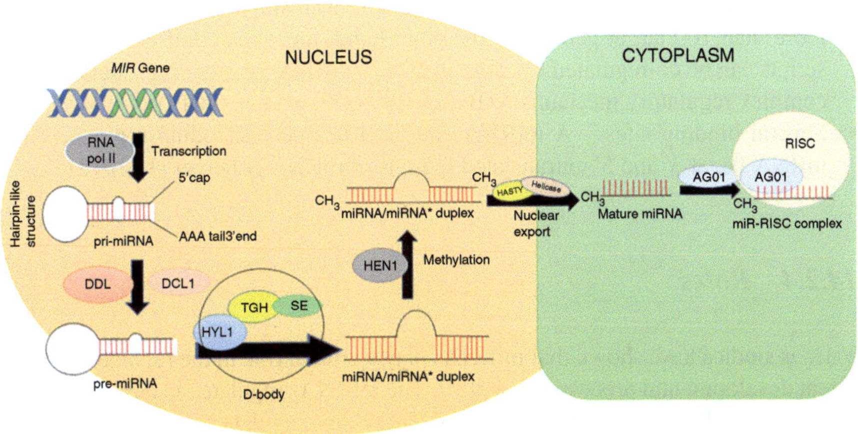

Fig. 14.1 Biogenesis of plant miRNA. The process of biogenesis starts in the nucleus with the transcription of the *MIR* gene with the action of DNA-dependent RNA polymerase II, which results in the formation of pri-miRNA. The pri-miRNA has a guanine cap at the 5′ end and a poly-adenine tail at the 3′ end, which are cleaved off by the action of DAWDLE (DDL) resulting in the formation of pre-miRNA. This pre-miRNA is then converted to miRNA/miRNA* duplex by the action of d-body that contains several proteins, including HYPONASTIC LEAVES 1 (HYL1), TGH, and SERRATE (SE). This miRNA/miRNA* duplex is methylated by a protein called HUA Enhancer 1 (HEN1). The methylated miRNA/miRNA* duplex is transported out of the nucleus to the cytoplasm with the help of HASTY (HST) protein and the two strands of miRNA/miRNA* duplex are separated by the action of helicase enzyme. After reaching the cytoplasm, miRNA incorporates into ARGONAUTE (AGO1) and combines with the RNA-induced silencing complex (RISC). These complexes can reduce gene expression by cleaving mRNAs or repressing their translation

which further combines with other proteins for forming of RNA-induced silencing complex (RISC). These complexes can reduce gene expression by cleaving mRNAs or repressing their translation (Gao et al., 2020). Figure 14.1 depicts a diagrammatic model of the biogenesis process in plants for better understanding.

14.2.3 Features

There are various notable features of miRNAs that are listed below:

- Small molecules – These are small in size (approximately 22 nucleotides long). Each miRNA contains a seed region/seed sequence (a conserved heptameric sequence, which is mostly situated at positions 2–7 from the miRNA 5′-end) (Berezikov et al., 2006).
- Vital – miRNAs play important roles in development, cell differentiation, apoptosis, innate immunity, molecular metabolism based on post-transcriptionally and dynamically negative regulation.

- High proficiency – These are highly efficient, as a single miRNA may regulate more than 100 target genes. On the other hand, more than half of mRNA transcripts can be co-regulated by more than two miRNAs, explaining a detailed and complex regulatory mechanism (Bartel, 2009; Zhuo et al., 2013).
- Several binding sites – A miRNA can bind to many encoding regions of the mRNA i.e. at 3′ and 5′ untranslated regions of mRNA (Zhuo et al., 2013).

14.2.4 Roles

Various studies have shown that miRNAs play a crucial role in the regulation of different developmental processes in a plant. The miRNAs regulate the development of anthers, vascular components and leaves. These also regulate root and flower morphogenesis. Moreover, miRNAs play a vital role in regulating biotic and abiotic stress responses in plants (Lotfi et al., 2017).

In *Arabidopsis* and maize, miR156 plays an important role in developing novel cultivars (Axtell & Bowman, 2008). It is present in high level during seedling stage and decreases as the plant age, this pattern is also observed in other plant species, including rice (*Oryza sativa*), tomato (*Solanum lycopersicum*), chinese cabbage (*Brassica rapa* subsp. *pekinensis*), maize (*Zea mays*), tobacco (*Nicotiana tabacum*), bittercress (*Cardamine flexuosa*) and alpine rockcress (*Arabis alpine*). miR156 is essential for the maintenance of juvenile plant phase, and if overexpressed, it delays flowering and prolongs the juvenile phase of a plant. In *Arabidopsis,* the various physiological and developmental processes regulated by miR156 are leaf shape, leaf complexity, leaf initiation rate, secondary metabolite accumulation, shoot regeneration capacity, flowering, stress response, embryo formation, trichome initiation and innate immunity (Yu et al., 2015).

It has been discovered that miRNAs are primary modulators in salinity stress conditions as these can control expression of genes responsible for salinity response (Shuai et al., 2013). Salinity prompts miRNAs expression/activity and down-regulates their respective target mRNAs, hence producing negative functional proteins that are engaged in salinity response. Contrarily, different down-regulations of other miRNAs result in the accumulation of positively contributing target mRNAs (Lotfi et al., 2017).

Furthermore, miRNAs play a key role in nutrient homeostasis in plants by transmuting target gene's expression. Different miRNAs have been reported for uptake and transportation of important minerals in plants. The miRNA families, miR1A69 and miR172 are solely found in plant nodules and are involved in providing tolerance to plants against nitrogen (N), phosphorous (P) and manganese (Mn) stresses; also, during nitrogen (N) and sulphur (S) starvation, upraised levels of miR167 and miR395 have been reported. Thus, nutritional homeostasis conditions can be regulated by certain miRNA through up-regulation or down-regulation of target genes expressions, suggesting that miRNA plays a common signaling role in regulating diverse nutrition stress responses (Paul et al., 2015).

In addition to these, miRNAs are also involved in antiviral defence during viral infections in plants. Turnip (*Brassica rapa*) infected by Turnip Mosaic Virus (TuMV) exhibited upregulated expressions of miR158 and miR1885 (Jatan & Lata, 2019).

14.3 Scope of Genetic Engineering in Crop Improvement

Biotechnology has revolutionised the era with its broad scope in traditional methods of crop enhancement. Genetic engineering, a branch of biotechnology, has enabled the incorporation of desired genes of interest in agricultural crops for the sole purpose of crop improvement. Genetic engineering is a complex procedure that involves isolating a desired gene, creating a recombinant DNA molecule by ligating the desired gene into a vector and then introducing the gene into host plant to develop desired functions (Thakur et al., 2012). Limited resources have made meeting agricultural demands of an increasing population challenging. However, genetic engineering with its promise for agricultural enhancement, and plant breeding have the ability to meet these challenges (John & Stewart, 2010).

Genetic engineering has also aided in the development of disease-resistant crop plants. Through the sole expression of TMV coat protein gene in plants, resistance to tobacco mosaic virus (TMV) infection, defined as protein coat mediated protection, has been obtained (Abel et al., 1986). Moreover, ripening, a highly synchronised and developing process, is accompanied by an ethylene burst in case of fruits. One of the key concerns was that 50 percent of fruits used to rot before shipping owing to ethylene-induced ripening. Genetic engineering, on the other hand, improved the fruit quality by lowering the ethylene concentration. A desired gene responsible for decomposing 1-amino cyclopropane-1-carboxylic acid, a forerunner of ethylene biosynthesis and an anti-sense version of the gene for ethylene producing enzyme was introduced in tomato plants, which considerably lowered ethylene concentration and slowed fruit ripening process by 2 weeks (Kahl & Winter, 1995).

Insect resistance has also been boosted by genetic engineering through introduction of genes of *Bacillus thuringiensis*, an entomocidal bacterium that kills some insect pests. *Bacillus thuringiensis* strains are toxic to a variety of dipteran, coleopteran and lepidopteran larvae, but not to humans, animals or useful insects. The insecticidal protein's mechanism of action is assumed to be interruption of ion transport across sensitive brush border membranes of an insect (Gasser & Fraley, 1989). Also, when it comes to crop output, abiotic and biotic stresses are key concerns. Genetic engineering technique is effective in fighting both crops biotic and abiotic stresses. CRISPR (clustered regularly interspaced short palindromic repeat) technology, a genetic technique is garnering a lot of attention these days, and has aided in the modification of genes that confer drought resistance (Sami et al., 2021).

14.4 microRNA: A Potential Tool in Genetic Engineering

miRNA, a short non-coding regulatory RNA with a length of 20–22 nucleotides, has been widely employed to produce genetically altered crops. miRNAs have been shown to influence the majority of metabolic and biochemical processes in plants (Khraiwesh et al., 2012). It is worth noting that miRNA is regarded as one of the master controllers, regulating a variety of gene networks (Jones-Rhoades et al., 2006). The analysis of their functions and regulatory network aids in investigating their role in conferring plant resistance to various stresses (Sun, 2012).

Plants have a highly complex and coordinated regulatory network of miRNAs that confer stress tolerant response through modification of gene expression involved in numerous physiological, metabolic and cellular processes (Vinocur & Altman, 2005). Under normal conditions, plants consume a variety of resources for growth and development, but growth is slowed when weather takes its toll on the plant, a stage in which conserved miRNA plays a critical role and target mRNA from a variety of transcriptional factors (Jones-Rhoades et al., 2006). It has been found that during such circumstances, the content of conserved miRNA increases in plants, as does the modulation of target genes, implying that miRNA regulates plant growth and development during stress conditions (Khraiwesh et al., 2012).

Furthermore, plants are vulnerable to a variety of climatic changes, which affect their product quality and yield, resulting in significant losses to the growers. It is imperative to offset these losses by incorporating resilience to climate stresses. miRNA has emerged as a possible genetic engineering tool for modifying gene expression in plants (Patil et al., 2021). For example, drought resistance was developed in barley (*Hordeum vulgare*) through overexpression of Hv-miR827 (Ferdous et al., 2017).

In the realm of functional genomics, artificial miRNAs have emerged as one of the potential molecules for mediating gene silencing. This technology can be employed to improve a lot of agronomic features as well as incorporate different nutritional qualities into crops. There are numerous examples that support artificial miRNA role in crop enhancement. For example, targeting of anther-specific TATA-binding protein (TBP) associated factor genes (*SmTAF10* and *SmTAF13*) in *Solanum melongena* with artificial miRNA resulted in full male sterility of pollen (Albright & Tjian, 2000; Tiwari et al., 2014; Toppino et al., 2011).

14.5 Strategies for microRNA-mediated Genetic Engineering

miRNAs provide an indispensable source for the novel transgenes as well as aid in the development of various strategies for genetic modification, thus allowing development of new crop cultivars having agronomically beneficial traits (Zhou & Luo, 2013). A better knowledge of its possible strategies and mechanisms involved would make it easier to incorporate desirable features into agricultural crops, while minimizing any harmful effects. Some of these strategies are discussed below:

14.5.1 Overexpression of miRNA

The overexpression of miRNA refers to the generation of mature miRNAs in large amounts, which result in suppression of negative regulatory genes under stress, making the crop resistant to stress. This helps in the development of superior trait when the miRNA targets are negative regulators of stress (Mandal et al., 2021).

For example, overexpression of miR1139 has improved the traits like biomass, phenotype, photosynthesis and inorganic phosphorous (Pi) acquisition in wheat (*Triticum aestivum).* Moreover, the elevated expression of miR408 in rice (*Oryza sativa*) resulted in improved grain yield by increasing the grain number and number of panicle branches (Zhang et al., 2017).

However, many side effects may arise due to overexpression of miRNA. These side effects mainly constitute the pleiotropic effects, which can be avoided by RNA interference (RNAi) technology (Mandal et al., 2021).

14.5.2 Down-Regulation of miRNA

Down-regulation of miRNA is taken into account when the miRNA functions as a negative regulator i.e. when the targets of miRNA are positive regulators of superior traits. This is also known as upregulation of the target genes. The best techniques for the down-regulation of miRNA are miRNA sponge (Ebert & Sharp, 2010), target mimicry (Tang et al., 2008), short tandem target mimic (STTM) etc. (Teotia & Tang, 2017).

For example, the knockdown of miR396 in rice by using the target mimic method resulted in increased yield. Similarly, the downregulation of miR399 in citrus resulted in male sterility, this was done using STTM method (Mandal et al., 2021).

14.5.3 Generation of Artificial miRNAs

Artificial miRNAs (amiRNAs) are an effective tool for the regulation of gene expression in plants. These are convenient as well as quick for reverse genetic technique (Liang et al., 2012). Just like miRNAs, amiRNAs are also single-stranded and 21 nucleotides long that are designed by altering the precursor miRNA complex by replacing the sequence of miRNA/miRNA* duplex, hence forming the artificial miRNA by natural biogenesis pathway that is used to target gene of interest (Khraiwesh et al., 2008). Figure 14.2 show miRNA-mediated mRNA silencing in plants.

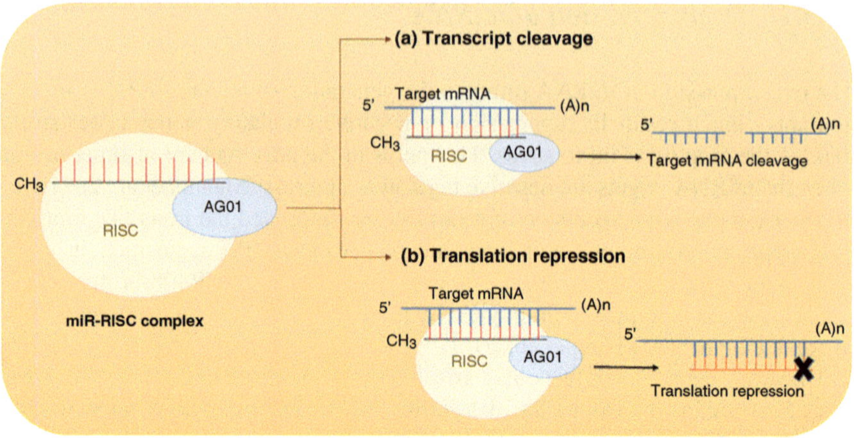

Fig. 14.2 miRNA-mediated mRNA silencing. The diagram depicts two methods of miRNA-mediated mRNA silencing. The first method is the cleavage of target mRNA transcript; the miR-RISC complex binds to the target mRNA and causes it to be cleaved at an appropriate site. The second method of repression is translational repression, in which miRNA attaches to the target mRNA, producing a stutter in translation, thereby silencing the target mRNA

14.6 Applications of microRNA-based Crop Improvement

miRNAs are a large family of small regulatory RNAs that play critical roles in all physiological and metabolic plant functions, including not just plant development and growth, but also compound biosynthesis and stress response. miRNA-based technology is emerging as a novel crop improvement strategy that will play a key role in agricultural sustainability in the future. Table 14.1 illustrates the potential importance of plant miRNAs in crop enhancement.

More than half of the world's population eats rice as their primary source of nutrition. Like other crops, rice is subjected to a variety of environmental stresses, including drought, salt, cold, heat and nutrient deficiency, all of which are key limiting factors for crop's development and output. Therefore, increasing rice production becomes critical using efficient technologies. Increasing crop productivity is a major challenge for modern agriculture, and as a result miRNA-based crop improvement has been used to increase grain yield in rice through manipulation of plant architecture. miR156-mediated regulation of *OsSPL14* (Squamosa Promoter Binding Protein) determines ideal plant architecture in rice. miR156 positively regulated the expression of *OsSPL14* and resulted in drastic enhancement in yield of rice grains. Overexpression of *OsSPL14* has also been found to modify rice plant architecture with decreased tiller number (with fewer unproductive tillers), increased grain yield and elevated lodging resistance (Jiao et al., 2010).

Biofuels derived from biomass crops have the potential to meet a substantial chunk of our transportation fuel requirements. miRNAs have emerged as an important class of gene regulatory factors with the ability to improve complicated features

Table 14.1 miRNA-based genetic engineering of crops for improvement of different traits

S. no.	Improved trait	miRNA(s) involved	Gene targeted	Crop	Reference(s)
1	Resistance to fruit borer	amiR319 (artificial miRNA)	Ecdysone receptor gene (*HaEcR*)	Tomato (*Solanum lycopersicum*)	Yogindran and Rajam (2021)
2	Tolerance to cold	miR398	Cu-Zn superoxide dismutase gene (*CSD1*)	Wheat (*Triticum aestivum*)	Lu et al. (2020)
3	Tolerance to drought, abscisic acid and poly-ethylene glycol	miR535	Squamosa promoter binding protein like gene (*SPL19*)	Rice (*Oryza sativa*)	Yue et al. (2020)
4	Enhanced anthocyanin biosynthesis	miR828	Trans acting siRNA4 (*TAS4*), activation of anthocyanin pigment 1 gene (*PAP1*)	Turnip (*Brassica rapa*)	Zhou et al. (2020)
5	Tolerance to drought	miR169	Transcription factors (*AtNF-YA1* and *AtNF-YA5*)	*Arabidopsis thaliana*	Yu et al. (2019)
6	Ripening regulation	miR156 and miR157	Colourless non-ripening gene (*CNR*)	Tomato (*Solanum lycopersicum*)	Moxon et al. (2008)
		miR1917	Constitutive triple response 4 genes (*LeCTR4sv1* and *LeCTR4sv2*)		
7	Tolerance to cadmium accumulation	miR166	Homeodomain containing protein 4 gene (*OsHB4*)	Rice (*Oryza sativa*)	Ding et al. (2018)
8	Tolerance to drought	miR827	Aberrant pollen transmission 1 gene (*APT1*), nucleotide-binding site – leucine-rich repeat (NBS-LRR) proteins	Barley (*Hordeum vulgare*)	Ferdous et al. (2017)

(continued)

Table 14.1 (continued)

S. no.	Improved trait	miRNA(s) involved	Gene targeted	Crop	Reference(s)
9	Resistance against viral diseases	miR528	L-ascorbate oxidase messenger RNA gene (*AO*)	Rice (*Oryza sativa*)	Wu et al. (2017); Fan et al. (2016); Yan et al. (2014); Guo et al. (2013)
10	Induction of male sterility	miR2118	Photoperiod sensitive genetic male sterility 1 gene (*PMS1T*)		
11	Enhanced tillering	miR444	MADS-box transcriptional factors (*MADS57* and *MADS23*)		
12	Tolerance to salt stress	miRNVL5 (miRNA ovual line 5)	Zinc-finger domain transcriptional factor (*GhCHR*)	Cotton (*Gossypium hirsutum* L.)	Gao et al. (2016)
13	Enhanced growth and development of primary roots	miR160	Auxin response factors (*ARF10*, *ARF16* and *ARF17*)	*Arabidopsis thaliana*	Zhang and Wang (2015); Wang et al. (2010)
14	Improvement in crop productivity	miR156	Squamosa promoter binding protein like genes (*SPL6*, *SPL12* and *SPL13*)	Alfa alfa (*Medicago sativa*)	Aung et al. (2015)
15	Tolerance to drought	miR408	Dehydration responsive element binding protein 2A and 1A genes (*DREB2A* and *DREB1A*)	Chickpea (*Cicer arietinum* L.)	Hajyzadeh et al. (2015)
16	Tolerance to Vericillium wilt	miR482	Nucleotide-binding-site-leucine-rich repeat gene (*NBS-LRR*)	Potato (*Solanum tuberosum*)	Yang et al. (2015)
17	Resistance to turnip mosaic and turnip yellow mosaic virus	amiR159	Viral silencing repressor proteins (HC-Pro and P69)	*Arabidopsis thaliana*	Kamthan et al. (2015)
18	Improved development of cotton fibre	miR828	D genome species (*GhMYB2D*)	Cotton (*Gossypium hirsutum* L.)	Guan et al. (2014)

(continued)

Table 14.1 (continued)

S. no.	Improved trait	miRNA(s) involved	Gene targeted	Crop	Reference(s)
19	Resistance against stripe rust	miR2013	Monodehydro-ascorbate reductase gene (*TAMDHAR*)	Wheat (*Triticum aestivum* L.)	Feng et al. (2014); Wang et al. (2012)
20	Tolerance to heat	miR159	Transcriptional factor (*TaGAMYB*)		
21	Resistance against pathogens	miR7695	Natural resistance-associated macrophage protein 6 (Nramp6)	Rice (*Oryza sativa*)	Campo et al. (2013)
22	Tolerance to salt and drought	miR319	Proliferating cell factors genes (*PCF5*, *PCF6* and *PCF8*)	Creeping bent grass (*Agrostis stolonifera*)	Zhuo et al. (2013)
23	Improved floral meristem determinacy	miR171	Scarecrow transcriptional factor gene (*HvSCL*)	Barley (*Hordeum vulgare* L. *cv. Golden promise*)	Curaba et al. (2013)
24	Resistance to leaf curl virus	amiR-AV1–1	Coat protein gene (*AV1*) and Pre-coat protein gene (*AV2*)	Tomato (*Solanum lycopersicum*)	Van Vu et al. (2013)
25	Increase in overall number of leaves	miR156	Squamosa promoter binding proteins (SBP)	*Arabidopsis thaliana*	Kim et al. (2012); Schwab et al. (2005)
26	Resistance to nematodes	miR396	Growth regulating factors (*GRF1*/*GRF3*)	*Arabidopsis thaliana*	Hewezi et al. (2012)
27	Flower development	miR156, miR159 and miR172	Meristem identity regulator (*APETALA2*)	*Arabidopsis thaliana*	Yamaguchi and Abe (2012)
28	Enhanced plant growth and development	miR156	Squamosa binding protein factors (SBP)	Peanut (*Arachis hypogea* L.)	Chi et al. (2011)
29	Tolerance to nutrient deficiency	miRNA399	Ubiquitin conjugating enzyme gene (*UBC*)	*Arabidopsis thaliana*	Zhao et al. (2011); Kawashima et al. (2009); Fujii et al. (2005)
		miRNA169	Nuclear factor Y sub-unit A (*NFYA*)		
		miRNA395	Sulphate transporter gene		
30	Increased grain productivity	miR156	Squamosa promoter binding protein like factor (*SPL14*)	Rice (*Oryza sativa*)	Miura et al. (2010)

(continued)

Table 14.1 (continued)

S. no.	Improved trait	miRNA(s) involved	Gene targeted	Crop	Reference(s)
31	Promotion of lateral root growth	miR390	Auxin response factors (*ARF2*, *ARF3* and *ARF4*)	*Arabidopsis thaliana*	Yoon et al. (2010); Marin et al. (2010)
32	Initiation and growth of adventitious roots	miR160 and miR167	Auxin response factors (*ARF6*, *ARF8* and *ARF17*)	*Arabidopsis thaliana*	Gutierrez et al. (2009)
33	Effective control over tuberization time	miR172	RAP1 protein (related to APETALA 2 and 1)	Potato (*Solanum tuberosum* L.)	Martin et al. (2009)
34	Tolerance to drought	miR169	Nuclear factors (*YA5* and *NFYA5*)	*Arabidopsis thaliana*	Li et al. (2008)
35	Resistance to cauliflower mosaic virus	amiR171	Viral suppressor of RNA silencing (*CMV2b*)	Tobacco (*Nicotiana tabacum cv. SR1*)	Qu et al. (2007)
36	Induction of parthenocarpy	miR167	Auxin response factor 8 (*ARF8*)	Tomato (*Solanum lycopersicum*) and *Arabidopsis thaliana*	Goetz et al. (2007)
37	Controlled bacterial speck disease	miR393	Transport inhibitor response 1 gene (*TIR1*)	Tomato (*Solanum lycopersicum*)	Fahlgren et al. (2007); Navarro et al. (2006)
38	Tolerance to mechanical stress	miR473	UV-B resistant gene (*UVR8*)	*Populus trichocarpa*	Lu et al. (2005)
39	Tolerance to alkali and salts	miR396	Gene encoding putative transcriptional factors (*LOC_Os01g32750*, *LOC_Os02g45570* and *LOC_Os04g51190*)	Rice (*Oryza sativa*) and *Arabidopsis thaliana*	Gao et al. (2010)

like biomass output. miRNA-based technologies for increasing biomass production have been explored in switchgrass (*Panicum virgatum*), an important biofuel crop. One miR156 precursor was overexpressed in switchgrass that resulted in an overall increase in tiller number, which was responsible for a significant rise in biomass yield. Targeted overexpression of miR156 enhanced solubilized sugar yield and fodder digestibility, as well as provided an effective method for transgene containment (Fu et al., 2012). These transgenic plants were also found to contain up to 250 percent more starch and better digestibility than control plants. In maize, overexpression of *corngrass1* (*Cg1*) encoding miR156 resulted in an extended vegetative phase and delayed blooming, both of which are beneficial features for bioenergy crops due to enhanced biomass output. Moreover, the maize *Cg1* gene was overexpressed in different plants to fix it during juvenile phase of development and examine its impact on biofuel production. *Cg1* is a unique tandem miRNA gene that has been identified in grass species till date (Chuck et al., 2011).

Global warming, climate change as well as environmental pollution are important irreversible problems that humans confront, resulting in serious concerns, such as drought, salinity, pollution in soil, besides high and low temperatures. These abiotic stresses have a considerable impact on plant growth and development, posing a serious challenge to sustainable agricultural production. miRNAs have emerged as a new key player in plant response to abiotic stress and increase in plant tolerance to these stimuli. Plant tolerance to abiotic stresses is considerably influenced by manipulating a single miRNA. For example, overexpression of miR319 greatly improved creeping bentgrass (*Agrostis stolonifera*) resistance to drought and salinity stress. In rice, however, miR319 was responsible for not only boosting leaf blade size, but also significantly improving cold stress resistance. Table 14.1 lists several other miRNAs implicated in conferring resistance to abiotic stresses in plants, namely miR169, miR398, miR827, miR408, miR535 and miR159 (Yang et al., 2013; Zhang, 2015; Zhuo et al., 2013).

To the best of our knowledge, there has been no example of miRNA being used directly to engineer crop nutrient quality. However, there have been an increasing number of successful examples of using RNAi technology to adjust the nutritional composition of crops. For example, high lysine maize was developed by employing RNAi technique to decrease the accumulation of zein proteins (Houmard et al., 2007). A synergistic technique combining seed-specific expression and RNAi knockdown of target genes increased lysine content in Arabidopsis seeds considerably (Angelovici et al., 2011). The nutritional value of tomatoes was enhanced by employing fruit-specific RNAi-mediated inhibition of an endogenous photomorphogenesis regulatory gene (*DET1*), which resulted in increased carotenoid and flavonoid content (Davuluri et al., 2005). RNAi was also utilized to inhibit the starch-branching enzyme, resulting in high-amylose wheat, which has the potential to benefit human health (Regina et al., 2006). Figure 14.3 shows miRNA-mediated genetic transformation of *Oryza sativa*.

14.7 Functional Foods and Their Production

Functional foods are substances that provide health advantages in addition to their nutritional value. These foods are often high in essential nutrients such as vitamins, minerals, healthy fats, fibre and a variety of bioactive substances, and are therefore linked to a number of substantial health benefits. They not only prevent vitamin deficiencies, but also defend against different diseases and support appropriate growth and development. So, production of these foods becomes vital. Cereal grains contain a variety of phytochemicals including phenolic acids, flavonoids, phytic acid, coumarins and terpenes, a significant amount of ferulic acid, glutathione and

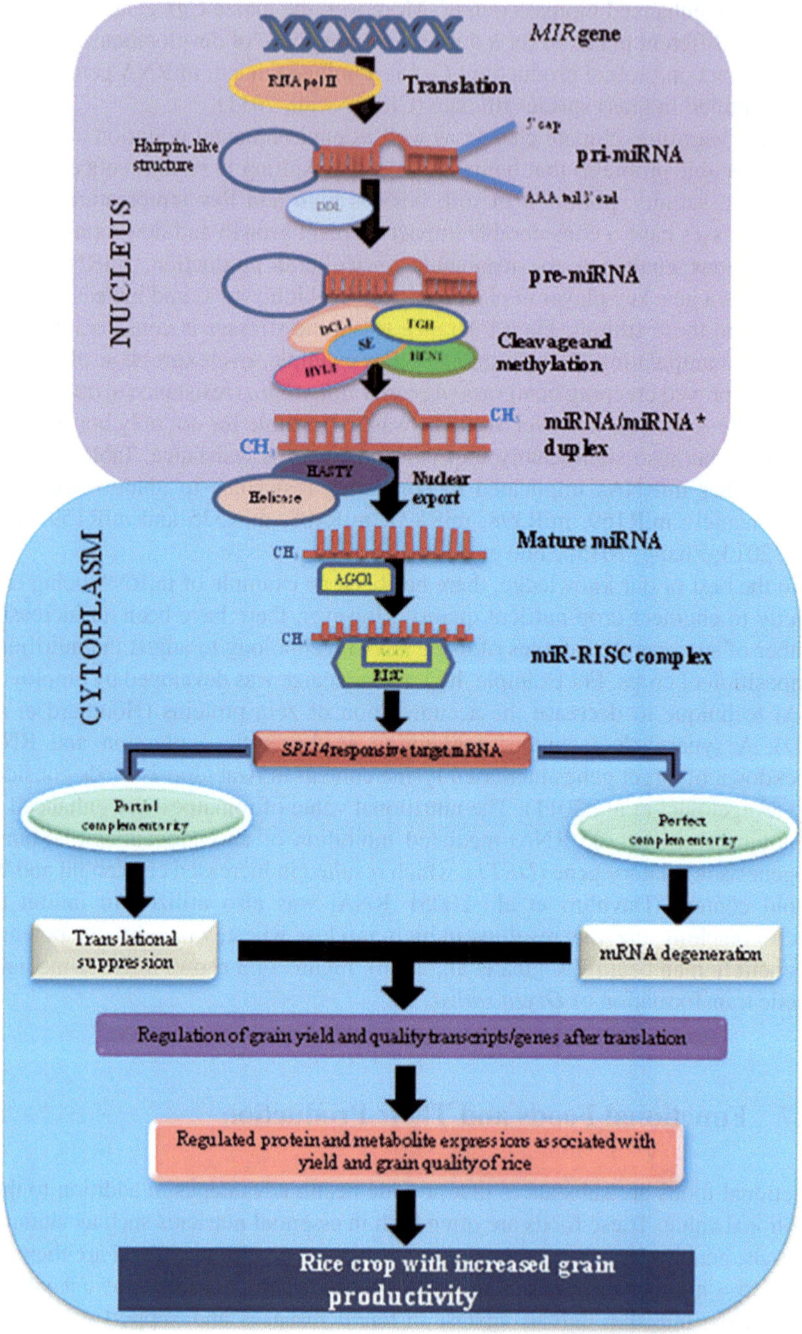

Fig. 14.3 miRNA-mediated genetic transformation of *Oryza sativa* to increase grain productivity and quality is shown in the figure. The mature miRNA formed by plant miRNA biogenesis pathway forms the miR-RISC complex. This complex attaches to the Squamosa promoter binding

phytosterols. Cereals constitute a crucial food component for the production of functional foods due to their high nutritional content (Sidhu et al., 2007).

Plant genetic engineering has been employed to create foods having increased quantity of endogenous nutrientsor to introduce foreign molecules in plants. The present biotechnological techniques has enabled purposeful manipulation ofmetabolic pathwaysin order to improve poly-unsaturated fatty acids (PUFA) and carotenoids production in plants. Cereals, as the primary food source, are low in key elements like PUFA and various essential pigments. PUFA and carotenoid enriched cereals have a unique potential for increasing the nutritional content of cereal crops, resulting in a variety of health benefits (Čertík et al., 2013).

RNA interference (RNAi) is an effective technique for gene silencing in plants and other organisms. It can be used in functional genomics to decipher gene function. This technique has also seen a number of possible applications in agriculture for improvement of crop nutrient content. For example, RNAi has been employed to increase amylose starch content in wheat, where starch branching enzymes of class II (*SBEIIa*) in two durum wheat cultivars utilising two different transformation methods (biolistic and *Agrobacterium*) were silenced. Amylose starch has gathered a huge attention because of its relationship with quality resistant starch, which provides plethora of health benefits to humans (Sestili et al., 2010).

Besides RNAi, miRNAs have been used as a potential candidate for enhancing iron (Fe) and zinc (Zn) content of cereal grains. The expression of miRNAs involved in miRNA biogenesis and nutritional homeostasis revealed that these small RNAs may play a role in the control of Fe and Zn metabolism in rice. The participation of miRNA-mediated regulation of Fe homeostasis was demonstrated by the downregulation of miRNAs and subsequent overexpression of their putative target genes (Agarwal et al., 2018). However, in rice grains the involvement of miRNAs in control of Fe and Zn has rarely been studied.

Phytosterols and brassinosteroids (BR) are considered key functional nutrients, with a considerable increase seen in rice grains. In rice, miR1848 post-transcriptionally modulates the expression of demethylase gene (*OsCYP51G3*) and thereby mediates phytosterol and BR biosynthesis. miR1848 and *OsCYP51G3* have the potential to be used in rice breeding programs to control leaf angle, seed size and quality (Zhang & Wang, 2016).

Studies on miRNAs in genome biology have provided a better understanding of the transcriptional control mediated by miRNAs in plants, but a lot is yet to be discovered. Furthermore, as miRNA is an emerging technology, its significance in crop nutritional improvement is relatively limited. However, computational and experimental methodologies will enable further discoveries in identifying potential role of

Fig.14.3 (continued) protein-like factor (*SPL14*) responsive mRNA to which it can be partially or completely complimentary. The partial complementarity leads to suppression of translation, whereas perfect complementarity leads to degradation of target mRNA, both of these result in post-transcriptional regulation of grain productivity and grain quality transcripts. This further leads to the regulation of protein and metabolite expressions associated with grain productivity and quality, hence resulting in increased grain productivity in rice crop

plant miRNAs in production of functional foods in future. This would eventually broaden the scope of miRNAs and aid in the development of functional foods with increased nutritional content (Challam et al., 2019).

14.8 Role of microRNAs in Production of Plant Secondary Metabolites

Secondary metabolites are organic compounds found in extremely low quantities in plants. Despite the fact that they are not directly engaged in plant growth and development, they are regarded as critical elements, since their absence could result in long-term damage, and even plant death (Bartwal et al., 2013). Terpenes, alkaloids, phenolics, glycosides, tannins and saponins are among the main phytochemical substances found in secondary metabolites. These metabolites are expected to function in a plant-environment interaction, protecting plants from a variety of environmental stresses. In recent years, miRNAs have been used as a potential technique to regulate metabolite production, since they can positively or negatively regulate the production of predicted metabolites, while limiting the creation of harmful metabolites and increasing production of novel compounds. Various computational and experimental methods are being employed for miRNAs identification (Sabzehzari & Naghavi, 2019). Table 14.2 summarises several plant miRNAs that have been found and their important contribution in the enhancement of plant secondary metabolites.

14.9 Advantages of microRNAs

miRNAs offer following advantages in crop improvement:

I. miRNAs play an inevitable role in plant development as well as provide resistance to diverse biotic and abiotic stresses through metabolic and biochemical control (Zhang and Unver, 2018; Zhang & Wang, 2015).
II. Plant miRNAs control differentiation and development of plant organs. It has been shown that many miRNAs and protein coding genes interact and express throughout organ differentiation and development of plants. miRNA156, for example, regulates flower development in Arabidopsis, a key organ in plant's life cycle (Couzigou & Combier, 2016; Li & Zhang, 2016).
III. To meet fundamental food needs of an expanding global population necessitates the production of more crop yields. Maintaining and improving agricultural output in heavy metal-stressed soils is complicated. miRNAs have emerged as a viable technique in the development of an intrinsic system in plants to deal with heavy metal stress (Gupta et al., 2014).

Table 14.2 A representative list of miRNA targets and their roles in plant metabolite accumulation

S. no.	Secondary metabolite	miRNA involved	Gene targeted	Crop	Reference
1	Enhanced lignin build-up	miR1438	Caffeoyl-CoA O-methyltransferase (*CCoAMT*)	Himalayan mayapple (*Podophyllum hexandrum*)	Biswas et al. (2016)
2	Enhanced accumulation of phenolic compound (gingerol)	miR1873	Phenylalanine ammonia lyase gene (*PAL*)	Ginger (*Zingiber officinale*)	Singh et al. (2016)
3	Enhanced proanthocyanidin accumulation	miR1438	Myeloblastosis proteins (MYB)	Persimmon (*Diospyros kaki*)	Luo et al. (2015)
4	Enhanced terpenoid (Secologanin) accumulation	miR396	Secologanin synthase gene	*Rauvolfia serpentina*	Prakash et al. (2016)
5	Enhanced benzylisoquinoline alkaloid synthesis	miR13	7-O-methyltransferase gene (*7-OMT*)	Opium poppy (*Papaver somniferum*)	Boke et al. (2015)
6	Enhanced glucosinolate biosynthesis	miR24	Branched-chain amino acid transaminase 3 gene (*BCAT3*)	*Nicotiana tabacum*	Gou et al. (2011)
7	Enhanced anthocyanin biosynthesis	miR156	Squamosa promoter binding protein like gene (*SPL9*)	*Arabidopsis thaliana*	Gou et al. (2011)

IV. Moreover, artificial miRNAs when compared to RNAi techniques like dsRNA or siRNA, were found to be more precise and effective towards target mRNA modification.

V. Viral vector-mediated miRNA delivery has quickly evolved into a widely used and simple-to-implement technology that allows for a new level of expression control (Basso et al., 2019).

VI. When an indigenous target gene has a negative effect, constitutive overexpression of the regulatory miRNA suppresses the matching mRNA, an approach that could be employed to boost agricultural yields when miRNAs of interest act as positive stress regulators (Djami-Tchatchou et al., 2017).

VII. The advances in genetic engineering, particularly miRNA-based technology, have shown promise in alleviating food poverty by incorporating improved agronomic traits into different crop varieties, resulting in higher yields and food security, which benefits both growers and consumers. Various miRNAs have shown promise in improving agricultural traits responsive to different environmental challenges and increasing biomass production. As a result, there has been an increase in use of marginal soils for plant production resulting in improved agricultural yields (Buiatti et al., 2013; Zhou & Luo, 2013).

VIII. miRNA technique can be used to suppress transgene expression, thereby complementing existing targeting methods, and adding another tool to the genetic engineer's arsenal (Basso et al., 2019).

14.10 Disadvantages of microRNAs

(i) Although miRNA and their targets are a novel source for development of transgenics and also wonderful tools for crop improvement, but there are many side effects that may cause problems. These side effects include undesirable pleiotropic changes in plant morphology and development (Zhou & Luo, 2013).
(ii) Discovering the roles of non-conserved and young miRNAs is a great challenge as low expression level of these miRNAs makes the identification process slow (Garcia, 2008). Also, the regulation of all miRNAs, including conserved miRNAs tend to vary in different species, thus it is not necessary that the same strategies will work on all species (Li et al., 2008; Zhang et al., 2011; Zhou & Luo, 2013).
(iii) Furthermore, there are several challenges in the development and commercialization of genetically modified crops developed using miRNA-based techniques. These challenges include enormous efforts and skills required to achieve a scientific breakthrough and the promotion for public acceptance of these modified crops, along with various economic and political issues (Zhou & Luo, 2013).

14.11 Computational Tools: An *In Silico* Approach

Various computational techniques and databases have emerged over the past decade for the purpose of discovering miRNAs and their targets from deep sequencing, which generates massive datasheets as well as traditional nucleotide sequences for combining expression and functional analyses. Table 14.3 lists most often used computational tools for identifying plant miRNA and its targets.

14.12 Conclusion and Future Perspective

Crops are subjected to a variety of stresses that impede their natural growth and development, impacting a variety of agronomic variables that have a significant influence on crop quality and yield. Agriculture, which is crucial to the global economy, necessitates the development of effective molecular tools and building intrinsic resistance in crops for preventing losses against diverse pressures. In this regard, research has been performed to understand the mechanism of efficacious regulators

Table 14.3 A representative list of various databases for miRNA identification

S. no.	Database	Remarks	Reference(s)
1	TarDB (plant miRNA database)	miRNA targets supporting degradome or parallel analysis RNA ends (PARE) sequencing, as well as miRNA triggered small interfering RNAs (phasiRNA) loci, are all included in this online database. An incredibly user-friendly interface that allows users to quickly find, explore and download miRNA targets and miRNA-induced phasiRNA in a wide range of plants.	Liu et al. (2021)
2	MepmiRDB (medicinal plant miRNA-based database)	Thousands of miRNA candidates from 29 medicinal plant species are stored in this database. The functional modules of this database incorporate miRNA information on expression, sequence patterns and regulatory networks. All the data available in this database are easily accessible.	Yu et al. (2019)
3	Chimira	It involves miRNA analysis using a browser system. Chimira beats all the other tools in terms of overall computational efficiency, and seeks to make small RNA-seq data processing easy, quick and accurate through the use of global miRNA alteration patterns in a user-friendly interface.	Vitsios and Enright (2015)
4	PMTED (plant miRNA target expression database)	PMTED makes use of value-added sequencing data to investigate the relevance of miRNA target genes, which could aid functional research on both miRNAs and their targets.	Sun et al. (2013)
5	miRNEST (database for plant and animal miRNAs)	This database gathers information from various attributes (novel miRNAs, targets and miRNA gene architectures), as well as from many other databases and articles. Data on plant degradome can be found in this database.	Szcześniak et al. (2012)
6	miRTarBase	The largest collection of authenticated miRNA target interactions may be found in this database. The miRTarBase provides a large number of positive data for developing computational approaches for detecting miRNA–target interactions.	Hsu et al. (2011)
7	PMRD (plant miRNA database)	The PMRD database contains a wealth of information about plant miRNAs, including miRNAs and their targets, genome editors, secondary structure analysis and expression profiling.	Zhang et al. (2010)
8	TransmiR	By searching for either a miRNA or transcriptional factors, interested researchers can quickly retrieve transcriptional factors–miRNA regulatory pairs using TransmiR's easy to use interface.	Wang et al. (2010)
9	µPC (Micro PC)	It helps in comparing and forecasting various plant miRNAs, with conserved miRNAs being discovered using homologous analysis system. This database provides a comprehensive overview of plant miRNA function, conservation and evolution.	Mhuantong and Wichadakul (2009)

(continued)

Table 14.3 (continued)

S. no.	Database	Remarks	Reference(s)
10	miRbase (miRNA registry)	All reported miRNA sequence and descriptions are available on this online searchable database. It is built on the names, annotations, keywords and references for miRNAs. It also includes target predictions based on a number of frequently used target prediction techniques.	Griffiths-Jones et al. (2007); Griffiths-Jones (2004)

in crops against diverse stress conditions, and in this direction, miRNAs have gathered a lot of attention for their role as master regulators in crop improvement. In this chapter, we have elucidated miRNAs and their unique target mRNAs to exemplify unexpected function of miRNAs in crop improvement, proving these as a potent tool in genetic engineering.

A significant advancement in the field of bioinformatics in the development of various databases has allowed us to utilise data for miRNA characterization, targets and discovery of miRNA families that could provide resistance to numerous stresses in crops through gene modification. There are numerous miRNAs that have demonstrated their promising potential in improving crop traits. Through targeting a particular mRNA, miR535 induced tolerance to salinity, salt, abscisic acid and polyethylene glycol environmental stresses in rice (Yue et al., 2020). Also, miR408, miR156, miR393 and a few other miRNAs have been found to regulate abiotic stresses in a variety of crops. In addition, artificial miRNAs have also been proposed as a potential crop improvement tool. amiR171, an artificial miRNA conferred resistance to cauliflower mosaic virus in tobacco (*Nicotiana tabacum* cv. *SR1*) (Qu et al., 2007). The techniques developed by bioinformatics experts have paved the path for deep learning that has made miRNA investigations quicker, more accurate and effective.

miRNAs have been demonstrated to play important roles in a variety of activities, including yield improvement, plant architectural modulation and organ growth. However, the relevance of miRNAs in the development of functional foods has been minimal. Through the down-regulation of miRNAs, the iron and zinc content of rice grains was significantly boosted (Agarwal et al., 2018). Various computational and experimental approaches will aid in identifying the potential role of miRNAs in functional food production, and boosting the food industry business in future. In many cases, altered miRNA activity results in a number of pleiotropic abnormalities. Therefore, it is critical to apply precise genome editing technologies like CRISPR-Cas9 to fine-tune the quantity of miRNAs and their target genes. As a result, determining the editing feasibility and effectiveness of current CRISPR–Cas9 systems or developing additional CRISPR–Cas9-derived methods that target plant miRNA locus and miRNA-binding sites, is necessary and important. A very efficient miRNA-related genome-editing method would greatly broaden the variety of miRNA uses for crop breeding programs.

References

Abel, P. P., Nelson, R. S., De, B., Hoffmann, N., Rogers, S. G., Fraley, R. T., & Beachy, R. N. (1986). Delay of disease development in transgenic plants that express the tobacco mosaic virus coat protein gene. *Science, 232*(4751), 738–743.

Agarwal, S., Mangrauthia, S.K. and Sarla, N. (2018). Genomic approaches for micronutrients biofortification of rice. In *Plant Micronutrient Use Efficiency* (pp. 245–260). Academic Press.

Albright, S. R., & Tjian, R. (2000). TAFs revisited: More data reveal new twists and confirm old ideas. *Gene, 242*(1–2), 1–13.

Angelovici, R., Fait, A., Fernie, A. R., & Galili, G. (2011). A seed high-lysine trait is negatively associated with the TCA cycle and slows down Arabidopsis seed germination. *New Phytologist, 189*(1), 148–159.

Aung, B., Gruber, M. Y., Amyot, L., Omari, K., Bertrand, A., & Hannoufa, A. (2015). Micro RNA 156 as a promising tool for alfalfa improvement. *Plant Biotechnology Journal, 13*(6), 779–790.

Axtell, M. J., & Bowman, J. L. (2008). Evolution of plant microRNAs and their targets. *Trends in Plant Science, 13*(7), 343–349.

Bartel, D. P. (2009). MicroRNAs: Target recognition and regulatory functions. *Cell, 136*(2), 215–233.

Bartwal, A., Mall, R., Lohani, P., Guru, S. K., & Arora, S. (2013). Role of secondary metabolites and brassinosteroids in plant defense against environmental stresses. *Journal of Plant Growth Regulation, 32*(1), 216–232.

Basso, M. F., Ferreira, P. C. G., Kobayashi, A. K., Harmon, F. G., Nepomuceno, A. L., Molinari, H. B. C., & Grossi-de-Sa, M. F. (2019). Micro RNAs and new biotechnological tools for its modulation and improving stress tolerance in plants. *Plant Biotechnology Journal, 17*(8), 1482–1500.

Berezikov, E., Cuppen, E., & Plasterk, R. H. (2006). Approaches to microRNA discovery. *Nature Genetics, 38*(6), S2–S7.

Biswas, S., Hazra, S., & Chattopadhyay, S. (2016). Identification of conserved miRNAs and their putative target genes in *Podophyllum hexandrum*(Himalayan Mayapple). *Plant Gene, 6*, 82–89.

Boke, H., Ozhuner, E., Turktas, M., Parmaksiz, I., Ozcan, S., & Unver, T. (2015). Regulation of the alkaloid biosynthesis by mi RNA in opium poppy. *Plant Biotechnology Journal, 13*(3), 409–420.

Buiatti, M., Christou, P., & Pastore, G. (2013). The application of GMOs in agriculture and in food production for a better nutrition: Two different scientific points of view. *Genes & Nutrition, 8*(3), 255–270.

Campo, S., Peris, C., Siré, C., Moreno, A. B., Donaire, L., Zytnicki, M., Notredame, C., Llave, C., & San Segundo, B. (2013). Identification of a novel microRNA (mi RNA) from rice that targets an alternatively spliced transcript of the *Nramp6* (natural resistance-associated macrophage protein 6) gene involved in pathogen resistance. *New Phytologist, 199*(1), 212–227.

Čertík, M., Klempová, T., Guothová, L., Mihálik, D., & Kraic, J. (2013). Biotechnology for the functional improvement of cereal-based materials enriched with PUFA and pigments. *European Journal of Lipid Science and Technology, 115*(11), 1247–1256.

Challam, C., Nandhakumar, N., & Kardile, H. B. (2019). Advances in crop improvement: Use of miRNA technologies for crop improvement. In *OMICS-based Approaches in Plant Biotechnology* (pp. 55–74). Scrivener Publishing LLC.

Chi, X., Yang, Q., Chen, X., Wang, J., Pan, L., Chen, M., Yang, Z., He, Y., Liang, X., & Yu, S. (2011). Identification and characterization of microRNAs from peanut (*Arachis hypogea* L.) by high-throughput sequencing. *PLoS One, 6*(11), e27530.

Chuck, G. S., Tobias, C., Sun, L., Kraemer, F., Li, C., Dibble, D., Arora, R., Bragg, J. N., Vogel, J. P., Singh, S., & Simmons, B. A. (2011). Overexpression of the maize *Corngrass1* microRNA prevents flowering, improves digestibility, and increases starch content of switchgrass. *Proceedings of the National Academy of Sciences, 108*(42), 17550–17555.

Couzigou, J. M., & Combier, J. P. (2016). Plant microRNAs: Key regulators of root architecture and biotic interactions. *New Phytologist, 212*(1), 22–35.

Curaba, J., Talbot, M., Li, Z., & Helliwell, C. (2013). Over-expression of microRNA171 affects phase transitions and floral meristem determinancy in barley. *BMC Plant Biology, 13*(1), 1–10.

Davuluri, G. R., Van Tuinen, A., Fraser, P. D., Manfredonia, A., Newman, R., Burgess, D., Brummell, D. A., King, S. R., Palys, J., Uhlig, J., & Bramley, P. M. (2005). Fruit-specific RNAi-mediated suppression of DET1 enhances carotenoid and flavonoid content in tomatoes. *Nature Biotechnology, 23*(7), 890–895.

Ding, Y., Gong, S., Wang, Y., Wang, F., Bao, H., Sun, J., Cai, C., Yi, K., Chen, Z., & Zhu, C. (2018). MicroRNA166 modulates cadmium tolerance and accumulation in rice. *Plant Physiology, 177*(4), 1691–1703.

Djami-Tchatchou, A. T., Sanan-Mishra, N., Ntushelo, K., & Dubery, I. A. (2017). Functional roles of microRNAs in agronomically important plants—Potential as targets for crop improvement and protection. *Frontiers in Plant Science, 8*, 378.

Ebert, M. S., & Sharp, P. A. (2010). Emerging roles for natural microRNA sponges. *Current Biology, 20*(19), R858–R861.

Fahlgren, N., Howell, M. D., Kasschau, K. D., Chapman, E. J., Sullivan, C. M., Cumbie, J. S., Givan, S. A., Law, T. F., Grant, S. R., Dangl, J. L., & Carrington, J. C. (2007). High-throughput sequencing of Arabidopsis microRNAs: Evidence for frequent birth and death of MIRNA genes. *PLoS One, 2*(2), e219.

Fan, Y., Yang, J., Mathioni, S. M., Yu, J., Shen, J., Yang, X., Wang, L., Zhang, Q., Cai, Z., Xu, C., & Li, X. (2016). PMS1T, producing phased small-interfering RNAs, regulates photoperiod-sensitive male sterility in rice. *Proceedings of the National Academy of Sciences, 113*(52), 15144–15149.

Feng, H., Wang, X., Zhang, Q., Fu, Y., Feng, C., Wang, B., Huang, L., & Kang, Z. (2014). Monodehydroascorbate reductase gene, regulated by the wheat PN-2013 miRNA, contributes to adult wheat plant resistance to stripe rust through ROS metabolism. *Biochimica et Biophysica Acta (BBA)-Gene Regulatory Mechanisms, 1839*(1), 1–12.

Ferdous, J., Whitford, R., Nguyen, M., Brien, C., Langridge, P., & Tricker, P. J. (2017). Drought-inducible expression of Hv-miR827 enhances drought tolerance in transgenic barley. *Functional & Integrative Genomics, 17*(2–3), 279–292.

Fu, C., Sunkar, R., Zhou, C., Shen, H., Zhang, J. Y., Matts, J., Wolf, J., Mann, D. G., Stewart, C. N., Jr., Tang, Y., & Wang, Z. Y. (2012). Overexpression of miR156 in switchgrass (Panicum virgatum L.) results in various morphological alterations and leads to improved biomass production. *Plant Biotechnology Journal, 10*(4), 443–452.

Fujii, H., Chiou, T. J., Lin, S. I., Aung, K., & Zhu, J. K. (2005). A miRNA involved in phosphate-starvation response in Arabidopsis. *Current Biology, 15*(22), 2038–2043.

Gao, S., Yang, L., Zeng, H. Q., Zhou, Z. S., Yang, Z. M., Li, H., Sun, D., Xie, F., & Zhang, B. (2016). A cotton miRNA is involved in regulation of plant response to salt stress. *Scientific Reports, 6*(1), 1–14.

Gao, Z., Nie, J., & Wang, H. (2020). MicroRNA biogenesis in plant. *Plant Growth Regulation, 93*(5), 1–12.

Garcia, D. (2008). A miRacle in plant development: Role of microRNAs in cell differentiation and patterning. *Seminars in Cell & Developmental Biology, 19*(6), 586–595.

Gao, P., Bai, X., Yang, L., Lv, D., Li, Y., Cai, H., Ji, W., Guo, D., Zhu, Y. (2010). Over-expression of osa-MIR396c decreases salt and alkali stress tolerance. *Planta, 231*(5), pp. 991–1001.

Gasser, C. S., & Fraley, R. T. (1989). Genetically engineering plants for crop improvement. *Science, 244*(4910), 1293–1299.

Ghosh, S. (2011). Micro RNA-biogenesis, mechanism of action and applications – A review. *International Journal of Research in Biotechnology and Biochemistry, 1*(1), 11–36.

Goetz, M., Hooper, L. C., Johnson, S. D., Rodrigues, J. C. M., Vivian-Smith, A., & Koltunow, A. M. (2007). Expression of aberrant forms of AUXIN RESPONSE FACTOR8 stimulates parthenocarpy in Arabidopsis and tomato. *Plant Physiology, 145*(2), 351–366.

Gou, J. Y., Felippes, F. F., Liu, C. J., Weigel, D., & Wang, J. W. (2011). Negative regulation of anthocyanin biosynthesis in Arabidopsis by a miR156-targeted SPL transcription factor. *The Plant Cell, 23*(4), 1512–1522.

Griffiths-Jones, S. (2004). The microRNA registry. *Nucleic Acids Research, 32*(suppl_1), D109–D111.

Griffiths-Jones, S., Saini, H. K., Van Dongen, S., & Enright, A. J. (2007). miRBase: Tools for microRNA genomics. *Nucleic Acids Research, 36*(suppl_1), D154–D158.

Guan, X., Pang, M., Nah, G., Shi, X., Ye, W., Stelly, D. M., & Chen, Z. J. (2014). miR828 and miR858 regulate homoeologous MYB2 gene functions in Arabidopsis trichome and cotton fibre development. *Nature Communications, 5*(1), 1–14.

Guo, S., Xu, Y., Liu, H., Mao, Z., Zhang, C., Ma, Y., Zhang, Q., Meng, Z., & Chong, K. (2013). The interaction between OsMADS57 and OsTB1 modulates rice tillering via DWARF14. *Nature Communications, 4*(1), 1–12.

Gupta, O. P., Sharma, P., Gupta, R. K., & Sharma, I. (2014). MicroRNA mediated regulation of metal toxicity in plants: Present status and future perspectives. *Plant Molecular Biology, 84*(1–2), 1–18.

Gutierrez, L., Bussell, J. D., Pacurar, D. I., Schwambach, J., Pacurar, M., & Bellini, C. (2009). Phenotypic plasticity of adventitious rooting in Arabidopsis is controlled by complex regulation of AUXIN RESPONSE FACTOR transcripts and microRNA abundance. *The Plant Cell, 21*(10), 3119–3132.

Ha, M., & Kim, V. N. (2014). Regulation of microRNA biogenesis. *Nature Reviews Molecular Cell Biology, 15*(8), 509–524.

Hajyzadeh, M., Turktas, M., Khawar, K. M., & Unver, T. (2015). miR408 overexpression causes increased drought tolerance in chickpea. *Gene, 555*(2), 186–193.

Hewezi, T., Maier, T. R., Nettleton, D., & Baum, T. J. (2012). The Arabidopsis microRNA396-GRF1/GRF3 regulatory module acts as a developmental regulator in the reprogramming of root cells during cyst nematode infection. *Plant Physiology, 159*(1), 321–335.

Houmard, N. M., Mainville, J. L., Bonin, C. P., Huang, S., Luethy, M. H., & Malvar, T. M. (2007). High lysine corn generated by endosperm-specific suppression of lysine catabolism using RNAi. *Plant Biotechnology Journal, 5*(5), 605–614.

Hsu, S. D., Lin, F. M., Wu, W. Y., Liang, C., Huang, W. C., Chan, W. L., Tsai, W. T., Chen, G. Z., Lee, C. J., Chiu, C. M., & Chien, C. H. (2011). miRTarBase: A database curates experimentally validated microRNA–target interactions. *Nucleic Acids Research, 39*(suppl_1), D163–D169.

Jatan, R., & Lata, C. H. A. R. U. (2019). Role of MicroRNAs in abiotic and biotic stress resistance in plants. *Proceedings of the Indian National Science Academy, 85*, 553–567.

Jiao, Y., Wang, Y., Xue, D., Wang, J., Yan, M., Liu, G., Dong, G., Zeng, D., Lu, Z., Zhu, X., & Qian, Q. (2010). Regulation of OsSPL14 by OsmiR156 defines ideal plant architecture in rice. *Nature Genetics, 42*(6), 541–544.

John, M. E., & Stewart, J. M. (2010). Genetic engineering applications in crop improvement. In *Physiology of cotton* (pp. 394–403). Springer.

Jones-Rhoades, M. W., Bartel, D. P., & Bartel, B. (2006). MicroRNAs and their regulatory roles in plants. *Annual Review of Plant Biology, 57*, 19–53.

Kahl, G., & Winter, P. (1995). Plant genetic engineering for crop improvement. *World Journal of Microbiology and Biotechnology, 11*(4), 449–460.

Kamthan, A., Chaudhuri, A., Kamthan, M., & Datta, A. (2015). Small RNAs in plants: Recent development and application for crop improvement. *Frontiers in Plant Science, 6*, 208.

Kawashima, C. G., Yoshimoto, N., Maruyama-Nakashita, A., Tsuchiya, Y. N., Saito, K., Takahashi, H., & Dalmay, T. (2009). Sulphur starvation induces the expression of microRNA-395 and one of its target genes but in different cell types. *The Plant Journal, 57*(2), 313–321.

Khraiwesh, B., Ossowski, S., Weigel, D., Reski, R., & Frank, W. (2008). Specific gene silencing by artificial MicroRNAs in Physcomitrella patens: An alternative to targeted gene knockouts. *Plant Physiology, 148*(2), 684–693.

Khraiwesh, B., Zhu, J. K., & Zhu, J. (2012). Role of miRNAs and siRNAs in biotic and abiotic stress responses of plants. *Biochimica et Biophysica Acta (BBA)-Gene Regulatory Mechanisms, 1819*(2), 137–148.

Kim, J. J., Lee, J. H., Kim, W., Jung, H. S., Huijser, P., & Ahn, J. H. (2012). The microRNA156-SQUAMOSA PROMOTER BINDING PROTEIN-LIKE3 module regulates ambient temperature-responsive flowering via FLOWERING LOCUS T in Arabidopsis. *Plant Physiology, 159*(1), 461–478.

Kozomara, A., & Griffiths-Jones, S. (2014). miRBase: Annotating high confidence microRNAs using deep sequencing data. *Nucleic Acids Research, 42*(D1), D68–D73.

Li, C., & Zhang, B. (2016). MicroRNAs in control of plant development. *Journal of Cellular Physiology, 231*(2), 303–313.

Li, W. X., Oono, Y., Zhu, J., He, X. J., Wu, J. M., Iida, K., Lu, X. Y., Cui, X., Jin, H., & Zhu, J. K. (2008). The Arabidopsis NFYA5 transcription factor is regulated transcriptionally and posttranscriptionally to promote drought resistance. *The Plant Cell, 20*(8), 2238–2251.

Liang, G., He, H., Li, Y., & Yu, D. (2012). A new strategy for construction of artificial miRNA vectors in Arabidopsis. *Planta, 235*(6), 1421–1429.

Liu, J. M., Livny, J., Lawrence, M. S., Kimball, M. D., Waldor, M. K., & Camilli, A. (2009). Experimental discovery of sRNAs in *Vibrio cholerae* by direct cloning, 5S/tRNA depletion and parallel sequencing. *Nucleic Acids Research, 37*(6), e46–e46.

Liu, J., Liu, X., Zhang, S., Liang, S., Luan, W., & Ma, X. (2021). TarDB: An online database for plant miRNA targets and miRNA-triggered phased siRNAs. *BMC Genomics, 22*(1), 1–12.

Liu, C., Xin, Y., Xu, L., Cai, Z., Xue, Y., Liu, Y., Xie, D., Liu, Y. and Qi, Y. (2018). *Arabidopsis* ARGONAUTE 1 binds chromatin to promote gene transcription in response to hormones and stresses. *Developmental Cell, 44*(3), pp. 348–361.

Lotfi, A., Pervaiz, T., Jiu, S., Faghihi, F., Jahanbakhshian, Z., Khorzoghi, E. G., & Fang, J. (2017). Role of microRNAs and their target genes in salinity response in plants. *Plant Growth Regulation, 82*(3), 377–390.

Lu, S., Sun, Y. H., Shi, R., Clark, C., Li, L., & Chiang, V. L. (2005). Novel and mechanical stress–responsive microRNAs in *Populus trichocarpa* that are absent from *Arabidopsis*. *The Plant Cell, 17*(8), 2186–2203.

Lu, Q., Guo, F., Xu, Q., & Cang, J. (2020). LncRNA improves cold resistance of winter wheat by interacting with miR398. *Functional Plant Biology, 47*(6), 544–557.

Luo, Y., Zhang, X., Luo, Z., Zhang, Q., & Liu, J. (2015). Identification and characterization of microRNAs from Chinese pollination constant non-astringent persimmon using high-throughput sequencing. *BMC Plant Biology, 15*(1), 1–18.

Mandal, K., Boro, P., & Chattopadhyay, S. (2021). Micro-RNA based gene regulation: A potential way for crop improvements. *Plant Gene, 27*, 100312.

Marin, E., Jouannet, V., Herz, A., Lokerse, A. S., Weijers, D., Vaucheret, H., Nussaume, L., Crespi, M. D., & Maizel, A. (2010). miR390, Arabidopsis TAS3 tasiRNAs, and their AUXIN RESPONSE FACTOR targets define an autoregulatory network quantitatively regulating lateral root growth. *The Plant Cell, 22*(4), 1104–1117.

Martin, A., Adam, H., Díaz-Mendoza, M., Zurczak, M., González-Schain, N. D., & Suárez-López, P. (2009). Graft-transmissible induction of potato tuberization by the microRNA miR172. *Development, 136*(17), 2873–2881.

Meijer, H. A., Smith, E. M., & Bushell, M. (2014). Regulation of miRNA strand selection: Follow the leader? *Biochemical Society Transactions, 42*(4), 1135–1140.

Mhuantong, W., & Wichadakul, D. (2009). MicroPC (μPC): A comprehensive resource for predicting and comparing plant microRNAs. *BMC Genomics, 10*(1), 1–8.

Miura, K., Ikeda, M., Matsubara, A., Song, X. J., Ito, M., Asano, K., Matsuoka, M., Kitano, H., & Ashikari, M. (2010). OsSPL14 promotes panicle branching and higher grain productivity in rice. *Nature Genetics, 42*(6), 545–549.

Moxon, S., Jing, R., Szittya, G., Schwach, F., Pilcher, R. L. R., Moulton, V., & Dalmay, T. (2008). Deep sequencing of tomato short RNAs identifies microRNAs targeting genes involved in fruit ripening. *Genome Research, 18*(10), 1602–1609.

Navarro, L., Dunoyer, P., Jay, F., Arnold, B., Dharmasiri, N., Estelle, M., Voinnet, O., & Jones, J. D. (2006). A plant miRNA contributes to antibacterial resistance by repressing auxin signaling. *Science, 312*(5772), 436–439.

Patil, S., Joshi, S., Jamla, M., Zhou, X., Taherzadeh, M. J., Suprasanna, P., & Kumar, V. (2021). MicroRNA-mediated bioengineering for climate-resilience in crops. *Bioengineered, 12*(2), 10430–10456.

Paul, S., Datta, S. K., & Datta, K. (2015). miRNA regulation of nutrient homeostasis in plants. *Frontiers in Plant Science, 6*, 232.

Phelps-Durr, T. L. (2010). MicroRNAs in Arabidopsis. *National Education, 3*, 51.

Prakash, P., Rajakani, R., & Gupta, V. (2016). Transcriptome-wide identification of *Rauvolfia serpentina* microRNAs and prediction of their potential targets. *Computational Biology and Chemistry, 61*, 62–74.

Qu, J., Ye, J., & Fang, R. (2007). Artificial microRNA-mediated virus resistance in plants. *Journal of Virology, 81*(12), 6690–6699.

Regina, A., Bird, A., Topping, D., Bowden, S., Freeman, J., Barsby, T., Kosar-Hashemi, B., Li, Z., Rahman, S., & Morell, M. (2006). High-amylose wheat generated by RNA interference improves indices of large-bowel health in rats. *Proceedings of the National Academy of Sciences, 103*(10), 3546–3551.

Rogers, K., & Chen, X. (2013). Biogenesis, turnover, and mode of action of plant microRNAs. *The Plant Cell, 25*(7), 2383–2399.

Sabzehzari, M., & Naghavi, M. R. (2019). Phyto-miRNAs-based regulation of metabolites biosynthesis in medicinal plants. *Gene, 682*, 13–24.

Sami, A., Xue, Z., Tazein, S., Arshad, A., He Zhu, Z., Ping Chen, Y., Hong, Y., Tian Zhu, X., & Jin Zhou, K. (2021). CRISPR–Cas9-based genetic engineering for crop improvement under drought stress. *Bioengineered, 12*(1), 5814–5829.

Schwab, R., Palatnik, J. F., Riester, M., Schommer, C., Schmid, M., & Weigel, D. (2005). Specific effects of microRNAs on the plant transcriptome. *Developmental Cell, 8*(4), 517–527.

Sestili, F., Janni, M., Doherty, A., Botticella, E., D'Ovidio, R., Masci, S., Jones, H. D., & Lafiandra, D. (2010). Increasing the amylose content of durum wheat through silencing of the *SBEIIa* genes. *BMC Plant Biology, 10*(1), 1–12.

Shuai, P., Liang, D., Zhang, Z., Yin, W., & Xia, X. (2013). Identification of drought-responsive and novel *Populus trichocarpa* microRNAs by high-throughput sequencing and their targets using degradome analysis. *BMC Genomics, 14*(1), 1–14.

Sidhu, J. S., Kabir, Y., & Huffman, F. G. (2007). Functional foods from cereal grains. *International Journal of Food Properties, 10*(2), 231–244.

Singh, N., Srivastava, S., & Sharma, A. (2016). Identification and analysis of miRNAs and their targets in ginger using bioinformatics approach. *Gene, 575*(2), 570–576.

Sun, G. (2012). MicroRNAs and their diverse functions in plants. *Plant Molecular Biology, 80*(1), 17–36.

Sun, X., Dong, B., Yin, L., Zhang, R., Du, W., Liu, D., Shi, N., Li, A., Liang, Y., & Mao, L. (2013). PMTED: A plant microRNA target expression database. *BMC Bioinformatics, 14*(1), 1–7.

Sunkar, R. (2010). MicroRNAs with macro-effects on plant stress responses. In *Seminars in Cell & Developmental Biology, 21*(8) (pp. 805–811). Academic Press.

Szcześniak, M. W., Deorowicz, S., Gapski, J., Kaczyński, Ł., & Makałowska, I. (2012). miRNEST database: An integrative approach in microRNA search and annotation. *Nucleic Acids Research, 40*(D1), D198–D204.

Tang, J., & Chu, C. (2017). MicroRNAs in crop improvement: Fine-tuners for complex traits. *Nature Plants, 3*(7), 1–11.

Tang, G., Xiang, Y., Kang, Z., Mendu, V., Tang, X., Jia, X., Chen, Q. J., & Tang, X. (2008). Small RNA technologies: siRNA, miRNA, antagomiR, target mimicry, miRNA sponge and miRNA profiling. In *Current perspectives in microRNAs (miRNA)* (pp. 17–33). Springer.

Teotia, S., & Tang, G. (2017). Silencing of stress-regulated miRNAs in plants by short tandem target mimic (STTM) approach. In *Plant stress tolerance* (pp. 337–348). Humana Press.

Thakur, A. K., Chauhan, D. K., Parmar, N., & Verma, V. (2012). Role of genetic engineering in horticultural crop improvement-a review. *Agricultural Reviews, 33*(3), 248–255.

Tiwari, M., Sharma, D., & Trivedi, P. K. (2014). Artificial microRNA mediated gene silencing in plants: Progress and perspectives. *Plant Molecular Biology, 86*(1–2), 1–18.

Toppino, L., Kooiker, M., Lindner, M., Dreni, L., Rotino, G. L., & Kater, M. M. (2011). Reversible male sterility in eggplant (*Solanum melongena* L.) by artificial microRNA-mediated silencing of general transcription factor genes. *Plant Biotechnology Journal, 9*(6), 684–692.

Van Vu, T., Choudhury, N. R., & Mukherjee, S. K. (2013). Transgenic tomato plants expressing artificial microRNAs for silencing the pre-coat and coat proteins of a begomovirus, tomato leaf curl New Delhi virus, show tolerance to virus infection. *Virus Research, 172*(1–2), 35–45.

Vinocur, B., & Altman, A. (2005). Recent advances in engineering plant tolerance to abiotic stress: Achievements and limitations. *Current Opinion in Biotechnology, 16*(2), 123–132.

Vitsios, D. M., & Enright, A. J. (2015). Chimira: Analysis of small RNA sequencing data and microRNA modifications. *Bioinformatics, 31*(20), 3365–3367.

Voinnet, O. (2009). Origin, biogenesis, and activity of plant microRNAs. *Cell, 136*(4), 669–687.

Wang, Q. L., & Li, Z. H. (2007). The functions of microRNAs in plants. *Frontiers in Bioscience, 12*, 3975–3982.

Wang, J., Lu, M., Qiu, C., & Cui, Q. (2010). TransmiR: A transcription factor–microRNA regulation database. *Nucleic Acids Research, 38*(suppl_1), D119–D122.

Wang, J., Mei, J., & Ren, G. (2019). Plant microRNAs: Biogenesis, homeostasis, and degradation. *Frontiers in Plant Science, 10*, 360.

Wang, Y., Sun, F., Cao, H., Peng, H., Ni, Z., Sun, Q. and Yao, Y. (2012). TamiR159 directed wheat *TaGAMYB* cleavage and its involvement in anther development and heat response. *PloS ONE, 7*(11), e 48445.

Willmann, M.R. and Poethig, R.S. (2007). Conservation and evolution of miRNA regulatory programs in plant development. *Current Opinion in Plant Biology*, *10*(5), pp. 503–511.

Wightman, B., Ha, I., & Ruvkun, G. (1993). Posttranscriptional regulation of the heterochronic gene lin-14 by lin-4 mediates temporal pattern formation in *C. Elegans*. *Cell, 75*(5), 855–862.

Wu, J., Yang, R., Yang, Z., Yao, S., Zhao, S., Wang, Y., Li, P., Song, X., Jin, L., Zhou, T., & Lan, Y. (2017). ROS accumulation and antiviral defence control by microRNA528 in rice. *Nature Plants, 3*(1), 1–7.

Yan, Y., Wang, H., Hamera, S., Chen, X. & Fang, R. (2014). MiR444a has multiple functions in the rice nitrate-signaling pathway. *The Plant Journal, 78*(1), pp. 44–55.

Yamaguchi, A. and Abe, M. (2012). Regulation of reproductive development by non-coding RNA in Arabidopsis: to flower or not to flower. *Journal of Plant Research, 125*(6), pp. 693–704.

Yang, C., Li, D., Mao, D., Liu, X. U. E., Ji, C., Li, X., Zhao, X., Cheng, Z., Chen, C., & Zhu, L. (2013). Overexpression of microRNA 319 impacts leaf morphogenesis and leads to enhanced cold tolerance in rice (*Oryza sativa* L.). *Plant, Cell & Environment, 36*(12), 2207–2218.

Yang, L., Mu, X., Liu, C., Cai, J., Shi, K., Zhu, W., & Yang, Q. (2015). Overexpression of potato miR482e enhanced plant sensitivity to *Verticillium dahliae* infection. *Journal of Integrative Plant Biology, 57*(12), 1078–1088.

Yogindran, S., & Rajam, M. V. (2021). Host-derived artificial miRNA-mediated silencing of ecdysone receptor gene provides enhanced resistance to *Helicoverpaarmigera* in tomato. *Genomics, 113*(1), 736–747.

Yoon, E.K., Yang, J.H., Lim, J., Kim, S.H., Kim, S.K. & Lee, W.S. (2010). Auxin regulation of the microRNA390-dependent transacting small interfering RNA pathway in Arabidopsis lateral root development. *Nucleic Acids Research, 38*(4), pp.1382–1391.

Yu, S., Lian, H., & Wang, J. W. (2015). Plant developmental transitions: The role of microRNAs and sugars. *Current Opinion in Plant Biology, 27*, 1–7.
Yu, Y., Jia, T., & Chen, X. (2017). The 'how' and 'where' of plant microRNA s. *New Phytologist, 216*(4), 1002–1017.
Yu, Y., Ni, Z., Wang, Y., Wan, H., Hu, Z., Jiang, Q., Sun, X., Zhang, H. (2019). Overexpression of soybean miR169c confers increased drought stress sensitivity in transgenic *Arabidopsis thaliana. Plant Science, 285*, pp. 68–78.
Yue, E., Cao, H., Liu, B. (2020). OsmiR535, a potential genetic editing target for drought and salinity stress tolerance in *Oryza sativa. Plants, 9*(10), p. 1337.
Zhang, B. (2015). MicroRNA: A new target for improving plant tolerance to abiotic stress. *Journal of Experimental Botany, 66*(7), 1749–1761.
Zhang, B., & Wang, Q. (2015). MicroRNA-based biotechnology for plant improvement. *Journal of Cellular Physiology, 230*(1), 1–15.
Zhang, B., & Wang, Q. (2016). MicroRNA, a new target for engineering new crop cultivars. *Bioengineered, 7*(1), 7–10.
Zhang, X., Zou, Z., Gong, P., Zhang, J., Ziaf, K., Li, H., Xiao, F., & Ye, Z. (2011). Over-expression of microRNA169 confers enhanced drought tolerance to tomato. *Biotechnology Letters, 33*(2), 403–409.
Zhang, J. P., Yu, Y., Feng, Y. Z., Zhou, Y. F., Zhang, F., Yang, Y. W., Lei, M. Q., Zhang, Y. C., & Chen, Y. Q. (2017). MiR408 regulates grain yield and photosynthesis via a phytocyanin protein. *Plant Physiology, 175*(3), 1175–1185.
Zhang, Z., Yu, J., Li, D., Zhang, Z., Liu, F., Zhou, X., Wang, T., Ling, Y. and Su, Z. (2010). PMRD: plant microRNA database. *Nucleic Acids Research, 38*(suppl_1), pp. D806–D813.
Zhang, B. & Unver, T. (2018). A critical and speculative review on microRNA technology in crop improvement: Current challenges and future directions. *Plant Science, 274*, pp. 193–200.
Zhao, M., Ding, H., Zhu, J. K., Zhang, F., & Li, W. X. (2011). Involvement of miR169 in the nitrogen-starvation responses in Arabidopsis. *New Phytologist, 190*(4), 906–915.
Zhou, M., & Luo, H. (2013). MicroRNA-mediated gene regulation: Potential applications for plant genetic engineering. *Plant Molecular Biology, 83*(1–2), 59–75.
Zhuo, Y., Gao, G., Shi, J., Zhou, X., & Wang, X. (2013). miRNAs: Biogenesis, origin, and evolution, functions on virus-host interaction. *Cellular Physiology and Biochemistry, 32*(3), 499–510.
Zhou, B., Leng, J., Ma, Y., Fan, P., Li, Y., Yan, H., Xu, Q. (2020). BrmiR828 targets *Brpap1, Brmyb82* and *Brtas4* involved in the light induced anthocyanin biosynthetic pathway in *Brassica rapa. International Journal of Molecular Sciences, 21*(12), p. 4326.

Index

S. Punia Bangar, A. Kumar Siroha (eds.), *Functional Cereals and Cereal Foods*, https://doi.org/10.1007/978-3-031-05611-6

D

E

MIX
Papier aus verantwortungsvollen Quellen
Paper from responsible sources
FSC® C105338

If you have any concerns about our products,
you can contact us on
ProductSafety@springernature.com

In case Publisher is established outside the EU,
the EU authorized representative is:
Springer Nature Customer Service Center GmbH
Europaplatz 3, 69115 Heidelberg, Germany

Printed by Libri Plureos GmbH
in Hamburg, Germany